# Lecture Notes in Nanoscale Science and Technology

Volume 49

Lecture Notes in Nanoscale Science and Technology (LNNST) aims to report latest developments in nanoscale science and technology research and teaching – quickly, informally and at a high level. Through publication, LNNST commits to serve the open communication of scientific and technological advances in the creation and use of objects at the nanometer scale, crossing the boundaries of physics, materials science, biology, chemistry, and engineering. Certainly, while historically the mysteries in each of the sciences have been very different, they have all required a relentless step-by-step pursuit to uncover the answer to a challenging scientific question, but recently many of the answers have brought questions that lie at the boundaries between the life sciences and the physical sciences and between what is fundamental and what is application. This is no accident since recent research in the physical and life sciences have each independently cut a path to the edge of their disciplines. As both paths intersect one may ask if transport of material in a cell is biology or is it physics? This intersection of curiosity makes us realize that nanoscience and technology crosses many if not all disciplines. It is this market that the proposed series of lecture notes targets.

Shrikaant Kulkarni • Susanta Das
Editors

# Microplastics and Nanoplastics

## Environmental Risks, Impact Assessment, and Remediation Technologies

Springer

*Editors*
Shrikaant Kulkarni
Victorian Institute of Technology
Melbourne, VIC, Australia

Susanta Das
School of Science
Ajeenkya DY Patil University
Pune, Maharashtra, India

ISSN 2195-2159 ISSN 2195-2167 (electronic)
Lecture Notes in Nanoscale Science and Technology
ISBN 978-3-032-17962-3 ISBN 978-3-032-17963-0 (eBook)
https://doi.org/10.1007/978-3-032-17963-0



This Springer imprint is published by the registered company Springer Nature Switzerland AG
The registered company address is: Gewerbestrasse 11, 6330 Cham, Switzerland

# Preface

*Microplastics and Nanoplastics: Environmental Risks, Impact Assessment, and Remediation Technologies* is a comprehensive book that delves into the intricate world of tiny plastic particles and their profound implications on the environment. This book explores the origins, sources, and distribution of microplastics and nanoplastics, shedding light on their pervasive presence in ecosystems. With a focus on environmental risks, this book discusses the potential threats these minuscule pollutants pose to marine and terrestrial ecosystems, wildlife, and human health. Impact assessment methodologies are critically examined, offering insights into the complex interactions between microplastics/nanoplastics and various environmental components. Furthermore, this book addresses innovative and emerging remediation technologies designed to mitigate the adverse effects of plastic pollution. From novel cleanup strategies to advanced monitoring techniques, the authors provide a forward-looking perspective on combating the challenges posed by these synthetic particles.

Authored by experts in the field, this book serves as a valuable resource for researchers, policymakers, and environmental enthusiasts seeking a deeper understanding of the intricate issues surrounding microplastics and nanoplastics. It bridges the gap between scientific knowledge and practical solutions, making it an indispensable guide in the ongoing efforts to safeguard the health of our planet.

This edited volume has 13 chapters: Chap. 1 provides a foundational overview of microplastics (MPs) and nanoplastics (NPs), exploring their origins, characteristics, and the growing global concern surrounding their environmental and health impacts. Chapter 2 delves into the behavior and fate of these particles within ecosystems, emphasizing their pervasive distribution, potential bioaccumulation, and the urgent need for effective remediation technologies. Chapter 3 discusses advanced detection and analytical methods. It highlights both established and emerging techniques for identifying and characterizing these contaminants with accuracy and precision. Chapter 4 investigates the challenges MPs pose to water management and treatment systems. It emphasizes the urgent need for advanced technologies and policy integration. Chapter 5 explores both natural and human-induced sources of microplastic pollution and their cascading impacts on global ecosystems. Chapter 6 discusses the toxicological effects of MPs and NPs on living organisms, linking their bioaccumulation to serious physiological disorders and diseases. Chapters 7–9 extend the discussion to environmental and terrestrial domains. These two chapters analyze the risks MPs and NPs impose on ecosystems, agricultural soils, and food safety. Chapter 10 focuses on the human health dimension, detailing the pathways of MP ingestion and the subsequent physiological and genetic risks, alongside strategies for reducing exposure and improving waste management. Chapter 11 examines the routes of human exposure to MPs through food, water, and everyday plastic use. It highlights their bioaccumulation and harmful physiological effects, including oxidative stress, DNA damage, and other disorders. Chapter 12 offers a holistic assessment of MP pollution in freshwater systems. It discusses their sources, transport dynamics, and biological consequences. It further emphasizes the need for integrated scientific and policy approaches to safeguard aquatic ecosystems. The last and final chapter, Chap. 13, introduces the emerging role of AI in detecting, assessing, and remediating MP contamination, showcasing how ML, computer vision, and predictive modeling can revolutionize environmental monitoring and management.

Melbourne, Australia Shrikaant Kulkarni
Pune, Maharashtra, India Susanta Das

# Contents

# Editors and Contributors

## About the Editors

**Shrikaant Kulkarni, Ph.D.**, works as an Adjunct Professor at the Faculty of Business, Victorian Institute of Technology, Melbourne, Australia; Adjunct Professor, Lincoln University College, Malaysia; and Adjunct Professor at Chitkara University, Punjab, India. Dr. Kulkarni has been a senior academician and researcher for 42 years. He has delivered invited lectures and conducted sessions at national and international conferences and faculty development programs. He has guided many major and minor projects in areas, such as engineering chemistry, green chemistry, nanotechnology, analytical chemistry, catalysis, chemical engineering materials, industrial organization and management, advanced AI, and educational technology. He has published over 100 research papers in national and international journals and conferences of repute. He has edited more than 50 books, including four in advanced AI, and authored over 100 book chapters in books published by CRC, Springer Nature, Apple Academic Press, Elsevier, Wiley, and the IET. He has authored four textbooks in engineering chemistry. Dr. Kulkarni holds M.Sc., M.Phil., and Ph.D. degrees in chemistry, as well as master's degrees in economics, business management, and political science. Two of his patents have been published, and two more are pending grants. He was invited by UNESCO to give an invited talk on "Green Chemistry Education for Sustainable Development" at the International Conference on Green Chemistry, Bangkok, and was well received by international jury members and dignitaries from across the world.

**Susanta Das** is a seasoned educator with a proven record of teaching and mentoring diverse students, working at multiple institutions, and providing service to various initiatives of organizations. He received his Ph.D. and M.A. degrees from Western Michigan University (WMU), USA, and his M.Sc. degree from Banaras Hindu University (BHU), India, all in Physics. He continued his research as a Marie-Curie post-doctoral fellow at Stockholm University, Stockholm, Sweden, on beam diagnostics for the DESIREE (Double ElectroStatic Ion Ring ExpEriment) in the project DITANET (Diagnostic Techniques for particle Accelerators—a Marie-Curie initial training NETwork), at the Indian Institute of Science Education and Research-Kolkata, India, on high-pressure physics as a Project Scientist-B, and at the University of Electro-Communications, Tokyo, Japan, on ion-surface interactions as a postdoctoral fellow. Throughout his career, Dr. Das has worked and collaborated with many researchers, Ph.D., master's, and visiting students from different countries. He has further visited many countries to discuss research and participated in international conferences (UK, Italy, Germany, Greece, Belgium, Bulgaria, Romania, Brazil, etc.). He worked at several institutions before assuming his current role at Ajeenkya D Y Patil University. He received the Marie-Curie postdoctoral fellowship in Sweden, the Science Academies' Summer Research Fellowship in India, the Gowen Frostic Doctoral Fellowship, the Department Graduate Research and Creative Scholar Award from WMU, and the Leo R. Parpart Doctoral Fellowship from the Dept. of Physics, WMU, among many others, throughout his academic journey. He has co-authored several research articles in WoS- and Scopus-indexed international journals, conference proceedings, and scholarly book chapters and has co-edited several books published by renowned international publishing houses (IGI-Global, Springer Nature, Nova Science Publishers, Wiley, Scrivener-Wiley, and Notion Press). At present, Dr. Das continues to delve into cutting-edge fields, with a keen interest in nanotechnology, the UN SDGs, and data science. His multifaceted contributions, spanning teaching, research, and administrative leadership, showcase a dedicated professional who is instrumental in advancing the vision and mission of the institutions he serves.

## About the Contributors

**Manish Baboo Agarwal** is an Assistant Professor in the Department of Applied Science (Physics), Sharda University, Agra, with over 22 years of teaching, research, and administrative experience. He earned his Ph.D. in Physics from IIT (ISM) Dhanbad and M.Tech. in Materials Science from IIT Bombay. His research interests include thin film technology, solar cells, nanotechnology, material characterization, and environmental sustainability. He has published extensively in SCI/Scopus-indexed journals, authored books and chapters, and holds several national and international patents.

**Seema Agarwal** is an Assistant Professor in the Department of Mathematics at the SRM Institute of Science and Technology, Ghaziabad, with over 12 years of teaching, research, and administrative experience. She earned her Ph.D. and M.Phil. in Mathematics from Dr. B. R. Ambedkar University, Agra. Her research expertise includes queuing theory, nanotechnology, and groundwater quality assessment. She has published widely in SCI-and Scopus-indexed journals, authored books and book chapters, and holds several national and international patents.

**Jayesh Baldota** is a distinguished professor and researcher in the field of pharmaceutical sciences, specializing in pharmaceutics. With over 6 years of academic and industry experience, he has contributed significantly to the advancement of drug formulation, delivery systems, and pharmaceutical analysis. He is working at Vishwakarma University, Pune. He is pursuing Ph.D. in Pharmaceutical Sciences from Dr. A.P.J Abdul Kalam University, Indore, and has published numerous research articles, book chapters, and patents related to pharmaceutical technology. As an esteemed educator, he has mentored UG and PG level students, making complex pharmaceutical principles accessible and applicable. In addition to his academic contributions, he has collaborated with the leading pharmaceutical sector, focusing on innovative drug delivery system development. Their passion for bridging the gap between theory and practice is reflected in this textbook, which serves as a comprehensive guide for students and practitioners in the field of physical pharmacy.

**Kazi Sabnam Banu** completed her B.Sc. in Chemistry in 2001 from the University of Calcutta, West Bengal. She earned her M.Sc. degree from Visva-Bharati University, West Bengal, in 2004 and her Ph.D. degree from the University of Calcutta in 2010. Her area of specialization was inorganic chemistry. She has 17 years of teaching experience. She served as a part-time lecturer at Ram Mohan College, affiliated with the University of Calcutta, from 2008 to 2010. Subsequently, she was an Assistant Professor at the government-aided Murshidabad College of Engineering and Technology from 2010 to 2017. Currently, she serves as an Assistant Professor of Chemistry at Netaji Nagar Day College, affiliated with the University of Calcutta, West Bengal. She has 9 years of research experience. Her areas of expertise are bioinorganic chemistry. Her thesis title was "Compartmental Ligand Complexes of d- and f-Block Elements: Synthesis, Characterization and Their Reactivity." Dr. Kazi Sabnam Banui has 9 years of experience in inorganic synthesis and the characterization of a wide range of inorganic complexes with rational design and synthesis of small target molecules. She is experienced in the development of new methodologies for the synthesis of important inorganic compounds and ligand design study. She has expertise in traditional and modern techniques for separating complex mixtures of organic compounds, including normal-phase, preparative, reverse-phase, and flash column chromatography, fractional crystallization, vacuum sublimation, and fractional distillation, including high vacuum technique. She has thorough knowledge of modern analytical and characterization techniques, such as FT-IR, LC, MS, ESI, UV-Vis spectrophotometer, and polarimeter. She has published 21 papers in reputed international journals, as well as nine book chapters. She has attended and presented papers at 23 conferences, symposia, seminars, and workshops. Her awards and achievements include: (a) GATE (Graduate Aptitude Test in Engineering) qualified in 2004 with a 97.79 percentile and an All India Rank of 71 and (b) qualified for the CSIR-JRF in June 2004 (conducted by UGC, India).

**Sonali Bhosale** is currently serving as an Assistant Professor and the Head of the Department of Environmental Science at Dr. D. Y. Patil, Arts, Commerce and Science College, Pimpri, Pune, Maharashtra. She has 2 years of teaching experience, and her academic interests include fundamentals of environmental biology and biodiversity, environmental impact assessment and environmental audit, environmental law, and solid and hazardous waste management. She presented a poster titled "Reclaiming the Roots: The Role of Indigenous Ecological Knowledge in Climate Adaptation and Biodiversity Conservation" at a National Conference organized by the Indian Society of Analytical Scientists (ISAS), Pune, Maharashtra, on 2nd August 2025. Miss Sonali Bhosale also participated in the Young Scientist Meet titled "Earth's Changing Contours: The Interplay of Forces and Time" held from 27 to 29th August 2025 and organized by TWAS-CASAREP at JNCASR, Bengaluru, India. She has published three research papers in peer-reviewed journals, focusing on themes, such as environmental sustainability, biodiversity conservation, and climate resilience.

**Kashvi Chaturvedi** is studying B.Tech., specialization in artificial intelligence and data science, at Ajeenkya DY Patil University. She started solidifying her programming skills by completing various certification courses like Data Structures in C, Tableau functions, and Operating Systems along with doing a few projects using languages like Python, C, C++, and creating web pages using HTML, CSS, and JavaScript. She had also attended an Arduino-Based Automation course at Dolphin Labs to enhance her software knowledge. There, she learned how to work with sensors, microcontrollers, and circuit design. She has also written multiple review papers on topics, such as Geographic Information Systems, Comparative Analysis of BJT Configurations, and Swiss Bank vs. European Central Bank: Data Analysis. She is also researching and learning many topics such as small modular reactors, bibliometric analysis, UN SDGs, etc. Because of her interest in the latest technologies, she also participated in Global Hyperloop Competition hosted by IIT Madras, representing her college as part of team Vac-Vectoor, where she learned more about various subtopics from the mentors.

**Tamanna Chaturvedi** is a student at the University Institute of Chemical Technology, Kavayitri Bahinabai Chaudhari North Maharashtra University, Jalgaon. Tamanna is working on nanoparticle synthesis using plant-based methodologies. She has carried out investigations on effects and origins of microplastics and detection methods.

**Jeba Sweetly Dharmadhas** completed her Doctor of Philosophy (Ph.D.) in Biotechnology and is currently working as an Assistant Professor in the Department of Biochemistry and Biotechnology, SF Programs, Campus II. Her research area is environmental biotechnology, with specific interests in nanotechnology, marine biology, and wastewater management. Previously, she served as an Assistant Professor in the Department of Biotechnology at Karpagam Academy of Higher Education, Coimbatore, for 4.5 years. She has published more than 20 research papers, 8 book chapters, and 1 full-length book. She has presented her research findings at various national and international conferences. Additionally, she has participated in several workshops, faculty development programs (FDPs), and refresher courses.

**Ajaygiri Goswami, Ph.D.**, has completed his Master's and doctorate in Chemical Engineering and Technology and is presently working as Professor and Head of Pharmaceutical Technology, KBC North Maharashtra University, Jalgaon. He has presented more than fifty research articles at international conferences and has also published more than one hundred research papers in national and international journals of repute. He is working in various academic bodies of the university, such as the Board of Studies, the Board of Governors, and as a Coordinator for the IIC, Start-up, KCIL cell, etc. He has more than 25 years of teaching and research experience in engineering colleges and university departments, and his research areas are pharmaceutical technology, chemical engineering, oil technology, biofuel, nanocatalysts, etc.

**Aparna Gunjal** completed her B.Sc. in Industrial Microbiology in 2003 from Annasaheb Magar Mahavidyalaya, Hadapsar, Pune. Currently, she is working as a Mentor and Assistant Professor, Department of Environmental Science, Dr. D. Y. Patil, Arts, Commerce and Science College, Pimpri, Pune, Maharashtra. She has 9 and 13 years of teaching and research experience, respectively. Her areas of expertise are bioremediation, composting, plant growth-promoting rhizobacteria, solid state fermentation, and waste management. She has 200 publications to her credit, which include research papers in national and international journals, research papers in national and international proceedings, review articles, books, book chapters, articles in magazines, newsletters, and sequences submitted to databases. She has one German patent granted. She has received 12 national and international awards. Some of them are: (a) Award of Recognition as "Environmentalist" in 2019 by Scire Science at the International Conference of SciCon Series on In Sync—With Next Generation Biosciences held at Goa, India; (b) Appreciation Award for contribution to environmental conservation in 2019–2020; (c) Best Article of the Month Award in May 2020 from Vigyan Varta, an international e-magazine for science enthusiasts; (d) The Outstanding Scientist of the Year Award in 2020 from Vigyan Varta, an international e-magazine for science enthusiasts for outstanding contribution in the field of microbiology and environmental sciences; (e) Most Impactful Article of the Month Award in January 2021 from Vigyan Varta, an international e-magazine for science enthusiasts; (f) Senior Scientist Award in 2023 from Microbiologists Society, India; (g) Award of Recognition of ISTE-AMIar from Rollascripting Nigeria country, in 2023 for working on project related to free education for children; and (h) Best Ph.D. Thesis Award from Vigyan Varta. She has also been selected as a mentor by the American Society for Microbiology (ASM) to mentor students in the aspects of research and teaching in microbiology and has been appointed as a trainer by the Australian Institute of Water Science (AIWS), Australia, to train students of microbiology and environmental sciences in wastewater treatment and research. She serves as an editor and reviewer for many international and national journals.

**Suganya Ilango** is an Assistant Professor of Biochemistry at the SRM Institute of Science and Technology. She completed her Ph.D. in Cancer Biology at the Avinashilingam Institute for Home Science and Higher Education for Women, where she was awarded the prestigious Avinashilingam Ayya Memorial Fellowship (2021) for the academic year 2020–2021. She has published over 15 research articles in reputed national and international journals and contributed to more than 10 book chapters in the fields of cancer biology, phytochemistry, and microplastic pollution. Her research expertise lies in cancer biology, the isolation and purification of plant metabolites, and the therapeutic applications of natural compounds. Dr. Suganya is skilled in advanced research methodologies, including flow cytometry, animal tissue culture, primary cell culture, column chromatography, high-performance thin layer chromatography (HPTLC), polyacrylamide gel electrophoresis (PAGE), and antioxidant activity assays. She has served as a resource person in workshops on animal tissue culture and basic techniques in biosciences research and has been part of the organizing team for demonstrations of advanced analytical techniques at Avinashilingam Institute. Her active involvement in academic and research training programs reflects her commitment to encouraging hands-on learning and promoting scientific excellence among young researchers. Her research interests continue to focus on cancer therapeutics, bioactive plant metabolites, and their molecular mechanisms, with the goal of advancing translational cancer research.

**Vinay Kumar Jadon** is an academician with more than 18 years of teaching, research, and administrative experience, currently serving as an Assistant Professor in the Department of Mathematics at Sharda University, Agra, Uttar Pradesh, India. He completed his Ph.D. in Mathematics from Dr. B. R. Ambedkar University, Agra, and specializes in fluid mechanics, nanotechnology, and groundwater quality assessment. He has published multiple research papers in SCI- and Scopus-indexed journals, authored books and book chapters, and holds several national and international patents.

**Priyanka Jayachandran** is an Assistant Professor of Biochemistry and Biotechnology at Avinashilingam Institute for Home Science and Higher Education for Women, where she also earned her Ph.D. in Biochemistry after completing her B.Sc. and M.Sc. in Biochemistry at PSG College of Arts and Science. With over a year of experience as an Assistant Professor at Sri Ramakrishna College of Arts and Science for Women, she has published 14 peer-reviewed articles and four book chapters. An active member of the Society of Biological Chemists, her research interests span plant biochemistry, recombinant DNA technology, nanotechnology, cancer biology, and immunology; additionally, she is an engaged participant in various national and international academic events.

**Ashish Katke**, a Bachelor of Pharmacy graduate from Vishwakarma University, is deeply passionate about integrating technology into everyday life to enhance health and promote sustainability. With a strong foundation in pharmaceutical sciences and a forward-looking mindset, he strives to bridge the gap between innovation and practical application. Driven by an insatiable curiosity and an unwavering commitment to progress, he continuously seeks out new opportunities to explore, learn, and create. His vision revolves around merging science, technology, and human well-being, ultimately aiming to contribute meaningfully to the advancement and betterment of society.

**Sunil Kulkarni, Ph.D.**, earned his Ph.D. from Sant Gadge Baba Amravati University and his Master's in Chemical Engineering from the Tatyasaheb Kore Institute of Engineering and Technology, Warananagar. He served as an Assistant Professor in the Chemical Engineering Department at Datta Meghe College of Engineering, Airoli, Navi Mumbai, India, for 13 years. He has 25 years of teaching and research experience. Dr. Sunil is involved in various academic activities across multiple universities in Maharashtra. His area of investigation in doctoral studies is phenol and heavy metal removal from wastewater. He has also worked on kinetics and isotherms of various sorption operations. His area of research includes adsorption, clean technology, and environmental biotechnology. He is on the reviewer board of many international journals and reviewed many international papers. He is also a governing body member, promotional editor, and executive member of many reputed international journals. Currently, he serves as Professor and Head of the Department of Chemical Engineering at the Gharda Institute of Technology, Lavel, India.

**M. Vijaya Kumar** earned his B.Tech in Civil Engineering from Y.S.R. Engineering College of Yogi Vemana University, Proddatur, Kadapa, in 2008 and completed his M.Tech with a specialization in Environmental Engineering from the National Institute of Technology (NIT) Warangal, Warangal, Telangana, in 2015. He possesses over 7 years of professional experience, including 6 years and 2 months in academia and 1 year and 5 months in industry. His research and academic contributions include the publication of three Scopus-indexed research papers, two Scopus-indexed book chapters, nine peer-reviewed journal articles, and two textbooks. His areas of interest and expertise encompass advanced water and wastewater treatment methods, air pollution control, and solid waste management. He remains actively engaged in teaching, research, and scholarly writing within the domains of civil and environmental engineering.

**Pranabi Maji** is an academician in the Department of Physics, JIS University. Her research interests lie in the synthesis of sustainable materials for environmental applications, such as waste management. She works on advanced nanocomposites, which are rigorously studied for their dielectric properties to develop sensitive platforms for detecting pollutants in water and soil. She actively integrates artificial intelligence and machine learning to model, predict, and optimize processes in waste management and environmental remediation.

**Manu Mehrotra** is an Associate Professor of Chemistry at the Anand School of Engineering and Technology, Sharda University, Agra, with over two decades of teaching and research experience. He earned his Ph.D. in Chemistry from Jai Narain Vyas University, Jodhpur, specializing in the kinetics and mechanisms of redox reactions. His expertise lies in nanomaterials, environmental chemistry, and material science applications. He has published extensively in reputed journals, authored books and chapters, and holds several patents, while also contributing to academic leadership and professional development through conferences and workshops.

**Dola Mukherjee** is currently an academician at Amity Institute of Biotechnology, Amity University, Kolkata. Her work focuses on evolutionary bioinformatics and plant systems biology, with an emphasis on understanding the genomic and regulatory mechanisms that shape plant metabolism, influence productivity, drive adaptability, and govern ecological interactions. She is currently integrating artificial intelligence (AI) and machine learning approaches to develop innovative strategies for sustainable agriculture and environmental management.

**Dibyarupa Pal** is an academician in the Department of Biosciences, JIS University. Her research interests are centered on advancing sustainable environmental remediation strategies by developing eco-friendly solutions with a dual focus on purifying water by removing industrial dyes and turbidity using natural resources. Complementing this work, she is also deeply invested in green synthesis, developing environmentally benign nanomaterials for the effective remediation of heavy metal pollutants contaminating both soil and water systems.

**Mahek Pathan** is currently pursuing her MBBS degree at Northern State Medical University, Arkhangelsk, Russia. Alongside her medical studies, she has a strong interest in pharmaceutical research, particularly in the areas of innovative formulations and healthcare solutions. Driven by curiosity and a passion for advancing patient care, she aspires to contribute to the integration of medicine and pharmacy for improved therapeutic outcomes and overall well-being.

**Krutika Patre** is an aspiring programmer and technology enthusiast currently pursuing engineering in artificial intelligence and data science. With a passion for learning, she is building experience in a variety of programming languages and technologies, including C, C++, Python, Data Structures (DS), HTML, CSS, JavaScript, and DBMS (MySQL). After learning these programming languages, she has developed interest in Full Stack Development. She also attended an Arduino-Based Automation course at Dolphin Labs to improve her software knowledge. There, she learned how to work with sensors, microcontrollers, and circuit design. She has written a review paper on sustainable engineering in the built environment. She is researching and learning about many topics related to AI, policy, and governance that align with her interests. She has taken part in hackathons, workshops, and tech events to explore many fields. She is also working with a group on Hyperloop. Being part of this project has really helped her grow, especially in areas like teamwork, innovation, and systems thinking.

**Ritam Paul** is a postgraduate student in the Department of Biosciences, JIS University. He is interested in the degradation of emerging pollutants, such as microplastics. He is actively engaged in remediation research for his project work using beneficial microbes.

**Sunil Sankathala** is pursuing a B. Tech in Computer Science and Engineering at Ajeenkya DY Patil University, with a diverse range of interests that bridge technology, sustainability, and innovation. His academic focus lies in computer science fundamentals like programming, data structures, algorithms, and software development. He actively supplements this with hands-on experience in web development and robotics. Beyond the core curriculum, he has a strong research inclination in emerging and impactful fields. His current research interests include artificial intelligence in healthcare, bioplastics, nanobioplastics, e-waste management, and solid waste management—areas where he believes technology can make a meaningful difference in solving real-world problems. He has been publishing research articles and book chapters in reputed national and international journals and books. He is also part of an ambitious Hyperloop project, which involves interdisciplinary collaboration and cutting-edge engineering aimed at revolutionizing transportation. Working on this project has enhanced his skills in teamwork, innovation, and systems thinking. In the tech domain, he has developed proficiency in web development using technologies, such as HTML, CSS, JavaScript, and React. His interests in robotics further reflect his goal of building smart, responsive systems that interact with the physical world. Throughout his course, he has strengthened his problem-solving and collaborative skills. He is proficient in languages such as Python, C, C++, and MySQL and is experienced with the Ubuntu operating system and machine learning. He is also familiar with frameworks and tools like React and Git. He actively participates in hackathons, coding competitions, and tech events to challenge himself and learn from peers. He is a quick learner and highly adaptable, with an eagerness to stay updated with the latest trends in the tech world. He aspires to become a skilled software engineer and contribute to building innovative solutions that have a positive impact on society. His long-term goal is to work in a dynamic environment where he can expand his knowledge, grow professionally, and make meaningful contributions to the field of technology. Driven by a spirit of exploration and sustainability, he aims to contribute to technological solutions that address both human and environmental needs. His long-term goal is to become a versatile engineer and researcher, working at the intersection of computer science, environmental innovation, and human well-being.

**K. V. Shalini** is an accomplished academic and researcher in biotechnology, currently serving as an Assistant Professor at Avinashilingam Institute for Home

Science and Higher Education for Women. Holding a Ph.D. in Biotechnology, she has accumulated 13 years of extensive teaching experience, including prior assistant professorships at Sri Ramakrishna College of Arts and Science and Rathnavel Subramaniam College of Arts and Science. Her significant research contributions encompass metallothioneins, environmental biotechnology, biofuels, and nanobiotechnology, supported by a UGC Minor Research Project. A recognized Ph.D. and M.Phil. guide at Bharathiar University, Dr. Shalini is a prolific author with 14 research papers published in Scopus and Web of Science indexed journals, 13 book chapters, and one book. She also serves as a reviewer for prestigious journals and has organized and presented at numerous national and international conferences. Her expertise extends to various academic services, including roles in NAAC and Green Audit committees, as well as being a question paper setter and external examiner for several universities.

**Tanishka Sharma** is a third-year Bachelor of Pharmacy student at Vishwakarma University, Pune, with a strong academic record and hands-on experience in pharmaceutical research. She has interned at GenePath Diagnostics and IIT Bombay, gaining expertise in molecular diagnostics, healthcare innovation, and clinical research. Passionate about advancing drug delivery systems and novel formulations, she has worked on projects ranging from herbal solutions to chewable drug-based gummies for children with autism. With several research publications and active involvement in social initiatives, Tanishka aspires to contribute meaningfully to pharmaceutical sciences and improve patient well-being through innovative healthcare solutions.

**Aditya Shrivastav** is an aspiring technologist currently pursuing a Bachelor of Technology (B. Tech.) in Computer Science and Engineering at Ajeenkya D. Y. Patil University. He holds solid programming expertise in C, C++, and Python, alongside strong command over data and business tools, such as Microsoft Excel and Tally software. Aditya is a member of Team Vac-Vectoor Hyperloop and participated with the team in the Global Hyperloop Competition organized by IIT Madras, contributing to cutting-edge transportation solutions. He actively engages in hackathons and various technical events to expand his skills and contribute to innovative problem-solving. He has authored and co-authored book chapters in the domains of solid waste management, bioplastics, nanotechnology applications, and artificial intelligence, showcasing his multidisciplinary approach and research acumen. Aditya is also working on an innovative project on predictive e-waste management. The project is targeted toward IT departments, schools, and colleges, aiming to support sustainable practices and resource optimization. In parallel with his academic journey, Aditya is serving as a faculty intern at Disha Computer Institute, where he teaches C, C++, basic Python, basic Java, Excel formulas, and Tally software. His dedication and teaching proficiency earned him a Letter of Appreciation and the opportunity to conduct a workshop on Advanced Excel at G. H. Raisoni College of Arts, Commerce and Science. With a strong foundation in technology, a passion for sustainability, and hands-on experience in both teaching and project development, Aditya Shrivastav is committed to driving impact through research, education, and innovation.

**Aishwarya Vetrivel** is an Assistant Professor of Biotechnology at Dr. G. R. Damodaran College of Science, Coimbatore. She completed her B.Sc. and M.Sc. in Biochemistry at PSG College of Arts and Science, Coimbatore, and her Ph.D. in Biochemistry at Avinashilingam Institute for Home Science and Higher Education for Women, Coimbatore. She has over 3 years of experience at Dr. G. R. Damodaran College of Science as an Assistant Professor in Biotechnology. She has published 16 peer-reviewed articles in various national and international journals and also 10 Book chapters. She has participated in various conferences, webinars, faculty development programs, and awareness camps at both the national and international level. Her fields of interest include microbial biotechnology, recombinant DNA technology, clinical biochemistry, and genetic engineering.

## List of Contributors

**Aminu Abdullahi** Department of Biotechnology, Modibbo Adama University, Gombi, Nigeria

**Abdulhalim Musa Abubakar** Department of Chemical Engineering, Faculty of Engineering, Modibbo Adama University, Yola, Adamawa, Nigeria

**Manish Baboo Agarwal** Sharda School of Basic Sciences and Research, Sharda University Agra, Keetham, Agra, Uttar Pradesh, India

**Seema Agarwal** Department of Mathematics, SRMIST, Ghaziabad, Uttar Pradesh, India

**Jayesh Baldota** Department of Pharmaceutics, School of Pharmacy, Vishwakarma University, Pune, Maharashtra, India

**Kazi Sabnam Banu** Department of Chemistry, Netaji Nagar Day College, University of Calcutta, Kolkata, India

**Sonali Bhosale** Department of Environmental Science, Dr. D. Y. Patil, Arts, Commerce & Science College, Pimpri, Pune, Maharashtra, India

**Kashvi Chaturvedi** School of Engineering, Ajeenkya D Y Patil University, Pune, Maharashtra, India

**Tamanna Chaturvedi** University Institute of Chemical Technology (UICT), Kavayitri Bahinabai Chaudhari North Maharashtra University (KBCNMU), Jalgaon, Maharashtra, India

**Susanta Das** School of Engineering, Ajeenkya D Y Patil University, Pune, Maharashtra, India

School of Science, Ajeenkya D Y Patil University, Pune, Maharashtra, India

**Gopi Devarajan** Department of Economics, Faculty of Science and Humanities, SRM Institute of Science and Technology, Chengalpattu, Tamil Nadu, India

**Jeba Sweetly Dharmadhas** Department of Biochemistry and Biotechnology, Avinashilingam Institute for Home Science and Higher Education for Women, Coimbatore, Tamil Nadu, India

**G. Flora** Department of Botany, St. Mary's College (Autonomous), Thoothukudi, India

**Ajaygiri Goswami** University Institute of Chemical Technology (UICT), Kavayitri Bahinabai Chaudhari North Maharashtra University (KBCNMU), Jalgaon, Maharashtra, India

**Aparna Gunjal** Department of Environmental Science, Dr. D. Y. Patil, Arts, Commerce & Science College, Pimpri, Pune, Maharashtra, India

**Suganya Ilango** Department of Biochemistry, Faculty of Science and Humanities, SRM Institute of Science and Technology, Chennai, Tamil Nadu, India

**Vinay Kumar Jadon** Sharda School of Basic Sciences and Research, Sharda University Agra, Keetham, Agra, Uttar Pradesh, India

**Priyanka Jayachandran** Department of Biochemistry and Biotechnology, Avinashilingam Institute for Home Science and Higher Education for Women, Coimbatore, Tamil Nadu, India

**Ashish Katke** School of Pharmacy, Vishwakarma University, Pune, Maharashtra, India

**Shrikaant Kulkarni** Victorian Institute of Technology, Melbourne, Australia

**Sunil Kulkarni** Chemical Engineering Department, Gharda Institute of Technology, Lavel, India

**Mamilla Vijaya Kumar** Department of Civil Engineering, KSRM College of Engineering, Kadapa, Andhra Pradesh, India

**Pranabi Maji** Department of Physics, JIS University, Kolkata, India

**K. Malathi** Department of Economics, University of Madras, Chennai, Tamil Nadu, India

**R. Beaulah Mary** Department of Allied Health Science, Ganga College of Nursing, Coimbatore, India

**Manu Mehrotra** Sharda School of Basic Sciences and Research, Sharda University Agra, Keetham, Agra, Uttar Pradesh, India

**Dola Mukherjee** Amity Institute of Biotechnology, Amity University, Kolkata, India

**Dibyarupa Pal** Department of Biosciences, JIS University, Kolkata, India

**Mahek Pathan** Northern State Medical University, Arkhangelsk, Russia

**Krutika Patre** School of Engineering, Ajeenkya D Y Patil University, Pune, Maharashtra, India

**Ritam Paul** Department of Biosciences, JIS University, Kolkata, India

**G. Ramesh** Department of Metallurgical and Materials Engineering, Rajiv Gandhi University of Knowledge Technologies, R K Valley Campus, Kadapa, India

**Sunil Sankathala** School of Engineering, Ajeenkya D Y Patil University, Pune, Maharashtra, India

**K. V. Shalini** Department of Biochemistry and Biotechnology, Avinashilingam Institute for Home Science and Higher Education for Women, Coimbatore, Tamil Nadu, India

**Tanishka Sharma** School of Pharmacy, Vishwakarma University, Pune, Maharashtra, India

**Aditya Shrivastav** School of Engineering, Ajeenkya D Y Patil University, Pune, Maharashtra, India

**S. Surendarnath** Department of Mechanical Engineering, DVR & Dr. HS MIC College of Technology (Autonomous), Vijayawada, India

**Aishwarya Vetrivel** Department of Biotechnology, Dr. G.R. Damodaran College of Science, Coimbatore, Tamil Nadu, India

# Chapter 1
# A Brief Overview of Microplastics and Nanoplastics

**Sunil Sankathala , Kashvi Chaturvedi , Krutika Patre , Aditya Shrivastav , Susanta Das , and Shrikaant Kulkarni**

## Introduction

A plastic is a synthetic substance that is largely composed of polymers, which are large molecules that are a combination of many smaller molecules called monomers [1]. The name plastic derives out of the Greek word plastikos, which means something that can be moulded or shaped, the fact that plastics can be softened in the heat or through other chemical processes and then worked into different shapes that hold their form when cooled down [2].

Plastics contain polymers that are primarily of two types: Thermosetting plastics and thermoplastics. Soft-plastic materials such as polyethylene (PE), polypropylene (PP), and polyvinyl chloride (PVC) are thermoplastics that can be reformed under heat, and recycled on several occasions [3]. The thermosetting plastics, e.g. epoxy resins, are crosslinked to become irretrievable in the process of curing, which give them thermal resistance and mechanical integrity, but makes them much harder to recycle [4].

Plastics have gained extensive application in industries because of their versatility, durability, chemical stability, and lightweight characteristics, which are necessary in the packaging, construction, healthcare, and electronics [5]. Nonetheless, the majority of the plastics are based on fossil fuels and are not biodegradable, which

S. Sankathala (✉) · K. Chaturvedi · K. Patre · A. Shrivastav
School of Engineering, Ajeenkya D Y Patil University, Pune, Maharashtra, India

S. Das (✉)
School of Engineering, Ajeenkya D Y Patil University, Pune, Maharashtra, India

School of Science, Ajeenkya D Y Patil University, Pune, Maharashtra, India

S. Kulkarni
Victorian Institute of Technology, Melbourne, Australia


S. Kulkarni, S. Das (eds.), *Microplastics and Nanoplastics*, Lecture Notes in Nanoscale Science and Technology 49,
https://doi.org/10.1007/978-3-032-17963-0_1

brings about a lot of environmental issues concerning pollution and waste disposal [6].

This chapter presents information on different aspects and important parts that are considered when researching on microplastics and nanoplastics. The chapter begins providing the necessary information from the basics, i.e. types of plastics, before we start researching on any topic, it is necessary to understand the basics of it. After understanding the different types of plastics and going through their respective examples, the chapter moves forward to the various applications of plastics and different types of plastics in various fields like packaging, construction, agriculture, etc. It is important to know about the chemical composition of these plastics, which will be represented in next section. The chapter will conclude itself by presenting bibliometric analysis on the topic "Microplastics and Nanoplastics" by illustrating different figures and table which helps in understanding different keywords, their relation, shift in research trend, increase and decrease in number of publications, and contribution of different countries and publishers.

## Types of Plastics

Plastics consist of long chains of molecules known as polymers which are formed by the combination of smaller molecules (monomers). Thermoplastics and thermosetting plastics are broadly categorized into two groups depending on their behavior on heating.

### *Thermoplastics*

Thermoplastics become soft in the presence of heat and hard in the presence of coolness. This is because this is a reversible change and therefore can be melted and reshaped many times without altering their chemical nature and hence it is easy to recycle. Due to this fact, the thermoplastics dominate the planet as the plastics [1, 3, 4].

**Common Examples Include [1, 3, 4]**
**Polyethylene (PE):** This is the most widely manufactured plastic that is usually utilized in grocery bags, bottles and containers. PE may be produced to possess varying densities and characteristics.

**Polypropylene (PP):** Polypropylene is a chemically resistant and tough material that is used in textiles, bottle tops, automobile components, and food packaging.

**Polyvinyl Chloride (PVC):** PVC is a multifunctional product. It is rigorously applied in pipeline and window frames, and as pliable plastic in cables and garments, depending on the additives.

**Polyethylene Terephthalate (PET):** PET is popular in bottles of plastic and clothing made by synthetic fibers. The thermoplastics are versatile materials, as they are easy to work with and recycle.

## *Thermosetting Plastics*

Plastics that are thermosetting begin as soft or liquid plastics. Heat, or other chemical curing, causes them to undergo permanent, crosslinked structures that cannot soften upon being heated again. This causes them to be vastly heat-resistant and mechanically resilient but extremely less recyclable.

**Examples Include**
**Epoxy resins:** Components of powerful adhesives and coats.

**Phenolic resins:** These are frequently used in electrical appliances and thermal resistant products.

**Polyurethanes:** Foam, furniture and insulation. Such plastics can be put into service in challenging industrial applications because of its stability and strength.

## *Microplastics and Nanoplastics [4]*

### Microplastics

Microplastics are small plastic fragments that take up less than 5 mm, smaller than a sesame seed, even microscopic.

**They Originate From**
Primary microplastics are purposeful microplastics such as microbeads in cosmetics or industrial pellets.

Secondary microplastics, are particles that occur when the larger plastic objects (bottles, bags, fishing nets, and others) are disintegrated by sunlight, waves, and wear.

The microplastics have been discovered all over the world in the ocean, rivers, soils, or even in the air. Microplastics have been the subject of concern to marine life and human health due to the food chain, as a result of them being ingested by many aquatic creatures accidentally [7].

#### Nanoplastics

Nanoplastics with even smaller plastic particles, usually less than 100 nm. Nanoplastics are primarily formed as a result of the further degradation of microplastics. Since nanoplastics are so small, they can enter cells and biological tissues and can be toxic to them, but their effects are still not completely understood [8].

### *Bio-plastics: Biodegradable and Non-biodegradable*

Plastics derived not only out of fossil fuels but out of renewable biological materials, like plants, are called bioplastics. Bioplastics may be either biodegradable or non-biodegradable:

#### Biodegradable Bioplastics

The microorganisms can break these materials down to harmless substances such as water, carbon dioxide and biomass.

**Examples Include**
**Polylactic acid (PLA):** This is a fermented corn starch or sugarcane derived product that finds application in packaging and disposable products.

**Polyhydroxyalkanoates (PHA):** It is a bacterial product that can be used in packaging and medicine.

**Starch plastics:** This is formed by altering plant starch with the help of chemicals to be used as packaging.

Many bioplastics, however, are biodegradable, but can only degrade in particular conditions, including industrial composting plants [9].

#### Non-biodegradable Bioplastics

They are produced using renewable materials but possess the same chemical structures as traditional plastics, which prevents them from biodegrading.

**Examples Include**
**Bio-polyethylene (bio-PE):** It is made of plant-based ethanol, which is not different in properties than fossil-based PE.

**Bio-polyethylene terephthalate (bio-PET):** Consumers Bottles and fibers are made with bio-PET, which is 100% renewable in source and 100% non-biodegradable. They help reduce the use of fossil fuel but still need to be recycled to control the waste [10].

## Applications of Plastics

Plastics are widely used in industries and our daily life. Plastics have brought revolution in the process of designing, production and distribution of products in the world. Plastics are a material that cannot be overlooked in many sectors because of their unique properties, i.e., they are lightweight, durable, corrosion-resistant, and easy to mold. This extended discourse explains the numerous large-scale applications of plastics, their advantages, the main material utilized and the difficulties with each application.

### *Packaging Industry*

Packaging is indeed the biggest industry that consumes plastics with an estimated percentage of around 40% of all plastic worldwide [5]. The benefits that plastic packaging presents are unparalleled because it is lightweight, has a good barrier capability that ensures that its products are not subject to moisture, oxygen and contamination, and can be made in diverse shapes and sizes and at cheap costs. The most common polymers used in packaging are Polyethylene (PE), polypropylene (PP) and polyethylene terephthalate (PET). PE and PP are also applicable in flexible films, bags, and containers where they are used to offer convenience in transportation and maintain the freshness of the product. PET is valued due to its strength, transparency, and recyclability, which make it the best choice in terms of bottle drinking beverages and food packaging [9].

Another innovation in the packaging field is the bioplastics like polylactic acid (PLA), which are potentially biodegradable products that can substitute the traditional plastics. Nevertheless, the issues of bioplastic packaging are the appropriate management of waste to achieve environmental gains [4].

### *Construction*

Plastics are durable, resistant to chemicals and versatile which are of great advantage to the construction industry. Polyvinyl chloride (PVC) enjoys popularity as a material in pipes, window frames and siding, and flooring because of its strength and resistance to the weather. Polyurethane (PU) foams have significance in relation to insulating panels that ensure better energy efficiency in buildings. Plastics can also be found in roofing and wall covering, and interior finishing. Moreover, composite materials, which consist of plastic matrices with reinforcing materials such as glass fiber, are presented with superior mechanical characteristics, which can be used in structural applications [3].

Plastic use minimizes the weight of construction materials, makes installations simpler, and in most cases, it is cheaper to use, and also ensures enhanced maintenance cycles as it is resistant to corrosion and water [4].

## *Automotive Sector*

Plastics have revolutionized production of vehicles, since they have greatly lowered the weight of cars and fuel consumption. Plastic products can take up about 50% of the volume of a vehicle and 10–15% of the car mass, including dashboards, door panels, bumpers, and under-hood components, all the way to lighting and assemblies [1]. The most common polymers that are used in the automotive industry include polypropylene (PP), polycarbonate (PC), polyurethane (PU), and acrylonitrile butadiene styrene (ABS) because they are impact resistant, thermally stable and can be easily molded.

Plastics result in weight reduction of vehicles, which leads directly to decreased fuel consumption and decreased $CO_2$ emissions, which is in line with environmental regulations in the world. In addition, plastics are used to enhance safety of vehicles through use of energy-absorbing bumpers & enhanced crash structures [5].

## *The Electronics and Electrical Industry*

Plastics are used in important purposes in the electronics and electrical devices with the ability to provide electrical insulation, mechanical strength, & heat resistance. Thermosetting plastics like epoxy resins play a vital role in the printed circuit board (PCBs). The smartphone, computer, and home appliance housing is made of thermoplastics, such as polycarbonate and ABS, that are durable & safe [4].

The growing sub-miniaturization of electronics and the need to have high strength, lightweight and fire-retardant materials have necessitated the use of plastics in this industry.

## *Healthcare Applications*

The plastics have been heavily integrated into the contemporary healthcare because of their sterility, disposability, and versatility. Syringes, IV tubing, surgical gloves, diagnostic equipment, implantable prostheses, and pharmaceutical package containers are largely manufactured in plastics of high standards of biocompatibility. Polylactic acid (PLA) polymers or polyhydroxyalkanoates (PHA) are also biodegradable polymers that offer safer resorbable alternatives to implantable devices and drug delivery systems [9].

Plastic clinical products that are used single-use lower the risk of infection and avert patient safety but present a problem of disposal that has to be solved by the healthcare system.

### *Agriculture*

Plastics are used in agriculture in a number of ways as a means of improving agricultural productivity and resource conservation. Mulching is done through the use of plastic films that retain soil moisture and weeds. Polyethylene greenhouse covers enhance all-year round growing activities. Plastic irrigation tubing and flexible pipes are made by using hard plastics that make delivery systems of water efficient. Protective packaging and silage wrap also use the plastics to spoilage feed and crops [7].

The negative aspect of using agricultural plastics is residues which can be deposited in the soil and cause microplastics hence studies are being made on the biodegradable options.

### *Consumer Goods and Textiles*

Plastics have an extremely diverse use in the manufacture of consumer products, including toys and dishware, sporting equipment and electronics accessories. The textile industry is dominated by synthetic fibers such as nylon, polyester and acrylic that are of plastic polymers and have good durability, wrinkle-free and easy-to-service properties [5].

Consumer goods that are packed with plastic make them affordable and diverse, yet also create the problem of waste because of their short durability.

## Chemical Composition of Microplastics and Nanoplastics Explanation

Micro and nanoplastics are small pieces of plastic materials that have disintegrated due to bigger plastic products or in some cases, are produced at very small sizes to be used in industries. Chemical knowledge of what those particles consist of can assist us in knowing how they will act in the environment and how they are likely to influence living things.

The majority of microplastics and nanoplastics are derived out of usual varieties of plastics used in daily products. These plastics consist of long chains of molecules known as polymers which are composed of smaller units known as monomers. The major polymers that are present in microplastics and nanoplastics are polyethylene (PE), polypropylene (PP), polystyrene (PS), polyvinyl chloride (PVC), and polyethylene terephthalate (PET). All these polymers possess distinctive chemical structures which influence the duration of their environmental persistence as well as their interaction with contaminants and organisms [4].

Polyethylene (PE) is made up of repeated units of ethylene. It is also hydrophobic, i.e. it is water resistant and is also chemically stable hence it does not easily decompose in nature. PE finds wide application in the case of plastic bags, bottles, and packaging films [11].

Polypropylene (PP) is a polymer that is produced using propylene monomers and propylene finds significant application as a bottle top, in the textile industry, and in the automobile industry. PP is durable and has a good chemical resistance, which causes it to break down into micro- and nanoplastics when subjected to sunlight and mechanical forces [12].

Polystyrene (PS) includes the monomers of styrene having aromatic rings, and thus it is rigid. PS is widely used in disposable cups and packaging foams and is not easily degraded in the environment because of its constant chemical bonds [13].

Polyvinyl chloride (PVC) is chemically different since it is resistant to weathering and chemicals due to the presence of chlorine atoms. Nonetheless, the chlorine concentration is also a matter of concern since it may produce harmful substances upon breakage. PVC is employed in cable, windows and pipes [14].

Polyethylene terephthalate (PET) refers to a polyester, a product of the combination of terephthalic acid and ethylene glycol. It is applied in the creation of beverage bottles and textile fibers, and it is strong, but it finally disintegrates into microplastic particles [15].

Besides these polymers, the plastics are usually mixed with additives such as plasticizers, flame retardants, dyes, and stabilizers to enhance the properties of the plastics. Micro- and nanoplastics may release some of these additives to the environment, which may be toxic to humans and sea creatures [16].

Nanoplastics are microscopic in size and their surface area is significantly high as compared to the volume. This would predispose them to environmental pollutants like heavy metals and persistent organic compounds, thus making them toxic and more influential on ecosystems and health [17].

More so, sunlight, oxygen, and microorganisms are environmental factors that alter the chemical structure of micro- and nanoplastics over time—a process termed weathering. This alters their surface chemistry and makes them brittle and reactive, thereby affecting their behavior in the environment and their harmfulness. Microplastics & nanoplastics are a combination of widely used plastic polymers such as PE, PP, PS, PVC, & PET, & chemical additives. They are long-lasting pollutants with their design of enduring chemical compositions. Their surface properties are altered by weathering and this impacts the interaction with pollutants and living organisms [15].

## Bibliometric Analysis

The following figures and context are related to research on microplastics and nanoplastics. These figures help in understanding important key terms that are being used for the research or which receive maximum focus due to different reasons. This can be understood from Figs. 1.1, 1.2, and 1.3 where Fig. 1.1 explains how keywords are segregated into different clusters and how these clusters are interconnected showcasing multidisciplinary research. Figure 1.2 helps in understanding the trend and how the focus has shifted from different keywords along with time whereas Fig. 1.3 helps in understanding the keywords that received maximum attention irrespective of time. Figure 1.4 shows the increase of number of publications on this topic, Fig. 1.5 indicates the contribution of different countries in the research and Fig. 1.6 represents the number of times a keyword has appeared in various publications. Table 1.1, in the end, represents the publishers with maximum contribution.

Figure 1.1 is a network visualization map using VosViewer [18] for data taken from PubMed [19] offers a closer look at the connections of the research on

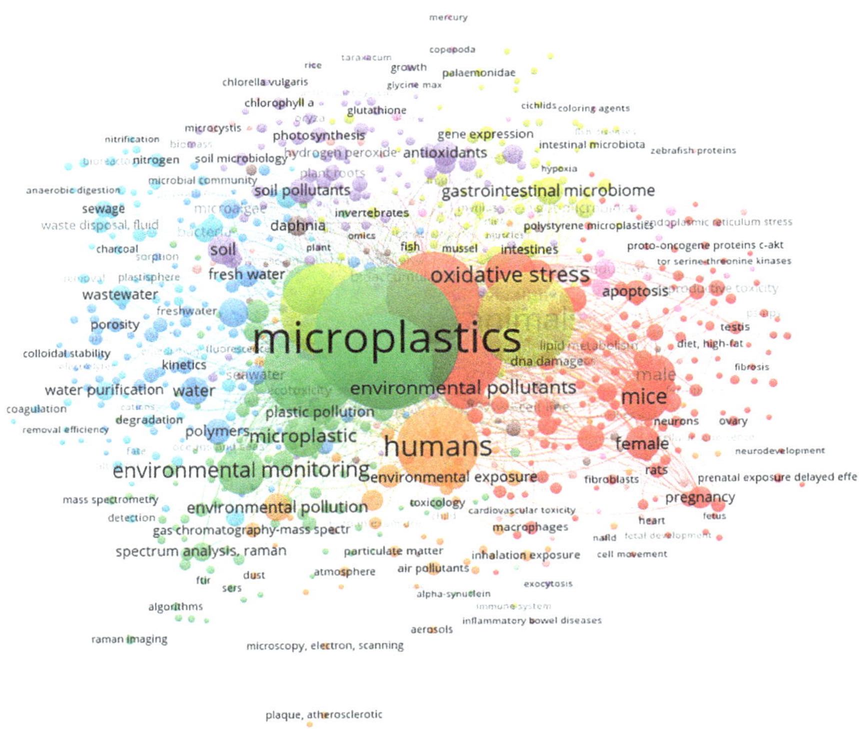


**Fig. 1.1** Network visualization of microplastic and nanoplastic. (Source: PubMed Database; created with VOSviewer) [18, 19]

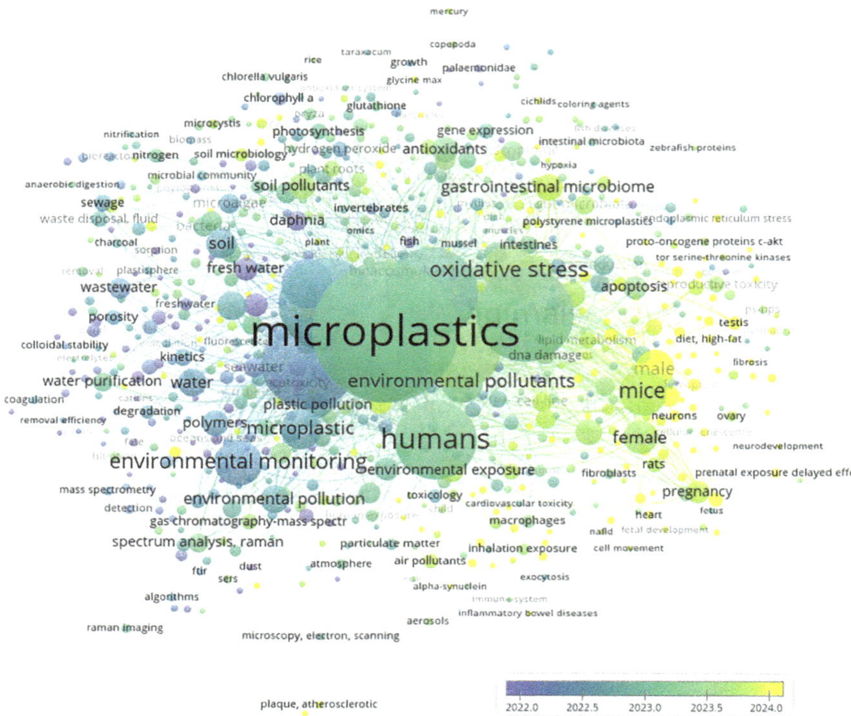


**Fig. 1.2** Overlay visualization of microplastic and nanoplastic. (Source: PubMed Database; created with VOSviewer) [18, 19]

microplastics in various scientific spheres. Every circle (node) is a keyword obtained in the research papers and the circle size indicates the frequency of occurrence of that word in the published research. The various colors are used to denote various clusters and each of these clusters showcase a research area or a field. This image collectively indicates how the research on the topic of microplastic has become a broad and interconnected system of environmental, toxicological, and biological topics.

The biggest and most dominating keyword is the term microplastics that is at the center of the visualization and is the primary connecting factor among all other areas of research. It is surrounded by the large corresponding terms like "humans," "environmental pollutants," "oxidative stress," "environmental exposure," and "environmental monitoring." This center keyword, microplastic, is linked with human and environmental health, which can be inferred from the neighboring keywords that it is connected to. The proximity of the term microplastics to the words humans and oxidative stress demonstrate the scientific interest of analyzing the occurrence of pollution, to the biological and toxicological impacts of microplastics on living organisms, and specifically, on humans.

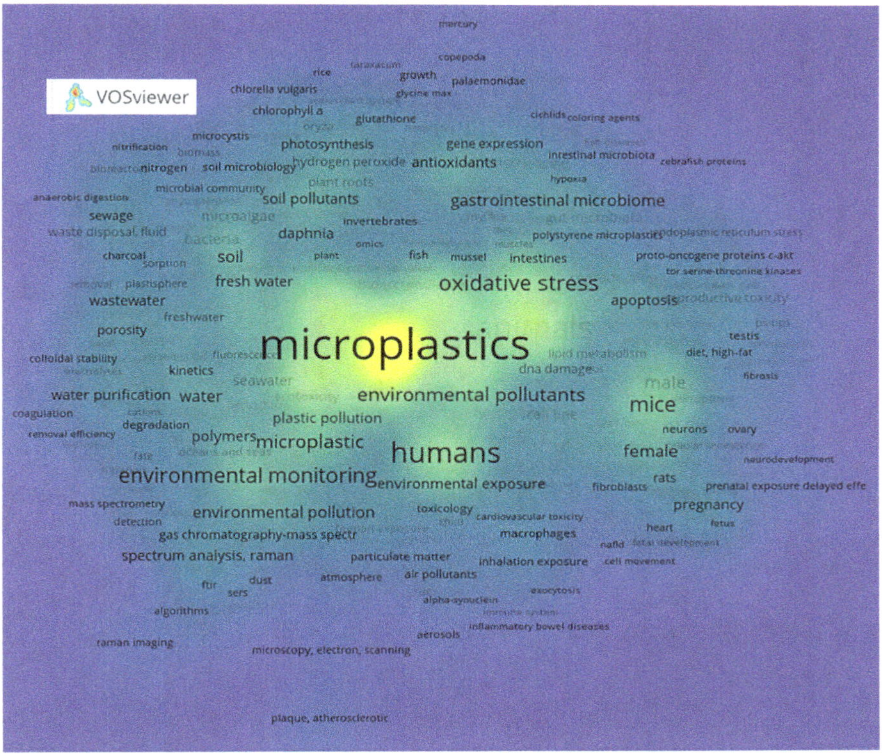


**Fig. 1.3** Density visualization of microplastic and nanoplastic. (Source: PubMed Database; created with VOSviewer) [18, 19]

The blue grouping on the left part of the map is more of the environmental and analytical research domain. In this case, the keywords include "soil," "fresh water," "wastewater," "water purification," "soil microbiology," "bacteria," "degradation," "filtration efficiency," and "mass spectrometry." This cluster demonstrates that much of the initial studies dealt with the way in which microplastics are dispersed across various environmental systems soil, water and air, and how these can be detected and eliminated. The fact that such words as gas chromatography, Raman spectroscopy and electron microscopy are technical shows that these works focus on the creation of analysis tools and measurement methods of microplastic pollution. This blue section symbolizes the background period of the field, during which the focus was on finding and determining the abundance of microplastics in nature.

Keywords including "air pollutants," "particle matter," "toxicity," "bioaccumulation," and "environmental contaminants" are present in bottom and central green cluster. This cluster connects environmental monitoring and biological research to link pollution to toxicological consequences. It demonstrates how researchers began to look at the locations of microplastics' occurrence as well as how they build up in living things & what potential harm they may cause. It is the intersection of

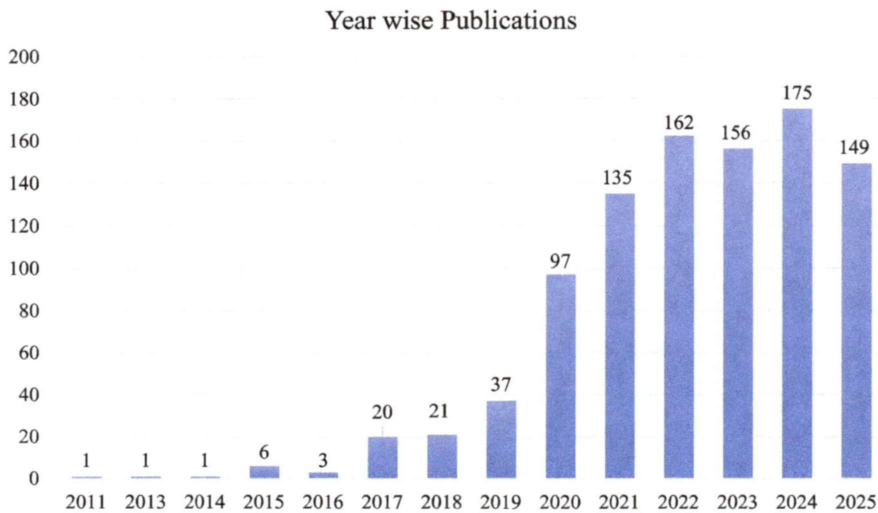


**Fig. 1.4** Year-wise publications on microplastic and nanoplastic research from 2011 to 2025, showing a steady increase in global research output. (Data source: Lens [20])

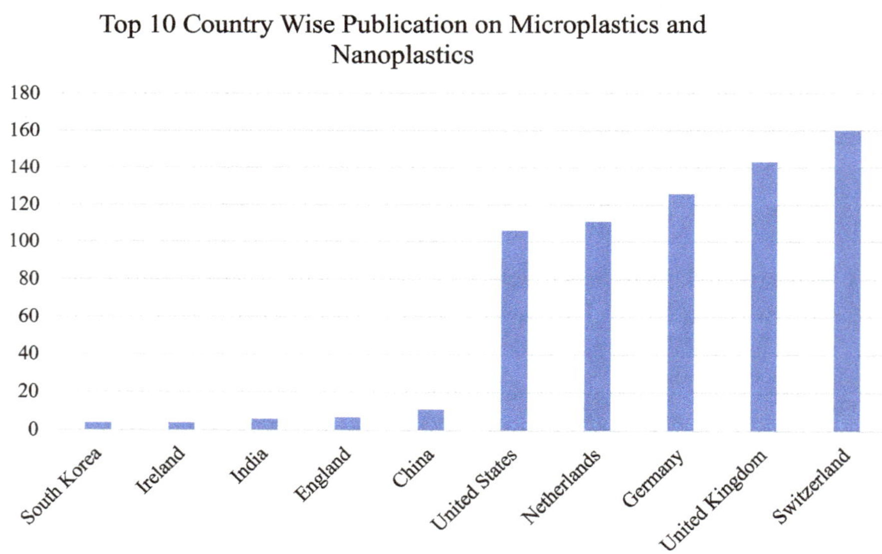


**Fig. 1.5** Top 10 country-wise publications on microplastic and nanoplastic research, highlighting leading contributors such as Switzerland, the United Kingdom, and Germany. (Data source: Lens [20])

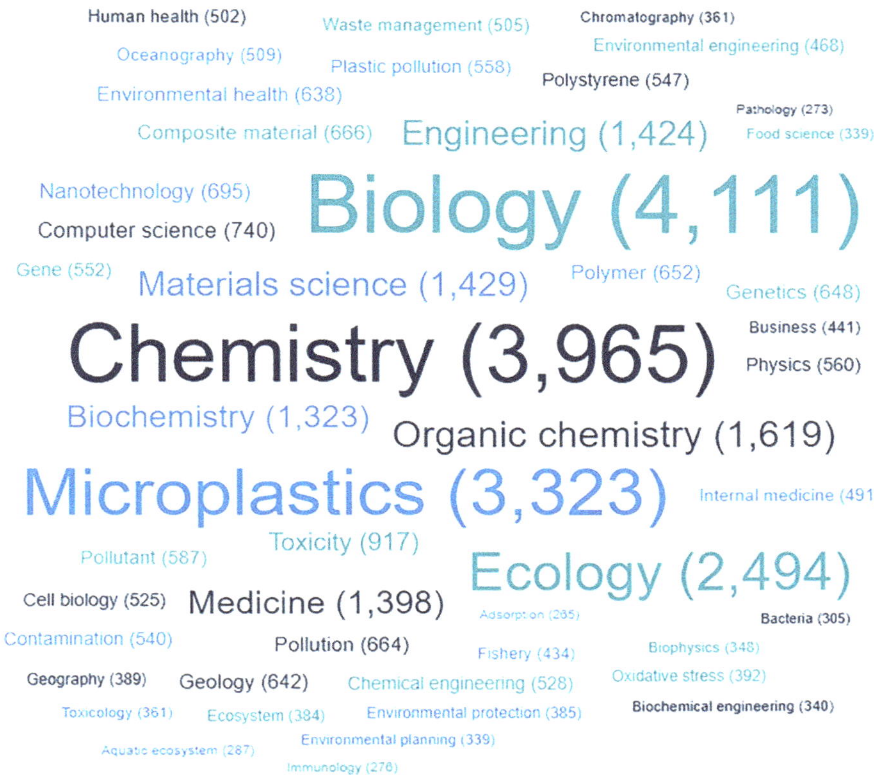


**Fig. 1.6** Word cloud of major research themes and subject areas related to microplastic and nanoplastic studies, emphasizing dominant fields such as biology, chemistry, ecology, and engineering. (Data source: Lens [20])

ecotoxicological and environmental study as scientists started looking at risk factors and microplastic exposure pathways.

The biological and health impact cluster of the right side is dominated by red-colored nodes and it has emerged as one of the rapidly advancing fields of research on microplastics. Major keywords in this case are "mice," "rats," "female," "male," "pregnancy," "neurons," "apoptosis," "intestinal microbiota," "fibrosis," "inflammation," "heart," "fetus," and "neurodevelopment." These expressions suggest a great interest in experimental research of the impact of exposure to microplastic on reproduction, development, brain health, and cellular processes through the use of animal models. The use of such words as oxidative stress, DNA damage, and cell death (apoptosis) indicate that the researchers are examining how microplastics can cause oxidative and inflammatory cells at a cellular scale. The terms "diet," "high-fat," and "gut microbiome" suggest that scholarly investigations into intestinal health and metabolism are currently linking microplastics to more serious health issues including immune system dysfunction, obesity, and digestive malignancies.

**Table 1.1** Major publishing organizations contributing to microplastic and nanoplastic research, with Springer, MDPI, Elsevier, and ACS leading in publication volume

| | |
|---|---|
| Cambridge University Press (CUP) | 5 |
| Oxford University Press (OUP) | 6 |
| Springer International Publishing | 10 |
| Multidisciplinary Digital Publishing Institute (MDPI) | 10 |
| Elsevier Limited | 13 |
| CRC Press | 17 |
| Royal Society of Chemistry (RSC) | 22 |
| Wiley | 32 |
| Springer Science + Business Media | 35 |
| Frontiers Media SA | 38 |
| Elsevier | 44 |
| American Chemical Society (ACS) | 86 |
| Elsevier BV | 98 |
| MDPI AG | 117 |
| Springer Science and Business Media LLC | 293 |

Data source: Lens [20]

On the upper portion, smaller clusters in purple and yellow represent unique subfields. "Antioxidants," "gene expression," "oxidative stress," and "zebrafish" are among the keywords found in the yellow cluster, which also displays molecular and genetic-level studies on the impact of microplastics on cells, genes, and biological processes. Research on the ecological and photosynthetic effects of microplastics on aquatic plants and microorganisms is included in the purple cluster, which also covers keywords like "photosynthesis," "chlorophyll," "algae," and "aquatic plants." Beyond the traditional pollution research, these clusters show that studies of microplastics have expanded into various biological and ecological subfields.

On the whole, the given network visualization is a vivid example of how the sphere of microplastics research has developed and became one of the most interdisciplinary ones integrating environmental science, toxicology, ecology and human health. The blue and green clusters indicate the preliminary environmental and detection research, whereas the red and yellow ones indicate the developing transition to toxicological and biomedical effects. The high level of interconnection between clusters indicates that the fields are not studied separately anymore, researchers are striving to learn more about the routes of microplastics further through the environment to the living organisms and what they entail. Simply put, this visualization represents the entire process of microplastic research, namely, how it detects the contamination, to how such invisible particles influence the planet, ecosystems, and, finally, our lives.

Figure 1.2 is a VosViewer [18] overlay visualization which helps in getting a clear view of the research terrain of microplastics and how this field of science has developed and is developing, and which areas are now receiving the largest attention. The circles (or nodes) on the image correspond to English words which are used in the scientific publications on microplastics and the diameter of each circle shows the frequency of the use of the corresponding word in the literature on the research topic—the bigger the circle, the more thoroughly the issue is discussed. The lines (or relations) between nodes indicate the strength of two topics in relation to each other i.e. they tend to be used interchangeably in the same papers. The color gradient is the most significant part of this overlay map, as it demonstrates an average of the year of publication of each keyword according to the scale on the bottom right. The older studies (around 2022), middle research (around 2023), and most recent researches (around 2024) are blue, green, and yellow respectively. This color coding allows to determine not only the most studied areas but also the way in which the interests of research have changed during the time.

The biggest and most interrelated keyword in the whole picture is the name of the "microplastics," which is the subject of the main research and connects several scientific subjects. Other important subjects, which are surrounding it, include; "humans," "environmental pollutants," "environmental monitoring," "oxidative stress," "animals," and "environmental exposure." These words constitute the central cluster, which is the confluence of environmental science, toxicology, and the health of the population. The proximity of the terms microplastics, humans, and oxidative stress suggests that numerous works have been done on how an exposure to microplastic could cause an oxidative stress and health-related issues in living organisms. The map visually validates the fact that the issue of human and ecological effects of microplastics has occupied the international research agenda.

On the left-hand side, with a majority of the colors being blue and turquoise, we observe words that are related to previous studies—most of them indicate the year 2021–2022 when they were on focus for research. The keywords are "soil," "fresh water," "wastewater," "water purification," "bacteria," "soil microbiology," "degradation," "porosity," and "mass spectrometry." This group depicts the environmental and analytical stage of the microplastic study, in which researchers were mostly exploring methods of transmission of microplastics within the natural environment, such as water, soil, and air, and how they could be identified or eliminated. The use of technical jargon in many expressions, including gas chromatography, Raman spectroscopy, microscopy, and filtration efficiency, demonstrates that previous research was concentrated on the detection procedures and methods of pollution control. Basically, this element of the visualization is the foundational research of the environment and analysis that characterized the initial phases of the discipline.

Conversely, on the right-hand side of the picture, which is painted in green and yellow colors, the future research directions of 2023–2024 are indicated. The keywords present in this area include: "mice," "rats," "female," "male," "pregnancy," "neurons," "apoptosis," "intestinal microbiota," "gastrointestinal microbiome," "inflammation," and "fibrosis." These words signify the fact that there is an increasing trend of environmental detection to be replaced with biological and health impact studies. Scientists currently examine the impact of microplastics on the reproductive systems, lethal development, brain and nervous biology, intestinal health, and even cellular tissue functions, including apoptosis (cell loss) and oxidative stress. The presence of the terms such as diet, high-fat, and intestines indicates the emergence of studies on the interactions of microplastics with the metabolism, diet, and the digestive system. This new emphasis points to a shift in the research of microplastic toxicity and bioaccumulation in living organisms, including human beings.

The green center zone is the place that will be an interface between both environmental and biomedical research relating the older pollution related studies to the new studies of how pollution affects health. The keywords such as "environmental exposure," "bioaccumulation," "toxicology," "cardiovascular toxicity," "inhalation exposure" mean that the researchers already study how the microplastics are spread in the environment and accumulated in the biological systems that may lead to the emergence of chronic diseases. This intersection shows how the field is increasingly becoming interdisciplinary with more and more environmental science, analytical chemistry, biology, and medicine being introduced.

In general, this overlay graphical representation shows how trends and alterations in the evolution of the research on global microplastics. The blue-green clusters foreshadow the previous research that galloped on the pollution and detection of the environment and the yellow-colored clusters demonstrate the visible shift of the research on to the health impacts and the biological mechanisms of the exposure of microplastics. According to the map, microplastics are no longer regarded as the origin of environmental pollution but the potential threat to human and animal health. Overall, the visualization is an emerging field of study at the crossroads of ecology, toxicology, and biomedical science with the point where the entire world is currently interested in the concept that microplastics in the environment are not necessarily dramatic, but their effects on health and life.

The density visualization map, as shown in Fig. 1.3, produced with the assistance of VOSviewer [18] is the clear illustration of how the research on the issue of microplastics has developed and expanded in a broad spectrum of the scientific fields. The intensity of color employed in this kind of map is level of research; activity with the bright yellow areas showing high level of research, the green and blue areas represent the moderate and low density of research respectively. And, the most noticeable word in this visualization is that of "microplastics," and surrounded by a fiery yellow glow, as it symbolizes that it is the center of the entire research network. The most significant ones are also its related terms i.e. "humans," "oxidative stress," "animals," "environmental pollutants," and "environmental monitoring" and form the major clusters of research which dominate the modern literature on the topic under consideration.

The bright yellow position in the middle of the map provides the emphasis on the reality that the most significant research focus is given to the issue of contacting the microplastic and people, animals, and the environment overall. One of the key words used is the term "human" beings since it demonstrates the growing global problem of exposing human beings to microplastics in food, water, and air. Microplastics have been discovered to have entered into the human tissue, lungs and even blood hence making researchers consider their toxicity to the cellulose and body organs. Similarly, the term "animals" is used to represent the vast research done on experimentation on marine animals, aquatic wildlife, and animals in the laboratory such as mice and rats. The research studies are aimed at finding out the mechanism of how the microplastics enter biological systems, how they accumulate in the tissues and what type of physiological or genetic damage they can cause.

One of the most common groups with regard to such central themes is that of "oxidative stress" which is a biological condition, mediated by the deficit of equilibrium between the reactive oxygen species and the antioxidants in the cells. This word was next to the center field, which implies that the effect of oxidative stress is one of the most examined results of exposure to microplastics. The pressure can cause cell damage, inflammation, DNA alteration, and cell death (apoptosis), and these are some of the significant processes that cause microplastics toxicity. Other related theories like the gene expression, the mitochondria and lipid metabolism are yet another pointer that researchers are taking the molecular level studies a step further, in an attempt to understand how microplastics are interfering with the normal biological processes.

In addition to human and biological health, the visualization also provides the environmental aspect of the microplastic study. The notion that the terms like "environmental monitoring," "environmental pollution," "wastewater," "soil," "freshwater," and "seawater" exist is an indication that the scientific community is striving to learn how microplastics originated, their distribution, and their concentration in the natural ecosystem. The aim of such studies is to quantify the pollution of microplastics and identify its destiny in nature using the tools of analysis of high quality. Other methods such as "Raman spectroscopy, "Fourier-Transform infrared (FTIR) spectroscopy," and "gas chromatography-mass spectrometry (GC-MS)" are also present as support keywords, which implies that methods are commonly used in the detection and characterization of microplastics in water, air and soil samples.

The external and colder regions of the density map that is represented in green and blue color show the new areas of research or least explored. Such terms are "pregnancy," "neurodevelopment," "fertility," "apoptosis," and "immune system." The latter are arranged in such a manner that such issues are not the first priority but have been getting an increasing amount of attention as scientists begin to analyze the possible effects of microplastic exposure on reproductive health, fatal development, and neurological capabilities. Such a study is a reflection of an increasing perception of the fact that microplastics is not only an environmentally problematic issue, but it is also a serious human health issue with biological implications with far-reaching consequences in the long run. On balance, this density visualization creates a general picture of the scientific situation on the topic of microplastic. The concentration of yellow spots in the center suggests that the core of research on microplastic contamination, toxicological effects, and monitoring has been established, while the gradients around it show that the research has expanded into additional areas such as biology, chemistry, and the environment. It demonstrates how study has evolved from just identifying microplastics in the environment to comprehending the complex network of interactions in ecosystems and living things. The map also shows a change in detection and analysis, as well as a movement in knowledge of toxicity, oxidative damage, and molecular reactions.

Overall, the provided graphic effectively illustrates how environmental research, toxicology, molecular biology, and even public health have come together to make microplastics a global scientific concern. The degree of interconnectedness between these problems is demonstrated, and the necessity of interdisciplinary research and long-term solutions to the growing threat of microplastic contamination is emphasized.

The Fig. 1.4 shows the annual publication on microplastics and Nanoplastics between 2011 and 2025. A vivid and steady rise in research activity is observed during the last ten and a half years. The data were taken from Lens.Org [20]. The information shows a great change in the scientific interest and global attention to the issue of plastic pollution, which has ceased to be an almost neglected issue and has become a large field of interdisciplinary research, uniting environmental science, materials science, chemistry, toxicology, and the health of the population. The first phase (2011–2015) showed very low efficiency in the field of research, as one or two publications were released each year. It was the period of the conceptual formation of microplastic pollution as a research issue, which cantered mainly on marine litter and the detection of plastic particles in oceans. The majority of research then was done on the visible or larger microplastics because analytical skills used to detect the smaller ones, particularly nanoplastics, were not yet developed.

The number of publications started to increase gradually around 2016, with three articles published in 2016 and 20 in 2017. This development portrays the increased scientific understanding and societal worry on the continued presence of plastics in the environment. This was especially the year of 2017, when various international agencies, such as the United Nations Environment Programme (UNEP) and the European Union, began to take microplastics as a high-priority environmental concern. Scientists started to extend their study past marine to freshwater and soil, and air to define microplastics as a ubiquitous pollutant. This wider perception led to the shift of the discipline to observational studies to the research based on mechanics and effect.

The number of publications in 2018 (21) and 2020 (97) exhibit a strong rising trend. Alongside this influx, technical advancements in analytical techniques including as scanning electron microscopy (SEM), Fourier-transform infrared (FTIR) spectroscopy, and Raman spectroscopy were made possible by the ability to view and measure tiny plastic particles. Research on the behavior, movement, and destiny of microplastics and nanoplastics in various environmental matrices was made possible in significant part by these methodological advancements. As the issue of plastic pollution became recognized as a worldwide environmental and human health concern, the sharp increase in 2020 also shows the growth of multidisciplinary collaborations between environmental scientists, chemists, biologists, and toxicologists.

The number of publications in the subject increased exponentially between 2021 and 2024, reaching 175 in 2024 from a baseline of 135 in 2021. Because of the growing geographic engagement of researchers worldwide and the range of themes, microplastic and nanoplastic research has reached the mature stage. Research was done not only on detection and quantification but also the ecotoxicological impact, bioaccumulation mechanisms as well as the possible health effects on persons of these particles. The finding of microplastics in human blood, lungs and even placental tissues led to a heightened reaction in research. At the same time, there was an increased momentum in research of remediation methods, including, though not limited to: biodegradable materials, filtration systems, and policy-based waste reduction. Governments and scientific institutions began to finance more and more projects to deal with plastic pollution and facilitate the principles of the circular economy. Even though 2025 indicates a minor decrease to 149 publications, it is probably related to the unfinished data on the current year than the reduction in research output. All in all, Fig. 1.4 can be considered clear evidence of the exponential increase in microplastic and nanoplastic research and its worldwide importance. The trend has made the study and alleviation of the effects of plastic pollution a pressing issue once practically unknown in 2011 but then a top environmental research focus in 2024. This trend shows that society at large is starting to recognize the serious and pervasive risks that plastic particles pose to human health and the environment, in addition to the scientific community's growing dedication to finding a solution to this environmental crisis.

Figure 1.5 presents ten leading countries that have authored most studies on micro and nanoplastics. It provides a clear insight into the most active countries in the area of the study of this increasing environmental issue. As the figure demonstrates, some countries such as Switzerland, United Kingdom, Germany, Netherlands, and United States are the top contributors in the world, with other countries such as China, India, Ireland and South Korea making their contributions towards the same slightly uphill.

Switzerland is on the first line of the list with the highest number of publications which demonstrates its serious interest in environmental science and high level of research. Swiss researchers have achieved significant advances in the research of the mobility of small amount plastic particles in the environment and their impact on living organisms. The United Kingdom follows, and many studies have been conducted about the issue of marine pollution and microplastics impacts on sea life. Some of the earliest studies on the spread of microplastics in oceans and rivers have been carried out in British universities and research institutes, including the University of Plymouth and the University of Exeter.

Germany is third, and it has elaborated and technical studies which are a combination of chemistry, environmental engineering and biology. German scientists have assisted in the explanation of the breakdown of bigger plastics to micro- and nanoplastics and the possible entrance of these small particles into food chains. Close behind, in the Netherlands and the United States, there are more than a hundred papers produced. The Netherlands has gained experience in water management and therefore is one of the important participants in the research on pollution in rivers and coastal waters, and U.S. research has been directed at health impacts of microplastics and how they may be solved through biodegradable plastics.

China has experienced an increasing research activity among the Asian countries as increasing numbers of people are becoming aware of environmental pollution. This has become a particularly important issue to China considering the large population and the manufacturing of plastic. India, with fewer publications, however, is rapidly catching up with the trend since scientists have started examining the impact of plastic pollution on soil, water, and agriculture. Ireland and South Korea are also found on the list, which indicates that the interest in the field is becoming widespread among the regions.

On the whole, (Fig. 1.5) demonstrates that the study of microplastics and nanoplastics is an international task, with developed nations primarily causing it, but developing countries increasingly contribute to it. The rise in the number of studies all over the world can be attributed to the fact that plastic pollution has become a serious issue and scientists in various continents are trying to figure out its origins, effects and the solution. This international cooperation will help in coming up with policies and technologies that will reduce plastic pollution and safeguard the environment and human health.

Figure 1.6 presents word cloud, and it shows the most frequent theme and discipline of research that studies micro plastics and nanoplastics. The word cloud indicates the frequency of use of some keywords in scientific literature, which demonstrates the areas that have been the most popular among researchers. This graphic overview shows that microplastic studies cut across a variety of disciplines, integrating interdisciplinary perspectives of biology, chemistry, ecology, medicine, and engineering in order to comprehend the causes, impact, and potential remedies to plastic pollution.

The biggest words in the middle of the picture are "Biology," "Chemistry," "Microplastics" and "Ecology" that reveal the leading research directions. Biology is in the forefront with the largest number (4111 occurrences) in terms of microplastic and nanoplastic studies. The main focus of the study in this field is how plastic particles affect aquatic life, plants, and humans. Researchers investigate the biological absorption, accumulation, and physiological impacts of microplastics on tissues and cells. Chemistry, which focuses on the chemical composition, dissolution, and reactivity of plastic particles with other environmental contaminants, ranks only below biology (3965). Moreover, chemists research the chemical additives in plastics and how they seep into the environment.

The keyword "Microplastics" has been used and appeared frequently in various publications, as can be understood by its repective number 3323, which illustrates the main idea of the research. How these tiny plastic particles are made, how they move through the air, water, and soil, and how long they stay in the environment are the subjects of several studies. Ecology is another important discipline that focuses on the effects of microplastics on the ecosystem and the environment (2494). Researchers describe how plastic pollution affects biodiversity, food chains, & ecological stability. In the field of microplastic and nanoplastic research, these four domains represent the world's strongest points.

In addition to these, other important entities like; Materials science, Engineering, Medicine & Biochemistry indicate that micro plastics are interdisciplinary research. Materials science and engineering (both of which have more than 1400 mentions) is concerned with coming up with alternative biodegradable material, enhancing waste management systems, and inventing new detection systems. The increased interest in the health hazards of microplastics and nanoplastics, such as oxidative stress, inflammation, and possible toxicity to humans and animals is reflected in medicine and biochemistry.

Subfields are represented by smaller, yet not unimportant terms, such as Environmental engineering, Pollution, Toxicity, Genetics and Nanotechnology. As an example, nanotechnology can be used to identify nanoplastics as well as develop sustainable materials, whereas environmental engineering can be used to provide solutions involving pollution management and recycling. Human health, waste management and toxicology are also emerging as major themes as more people become conversant with plastic particles finding their way into the food, water, and even human body.

In general, Fig. 1.6 indicates that studies of microplastics and nanoplastics are broad in terms of disciplines they can be applied to. It emphasizes the interdisciplinary aspect of this discipline, in which biologists, chemists, ecologists, medics, and engineers come together in order to comprehend the multi-faceted issues of plastic pollution. The number represents the full range of scientific diversity of the study as well as the international urgency to work on one of the most burning environmental problems of the current age.

Table 1.1 shows the list of key publishers working on the topic of microplastics and Nanoplastics and the number of publications they created. The figure brings out the publishing landscape of this research area and has indicated what academic and scientific publishers have had the most important roles in creating knowledge about plastic pollution. It gives an idea about the multidisciplinary nature of the issue as studies on microplastics and nanoplastics have diffused throughout various scientific publications and publishing houses.

Springer Science and Business Media LLC are the most active publisher in this sphere, and thus it takes the first place in the list with 293 publications. It is no wonder that Springer dominates since it releases numerous environmental science and materials journals such as Environmental Science and Pollution Research, Marine Pollution Bulletin, and Scientific Reports. The journals usually rank as some of the major platforms among researchers on pollution, sustainability, and environmental health.

After Springer, MDPI AG is the best with 117 publications. MDPI (Multidisciplinary Digital Publishing Institute) has a reputation of open-access model and a wide range of journals such as Polymers, Toxics, and Sustainability, which also regularly publish articles on the subject of microplastic pollution, nanomaterials, and environmental contamination. The large number of publications in the MDPI has indicated how easy and fast the environmental research has been able to spread through open access.

Elsevier BV follows it with 98 publications indicating that it is a powerhouse in publishing scientific journals worldwide. The most frequent journals related to the topic that are cited in this scientific area are Elsevier journals like the Science of the Total Environment, the Environmental Research, and the Marine Pollution Bulletin. Combining the numbers of Elsevier Limited (13) and Elsevier (44), the Elsevier group is a cumulative publisher of more than 150 publications, and therefore, one of the largest contributors in total.

A significant presence of the American Chemical Society (ACS) is also made with 86 publications and the necessity of chemical science in examining the makeup, deterioration, and poisonous effects of plastics. The journals like the Environmental Science Technology and ACS Sustainable chemistry and engineering are pivotal in the research of plastic pollutants, chemistry of polymers, and sustainable solutions.

Other non-commercial sponsors include Frontiers Media SA (38) that encourages open-access interdisciplinary research and Wiley (32) with reputable journals on environmental & materials science. The Royal Society of Chemistry (RSC) is another stakeholder that publishes 22 journals, including Environmental Science, Nano and Environmental Science, Processes and Impacts where micro- and nanoplastic toxicity & environmental behaviour regularly appear at the cutting edge.

The smaller but strong players include the CRC Press, Springer International Publishing and the Cambridge University Press that publish smaller academic literature or reviews. Overall, points to the suggestion that research on microplastics & nanoplastics is published in a vast array of academic materials, & this aspect reveals the interdisciplinary status of the area. The main manufacturers of the output are the large-scale international publishers like Springer, MDPI, Elsevier and ACS that enable a global effort to create knowledge on plastic pollution. This wide dissemination of publications also refers to a good cooperation of scientists, policymakers & environmental agencies who are trying to address one of the most pressing environmental problems of the days.

## Conclusion

The plastics has transformed the industrial & everyday life with its lightweight, cheap, durable, & easy-to-work materials. They are used in practically every application: polyvinyl chloride is essential to the building & plumbing industries; polyethylene & polypropylene are used in consumer goods and packaging; and various engineering plastics are used in electronics, auto parts, and medical equipment. With the utilization of renewable raw materials and similar functional properties to conventional plastics, the development of bioplastics, such as polylactic acid (PLA) and bio-polyethylene, can be viewed as a major step toward sustainability. These are the benefits that have rendered plastics inevitable to the economy and the growth of technologies. However, the accumulation and persistence of plastic waste are a threat in form of a global environmental crisis. The plastics are resistant to decomposition and thus they break down to form microplastics and nanoplastics which pollute the oceans, freshwater reservoirs, soils, and even the air. These minute particles absorb toxic substances and can be readily consumed by the organisms bypassing food webs and may end up in the human body. The growing scientific literature is suggesting exposure to micro- and nanoplastics, as the source of oxidative stress, inflammation, and impacts on reproduction and development in animals and humans as detrimental. Their gigantic proliferation illuminates the necessity to possess more sustainable production and waste treatment processes. People are becoming increasingly aware of the negative effects plastic pollution has on the environment and human health, as seen by the sharp rise in research on the topic since 2011. The global academic community and governments are now focusing on ways to reduce the amount of plastic that is leaking into the environment and improve recycling. In order to tackle this problem, the circular economy concept—which aims to reduce plastic manufacture, recycle resources, and use biodegradable or biobased alternatives—is necessary. A significant transformation requires the cooperation of industrial progress, consumer sensitization, and government policy. Overall, plastics have greatly aided human progress, yet they are currently one of the world's most difficult trash issues. Through scientific advancements, sound policies, and international collaboration, it is possible to balance the benefits of plastics with the wellbeing of the environment and people, paving the way for a cleaner and better future.

## References


1. Cuthbertson, A. A., Lincoln, C., Miscall, J., Stanley, L. M., Maurya, A. K., Asundi, A. S., Tassone, C. J., Rorrer, N. A., & Beckham, G. T. (2024). Characterization of polymer properties and identification of additives in commercially available research plastics. *Green Chemistry, 26*(12), 7067–7090. https://doi.org/10.1039/d4gc00659c
2. Plastics Explained. (2023). *Plastics Europe*. Retrieved October, 2025, from https://plasticseurope.org/plastics-explained/
3. Grigore, M. (2017). Methods of recycling, properties and applications of recycled thermoplastic polymers. *Recycling, 2*(4), 24. https://doi.org/10.3390/recycling2040024
4. Olatunji, O. (2024). *Re-envisioning plastics role in the global society*. Springer. https://doi.org/10.1007/978-3-031-48945-7
5. Geyer, R., Jambeck, J. R., & Law, K. L. (2017). Production, use, and fate of all plastics ever made. *Science Advances, 3*(7). https://doi.org/10.1126/sciadv.1700782
6. Oluwatobi, F. B., Iyanda, O. J., & Akpor, O. B. (n.d.). The evolution and environmental impact of plastics: Challenges and sustainable solutions. In *IEEE Xplore. IEEE 5th International Conference on Electro-Computing Technologies for Humanity, Ado Ekiti, Nigeria*. https://doi.org/10.1109/nigercon62786.2024.10926981
7. Zarfl, C. (2019). Promising techniques and open challenges for microplastic identification and quantification in environmental matrices. *Analytical and Bioanalytical Chemistry, 411*(17), 3743–3756. https://doi.org/10.1007/s00216-019-01763-9
8. Tsochatzis, E. D., Gika, H., Theodoridis, G., Maragou, N., Thomaidis, N., & Corredig, M. (2024). Microplastics and nanoplastics: Exposure and toxicological effects require important analysis considerations. *Heliyon, 10*(11), e32261. https://doi.org/10.1016/j.heliyon.2024.e32261
9. Ding, Y., Wang, H., Liu, Y., & Lei, X. (2024). Urban waterlogging structure risk assessment and enhancement. *Journal of Environmental Management, 352*, 120074. https://doi.org/10.1016/j.jenvman.2024.120074
10. Emadian, S. M., Onay, T. T., & Demirel, B. (2017). Biodegradation of bioplastics in natural environments. *Waste Management, 59*, 526–536. https://doi.org/10.1016/j.wasman.2016.10.006
11. Lambert, S., & Wagner, M. (2016). Characterisation of nanoplastics during the degradation of polystyrene. *Chemosphere, 145*, 265–268. https://doi.org/10.1016/j.chemosphere.2015.11.078

12. Hidalgo-Ruz, V., Gutow, L., Thompson, R. C., & Thiel, M. (2012). Microplastics in the marine environment: A review of the methods used for identification and quantification. *Environmental Science & Technology, 46*(6), 3060–3075. https://doi.org/10.1021/es2031505
13. Cole, M., Lindeque, P., Halsband, C., & Galloway, T. S. (2011). Microplastics as contaminants in the marine environment: A review. *Marine Pollution Bulletin, 62*(12), 2588–2597. https://doi.org/10.1016/j.marpolbul.2011.09.025
14. Andrady, A. L. (2011). Microplastics in the marine environment. *Marine Pollution Bulletin, 62*(8), 1596–1605. https://doi.org/10.1016/j.marpolbul.2011.05.030
15. Gewert, B., Plassmann, M. M., & MacLeod, M. (2015). Pathways for degradation of plastic polymers floating in the marine environment. *Environmental Science: Processes & Impacts, 17*(9), 1513–1521. https://doi.org/10.1039/C5EM00207A
16. Teuten, E. L., Saquing, J. M., Knappe, D. R. U., Barlaz, M. A., Jonsson, S., Björn, A., Rowland, S. J., Thompson, R. C., Galloway, T. S., Yamashita, R., Ochi, D., Watanuki, Y., Moore, C., Viet, P. H., Tana, T. S., Prudente, M., Boonyatumanond, R., Zakaria, M. P., Akkhavong, K., … Takada, H. (2009). Transport and release of chemicals from plastics to the environment and to wildlife. *Philosophical Transactions of the Royal Society B Biological Sciences, 364*(1526), 2027–2045. https://doi.org/10.1098/rstb.2008.0284
17. Koelmans, A. A., Besseling, E., & Foekema, E. M. (2014). Leaching of plastic additives to marine organisms. *Environmental Pollution, 187*, 49–54. https://doi.org/10.1016/j.envpol.2013.12.013
18. *VOSviewer—Visualizing scientific landscapes*. VOSviewer. Retrieved October, 2025, from https://www.vosviewer.com/
19. *Microplastic and nanoplastic—Search results—PubMed*. PubMed. Retrieved October, 2025, from https://pubmed.ncbi.nlm.nih.gov/?term=Microplastic+and+Nanoplastic&sort_order=asc
20. *Microplastic and Nanoplastic—Search results—Lens*. Lens. Retrieved October, 2025, from https://www.lens.org/lens/search/scholar/list?q=Microplastic%20and%20Nanoplastic

# Chapter 2
# Understanding Behavior, Fate of Microplastics and Nanoplastics

**Dibyarupa Pal, Pranabi Maji, Ritam Paul, and Dola Mukherjee**

## Introduction

The discovery of plastic was a game changer event, compounded with its widespread and uncontrolled use, has become a paramount nuisance threatening the survivality of life on Earth. Common polymer types found in soil environments include polyethylene (PE), polypropylene (PP), high-density polyethylene (HDPE), polystyrene (PS), polyurethane (PUR), polyvinyl chloride (PVC), low-density polyethylene (LDPE), polyethylene terephthalate (PET), polyamide (PA) and cellulose acetate (CA). Notably, these collections collectively account for 90% of the huge plastic production, and weathering of larger plastic give rise to the microplastics (MPs) and nanoplastics (NPs). The majority of the plastic products consist of bottles, containers, medical supplies, food packets, electronics and others. The chemical composition, shape and size vary greatly and are available as fragments, pellets, fibers, films, and foams. They are very persistent and stabilized in the environment and take millions of years to deteriorate or disintegrate into the atmosphere. The vast majority of the 360 million metric tons produced each year are synthetic plastics, which include polymers generated from petrochemicals such as PE, PP and PET.

MPs and NPs are omnipresent encompassing all three spheres of the ecosystem. Not only are they abundant in soil and water due to direct insinuation but indirectly

D. Pal · R. Paul
Department of Biosciences, JIS University, Kolkata, India

P. Maji
Department of Physics, JIS University, Kolkata, India

D. Mukherjee (✉)
Amity Institute of Biotechnology, Amity University, Kolkata, India
e-mail: dmukherjee5@kol.amity.edu



S. Kulkarni, S. Das (eds.), *Microplastics and Nanoplastics*, Lecture Notes in Nanoscale Science and Technology 49,
https://doi.org/10.1007/978-3-032-17963-0_2

are omnipresent in the atmosphere as a result of sea spray/evaporation. MPs and NPs originate from breakdown products of large plastics due to physical abrasion, weathering, sunlight and UV action on them. Being light weight they are found in non-point sources migrating far away from point of origin and they infiltrate into groundwater, present in the glaciers, ocean trench; and are uptaken by plants, insects and animals and eventually enter the human food chain [1–4].

## *Vertical Behaviour of MPs and NPs*

The complex process of microplastic and nanoplastic vertical mobility in soil and water is impacted by a myriad of elements, including the intrinsic physicochemical features of the particles, environmental factors, and biological processes. Smaller MPs and NPs permeate deeper into soil sediment or groundwater and provide a greater ecological danger because of quicker uptake by organisms. Plastic pellets migrate more easily than fiber plastics because of their mass, but fiber plastics stick to soil pores, enabling deeper flow with less transfer rate; therefore even the different form of the plastic has serious implications. While hydrophobic microplastics and nanoparticles are less likely to wander off in water, hydrophilic ones are more likely to do so. Rougher plastic surfaces also have a tendency to engage with soil particles, which decreases infiltration and vertical mobility. The surface charge and functional groups of MPs regulate electrostatic interactions with soil media and play an important role in the early phases of migration. Hydrological processes like rain, flood or irrigation help in MP movement from point to non-point sources like transport vertically to deeper groundwater or stuck in soil sediments. On the other hand, migration might be hindered by extremely high flow rates because of the greater likelihood of accidents and MP capture. Vertical dispersion is accelerated by dry-wet cycles, and MP transport is further enhanced by freeze-thaw cycles, which release colloids and generate preferential flow pathways.

## *Implications on Different Life Forms*

MPs and NPs are accessible to several marine organisms, including corals, zooplanktons, lobsters, turtle, fish and whales due to their small size. During filtering saltwater, macro organisms like Baleen whales, are susceptible to microplastic pollution as they enter into their respiratory system. Through the food chain these polymers reach greater tropical levels and bioaccumulate, for instance MPs and NPs are found in the stomach and intestine of whales due to their high fat and lipid content. The polymer waste clogs respiratory organs of sea turtles, found in the internal system of seals, sea lions and even polar bear. Internally, these micro and nano range plastics inhibit gastric enzyme release, cause hormonal imbalance, delays in ovulation and impairs reproduction. In the human system, either by direct contact or indirectly through trophic levels can cause both immediate and delayed toxicity in living organisms. Skin care products, hot food delivered in plastic containers, bottled water and beverages give direct exposure to NPs through dermal contact whereas devouring micro-nano plastics infiltrated livestock meat, milk or plant products gives indirect exposure. Various symptoms like stunted growth, mental health issues, intestinal dysfunction, impaired digestion, lung and reproductive disorder, cancer and even mortality occur in the worst-case scenario. Moreover, MPs and NPs can harbour toxic compounds, heavy metals, persistent organic pollutants (POPs), and diseases causing germs on them which may influence secondary health factors in humans like chromosomal changes and severe inflammation [5, 6].

## Outline of the Technologies for Removal of MPs and NPs

### *Physical Processes*

Filtration technology initially separates out and concentrates MPs depending on the parameters like size and density followed by adsorption technique. Advanced processes like membrane based filtration techniques, ultrafiltration, microfiltration, reverse osmosis and membrane bioreactors are utilized to clean up MPs and NPs. All these advanced technologies though efficiently remove MPs, but are costly and with constraints like membrane fouling and replacement, low flux and requirement for trans-membrane pressure. Generation of toxic intermediates or smaller fragments and mobility issues are also quite common constraints [7].

***Coagulation Technique*** Coagulation performed by addition of common chemical salts of Aluminium and Iron, remove chemical pollutants from water by the incorporation of highly charged cations that neutralizes colloidal surface charge and sediments them down. Particles aggregate into larger flocs due to strong van der Waals forces which eliminate them. In similar fashion, MPs and NPs interact with coagulant or coagulant aids and through sweep coagulation removes the pollutants. Coagulant nature, dosage and solution pH also influences the settling floc and its removal [8–10]. Recently Aluminium salts along with *Moringa* have been used for successful removal of MPs common in wastewater like PE, PA and PS respectively [11]. The investigators used a combination of Aluminium salt (40 mg/L) and *Moringa* (60 mg/L) and achieved the highest (92.99%) removal of PA. The determining factor for such high efficiency MP removal was additive effects of charge neutralization, agglomeration and adsorption on *Moringa* surface, validated by zeta potential and scanning electron microscopy (SEM) assessments. Researchers have also combined the processes of electrocoagulation with coagulation-flocculation for removal of MPs and NPs. Recently more focus is being given on Plant-based coagulants/flocculants to negate the challenges of using only chemical coagulants. Incorporating Plants like *Moringa, Cucurbita, Cicer* can also aid in coagulation—flocculation of MPs and NPs and if used as coagulant aid can help in removal of the polymer waste. Green coagulants are safer, biodegradable and easily available, compared to chemical inorganic and organic polymers which generate toxic sludge and impart neurological disease upon repeated use [11, 12].

## *Chemical Processes*

***Advanced Oxidation Process (AOPs)*** MPs and NPs can be effectively degraded using Advanced Oxidation Processes (AOPs), which use powerful oxidative radicals ($^{\bullet}OH$, $SO_4^{\bullet -}$, $O_2^{\bullet -}$) to split polymer chains into smaller, less toxic components. Generally AOPs include photocatalysis, ozonation and electrochemical reactions; each of these processes has its own set of benefits. As an example, electro-Fenton processes increase efficiency through electrochemically driven reactions, while $UV/H_2O_2$ systems generate hydroxyl radicals, which lead to high degradation rates. Optimal conditions allow AOPs to achieve near-complete mineralization; nonetheless, there are still obstacles to overcome, including high energy intake, the cohort of hazardous intermediates and the scalability of treatments. To increase cost-effectiveness and environmental compatibility, recent advancements focus on hybrid AOPs, such as solar-driven photocatalysis coupled with persulfate activation [13–16]. These technologies have the potential to address MP and NP contamination on a broad range, but still need fine-tuning to ensure efficiency, scalability and environmental safety. The inherent principle lies in the *in situ* generation of numerous reactive oxygen species (ROS) which catabolize the plastic components into non-toxic biodegradable materials [17]. For instance hydroxyl radicals (•OH), sulfate radicals ($SO_4^{\bullet -}$), singlet oxygen ($^1O_2$), superoxide anions ($O_2^{\bullet -}$), and hydroperoxyl radicals ($HO_2^{\bullet}$), all contribute to the degradation of MPs and NPs into harmless by-products like water ($H_2O$) and carbon dioxide ($CO_2$) by initiating non-selective oxidation reactions.

***Photocatalytic Oxidation*** A promising, sustainable and cost-effective approach relies on light-induced reactions that cause macromolecule chain breakdown, surface oxidation and cohort of ROS. Wide band gap metal oxide semiconductors like Titania ($TiO_2$) or zinc oxide (ZnO) are commonly used. When these semiconductors interact with light, electrons move from the valence to conduction band, leaving holes in the valence shell. These holes react with water to produce hydroxyl radicals, while electrons in the conduction band counter with oxygen to generate superoxide radicals ($O_2^{\bullet -}$), both of which are strong oxidizing agents that convert plastics into $H_2O$ and $CO_2$. In the Photolysis process, there is direct breakdown of plastics upon exposure to UV or visible light. This process is generally limited, with 17.1% mineralization achieved for $^{14}C$-labeled polystyrene NPs after 48 h of UV irradiation at 230 nm. It often results in oxygenated polymer fragments rather than full mineralization.

In another related phenomenon called as Homogeneous photocatalysis, MP and NP degradation is enhanced by utilizing light-activated oxidants such as hydrogen peroxide ($H_2O_2$), peroxymonosulfate (PMS), or chloride ions, which generate highly reactive radicals (•OH, $SO_4^{\bullet-}$, chlorine radicals) to break down plastic polymers. Hybrid systems like the UV/PMS system can achieve 63.90% mineralization for polystyrene NPs, demonstrating high efficiency in organic carbon reduction. However, a major drawback is the potential formation of more toxic by-products (e.g., chlorinated compounds or smaller reactive fragments) compared to untreated MPs, raising concerns about secondary pollution. Thus optimizing oxidant doses and reaction conditions is crucial to minimize harmful intermediates while ensuring effective degradation. Whereas in Heterogeneous photocatalysis, solid semiconductor catalysts (e.g., $TiO_2$, ZnO) are utilized under light irradiation to generate ROS that degrade MPs and NPs via redox reactions. For example, $TiO_2$-$P^{25}$ immobilized on β-SiC foam achieved 80% mineralization for 140 nm polystyrene NPs and 68% for 508 nm NPs, highlighting a size-dependent degradation efficiency. Recent advancements focus on composite semiconductor systems (e.g., doped $TiO_2$, ZnO/g-$C_3N_4$ hybrids) to improve solar light absorption and charge carrier separation, enhancing degradation rates. Unlike homogeneous methods, this approach minimizes toxic by-product formation but faces challenges like catalyst recovery and low visible-light activity. Optimizing catalyst design and reactor systems could make this a sustainable solution for microplastic pollution.

***Ozonation*** The process breaks down ozone ($O_3$) into ROS such as hydroxyl radicals (•OH). When $O_3$ comes in contact with electron-rich elements found in plastic structures, it causes ring opening and subsequent molecular reduction involving double bonds, amines, and others. Oxidative cleavage is triggered when the exposed surface of MPs and NPs interacts with oxygen from $O_3$. Earlier experimentation revealed ozonation of commercial PS-NPs catalyzed 42.7% breakdown after 240 min in clean water. But the possibility of formation of partially oxidized plastics has become a concern, despite the fact that the procedure can greatly simplify complex plastic components. Catalytic ozonation is another supplementary process that uses metal oxides and transition metal ions as catalysts to speed up the breakdown of $O_3$ and produce more ROS (like •OH and $O_2^{\bullet-}$). In one of the earlier investigations carried out, 70% mineralization of PS-NPs within 2 h was accomplished by a $Co^{2+}$-catalyzed ozonation, compared to single ozonation which took several hours more. However, owing to mass transfer restrictions, the method's efficiency may decrease at lower NP concentrations, despite its overall high efficiency. Hybrid approaches are also encouraged to reduce energy consumption and increase mineralization performance such as concentrating NPs with nanofiltration (NF) techniques as a pre-treatment step [18].

***Fenton Reaction*** The cleavage and mineralization of micro- and nanoplastics in water is an captivating AOP which liberates radicals (like •OH), that can degrade organic contaminants. Here is how the Fenton process operates:

Extremely reactive hydroxyl radicals are initially produced when hydrogen peroxide and ferrous ions combine. But accumulation of ferric ions ($Fe^{3+}$) in the reaction can reduce its efficiency over time. To overcome this, the Photo-Fenton process ($UV/H_2O_2/Fe^{3+}$) utilizes UV or visible light to regenerate $Fe^{2+}$ from $Fe^{3+}$, thereby enhancing the continuous production of $^{\bullet}OH$ and improving the efficiency of pollutant degradation. The core reaction is expressed as

$$Fe^{2+} + H^2O^2 + hv \rightarrow Fe^{3+} + 2 \bullet OH.$$

The generated hydroxyl radicals attack and oxidize plastic particles.

The Photo-Fenton process has demonstrated high effectiveness in degrading NPs. For instance, 100% mineralization and turbidity decrease was achieved for 140 nm PS-NPs after 60 min of treatment using the $UV/H^2O^2/Fe^{3+}$ Photo-Fenton process under optimal conditions. Although Fenton reactions are efficient in breaking down NPs, they can also produce free radicals, fragmented NPs with oxidized surfaces, and low molecular weights monomers, and oligomers. When contrasted with the original NPs, these intermediates may be more mobile via trophic transfer, enhanced bioavailability, and even be lethal to lives. To overcome the constraints, Photo-Fenton can be combined with other post-treatment technologies like adsorption or biological treatments which could further improve mineralization speed and environmental concerns by capturing and breaking down residual by-products.

While these AOPs demonstrate promising performance for plastic degradation in controlled laboratory settings, significant challenges remain for their large-scale applicability in real-world environmental conditions. These include the need for more energy-efficient systems, better understanding and control of degradation by-products, and the development of standardized protocols that account for complex water matrices and environmentally relevant plastic concentrations and types.

### *Biological Processes*

The carbon chains that make up synthetic plastics are not uniform; rather, they are complex mixtures of base polymers and a number of chemical additives to ensure material stability and durability. The enzymes derived from microbes possess metabolic machinery that has the potential to attack and break down varied types of plastic. Initially colonization over the plastic materials over time triggers their enzymatic repertoire to produce and attack the exposed and underlying polymer matrix. Sometimes biofilm formation also elicits extracellular enzyme production that can depolymerize the synthetic macromolecules with great ease.

Enzymes produced by microbes play an important part in the biodegradation of NPs and MPs, providing an environmentally friendly alternative to more traditional physical and chemical removal processes. Microbes such as bacteria, actinomycetes, fungi and algae along with insects are being studied as a source of enzymes that could degrade plastic. These organisms that are isolated from various plastic rich environments including oceans, organism guts and landfills secrete enzymes that are specialized in breaking these polymers into carbon dioxide and aqueous products. The enzymatic degradation process typically involves hydrolysis, oxidation, and cleavage of polymer chains, with different enzymes targeting specific plastic types. Among the most studied enzymes are microbial hydrolases (e.g., esterases, lipases, cutinases) and oxidoreductases (e.g., laccases, peroxidases), which act on the ester and carbon-carbon bonds in plastics.

### Bacteria

Bacteria are among the most extensively studied microorganisms in the context of plastic elimination due to their metabolic versatility and capacity to produce extracellular enzymes capable of attacking diverse synthetic polymers. A wide range of bacterial taxa have been reported to colonize and degrade recalcitrant plastics mainly PET, PE, PS, PUR and PVC. Several studies have highlighted the role of the Bacteroidetes phylum, particularly members of the Flavobacteriales order such as Flavobacterium and Tenacibaculum, in colonizing polymer surfaces. The enzymatic machinery employed by these bacteria primarily includes hydrolases, oxidoreductases, and depolymerases, which initiate the cleavage of long polymer chains into oligomers (3–6 subunits) and monomers (single subunit) suitable for microbial assimilation. Both Gram-negative and Gram-positive species are involved in polyethylene degradation, with genera such as *Pseudomonas*, *Bacillus*, and *Clostridium* frequently cited as potent polymer colonizers and degraders. Soil-derived *Bacillus* species including *B. cereus*, *B. amyloliquefaciens*, *B. pumilus*, *B. mycoides*, *B. firmus*, and *B. marisfavi* have been repeatedly isolated from landfill environments and shown significant activity against PE and PP [19, 20]. Interestingly, *Bacillus* spp. have also been identified in marine ecosystems where they contribute to the biodegradation of PVC, LDPE, and HDPE, and *Rhodococcus* spp., initiates PE degradation, highlighting the adaptability of bacterial plastic-degrading mechanisms to diverse ecological niches [21, 22]. Figure 2.1 illustrates the range of microbial enzymes implicated in bacterial plastic degradation.

The enzymatic arsenal of these bacteria often includes classes such as hydrolases, esterases, cutinases, PETases, and lipases capable of cleaving ester bonds in polyesters and attacking hydrocarbon backbones in polyolefins. Among the most significant discoveries in bacterial plastic biodegradation is the identification of *Ideonella sakaiensis*, which secretes PETase to hydrolyze PET into two by-products namely terephthalic acid (TPA) and ethylene glycol (EG). This bacterium, first isolated from sludge in a PET bottle recycling facility in Osaka, Japan, represents a breakthrough in enzymatic plastic valorization [23]. A companion enzyme, MHETase, further breaks PET into mono(2-hydroxyethyl) terephthalic acid (MHET) and thereafter into TPA and EG, both of which can be metabolized by microorganisms. PETase variants are genetically engineered by researchers with enhanced stability and activity, significantly improving PET degradation efficiency. Another key enzyme, cutinase, derived from fungi like *Fusarium solani* and

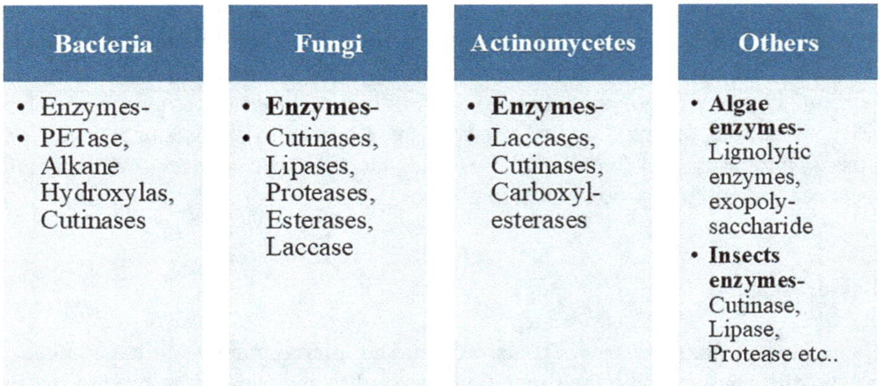


**Fig. 2.1** Microbial categories and enzyme repertoire for the polymer degradation

bacteria like *Thermobifida fusca*, also degrades PET and other polyesters by cleaving ester bonds. Similarly, lipases from *Pseudomonas* and *Bacillus* species have established the capacity to degrade PU and PE by breaking ester and urethane linkages. PETase shares structural similarities with cutinases and lipases but exhibits a relatively slow degradation rate compared to the massive accumulation of PET in the environment. To enhance the efficiency of this process, surface-modifying enzymes such as hydrolases and engineered PETases have been explored to increase substrate accessibility and catalytic turnover [24]. *Pseudomonas* and *Bacillus* spp. also generate polyethylene-degrading enzymes (PEases) and styrene monooxygenases, which are critical in depolymerizing polypropylene and aiding in the conversion of PP into monomeric forms [25]. Gut-associated bacteria from the larvae of *Galleria mellonella* (greater wax moth) have demonstrated the ability to hydrolyze polyethylene, suggesting a synergistic interaction between insect gut physiology and microbial enzymatic activity. Additionally; *Pseudomonas* sp. E4 and *P. aeruginosa* have been reported to produce alkane hydroxylase, rubredoxin, and rubredoxin reductase, enzymes crucial in PE degradation pathways [26, 27]. The available information highlights the complex process by which bacteria aid in the biodegradation of various plastic forms. This process includes biofilm-mediated colonization, surface modification by enzymes, breaking of polymer chains, and finally, assimilation of monomers. These bacterial enzymes are being identified and targeted for large-scale biotechnological applications taking in account the contributions from synthetic biology, metagenomics, and protein engineering.

### Fungi

For plastic biodegradation, fungi are an important microbial sections to consider because of the wide variety of enzymes they produce and the explicit ecological roles they play. Fungal hyphae allow easy and widespread fungal invasion on polymer surfaces compared to bacteria, which in turn allows mechanical disruption and enzymatic depolymerization. Initiating the breakdown of resistant polymers like PET, PE, and PP, is a challenging task, but due to inherent properties of the fungi, they are capable of combining surface colonization with enzymatic dissolution as effective agents [28, 29]. Fungi that break down plastics have an extensive arsenal of enzymes that can break down different types of synthetic polymers. These include cutinases, lipases, esterases, carboxylesterases, proteases, and lignocellulolytic enzymes. Complex polymers, such as PS and PP, can be effectively oxidized by laccases and peroxidases enzymes produced by white-rot fungi (*Phanerochaete chrysosporium*) and button mushrooms (*Pleurotus ostreatus*). This class of enzymes employ radical-mediated processes to degrade long-chain hydrocarbons for instance, the potential for manganese peroxidase (MnP) and lignin peroxidase (LiP) to break PS into smaller oligomers for further metabolism is worth noting. Fungal cutinases derived from *Fusarium solani pisi, Aspergillus, Fusarium* sp., and *Humicola insolens* ave demonstrated remarkable efficacy in degrading low-crystallinity PET films, achieving up to 97% weight loss within 96 h, indicating their potential for industrial applications [30]. These enzymes also target PLA, converting it into oligomers and monomers that can serve as carbon sources for fungal growth [31]. *Aspergillus fumigatus* has been shown to produce polyhydroxybutyrate (PHB) depolymerase, capable of degrading biodegradable polyesters such as polybutylene succinate (PBS), PHB, and polyethylene succinate (PES). *Phanerochaete chrysosporium* and *Trametes versicolor*, employ powerful oxidative manganese peroxidase enzymes to depolymerize PE and other polyolefins, highlighting the role of ligninolytic systems in attacking chemically inert plastics [32, 33]. Hydrophobins are secreted by fungi, small surface-active proteins that enhance hyphal adhesion to hydrophobic substrates like plastics. This adaptation, coupled with the mechanical penetration of hyphae, accelerates the initial colonization phase, allowing enzymes to act more efficiently on the polymer surface [34]. The rate of fungal degradation is significantly affected by environmental factors such as temperature, humidity, pH, and the presence of co-substrates.

Fungal-mediated plastic degradation, however, is highly dependent on the structural assets of the polymer, including crystallinity, surface roughness, and the presence of additives. Despite these constraints, fungi have emerged as promising candidates for bioaugmentation strategies, particularly in soil and compost environments where their enzymatic activity can be combined with bacterial consortia for synergistic degradation of complex polymer mixtures.

### Actinomycetes

Actinomycetes, a group of filamentous, Gram-positive bacteria, are well known for their ability to produce a wide variety of extracellular enzymes and secondary metabolites. They have become a focal point in the field of plastic biodegradation due to their inherent ecological function in breaking down complex polymers like lignin and chitin. These organisms combine bacterial metabolic versatility with fungal-like hyphal growth, enabling effective colonization and breakdown of synthetic polymers. Strains belonging to the genera Streptomyces and Micromonospora have validated the capacity to degrade synthetic aliphatic polyesters such as polyhydroxybutyrate. These actinomycetes were isolated from both up and down the stream environments, indicating their ecological adaptability in diverse habitats including soil, marine sediments, and landfill leachates [35]. Several studies have reported the breakdown of LDPE, HDPE, PVC, PS and PP films under the influence of actinomycetal enzymes. The enzymatic toolkit of these organisms includes laccases, cutinases, esterases, carboxylases, and lipases, which initiate oxidative and hydrolytic cleavage of the polymer backbone [36, 37]. For example, *Streptomyces scabies* and *Nocardiopsis* spp. produce esterases with broad substrate specificity capable of attacking PET and other polyesters isolated from natural environments. Marine-derived actinomycetal consortia have also shown the ability to degrade polyurethane and associated chemical additives, facilitated by the secretion of extracellular polymeric substances (EPS) which enhance biofilm formation and surface adhesion, a crucial step before enzymatic depolymerization [38]. One of the key features of actinomycetes in plastic biodegradation is their ability to form structured biofilms, which create microenvironments favorable for sustained enzymatic activity. These biofilms help in concentrating hydrolytic enzymes on the polymer surface, improving degradation efficiency compared to planktonic cells. Additionally, their filamentous growth pattern increases surface contact, enabling the colonization of larger areas of polymeric substrates compared to unicellular bacteria.

Actinomycetes are also of significant biotechnological interest due to their amenability to genetic manipulation, which allows the engineering of enhanced enzyme systems for targeted plastic depolymerization. This makes them promising candidates for the development of integrated bioprocesses for plastic waste valorization in industrial settings.

## Algae

Algae, particularly microalgae and cyanobacteria, have emerged as unconventional yet promising agents for plastic biodegradation. Unlike bacteria and fungi, which rely heavily on biofilm formation and enzyme secretion, algae can employ a combination of enzymatic, biochemical, and physicochemical mechanisms to attack plastic polymers. Their ability to secrete extracellular enzymes, organic acids, and exopolysaccharides, coupled with their natural presence in aquatic ecosystems, makes them significant players in the degradation of plastics in marine and freshwater environments [39]. Microalgae can colonize polymer surfaces in aquatic environments and initiate hydrolytic cleavage of plastics into smaller subunits that are then absorbed into the algal cell for further metabolic processing. Some algae produce biofilms that trap MPs, while their enzymatic secretions (ligninolytic oxidases and hydrolases) weaken the polymer structure over time. Additionally, the secretion of exopolysaccharides (EPS) enhances surface adherence and promotes biofilm formation, facilitating sustained enzymatic activity on otherwise inert polymer surfaces [40]. Studies have shown species-specific variations in algal degradation efficiency. For example, the blue-green cyanobacterium *Anabaena spiroides* demonstrated the highest LDPE breakdown rate at 8.18%, followed by the green alga *Scenedesmus dimorphus* at 3.74%, highlighting the role of algal taxonomy in plastic biodegradation. Similarly, freshwater cyanobacteria such as *Phormidium lucidum* and *Oscillatoria subbrevis* have been reported to colonize PE surfaces and degrade LDPE efficiently without the need for pre-treatments or pro-oxidant additives [41]. Marine algae also play a decisive role in the natural breakdown of oceanic plastic debris. Species such as *Picochlorum maculatum*, *Amphora* sp., and *Hymenomonas globosa* have been linked to LDPE degradation, with reported weight loss of up to 20% within 45 days, while *Spirulina* sp. demonstrated up to 48.61% degradation of PP and PET under controlled laboratory conditions. A unique aspect of algal-mediated plastic degradation is the synergistic interaction between algae and bacteria in mixed biofilms. In natural ecosystems such as sewage water and marine sediments, algal biofilms often coexist with heterotrophic bacteria, creating a consortium where algae provide oxygen and organic exudates while bacteria contribute specialized depolymerizing enzymes. This cooperation enhances overall polymer breakdown rates and suggests that algal-bacterial consortia may be an effective strategy for engineered biodegradation systems.

Despite these promising findings, algal plastic degradation faces challenges in scalability due to its dependency on environmental factors such as light intensity, nutrient accessibility, and temperature. Nevertheless, with advances in photobioreactor technology and synthetic biology, algae are increasingly being explored as green catalysts for eco-friendly plastic waste management in both natural and engineered systems.

### Insects

In recent years, insects and their gut microbiomes have gained significant attention as biological agents for plastic biodegradation. The insect gut represents a unique microenvironment populated by diverse microbial communities capable of producing specialized enzymes that can degrade/detoriate synthetic polymers. Depending on the host species and diet, the gut microbiota can adapt and evolve to degrade complex hydrocarbons, making insects potential allies in tackling plastic waste [42]. Among the most studied insect species is the yellow mealworm (*Tenebrio molitor*), whose larvae can consume and degrade PS, PE, PU and PVC. Studies report that mealworms can convert PS into $CO_2$ and low-molecular-weight compounds, incorporating a portion of the degraded polymer into their biomass. Superworms (*Zophobas atratus* larvae) and larger wax moth larvae (*Galleria mellonella*) have also demonstrated the ability to degrade plastics, with waxworms showing nearly 100% polyethylene breakdown under experimental conditions [43]. *Tenebrio molitor* larvae exhibit a particularly broad plastic substrate range. In controlled studies, mealworms achieved 67% degradation of polyester foams, showcasing their ability to process not only polyolefins but also polyurethane-based materials

[44–46]. Interestingly, the efficiency of insect-mediated plastic breakdown is inclined to polymer characteristics like molecular weight, degree of branching, and additive content. A striking example is *Zophobas atratus* larvae, which can degrade PS and PE effectively; however, their waste lacks detectable breakdown products, suggesting that the degradation process may involve near-complete assimilation of the polymer or alternative mineralization pathways. This property highlights insect gut biodegradation as a potentially eco-friendly and closed-loop process compared to surface-level microbial depolymerisation. Beyond beetle larvae, several other invertebrates have demonstrated plastic degradation potential. The lesser waxworm (*Achroia grisella*), the confused flour beetle (*Tribolium confusum*), and even terrestrial gastropods such as the giant African land snail (*Achatina fulica*) have been reported to attack PS, PE, and PVC polymers under laboratory conditions [47–49]. Furthermore, earthworms particularly anecic and endogeic species have been suggested to accelerate the decomposition of biodegradable plastics in soil ecosystems by altering soil structure and facilitating microbial activity. Termites, as major soil engineers, also show potential in plastic biodegradation through their symbiotic gut microbiota. While early studies indicate their ability to attack polyolefins, more research is required to elucidate the enzymatic pathways and microbial consortia involved in termite-mediated polymer breakdown [50].

Despite the promising results, the large-scale application of insect-based plastic biodegradation faces limitations such as ethical concerns, containment requirements, and the need to optimize gut microbial consortia for consistent performance. However, the ability of insects to integrate polymer-derived carbon into their biomass, coupled with their gut-specific microbial enzymes, positions them as a novel bioresource for sustainable plastic waste management and bioconversion (Table 2.1).

## Mechanistic Pathway of Microbial Plastic Degradation

The synthetic polymer disintegration by microorganisms is a complex, multi-step process that integrates microbial colonization, enzymatic activity, and metabolic assimilation. Plastic polymers, with high molecular weight, hydrophobicity, and chemical inertness, are not directly metabolizable by most microorganisms. Instead, they undergo a sequence of physical, chemical, and biological transformations before complete mineralization. Microbial enzymatic plastic degradation is generally divided into major stages: biodeterioration, biofragmentation, assimilation, and mineralization, each governed by specific enzymatic and environmental factors [54]. This is schematically shown in Fig. 2.2.

**Table 2.1** Microbial degradation of polymers

| Microbes/insects | Plastic types | Degradation time and efficiency | Detection methods | Degradation mechanism | References |
|---|---|---|---|---|---|
| **1. Bacteria** *Ideonella sakaiensis* (genetically engineered) | PET | Complete degradation after 72 h | AFM and SEM analyses | PETase, a cutinase like enzyme degrades PET into monomers | [19] |
| *Pseudomonas* spp. | PE, PP | >10% weight loss | SEM and FTIR | Extracellular enzymes | [51] |
| *P. protegens* | PU | Detectable weight loss, biofilm growth | Analysis by confocal Raman and μFTIR | Ester-bond cleavage by PueA and PueB hydrolase | [52] |
| **2. Fungi** *Aspergillus flavus, A. niger A. glaucus* | PET, PP | 28.78–40% Weight loss after 30 days | FTIR, Raman spectroscopy, SEM | Laccases, cutinase lipase and ureases | [53] |
| *Embarria clematidis* | PUR | 88.84% clearance of Impranil in liquid culture after 14 days | Sturm test, FTIR spectroscopy, (GC-MS) | Esterase, protease and urease | [29] |
| *Ascocoryne* sp., *S. chlorohalonata, Phoma* sp., *T. porosum, S. rugosoannulata* | PE, DBP and DEP | >50%, within 14 days | UPLC-QTOF-MS, enzyme activity, | Laccase, peroxidase, esterase, monooxygenase | [28] |
| **3. Actinomycetes** *Streptomyces scabies* | PE | Degradation efficiency increases when plastics are pretreated (UV, heat, oxidation), | FTIR spectroscopy, (GC-MS) | Lipases and esterases, | [36] |
| *Streptomyces gougerotti, N. prasina,* and *M. matsumotoense* | LDPE, PS,PLA | 0.56–1.27% degradation, aided by UV radiation and yeast extract | FTIR, epifluroscence microscopy | Plastic degrading enzyme | [35] |
| **4. Algae** *P. maculatum, C. saccharophilum, Amphora* sp., *Hymenomonas globosa* and *Limnospira indica* | LDPE | 20% weight loss within 45 days | SEM and ATR-FTIR spectroscopy | Plastic degrading enzyme | [40] |
| *Spirulina* sp | PP, PET | 36.7–48.61% for 112 days | EDX, FTIR | – | [21] |
| *Chlorella vulgaris* | PE | 84% | FE-SEM, FTIR, EDX | Extracellular enzymes | [53] |

(continued)

**Table 2.1** (continued)

| Microbes/insects | Plastic types | Degradation time and efficiency | Detection methods | Degradation mechanism | References |
|---|---|---|---|---|---|
| **5. Insects**<br>Waxworm (*Galleria mellonella*) | PE, PP | Visible surface oxidation of LDPE within hours–days ~40–60 mg degraded per 100 larvae in 12–24 h. Mass loss: a few % in short time | FTIR (carbonyl/ hydro xyl groups)<br>GPC, DSC (molecular weight, crystallinity changes)<br>GC-MS (short-chain hydrocarbons)<br>SEM (erosion) | Enzymatic and chemical degradation | [43, 44] |
| Mealworm (*Tenebrio molitor*) | PS, PE | Consumes PS, PE over weeks ~50% PS foam mass reduction in 1–2 months ~50% carbon mineralized to $CO_2$ | FTIR, NMR (oxygenated products)<br>TGA, DSC (thermal property shifts)<br>13C radio-label tracking into $CO_2$/ biomass | Gut microbial activity | [46] |

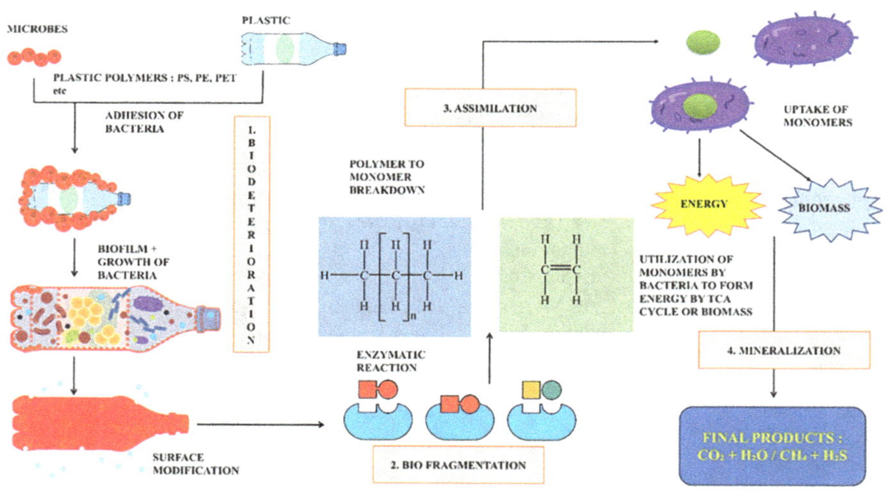


**Fig. 2.2** Illustrates the sequential stages of microbial plastic biodegradation, highlighting the transition from surface colonization to complete mineralization

### *Biodeterioration*

This preliminary phase comprises the surface conditioning and colonization of plastic by microbial cells. The formation of a biofilm plays a critical role in this stage. Microorganisms secrete extracellular polymeric substances (EPS) that facilitate adhesion and alter the physicochemical properties of the polymer surface. Enzymes or reactive oxygen species secreted by the microbes introduce polar functional groups, such as hydroxyls and carbonyls, into the hydrophobic polymer chains, thereby increasing surface wettability and initiating structural weakening. Environmental weathering factors such as UV light, heat, and mechanical stress can synergize with microbial biodeterioration to accelerate this stage.

### *Biofragmentation*

Once the polymer surface becomes more accessible, extracellular enzymes cleave the long polymer chains into oligomers and monomers. The type of enzyme involved depends on the polymer class:

- Hydrolases (e.g., esterases, cutinases, lipases) target ester linkages in polyesters such as PET and PUR.
- Oxidoreductases (e.g., laccases, peroxidases) are crucial for recalcitrant polyolefins like PE and PS, often requiring prior oxidative activation via UV or advanced oxidation treatments.

This enzymatic scission reduces molecular weight and generates water-soluble intermediates, which can diffuse into microbial cells for further metabolism.

### *Assimilation*

The smaller oligomers and monomers generated during biofragmentation are conveyed into microbial cells through specific membrane transport systems. Inside the cell, they are funnelled into central metabolic pathways such as the tricarboxylic acid cycle, β-oxidation, or glycolysis. These pathways convert the monomers into energy (ATP) and biomass precursors, enabling microbial growth and proliferation on plastic-derived carbon.

### *Mineralization*

The final stage is the complete oxidation of polymer-derived intermediates into inorganic end products such as $CO_2$, $H_2O$, $CH_4$, and microbial biomass. This stage signifies the full biodegradation of plastic, distinguishing it from partial depolymerization or surface erosion. The efficiency of mineralization depends on polymer type, microbial consortia, and environmental conditions [55].

## Biofilm-Assisted Trapping and Removal

The creation of biofilms also changes the physicochemical characteristics of MPs and NPs, which makes them more amenable to microbial colonization and enzyme activity. The extracellular polymeric substances (EPS) produced by biofilms increase the surface area and introduce functional groups to MPs and NPs, enhancing their ability to adsorb pollutants. This effectively traps contaminants within the biofilm matrix or MP aggregates. Although this does not directly eliminate MPs, the resulting aggregation reduces their mobility, concentrating them in specific zones. Such localized accumulation could facilitate targeted removal strategies if proper management approaches are implemented [56]. Multi-species biofilms create microenvironments with stable pH, nutrient gradients, and localized enzyme concentrations. This structured community architecture enhances polymer surface colonization and facilitates cooperative interactions, such as cross-feeding and quorum sensing, that optimize enzyme secretion and metabolic flux. Mixed biofilms of bacteria and fungi have shown particular promise in degrading polyolefins and polyurethane by combining fungal hyphal penetration with bacterial hydrolytic enzymes [57, 58].

### *Polymer-Specific Pathways and Influencing Factors*

The mechanism of microbial plastic degradation varies significantly depending on the chemical composition of the polymer:

- Polyesters (PET, PUR): More susceptible to hydrolytic cleavage via esterases and urethanases, enabling relatively direct biofragmentation.
- Polyolefins (PE, PP, PS): Chemically inert and require oxidative pretreatments (UV, ozone, or pro-oxidant additives) to introduce functional groups that make them susceptible to enzymatic attack.
- Polyvinyl chloride (PVC): Requires dechlorination steps before enzymatic degradation due to its halogen content.

Polymer crystallinity, surface roughness, and additive content also significantly affect degradation rates. Environmental variables such as pH, temperature, oxygen levels, nutrient availability, and light exposure further modulate microbial and enzymatic activity [59, 60].

Also Biofilm formation alters plastic density and promotes cell aggregation, facilitating sinking and sediment deposition, reducing their dispersion in water columns. As a result, MP/NP-biofilm complexes become denser, which changes their buoyancy when exposed to water. Thus such biofilm encased MPs/NPs tend to accumulate in sediments rather than surface waters. On land, water seepage and root movement cause biofilm-coated microplastics to bunch up and move to lower soil layers. Although there are various factors that might cause biofilm-induced sinking, such as organism ingestion or physical pressures, which can temporarily restore buoyancy, the sluggish clearance of biofilms in marine environments causes most MPs to settle in benthic zones. Together, these processes contribute to the long-term sequestration of MPs and NPs in aquatic and terrestrial environments, reducing their ecological mobility and persistence [61–63].

Different types of microbes adhere to one another in the EPS web to form biofilms. The EPS matrix traps microbial polysaccharides, proteins, e-DNA, peptides, enzymes and refractory substances which aids in survival of the trapped microbes [64]. These biofilm assemblages aid in the degradation of plastics like PE, PET and PP through biofragmentation and biodeterioration, where inherent microbial enzymes break complex polymers into smaller non-toxic compounds. Microplastic pollution is mitigated in a similar fashion by fungi such as *Penicillium simplicissimum* and *Phanerochaete chrysosporium*, as well as by couple of bacteria such as *Lysinibacillus*, *Ideonella*, *Serratia* and others. NPs impact the biofilm's structure via modifying the EPS hydrophilicity/hydrophobicity, surface charge, and surface roughness and creating oxidative stress, which modifies and senses the biofilm structure. This, in turn, alters the protein:carbohydrate ratio inside the EPS matrix.

As an example, hydrophobic surfaces, such as PE and PP, are more favored by bacteria for colonization than hydrophilic ones. The breakable plastic parts supply nutrients for microbial growth and cause biofilms to expand, thus biodegradable MPs and NPs break down easily when colonized by microbes. The type of plastic can also influence biofilm colonization, like *Neisseria*, was found to be augmented in PE biofilm but depleted in PVC, possibly due to PE's lesser toxicity. Presence of plastic additives like plasticizers, UV-stabilizers, certain heavy metals and flame retardants, can influence biofilm formation, as they are toxic and when they leach out due to weathering or photodegradation affects functioning of microbial communities, potentially decreasing their diversity and impacting the biofilm's structure and function [65].

## Hybrid Technologies to Determine Plastic Fate

### *Integration of Microbial Systems*

Plastic biodegradation in natural environments rarely occurs through the action of a single microorganism. Instead, it is typically mediated by microbial consortia, where different species perform complementary roles in breaking down complex polymer structures. These cooperative systems integrate bacteria, fungi, algae, and in some cases, insect gut microbiomes, creating a synergistic network of enzymatic activities and metabolic exchanges that enhances the overall degradation efficiency. In aquatic systems, algae often co-exist with heterotrophic bacteria in biofilms, creating phototrophic–heterotrophic consortia. Algae contribute oxygen and organic exudates, which stimulate bacterial growth, while bacteria produce enzymes that degrade polymer surfaces. This symbiotic relationship is especially relevant for marine plastic debris, where light-driven algal activity can sustain bacterial communities on floating polymers over extended periods [66]. Immobilization of consortia on carriers or biofilms can further enhance polymer contact and enzymatic activity. The combination of enzymatic pretreatment (e.g., oxidative modification) followed by microbial consortia action represents a promising hybrid strategy for tackling highly crystalline and chemically inert plastics [67].

### *Synergistic Enzymatic Action*

In consortia, one microbial group often initiates the process by producing enzymes that oxidize or hydrolyze high-molecular-weight polymers into oligomers. Other members subsequently metabolize these intermediates into monomers and funnel them into central metabolic pathways. For example, bacterial PETases can depolymerize PET into TPA and mono(2-hydroxyethyl) terephthalate (MHET), while secondary microbes with MHETases further hydrolyze MHET into EG and TPA, ensuring complete mineralization. Such division of labor minimizes metabolic bottlenecks and accelerates polymer turnover [68].

### *Metagenomics for Engineered Consortia*

Advances in synthetic biology and metagenomics have facilitated the strategy of artificial consortia tailored for plastic waste valorization. By combining engineered strains expressing optimized enzymes (e.g., PETase variants, laccases, cutinases) with naturally adapted degraders, researchers can build microbial systems capable of targeting mixed plastic streams in industrial settings. Enzyme secretion and substrate selectivity can be fine-tuned using CRISPR-based genome editing, and metabolic channeling can be used to decrease buildup of toxic intermediates [69, 70].

## Strategies to Enhance Plastic Biodegradability

Plastic surfaces are persistent and don't biodegrade on their own, moreover complex surface are not penetrable by microbes due to their extreme hydrophobic nature. This directs towards the use of a couple of physio-chemical pre-treatment technologies, such as gamma irradiation, plasma therapy, photocatalytic degradation not solely but in conjunction with microbial enzymes, to facilitate the destruction of MPs and NPs. Another option for removing these microscopic plastics is to use nanocomposite materials. Without harming the ecology, MPs and NPs will be eradicated in a sustainable manner, using these green nanocomposite technology. Optimal environmental conditions (pH, temperature, oxygen levels) and the slow degradation rates of highly crystalline plastics like HDPE are still obstacles in plastic mineralization. To improve the efficiency and stability of enzymes, researchers are investigating genetic engineering and enzyme immobilization approaches. To illustrate the point, fusion enzymes that combine PETase and MHETase have demonstrated a synergistic impact in hastening the breakdown of PET. Furthermore, microbial consortiums that comprise bacteria, fungus, and algae are currently under investigation for the purpose of enhancing degradation via supplementary enzymatic pathways.

## Conclusion

MPs and NPs pollution has drawn thoughtful attention from researchers, environmentalists, and scientists from all over the world. The ubiquitous and widespread presence of these polymers threatens the very existence of life on Earth. The immediate need is to investigate cutting-edge remedial techniques to facilitate and accelerate degradation of MPs and NPs. One potential strategy is the use of microbial enzymes, optimizing their production and scalability, and integrating them into wastewater treatment systems. Utilization of both naturally occurring and artificially produced microbes lessens ecological imprint compared to other available methods of deposing plastics. Synergistic use of different optimized techniques is needed to speed up and aid in degradation.

## References

1. Bullard, J. E., Ockelford, A., O'Brien, P., & Neuman, C. M. (2021). Preferential transport of microplastics by wind. *Atmospheric Environment, 245*, 118038. https://doi.org/10.1016/j.atmosenv.2020.118038
2. Gan, Q., Cui, J., & Jin, B. (2023). Environmental microplastics: Classification, sources, fates, and effects on plants. *Chemosphere, 313*, 137559. https://doi.org/10.1016/j.chemosphere.2022.137559
3. Maji, P., & Pal, D. (2022). Photodegradation processes of microplastics in aquatic medium. In *Trends in environmental biology* (Vol. 1, pp. 120–126. ISBN: 978-93-90847-55-6). Kripa-Drishti Publications.
4. Zhai, Y., Bai, J., Chang, P., Liu, Z., Wang, Y., Liu, G., & Vijver, M. G. (2024). Microplastics in terrestrial ecosystem: Exploring the menace to the soil-plant-microbe interactions. *TrAC Trends in Analytical Chemistry, 1176*. https://doi.org/10.1016/j.trac.2024.117667
5. Feng, Y., Tu, C., Li, R., Wu, D., Yang, J., Xia, Y., & Luo, Y. (2023). A systematic review of the impacts of exposure to micro-and nano-plastics on human tissue accumulation and health. *Eco-Environment & Health, 2*(4), 195–207. https://doi.org/10.1016/j.eehl.2023.08.002
6. Yang, X., Huang, G., Geng, X., Lyu, L., Bi, H., & An, C. (2025). Deciphering the behavior and fate of microplastics in coastal aquatic environments: A comprehensive review illuminating coastal dynamics and driving mechanisms. *Earth-Science Reviews, 270*, 105235. https://doi.org/10.1016/j.earscirev.2025.105235
7. Mustapha, S., Tijani, J. O., Elabor, R., Salau, R. B., Egbosiuba, T. C., Amigun, A. T., & Mohammed, A. K. (2024). Technological approaches for removal of microplastics and nanoplastics in the environment. *Journal of Environmental Chemical Engineering, 12*(2), 112084. https://doi.org/10.1016/j.jece.2024.112084
8. Rajala, K., Grönfors, O., Hesampour, M., & Mikola, A. (2020). Removal of microplastics from secondary wastewater treatment plant effluent by coagulation/flocculation with iron, aluminum and polyamine-based chemicals. *Water Research, 183*, 116045. https://doi.org/10.1016/j.watres.2020.116045
9. Skaf, D. W., Punzi, V. L., Rolle, J. T., & Kleinberg, K. A. (2020). Removal of micron-sized microplastic particles from simulated drinking water via alum coagulation. *Chemical Engineering Journal, 386*, 123807. https://doi.org/10.1016/j.cej.2019.123807
10. Badawi, A. K., Hasan, R., & Ismail, B. (2025). Sustainable coagulative removal of microplastic from aquatic systems: Recent progress and outlook. *RSC Advances, 15*(31), 25256–25273. https://doi.org/10.1039/D5RA04074D
11. Avazpour, S., & Noshadi, M. (2024). Enhancing the coagulation process for the removal of microplastics from water by anionic polyacrylamide and natural-based Moringa oleifera. *Chemosphere, 358*, 142215. https://doi.org/10.1016/j.chemosphere.2024.142215
12. Othmani, B., Rasteiro, M. G., & Khadhraoui, M. (2020). Toward green technology: A review on some efficient model plant-based coagulants/flocculants for freshwater and wastewater remediation. *Clean Technologies and Environmental Policy, 22*, 1025–1040. https://doi.org/10.1007/s10098-020-01858-3
13. Paul, B., Das, D., Aich, T., & Pal, D. (2024). Plant based biocoagulants from Cucurbita pepo and Cicer arietinum for improving water quality. *International Journal of Agriculture Environment and Biotechnology, 17*(01), 29–36. https://doi.org/10.30954/0974-1712.01.2024.5
14. Shen, M., Song, B., Zhou, C., Hu, T., Zeng, G., & Zhang, Y. (2022). Advanced oxidation processes for the elimination of microplastics from aqueous systems: Assessment of efficiency, perspectives and limitations. *Science of the Total Environment, 842*, 156723. https://doi.org/10.1016/j.scitotenv.2022.156723
15. Kim, S., Sin, A., Nam, H., Park, Y., Lee, H., & Han, C. (2022). Advanced oxidation processes for microplastics degradation: A recent trend. *Chemical Engineering Journal Advances, 9*, 100213. https://doi.org/10.1016/j.ceja.2021.100213

16. Fazli, A., Athanassiou, A., & Fragouli, D. (2025). Current approaches and challenges in advanced oxidation processes for nanoplastic degradation. *Advanced Science, 12*, e04352. https://doi.org/10.1016/j.ceja.2021.100213
17. Maji, P., & Pal, D. (2025). Nanotechnology in advanced oxidation processes for water remediation. In M. C. Garg, V. D. Rajput, T. Minkina, & S. K. Himanshu (Eds.), *Nano-solutions for sustainable water and wastewater management. Nanotechnology in the life sciences*. Springer. https://doi.org/10.1007/978-3-031-82794-5_10
18. Dhali, S., Parida, D., Kumar, B., & Kiran, B. (2024). Recent trends in microbial and enzymatic plastic degradation: A solution for plastic pollution predicaments. *Biotechnology for Sustainable Materials*. https://doi.org/10.1186/s44316-024-00011-0
19. Sevilla, M. E., Garcia, M. D., Perez-Castillo, Y., Armijos-Jaramillo, V., et al. (2023). Degradation of PET bottles by an engineered Ideonella sakaiensis PETase. *Polymers (Basel), 15*, 1779. https://doi.org/10.3390/polym15071779
20. Yao, Z., Seong, H. J., & Jang, Y. S. (2022). Degradation of low density polyethylene by Bacillus species. *Applied Biological Chemistry, 65*, 84. https://doi.org/10.1186/s13765-022-00753-3
21. Yang, X. G., Wen, P. P., Yang, Y. F., Jia, P. P., Li, W. G., & Pei, D. S. (2023). Plastic biodegradation by in vitro environmental microorganisms and in vivo gut microorganisms of insects. *Frontiers in Microbiology, 13*, 1001750. https://doi.org/10.3389/fmicb.2022.1001750
22. Venkatesh, S., Mahboob, S., Govindarajan, M., Al-Ghanim, K. A., et al. (2021). Microbial degradation of plastics: Sustainable approach to tackling environmental threats facing big cities of the future. *Journal of King Saud University-Science, 33*, 101362. https://doi.org/10.1016/j.jksus.2021.101362
23. Sagong, H. Y., Son, H. F., Seo, H., Hong, H., et al. (2021). Implications for the PET decomposition mechanism through similarity and dissimilarity between PETases from Rhizobacter gummiphilus and Ideonella sakaiensis. *Journal of Hazardous Materials, 416*, 126075. https://doi.org/10.1016/j.jhazmat.2021.126075
24. Kim, H. W., et al. (2021). Biodegradation of polystyrene by bacteria from the soil in common environments. *Journal of Hazardous Materials, 416*, 2021. https://doi.org/10.1016/j.jhazmat.2021.126239
25. Xiang, P., Zhang, Y., Zhang, T., et al. (2023). A novel bacterial combination for efficient degradation of polystyrene microplastics. *Journal of Hazardous Materials, 458*. https://doi.org/10.1016/j.jhazmat.2023.131856
26. Sanluis-Verdes, A., Colomer-Vidal, P., Rodriguez-Ventura, F., Bello-Villarino, M., et al. (2022). Wax worm saliva and the enzymes therein are the key to polyethylene degradation by Galleria mellonella. *Nature Communications, 13*, 1–11. https://doi.org/10.1038/s41467-022-33127-w
27. Kumari, S., & Das, S. (2023). Bacterial enzymatic degradation of recalcitrant organic pollutants: Catabolic pathways and genetic regulations. *Environmental Science and Pollution Research, 30*, 79676–79705. https://doi.org/10.1007/s11356-023-28130-7
28. Carstens, L., Cowan, A. R., Seiwert, B., & Schlosser, D. (2020). Biotransformation of phthalate plasticizers and bisphenol A by marine-derived, freshwater, and terrestrial fungi. *Frontiers in Microbiology, 11*, 317. https://doi.org/10.3389/fmicb.2020.00317
29. Khruengsai, S., Sripahco, T., & Pripdeevech, P. (2022). Biodegradation of polyester polyurethane by Embarria clematidis. *Frontiers in Microbiology, 13*, 874842. https://doi.org/10.3389/fmicb.2022.874842
30. Ormsby, M. J., Akinbobola, A., & Quilliam, R. S. (2023). Plastic pollution and fungal, protozoan, and helminth pathogens—A neglected environmental and public health issue? *Science of the Total Environment, 882*, 163093. https://doi.org/10.1016/j.scitotenv.2023.163093
31. Saritha, B., Sindgi, S. A., & Remadevi, O. K. (2021). Plastic degrading microbes: A review. *Microbiology Research Journal International, 31*, 22–28. https://doi.org/10.9734/mrji/2021/v31i630324
32. Okal, E. J., Heng, G., Magige, E. A., et al. (2023). Insights into the mechanisms involved in the fungal degradation of plastics. *Ecotoxicology and Environmental Safety, 262*, 115202. https://doi.org/10.1016/j.ecoenv.2023.115202

33. Saeed, S., Iqbal, A., & Deeba, F. (2022). Biodegradation study of polyethylene and PVC using naturally occurring plastic degrading microbes. *Archives of microbiology, 204*, 497. https://doi.org/10.1007/s00203-022-03081-8
34. Numoto, N., Kamiya, N., Bekker, G. J., et al. (2018). Structural dynamics of the PET-degrading cutinase-like enzyme from Saccharomonospora viridis AHK190 in substrate-bound states elucidates the Ca(2+)-driven catalytic cycle. *Biochemistry, 57*, 5289–5300. https://doi.org/10.1021/acs.biochem.8b00624
35. Oliveira, J., Almeida, P. L., Sobral, R. G., et al. (2022). Marine-derived actinomycetes: Biodegradation of plastics and formation of PHA bioplastics—A circular bioeconomy approach. *Marine Drugs, 20*, 760. https://doi.org/10.3390/md20120760
36. Soleimani, Z., Gharavi, S., Soudi, M., & Moosavi-Nejad, Z. (2022). A survey of intact low-density polyethylene film biodegradation by terrestrial actinobacterial species. *International Microbiology, 24*, 65–73. https://doi.org/10.1007/s10123-020-00142-0
37. Mukhopadhyay, R., & Biswas, K. (2024). Reports of actinomycetes to degrade microplastic. In *Microplastic pollution* (pp. 473–489). Springer. https://doi.org/10.1007/978-981-99-8357-5_28
38. Hosseini, F. S., Asoodeh, A., Ostad, M. S., et al. (2024). An integrated approach for plastic polymer degradation by the gut bacterial resident of superworm, Zophobas morio (Coleoptera:Tenebrionidae). *Environmental Science and Pollution Research, 31*, 60359–60370. https://doi.org/10.1007/s11356-024-35244-z
39. Chia, W. Y., Tang, D. Y. Y., Khoo, K. S., et al. (2020). Nature's fight against plastic pollution: Algae for plastic biodegradation and bioplastics production. *Environmental Science and Ecotechnology, 4*, 100065. https://doi.org/10.1016/j.ese.2020.100065
40. Gowthami, A., Marjuk, M. S., Raju, P., et al. (2023). Biodegradation efficacy of selected marine microalgae against Low-Density Polyethylene (LDPE): An environment friendly green approach. *Marine Pollution Bulletin, 190*, 114889. https://doi.org/10.1016/j.marpolbul.2023.1148896
41. Austin, H. P., Allen, M. D., Donohoe, B. S., et al. (2018). Characterization and engineering of a plastic-degrading aromatic polyesterase. *Proceedings of the National Academy of Sciences of the United States of America, 115*, E4350–E4357. https://doi.org/10.1073/pnas.1718804115
42. Luo, H., Liu, C., He, D., Xu, J., Sun, J., Li, J., & Pan, X. (2022). Environmental behaviors of microplastics in aquatic systems: A systematic review on degradation, adsorption, toxicity and biofilm under aging conditions. *Journal of Hazardous Materials, 423*, 126915.
43. Brandon, A. M., et al. (2018). Biodegradation of polyethylene and plastic mixtures in mealworms (Larvae of Tenebrio molitor) and effects on the gut microbiome. *Environmental Science & Technology, 52*, 6526–6533. https://doi.org/10.1021/acs.est.8b02301
44. Boschi, A., Scieuzo, C., Salvia, R., et al. (2024). Beyond microbial biodegradation: Plastic degradation by Galleria mellonella. *Journal of Polymers and the Environment, 32*, 2158–2177. https://doi.org/10.1007/s10924-023-03084-6
45. Brandon, A. M., Garcia, A. M., Khlystov, N. A., et al. (2021). Enhanced bioavailability and microbial biodegradation of polystyrene in an enrichment derived from the gut microbiome of Tenebrio molitor (mealworm larvae). *Environmental Science & Technology, 55*, 2027–2036. https://doi.org/10.1021/acs.est.0c04952
46. Liu, J., et al. (2022). Biodegradation of polyether-polyurethane foam in yellow mealworms (Tenebrio molitor) and effects on the gut microbiome. *Chemosphere, 304*. https://doi.org/10.1016/j.chemosphere.2022.135263
47. Kundungal, H., Amal, R., & Devipriya, S. P. (2024). Nature's solution to degrade long-chain hydrocarbons: A life cycle study of beeswax and plastic-eating insect larvae. *Journal of Polymers and the Environment, 33*, 1–14. https://doi.org/10.1007/s10924-024-03366-7
48. Song, Y., Qiu, R., Hu, J., et al. (2020). Biodegradation and disintegration of expanded polystyrene by land snails Achatina fulica. *Science of the Total Environment, 746*, 141289. https://doi.org/10.1016/j.scitotenv.2020.141289
49. Sanchez-Hernandez, J. C., Capowiez, Y., & Ro, K. S. (2020). Potential use of earthworms to enhance decaying of biodegradable plastics. *ACS Sustainable Chemistry & Engineering, 8*, 4292–4316. https://doi.org/10.1021/acssuschemeng.9b05450

50. Anuar, K. N. F. S., Huyop, F., Ur-Rehman, G., Abdullah, F., Normi, Y. M., Sabullah, M. K., & Abdul Wahab, R. (2022). An overview into polyethylene terephthalate (PET) hydrolases and efforts in tailoring enzymes for improved plastic degradation. *International Journal of Molecular Sciences, 23*, 12644. https://doi.org/10.3390/ijms232012644
51. Wrobel, M., Szymanska, S., Kowalkowski, T., & Hrynkiewicz, K. (2023). Selection of microorganisms capable of polyethylene (PE) and polypropylene (PP) degradation. *Microbiological Research, 267*, 127251. https://doi.org/10.1016/j.micres.2022.127251
52. Nadeau, L. J., Barlow, D. E., Hung, C. S., Biffinger, J. C., Crouch, A. L., Hollomon, J. M., & Crookes-Goodson, W. J. (2021). Colonization and degradation of polyurethane coatings by Pseudomonas protegens biofilms is promoted by PueA and PueB hydrolases. *International Biodeterioration & Biodegradation, 156*, 105121. https://doi.org/10.1016/j.ibiod.2020.105121
53. Nasrabadi, A. E., Ramavandi, B., & Bonyadi, Z. (2023). Recent progress in biodegradation of microplastics by Aspergillus sp. in aquatic environments. *Colloid and Interface Science Communications, 57*, 100754. https://doi.org/10.1016/j.colcom.2023.100754
54. Du, H., Xie, Y., & Wang, J. (2021). Microplastic degradation methods and corresponding degradation mechanism: Research status and future perspectives. *Journal of Hazardous Materials, 418*, 126377. https://doi.org/10.1016/j.jhazmat.2021.126377
55. Ganguly, A., Kaibarta, R., Das, S., Kundu, R., Paramanik, S., Mohapatra, S., Mukherjee, S. K., Mitra, D., & Chattaraj, S. (2025). Biodegradation and valorisation of plastic based food packets: A microbial solution for sustainability and circular economy. *Discover Sustainability, 6*(1). https://doi.org/10.1007/s43621-025-01257-y
56. Ventura, E., Marín, A., Gámez-Pérez, J., & Cabedo, L. (2024). Recent advances in the relationships between biofilms and microplastics in natural environments. *World Journal of Microbiology and Biotechnology, 40*(7), 220. https://doi.org/10.1007/s11274-024-04021-y
57. Debroy, A., George, N., & Mukherjee, G. (2022). Role of biofilms in the degradation of microplastics in aquatic environments. *Journal of Chemical Technology & Biotechnology, 97*(12), 3271–3282.
58. Gueboudji, Z. (2025). The degradation of microplastics. In *Handbook of microplastic pollution in the environment. Microplastic pollution in aquatic environments* (p. 358).
59. Yasin, N. M., Akkermans, S., & Van Impe, J. F. (2022). Enhancing the biodegradation of (bio) plastic through pretreatments: A critical review. *Waste Management, 150*, 1–12. https://doi.org/10.1016/j.wasman.2022.06.004
60. Kotova, I. B., Taktarova, Y. V., Tsavkelova, E. A., et al. (2021). Microbial degradation of plastics and approaches to make it more efficient. *Microbiology, 90*(6), 671–701. https://doi.org/10.1134/s0026261721060084
61. Pinto, M., Langer, T. M., Hüffer, T., Hofmann, T., & Herndl, G. J. (2019). The composition of bacterial communities associated with plastic biofilms differs between different polymers and stages of biofilm succession. *PLoS One, 14*(6), e0217165. https://doi.org/10.1371/journal.pone.0217165
62. He, S., Jia, M., Xiang, Y., Song, B., et al. (2022). Biofilm on microplastics in aqueous environment: Physicochemical properties and environmental implications. *Journal of Hazardous Materials, 424*, 127286. https://doi.org/10.1016/j.jhazmat.2021.127286
63. Chattopadhyay, I. (2022). Role of microbiome and biofilm in environmental plastic degradation. *Biocatalysis and Agricultural Biotechnology, 39*, 102263. https://doi.org/10.1016/j.bcab.2021.102263
64. Flemming, H. C., van Hullebusch, E. D., Neu, T. R., Nielsen, P. H., Seviour, T., Stoodley, P., & Wuertz, S. (2023). The biofilm matrix: Multitasking in a shared space. *Nature Reviews Microbiology, 21*(2), 70–86. https://doi.org/10.1038/s41579-022-00791-0
65. Tu, C., Chen, T., Zhou, Q., et al. (2020). Biofilm formation and its influences on the properties of microplastics as affected by exposure time and depth in the seawater. *Science of the Total Environment, 734*, 139237. https://doi.org/10.1016/j.scitotenv.2020.139237
66. Amirian, V., Russel, M., Yusof, Z. N. B., et al. (2025). Algae-and bacteria-based biodegradation of phthalic acid esters towards the sustainable green solution. *World Journal of Microbiology and Biotechnology, 41*(2), 24. https://doi.org/10.1007/s11274-024-04243-0

67. Rambabu, K., Bharath, G., Govarthanan, M., et al. (2023). Bioprocessing of plastics for sustainable environment: Progress, challenges, and prospects. *TrAC Trends in Analytical Chemistry, 166*, 117189. https://doi.org/10.1016/j.trac.2023.117189
68. Shokunbi, A. O., Ezeh, O. M., Fadipe, D. P., et al. (2025). Microbial consortia: Synergistic effects on plastic degradation and enzyme production. *Journal of Medical Science, Biology, and Chemistry, 2*(1), 149–156. https://doi.org/10.69739/jmsbc.v2i1.681
69. Adamu, K. S., Bichi, Y. H., Nasiru, A. Y., et al. (2023). Synthetic microbial consortia in bioremediation and biodegradation. *International Journal of Research and Innovation in Applied Science, 8*, 232–241. https://doi.org/10.51584/IJRIAS.2023.8727
70. Che, S., & Men, Y. (2019). Synthetic microbial consortia for biosynthesis and biodegradation: Promises and challenges. *Journal of Industrial Microbiology and Biotechnology, 46*(9–10), 1343–1358. https://doi.org/10.1007/s10295-019-02211-4

# Chapter 3
# The Detection Methods of Microplastics and Nanoplastics

**Tamanna Chaturvedi, Sunil Kulkarni, and Ajaygiri Goswami**

## Introduction

Microplastics and nano-plastics possess sophisticated elementary properties that affect their migration, toxicity, bioavailability, and interactions with surrounding pollutants [1]. The detection methods for these pollutants involve isolation of the particles from the habitat. The most widely used method is the microscopical technique involving dissecting microscopy, polarized microscopy, fluorescence microscopy, scanning electron microscopy, and atomic force microscopy [2]. Spectroscopic technique involves studying the motion of molecular vibration for the detection of material. This technique uses a monochromatic laser source for identification, along with exciting the molecules to analyze the sample.

Thermal Analysis involves the identification of MPs and NPs based on the products formed by the thermal degradation of the material [3]. Plastics are substances composed of a long polymeric chain of hydrocarbons. This material is very versatile, non-corrosive, strong, and has an extended shelf life. Plastic has become an integral part of our lifestyle and the manufacturing industry [4, 5]. The use of plastic is growing day by day, but the waste management is still not efficient enough to dispose of all the plastics being produced. These plastic pollutants, with time, start degrading as they are exposed to different chemical reactions and physical processes. The degradation eventually results in the formation of microplastics (MPs) and nanoplastics (NPs) [6]. The size of microplastics is less than 5 mm, while nanoplastics are less than 0.1 μm [7, 8]. There has been an estimation of about 80% of

T. Chaturvedi · A. Goswami
University Institute of Chemical Technology (UICT), Kavayitri Bahinabai Chaudhari North Maharashtra University (KBCNMU), Jalgaon, Maharashtra, India

S. Kulkarni (✉)
Chemical Engineering Department, Gharda Institute of Technology, Lavel, India



S. Kulkarni, S. Das (eds.), *Microplastics and Nanoplastics*, Lecture Notes in Nanoscale Science and Technology 49,
https://doi.org/10.1007/978-3-032-17963-0_3

the MPs and NPs originating from the land, while the rest 20% is from the sea. As these materials are uniquely light, indestructible, and able to float, they create an ease for the spread of pollution [9]. These eroded fragments of plastic pose a great threat to the environment as they reduce the fertility of the soil and choke aquatic animals.

MPs can be grouped into two types based on how they were formed [10]. The MPs that are created purposefully for industry are called primary MPs [11, 12]. The plastic formed by prolonged degradation of virgin plastic is called secondary NPs [11, 12]. Microplastics affect marine life [13]. Geyer et al. estimated that by the year 2017, the production of virgin plastic amounts to 8300 million metric tons [14]. The most prominently present polymers were polyethylene (PE) [15], polystyrene (PS), and polypropylene (PP) [13]. In the sample, the presence of additives such as phthalates, alkylphenols, bisphenol A, titanium dioxide, sulfur, and zinc were found in the form of NPs in the MPs [13]. It also contained additives in the "remnants of monomers, persistent organic pollutants (POPs) like polychlorinated biphenyl (PCBs), polycyclic aromatic hydrocarbons (PAHs), and organic chlorine pesticides" [13].

The study of the development of detection methods for controlling the contamination of nature due to MPs and NPs is becoming crucial with time, as these invisible particles may soon become a new threat to the life forms of this planet. In this chapter, we explore different types of methods developed and under development for detecting the presence of MPs and N Ps [14, 16].

## Microscopy Method of Detection

In the current state of time, the microscopy method is widely used and the most exploited method for detecting the presence of MPs and NPs. This method of detection is relatively based on plastic particle analysis, and facilitates gaining information in regards to the morphology of the sample, which includes the geometry of the plastic, surface characteristics, color, and size [17]. The microscopy method focuses on the structure of the particle for the identification. Factors such as presence of impurities, chipping off of the particle and adding of additives for sample preservation still holds a threat for misidentification of the particles.

The types of microscopy techniques are as follows:

### *Stereo- (or Dissect) Microscopy*

Number of optical and spectrum detection methods including stereo microscopy and other spectrum microscopy have been used by investigators [18]. Rhodamine B dye (RhB) staining method for microplastics can be used for labeling [19]. The comprehensive structural details about the sample hold great importance for identification of the plastic topology [17]. Stereomicroscopy is one of the most fundamental methods for the identification of particles.

The size of particles based on identification efficiency is divided into three categories as follows:

1. Particle size less than 100 μm.
2. Particle size between 100 and 500 μm.
3. Particle size greater than 1000 μm.

Here, the particles in the sample are identified based on their physical appearance. The quick screening technique allows identification of the shape, size, and color of the sample. When the size of the particle is smaller than 100 μm [20], they tend to be transparent or in particulate shape, making them difficult to recognize. The magnified image of the particles provides us with detailed morphology and typology for the sample. The presence of dense sediments in the sample, which are generally greater than 500 μm and cannot be eliminated by chemical digestion [17], also hinders the detection process [21]. Along with these, the presence of synthetic and natural fibers, which are very abundant in aquatic habitat, obstructs the identification of MPs [22]. These problems are solved by combining stereomicroscopy with methods of spectroscopy identification. An example of this can be considered the combination of stereomicroscopy and the FTIR identification method to detect samples for the sea surface micro layer and beach sand [17]. This combination allows identification of smaller MPs and NPs that cannot be detected alone with stereomicroscopy.

Stereomicroscopy fails to distinguish between natural polymers and synthetic particles created by humans [18, 22]. The insights gained from stereomicroscopy are largely influenced by the researcher's viewpoint. Micro Raman analysis enables to analyze hundreds of micro and nanoparticles [23]. FTIR can be used to analyze different shapes of microplastics [24].

## *Fluorescence Microscopy*

Fluorescence microscopy utilizes the gathering of fluorescence released from the fluorophore substance of the studied sample by excitation of specific wavelength [24]. The wavelength of light used for excitation is mainly of three types [17]:

1. Blue fluorescent (390 nm)
2. Green fluorescent (475 nm)
3. Red fluorescent (542 nm)

Traditionally, this method utilizes a laser that emits visible blue light and ultraviolet light continuously, which passes through a minute hole and focuses on a specific point of the sample to be studied [25]. The emitter laser is shorter by 50–200 nm in wavelength than what is emitted by the fluorophore. This allows photons from the source to excite the sample to emit fluorescence [17]. The excitation of the electron in the particles results in emission of fluorescence due to the gained energy. This method is primarily used for detecting white or transparent microplastics, as well as primary sub-micro and nano-plastics. This also facilitates the identification of particles which are smaller than the range of optical microscopic techniques. It is a cost-effective and time-saving method, generally used in combination with μ-Raman or μ-FTIR. It can measure particles of size less than 100 nm when combined with confocal scanning [26].

The presence of impurities in the sample can cause co-staining, additional fluorescent signals, and leaching of fluorescent dyes; hence, the sample pretreatment method holds great importance. The sample treatment allows the elimination of organic impurities, which can create hindrance in the identification process [9]. Unfortunately, pretreatment only helps remove surface impurities and contaminations. The pretreatment fails to remove plasticizer and dyes, which are blended into the sample [17, 27]. The addition of additives to induce fluorescence in the sample can affect the identification process by interfering with the natural fluorescence of the MPs. If a high high-powered laser is used for the identification process, it may also affect the particle by bleaching it, misleading the identification process [17]. The use of lasers also limits the exposure time for the identification [17]. Even though this method has proven to be quite efficient for detection of MPs and NPs, it still holds a great risk in generating the correct result.

## *Transmission Electron Microscopy (TEM)*

Transmission electron microscopy [17] uses an electron microscope with a very high resolving power [17] to observe samples at an infinitesimal size using the wave nature of electrons [17]. Thermo-ionic material such as tungsten or lanthanum hexaboride is used in the form of a slender filament [17]. This filament emits electrons which travel via a hole in the anode and then into a magnetic capacitor [17]. The intensity of the conjunction of the beam consisting of electrons is regulated by a capacitor [17]. Once the beam of the electron passes through the sample, the electrons get scattered. This scattering of electrons can be elastic or inelastic [18, 28].

The signals which are collected by the detectors are used to obtain chemical and imaging data at spatial resolutions for NPs. Since MPs and NPs are not electron dense, they are stained using heavy metals, but this can change the chemical composition of the plastic. When the elastic interaction with the electron is weak, it reflects a poor contrast in TEM [29]. The chemical characteristics of the particles present in the samples are obtained by adding fluorescence of the particles [17]. The fluorescent information is displayed in the form of a colored overlay over the TEM image [17]. The images obtained by TEM not only provide morphological information about the sample, but also portray the chemicals present in the sample.

To understand the toxic effects of polystyrene MPs, a study was conducted on Halomonas alkali philia, which is a bacterium present in the marine ecosystem [17, 30]. An increase in extracellular polymeric substances was observed under TEM, as a possible bacterial protective mechanism [17, 30]. Another study reported the TEM analysis to determine the size of polyethylene NP (<100 nm) and polystyrene NP (~500 nm), which were very heterogeneous [31, 32]. In a study, it was concluded that there is a need for multiple complementary technologies to provide accurate shape and size characterization. Along with this, it can be interpreted that no technology is completely capable of characterizing MPs and NPs [33].

Using the established protocol, Yang et al. in 2021 prepared PET NPs. He dissolved 1 g of the prepared PET into 100 mL of trifluoroacetic acid (90% v/v) [9]. This solution was then maintained at 50 °C, stirred at 700 rpm for 2 h and stored overnight [9]. In a dropwise manner, 10 mL aqueous TFA solution (20% v/v) and was then stirred overnight [9]. This suspension was then centrifuged at 2500 rpm for an hour to recover the PET in the form of pellets [9]. The pellets were then dispersed into 100 mL of 0.5% sodium dodecyl sulphate (SDS) solution [9]. The new suspension formed was stirred for about an hour and ultrasonicated thrice, for 5 min [9]. From the cylinder it was stored into, 20% of the solution from the top was taken for further inspection [9]. For preparation of the final result PS was drop-casted into the suspension onto the TEM grid [34]. This detailed study infers the methodology for the detection of nano-sized PET pellets and can be further used for detection of other MPs and NPs.

TEM is still in an early stage of development and has not detected NPs so far [17]. The development of the method for precise detection of the MPs and NPs creates an opportunity for the innovation of new technologies.

## *Scanning Electron Microscopy (SEM)*

Scanning electron microscopy is a technology that gives information about the morphology of the exterior of MPs and NPs [35]. The combination of SEM with energy-dispersive X-ray spectroscopy (EDS), provides information about the composition of the sample [36]. The electron beam is made up of primary and secondary electrons [37]. The primary electron penetrates through the solid sample to create elastic or inelastic scattering, creating signals which are detected by detectors, while the secondary electron is responsible for providing morphological information by accelerating across the sample surface, when applied the voltage of 10–20 kV and the back scattered electron passing through the sample provides information regarding topography and contrast based on atomic number [17, 38].

In this method, the electron emitter generates a ray of electrons, and then the primary ray of electrons collides with the surface of the sample to generate multiple interactions [17]. These interactions generate data about the element and the spatial distance, along with characterization of the sample [36]. For attaining proper visibility of the sample morphology, the sample is coated with a conductive path of metals to increase the conduction of electricity on the sample surface. It was observed that the addition of chemical additives and coating leads to susceptibility to misclassification of the particles [39]. The pre-treatment step does not involve any special digestion of organic matter; but needs to be coated with metal [40]. This metal coating increases the electron density of the sample, easing the detection process [40]. The organic matter present on the sample can hinder the identification process. Though this study can provide us with the detail of the exterior of the particles, the information about the inner material still remains unknown [40]. If the beam of electron falls strongly on the particle, it may alter the surface structure of the sample [40].

In a study done by Mariano et al. (2019) [2], they concluded that SEM is versatile, allowing NPs to be characterized as STEM holder integration to the sample provides analysis on a grid [40]. This method also allows visualization of small-size particles without increasing the energy and voltage as done in the TEM method [40]. Compared to TEM EDS, SEM EDS is more efficient due to larger sample space [41].

In another study, on MPs and NPs for indoor material for understanding the surface structure and fundamental constituents of the sample, providing a graphical representation of elemental analysis for household material [21, 42, 43]. The size of all the particles studied was less than 2.5 μm. The fourth-stage and first-stage filters, when compared, were very similar in composition [21]. In the study, the presence of plastic nature was confirmed along with elements such as carbon, oxygen, fluorine, sodium [44], chlorine, calcium, gold, magnesium, aluminum, silicon, potassium, iron, and zinc [44]. It was observed that the MPs and NPs had a smooth surface with irregular patterns, such as grooves, pits, cracks, and linear fractures, which are caused by chemical and mechanical deformation, collisions, frictions, and wind actions. The increased quantity of carbon, oxygen, and organic polymers helped in MPs [37, 45].

These elements are not chemically bonded and can be released into the environment due to degradation and weathering processes [21]. In another study, it was detected that 23% of NP particles present in fresh water were eliminated using an elimination-based strategy on the results obtained from SEM analysis [46]. SEM analysis has always been important for wastewater analysis [47].

The SEM method helps obtain high-resolution topography images for particle sizes greater than 0.5 nm and provides high magnification of about 200–10,000 times. This method provides very detailed information about the structure of the MPs and NPs. MPs and NPs in this method are differentiated based on surface features of the particles. This method can also visualize weathering effects on plastic particles, along with detecting the organic and inorganic elements present in the particles [18, 27].

## *Atomic Force Microscopy (AFM)*

Atomic force microscopy is used for visualizing dielectric materials without the need for surface plating with metal. Here a small rigid conductive tip is attached to the cantilever for the detection.

The exterior of the sample is pressed against the conductive tip during the process of measurement [17, 48]. The probe, using a piezoelectric mechanism, moves on the surface of the sample for all the Cartesian axes. The tip is maintained at a persistent height to gather the particulars about the variation in the texture of the sample's surface. An oscillation system is then used to record the slightest movements of the tip [49]. By processing the information about the relative motion of the cantilever and the sample's exterior, we then construct the image of the sample [50]. Alemansour has discussed probe-based devices and methods for applications in micro/nano characterization and fabrication [51].

The interaction between the cantilever tip and the surface of the sample can be carried out in mainly three ways, as classified below.

(a) **Contact mode:** Here, as the name suggests, the tip is physically touching the sample's exterior. The repulsive Van der Waals' forces make the lever bend accordingly and this electrostatic interaction is about $10^{-9}$ N [17]. The small elastic constant of the lever helps to avoid extra pressure and alteration to the surface of the sample [17]. It is generally used when the particle is a hard sample, which won't easily break on contact with the tip [52]. Even though the material is hard, it is necessary to make sure that no alterations due to the tip movement occurs to the sample's surface.

(b) **Non-Contact mode:** This mode maintains the cantilever a few nanometers above the surface of the sample and the cantilever resonates according to the sample's surface, resulting in continuous vibration [53]. Based on the distance of the cantilever from the exterior of the sample, the oscillation frequency of the vibration is obtained. This is then used to build the surface image for the particle [17]. It is used to analyze particles that are soft and can get altered due to contact with the tip of the probe [54]. This method provides a safer way of detection of malleable samples with accuracy.

(c) **Tapping mode:** Tapping mode provides images with maximum resolution but requires a lever with a high elastic constant. Here, the lever vibrates at a frequency which is very close to the natural oscillation of the sample. The interval between the lever and the sample is maintained such that only at the maximum amplitude of oscillation it makes ephemeral contact to the surface [17]. Force applied by tip gets absorbed by the sample with no damage, and the changes in the amplitudes are then recorded and used for the construction of the surface image [17].

AFMs used for the identification of MPs and NPs are of two types—Quasi-static AFM and Multifrequency AFM [55]. The most significantly used AFM is the quasi-static AFM, but fails when it comes to the detection of the mechanical properties of NPs [55]. It uses the lateral displacement method with forced application, while the multifrequency is an innovative method, but literary studies are yet to be done on it [55].

Akhatova (2022) [37] used AFM to obtain images of differently sized polystyrene particles (100, 200, 500, and 1000 nm) [37] and observed that all these particles were spherical and their dimensions corresponded to the data provided by the manufacturer [37]. Minor difference in shape as a result of the effects due to the presence of the tip and fragmentary removal of particles during scanning was noticed [37]. The polystyrene particles were identified and imaged based on modulus and deformation data. This study concluded that AFM provides accurate data about topography, adhesion, and deformation channels, for particles whose diameter ranged between 500 and 1000 nm, while it fails to detect particles less than 200 nm in diameter under fixed cells under air imaging conditions. By standardizing the data of known MPs, this method can help in identifying multiple polymer types present in the particle, allowing to widen the scope of standardization of parameters for the detection of MPs and NPs.

In the study carried out by by Galluzzi et al. (2025), quasi-static AFM and multifrequency were both utilized to observe images of polystyrene and sodium bentonite on substrate of glass or mica [55]. Here, the force vs. indentation curve was used to estimate the elastic property at the nano-scale level [55], and the multifrequency AFM uses a probe in bimodal tapping mode, where in the first oscillation, the probe amplitude is fixed. In the second oscillation, the probe frequency is fixed to get the image of the surface of the sample. It was observed that the polystyrene particles existed in micro-spherical shape while the bentonite aggregate was found to be irregular in morphology. The elastic modulus mapping, along with morphological overlay, displayed a sharp contrast in the mechanical properties and thus differentiated in the elastic properties of the mixture [55, 56].

AFM also provides information about additional characteristics such as stiffness, hydrophobicity, conductivity, and magnetization [1]. This method is advantageous in many ways as it provides high-resolution images for up to a few nanometers, which are superior to those obtained from SEM. It preserves the sample's surface, investigates non-conducting polymers, analyses nano-composite materials, and avoids damage from radiation. However, AFM has its limitations: the sample size is restricted, which can pose problems for larger MPs; only surface particles can be detected; the samples are prone to contamination; and the contact method for adhesive polymers can lead to the release of fragments, resulting in incorrect images [1]. Advanced materials such as lyotropic liquid crystals calls for advanced analysis methods [57].

## Fourier Transform Infrared (FTIR) Spectroscopy

FTIR spectroscopy integrates the Fourier transform with infrared spectroscopy to study chemical bonds, focusing on the interaction between infrared radiation and molecular vibration [1]. It is performed using an interferometer, which scans the overall frequency existing in the infrared spectrum emitted by the source [1]. Here, the difference in the optical path, which is created by a moving mirror facilitates the observations for the identification. The ray reflected from the fixed mirror results in the formation of constructive and destructive interferences; these interferences are then used for the processing of the particle image [1]. For the time domain of the study, the intensity of the radiations is represented in the form of an interferogram [58]. An infrared spectrum is plotted from the data obtained. The abscissa of the plot represents the wavelength of the incoming photon, while the effectiveness in transmitting the radiant energy is shown on the ordinate [17]. The vibrations observed in FTIR are caused by the stretching of chemical bonds and the deformation of bond angles [59].

For obtaining the image of the MPs and NPs, FTIR is used in four modes, namely transmission, reflectance, true specular reflectance, and attenuated total reflection. Various methods used for the analysis ranging from scanning electron microscopy to atomic force microscopies have been reported [60–62].

FTIR when used with atomic force microscopy (AFM) is called AFM-IR. In this setup, a pulse infrared source with a pulse length of 10 ns is coupled with a microlever AFM, which features a sharp tip for measuring the imperceptible movement of the sample's surface using an infrared laser beam. The images obtained from this have discernible fine details of 50–100 nm [35]. The infrared light absorbed by the sample leads to a slight expansion of it, which causes a consequent deflection of the microlever, resulting in the production of IR signals [35]. The absorbed pulse light generates heat leading to slight thermal expansion of the material. This creates an impulse in the reading of the AFM [37, 63].

Micro-FTIR (μ-FTIR) is another widely used method of using FTIR, where the IR instrument is combined with an optical microscope, giving imaging and chemical, and spatial information simultaneously, allowing concurrent identification and quantification information. μ-FTIR is generally used with a focal plane array detector or with point analysis through the transmission mode of instrumentation [18]. This method captures the morphological image of particles of size about 1 μm. The microscope diffraction limit strongly affects the detection limit of the particle.

Minimization of undesired interferences are overcome by employing suitable filters for the process. In this method, sample treatment plays an important role for the removal of undesired organic matter which poses a threat to misidentification of the particles [18, 64].

The ratio of signal to the background noise is way better in FTIR for tracing the IR spectrum, as FTIR provides higher energy availability and better performance of the system [17]. It is a non-destructive analysis method as the sample remains intact during analysis [65]. Along with this, particle ageing is also observed by studying surface oxidation, and it can measure MPs up to 50–500 μm with a high resolution of 10–20 μm; however, it has not yet been proven to detect NPs and MPs below 10 μm. Here, the sample pretreatment plays a crucial role in eliminating the impurities present in the sample. Thus, there is a probability of the vibrational bands being impacted by the chemical digestion process [39, 66]. Traces of moisture can also be inferred in the analysis of the sample [39, 67, 68]. Identification of MPs using FTIR has been successfully reported in the coastal environment [69]. The accuracy of FTIR makes this method approachable for the detection of MPs and provides a wide scope for researching for the detection of NPs via this method.

## Raman Spectroscopy

Raman Spectroscopy is another benign method of analysis, allowing sample identification based on the scattering of photons when the kinetic energy of the incident particle is not conserved, from a monochromatic source (laser) [35]. Integration of Raman spectrometer to an optical microscope forms the system for the detection of the particles [39]. Here, a monochromatic beam is shone upon the sample, where a part of the beam elastically diffuses through the sample at the initial frequency due to Rayleigh scattering, and the other part undergoes inelastic diffusion, creating a spectrum of scattered radiation, representing a series of lines with increasing and decreasing frequency than the Rayleigh line [70]. This method identifies the polymer and differentiates the types of additives present in the polymer. Small MPs up to the size of 20 μm and the upper range of NPs can be easily identified using Raman spectroscopy. When used as micro-Raman spectroscopy, it can identify particles up to 1 μm and can be combined with image mapping to detect NPs of 100 nm as well, but it does not facilitate the depiction of the true morphology of NPs [71, 72]. The three-dimensional morphological image contains "two spatial dimensions and one spectral dimension of the particle" [18]. The image obtained is of very high resolution, as each pixel in the image is thoroughly scanned for the spectrum [18, 72].

Sample purification is necessary for matrix removal to increase the correct identification of MPs [73]. This method involves recognizing laser spots, pixel image resolution, NP/MP positioning, and Raman mapping of signal intensity [74]. Optical tweezers are used to resolve the issue of low Raman spectra by capturing the particles of the same size [72]. Optimizing plasmonic hotspots also helps in enhancing the efficiency of detection [72].

Schmidt et al., [75] proposed a system where Raman imaging was combined with scanning electron microscopy. This combined micro-Raman for chemical identification and scanning electron microscopy to obtain high-resolution imaging. In his investigation Ivelva, [71], was able to obtain a three-dimensional image of PVC MPs from the gut area of Daphnia magna [71]. In another study, he explained the need for a longer acquisition time for correct identification, as he observed that pigmented plastic, having film formers which led to wrong identification as paint pigments [71]. Other studies are also reported for acrylic acid monomer detection and characterisation of the physico-chemical and solid-state behaviour of drugs [76, 77].

In a study by Schröder et al., he brought the human bias nature as a factor to misidentification of the MPs and NPs [78]. This study helps understand that human bias increases the occurrence error when the size of particles decreases [78]. The number of particles requiring study to "get a statistically meaningful result" was given by Anger et al. by using a random sampling approach [79]. Confocal Laser Scanning Microscopies becoming important for obtaining high resolution and in dept images [80].

Raman spectroscopy surpasses FTIR as it remains unaffected by sample thickness and moisture, making it suitable for aquatic and biological samples. It effectively assesses stereoregular characteristics in small amounts, reducing misidentification from agglomeration and overlapping particles [53]. Tip-enhanced Raman Spectroscopy (TERS) merges surface-enhanced Raman spectroscopy (SERS) Sensitivity with high-resolution microscopy, achieving nano-precision imaging [81]. However, challenges in quantitative analysis persist due to a lack of standardization in Raman signal intensity, varying magnifications, substrate variability, and the time-intensive sample preparation process. Identifying larger particles and multiple-colored fibers also presents difficulties [82].

The main obstruction in the identification process is the fluorescence caused by particles, as the Raman scattering and fluorescence both follow the same phenomenon of photon emission due to excitation in the absorption [17]. This fluorescence in the particle is caused by the presence of additives, pigments, and impurities present in the particle. Black fibers absorb Raman laser in a large amount and also emit intense background fluorescence, causing an error in the observation [18]. To minimize the error caused by interference from strong fluorescence, choosing parameters like wavelength, power, acquisition time, magnification of the object, and confocal mode becomes crucial [71].

Raman Spectroscopy is better suited for preliminary identification or confirmation of other detection methods [83]. It detects polyethylene, polypropylene from plastic containers and food packaging, polystyrene, polyethylene terephthalate, and PES. It also differentiates between synthetic and natural fibers [84].

## Thermal Analysis

Thermal analysis identifies plastic particles from the product obtained after pyrolysis and ion exploitation of the sample [85, 86]. MPs which are lightly insoluble and has inseparable additives are identified using this product [86, 87]. This method focuses on exploiting the physicochemical properties of the polymer for identification. The various methods in which thermal analysis can be carried out are mentioned below [88].

### *Differential Scanning Calorimetry (DSC)*

DSC identifies the particles present in the sample based on the transition occurring in the sample based on measure of heat flux and the temperature [30]. The information about the material is obtained from the continuous heating and cooling of the sample [89]. The methods consist of limitations for polymers which do not have a selective range of temperature for melting while it has successfully identified PE MPs with known characteristics [86].

### *Thermogravimetry Analysis (TGA)*

TGA offers quantitative thermal analysis for a certain temperature by quantifying the loss of weight from the sample [90]. This does not identify the chemical composition of the sample [90]. The pyrolysis reaction for the study, takes place in an oxygen deficient environment. The sample disintegration at low boiling temperatures provides structural information along with the observation of modifications the sample undergoes as the temperature rises [90]. The details of the sample from the thermo-gravimeter are obtained as graphs where the mass is represented as a function of time. The understanding of how the alteration in the environmental condition can affect the degradation mechanism is crucial to carry out the identification process [90]. For gaining the reliability in the information, detailed context of operation conditions related to the physical characteristics of the sample, rate at which the heating is carried out and the atmosphere of the experiment, and the standardization of the same helps in characterization and quantification of MPs/NPs [86].

The application of this method to both the organic and inorganic matrices make it a very versatile method [35]. This method also identifies MPs with low solubility [35]. When used with an air or water filtrate, it produces results in 2–3 h [91]. It can accommodate samples with higher mass and is suitable for small-scale heterogeneous samples. The sample preparation steps involve grinding and mixing, and the solvent elimination helps in preventing background contamination [91]. Unfortunately, the data obtained from this method is intricate, poses operational challenges, and provides no data regarding size and shape. In the thermal degradation of material, the loss of mass starts with the change in enthalpy, which is not measured in TGA, and hence a combination with DSC is suggested in the analysis of MPs [90]. Polymeric particles other than MPs cannot be detected using thermal analysis due to the overlapping phase transistor signal [92].

## Pyrolysis-Gas Chromatography-Mass Spectroscopy (Py-GC-MS)

Py-GC-MS is another method of thermal analysis for the simultaneous identification and quantification of MPs in complicated samples [93]. This method identifies the polymer by first degrading the sample to a low molecular weight product and then separating these by gas chromatography [91]. The substances obtained by separation are then identified via mass spectroscopy, and all the data from these results are then used to identify the composition of the sample [91]. It is a small-sized, fully automated system providing detailed analysis regarding chemical composition, endogenous additives, and environmental contaminants. Polymer and additive identification via this method is sped up using research software and libraries. In this method, double-shot pyrolysis is also used where the gas chromatography takes place at two temperatures, i.e. first shot being at a lower temperature range for low molecular weight composition and the second shot being at a higher temperature for actual identification and quantification. This method allows characterization of polymers and additives, but lacks particle size information [94].

## Receptor-Based Detection

Receptor-based detection is a miniature detection system, similar to a bio-receptor, allowing identification of MPs and NPs based on a "lock and key" mechanism. This method can detect NPs at the molecular level, as the interaction between plastic molecules and the receptor is at the molecular level [95]. As this method is highly specificity for identification of polymer molecules, it is not affected by contaminants present in the environmental sample [95]. Biological recognition elements are used for identification purposes [95]. These receptors get stuck transducers and create unique analytes [95]. The bio-receptors used here are bacteria [95], cells [95], proteins [95], short peptides [95], fluorescent dyes, polymers, and micro/nano structures [95] are discussed in detail below [95].

### *Receptor-Based Detection Using Bacteria*

Receptor-based detection using bacteria in the method incorporates live bacteria on transducers, which bind to MPs, resulting in activation or deactivation of biochemical reactions and a change in transducer signals [96]. The transducers used here are mainly of two types, electrochemical transducers and optical transducers [97, 98]. In an electrochemical transducer, the bacteria are coupled with an electrode for the movement of electrons, and the required exterior is functionalized, and the cells are inactivated for detection purposes. The types of electrochemical transducers used are potentiometric, amperometric, impedimetric, conductimetric, and capacitive. Among these, impedimetric is most widely used and is a reliable technique to detect bacterial activity [96]. In optical transducers, the genes of the bacteria are modified to produce bioluminescent and fluorescent signals in the reporter system [95], to provide optical and visual identification of MPs and NPs [99].In a study conducted by Raghavan et al. [99], whole-cell biosensors were developed to detect acrylic acid, which is a degraded product of polyacrylic acid [100, 101].For this method, many bacteria are not identified, but it creates an opportunity to identify more plastic-specific micro-biomes in nature [95]. To identify the bacteria that can be used as a receptor, high-throughput 16s rRNA gene sequence and commander was used on bacterial communities living on plastic particles [95, 102]. Out of 100s of species of bacteria, several were experimentally validated and found that marine hydrocarbon-degrading bacteria could also break down PET [103]. Studies have also shown that MPs can be toxic to the genes and cause the microorganism to mutate in order to develop resistance against them [104]. These mutated genes are further used towards receptor selection for the identification of MPs and NPs [95]. This technique is still in the laboratory stage and is not applicable for field analysis [95].

## *Receptor-Based Detection Using Proteins*

Receptor-based analysis utilizing proteins primarily employs antibodies, which identify distinct epitopes on the surface of analytes. This poses a challenge with MPs, as they lack unique epitopes, making it difficult to navigate morphological variations in MPs [95]. Antimicrobial materials also can be analysed by their resistance [104]. Consequently, various antibodies have been identified to capture plastic-related toxic chemicals [105–109]. In enzyme-based sensors, the enzyme either produces or consumes electro-active species and utilizes electrons as transducers [95, 110–112]. Based on the location of the sensor, it is divided as transmembrane receptor and intracellular receptors [95, 113, 114]. The transmembrane receptor participates in the internal placement of the MPs into the cell. Li et al. [115] developed a sensor which was functionalized as an estrogen receptor and called surface plasmon resonance (SPR) [95]. The SPR uses surface plasmons to create an oscillation condition of electrons and excites them with incoming ray of light at the interface of the gold film and the dielectric [95]. The interaction brought about by the plasmon at the surface and the incident light affects the plastic binding, leading to changes in the refractive index [95, 111]. This interaction alters the wavelength of the reflected ray, angle of reflection, intensity, phase and polarization of the reflected light [95, 111].Reflection from the dielectric is observed in the spectrophotometer and based on the findings, further analysis is made for understanding of the components in the sample. The main challenge this method holds is the need to study and identify a considerable number of microorganisms and receptors for identification of various polymer molecules as the variations in the morphology can alter the responsiveness towards the receptor [95].

## *Receptor-Based Detection Using Short Peptides*

Receptor-based detection is also performed using short peptides [95]. In 1995, Adey et al. discovered a class of bacteriophages that bound right to the top exterior during a phage display experiment with a random peptide library [112]. For the identification of appropriate peptides which can effectively bind to the plastic, highly accurate screening systems have been developed based on the structure of E. coli's surface [113–115]. Localized surface plasmon resonance (LSPR) has been known to detect MPs and fragmented debris of PS [95, 116]. By confining the incoming light within the metal NPs, stronger plasmon is generated. The incident light used for the analysis has the wavelength shorter than that of the particle size [111, 117].

HWGMWSY is isolated in various selection processes a s it has affinity to polystyrene [118]. Fluorescent tagged peptides with electrochemical impedance spectroscopy can be used for microplastics in water [119]. To optimize the trial-and-error efforts for exploring such peptides, advances in machine learning are needed [120].

## *Receptor-Based Detection Using Fluorescent Dyes*

Receptor-based detection is also implemented with the help of fluorescent dyes [95]. The fluorescent dyes used for staining of samples allows the plastic to emit light when excited by an energy source [95, 121]. The most widely used dye is the Nile red stain, which is a lipophilic dye for MPs [122]. In a study by Maes et al., [122] he invented a quick screening method by using Nile red stain [95]. When the sample was irradiated under fluorescence microscopy, the red signals were collected and the identification was further done on the basis of the density of the sample [95]. Counter-staining technology has also been reported and validated the use of calcofluor white and Evans blue staining prior to the application of Nile red to obtain green, red, or yellow color for plastic and bluish purple for ground biomass, facilitating easy identification [95, 123]. It also provides morphological information, helping to differentiate plastic from other biomaterials.

Polymer receptors have also been emerging as a prospective technique for the recognition of MPs and NPs. Polymers are made up of repeating units called monomers, and these monomers have a positive or a negative charge on them. These charges, when present in repetition, result in amplification of electrostatic attraction between polymer and MPs/NPs [95]. Poly-L-lysine (PLL) was used for MPs detection by functionalizing PLL with a terahertz (THz) metamaterial sensors capturing region, and the rest was coated with a layer of hexamethyl disilazane. The PLL having the ability to form stable complexes with negatively charged macromolecules, captures the negatively charged plastic and cause a shift in the meta-material's resonance frequency, thus allowing the identification of polymer by spectroscopy [95, 124].

Polymer-based chromogenic chemo-sensor was found capable of identifying primary aliphatic amines and thus displays a potential for detection of amine-coated plastic polymer synthesized from hyper-branched polyphenolic structures and are surrounded by large number of Trifluoracetyl azobenzene groups, which acts as reporter groups [95]. The trifluoro acetyl group causes a spectral change of blue shift in the presence of primary amines [95, 125]. This further reacts to form a hemiaminal group, which leads to a decrease in the length of the conjugated pi-bonds [95, 125].

## Challenges in Detection

MPs and NPs, being very small particles and present in very low concentrations, make their identification more challenging. The uneven distribution of particle debris in the sample reduces the reliability of the detection results. The increase in concentration also increases the risk of sample contamination. These particles vary in size and shape within their category, making identification more difficult. Their collection is also quite challenging, and upon interacting with environmental entities, they tend to accumulate and attract pollutant particles onto their surfaces. The wide range of physical features and count for the same plastic sample, combined with limited information about them, can restrict the usefulness of collected data for risk assessment" [23, 101]. Since every polymer is produced using various chemical additives—some semi-volatile, volatile, or toxic to [23] human health—this can also impact pretreatment success rates and complicate identification. Variations in aging states of polymer particles, surface charge, and hydrophobicity further influence the identification process [101].

## Conclusion

The existence of MPs and NPs in the environment can be of great ecological threat to animals and humans [2]. The development of detection methods has become important as the accumulation of particles is increasing with time. With increasing industrialization, the plastic materials are being modified and hence new contaminants are emerging, posing the challenging tasks for chemists to analyse them. The analysis technics are always evolving to accommodate for new types of pollutants and emerging contaminants. This chapter explores the detection methods that are developed and under development, along with quoting the studies done to understand the identification process. It has been observed that standardization of methods or developing a method that detects all types of MPs and NPs are not possible because of their versatile nature. Detection of MPs and NPs still needs a lot of research and study.

## References

1. Fu, W., Min, J., Jiang, W., Li, Y., & Zhang, W. (2020). Separation, characterization and identification of microplastics and nanoplastics in the environment. *Science of the Total Environment, 721*, 137561. https://doi.org/10.1016/j.scitotenv.2020.137561
2. Mariano, S., Tacconi, S., Fidaleo, M., Rossi, M., & Dini, L. (2021). Micro and nanoplastics identification: Classic methods and innovative detection techniques. *Frontiers in Toxicology, 3*(1). https://doi.org/10.3389/ftox.2021.636640

3. Adhikari, S., Kelkar, V., Kumar, R., & Halden, R. U. (2022). Methods and challenges in the detection of microplastics and nanoplastics: A mini-review. *Polymer International*. https://doi.org/10.1002/pi.6348
4. Choudhury, A., Faizan Zarreen Simnani, Singh, D., Patel, P., Sinha, A., Nandi, A., Ghosh, A., Saha, U., Kumari, K., Saravana Kumar Jaganathan, Nagendra Kumar Kaushik, Pritam Kumar Panda, Mrutyunjay Suar, & Verma, S. K. (2023). Atmospheric microplastic and nanoplastic: The toxicological paradigm on the cellular system. *Ecotoxicology and Environmental Safety, 259*, 115018. https://doi.org/10.1016/j.ecoenv.2023.115018
5. Parkar, S., Mulukh, R., Narhari, G., & Kulkarni, S. (2021, May). An insight into treatment, reuse, recycle and disposal of biodegradable and non-biodegradable solid waste. In *Proceedings of the 4th International Conference on Advances in Science & Technology (ICAST2021)*. https://doi.org/10.2139/ssrn.3867475
6. Andrady, A. L., & Neal, M. A. (2009). Applications and societal benefits of plastics. *Philosophical Transactions of the Royal Society B: Biological Sciences, 364*(1526), 1977–1984. https://doi.org/10.1098/rstb.2008.0304
7. Yong, C. Q. Y., Valiyaveetill, S., & Tang, B. L. (2020). Toxicity of microplastics and nanoplastics in mammalian systems. *International Journal of Environmental Research and Public Health, 17*(5), 1509. https://doi.org/10.3390/ijerph17051509
8. Hashemihedeshi, M. (2023). Exposure to (micro/nano)-plastics and their combustion products studied by cyclic ion mobility-mass spectrometry—Memorial University Research Repository. Library.mun.ca. https://research.library.mun.ca/16346/1/converted.pdf
9. Yang, T., Luo, J., & Nowack, B. (2021). Characterization of nanoplastics, fibrils, and microplastics released during washing and abrasion of polyester textiles. *Environmental Science & Technology, 55*(23), 15873–15881. https://doi.org/10.1021/acs.est.1c04826
10. Kulkarni, S. J., & Goswami, A. K. (2015). A review on wastewater treatment for petroleum industries and refineries. *International Journal of Innovative Research in Science, Engineering and Technology, 1*, 280–283.
11. Jakubowicz, I., Enebro, J., & Yarahmadi, N. (2021). Challenges in the search for nanoplastics in the environment—A critical review from the polymer science perspective. *Polymer Testing, 93*, 106953.
12. Kulkarni, S. J. (2020). Studies, efforts and investigations on various aspects of solid waste management with emphasis on developing countries. In *Sustainability concept in developing countries* (pp. 1–13) https://www.intechopen.com/chapters/71566
13. Ali, S. S., Alsharbaty, M. H. M., Al-Tohamy, R., Schagerl, M., Al-Zahrani, M., Kornaros, M., & Sun, J. (2024). Microplastics as persistent and vectors of other threats in the marine environment: Toxicological impacts, management and strategical roadmap to end plastic pollution. *Environmental Chemistry and Ecotoxicology, 7*, 229–251. https://doi.org/10.1016/j.enceco.2024.12.005
14. Geyer, R., Jambeck, J. R., & Law, K. L. (2017). Production, use, and fate of all plastics ever made. *Science Advances, 3*(7). https://doi.org/10.1126/sciadv.1700782
15. European Food Safety Authority. (2016). Presence of microplastics and nanoplastics in food, with particular focus on seafood. *EFSA Journal, 14*(6). https://doi.org/10.2903/j.efsa.2016.4501
16. Campanale, C., Dierkes, G., Massarelli, C., Bagnuolo, G., & Uricchio, V. F. (2020). A relevant screening of organic contaminants present on freshwater and pre-production microplastics. *Toxics, 8*(4), 100. https://doi.org/10.3390/toxics8040100
17. Veerasingam, S., Ranjani, M., Venkatachalapathy, R., Bagaev, A., Mukhanov, V., Litvinyuk, D., Mugilarasan, M., Gurumoorthi, K., Guganathan, L., Aboobacker, V. M., & Vethamony, P. (2020). Contributions of Fourier transform infrared spectroscopy in microplastic pollution research: A review. *Critical Reviews in Environmental Science and Technology, 51*(22), 1–63. https://doi.org/10.1080/10643389.2020.1807450

18. Ma, C., Ramachandraiah, K., & Jiang, G. (2024). Micro and nano plastics: Contaminants in beverages and prevention strategies. *Frontiers in Sustainable Food Systems, 8*. https://doi.org/10.3389/fsufs.2024.1491290
19. Tong, H., Jiang, Q., Zhong, X., & Hu, X. (2020). Rhodamine B dye staining for visualizing microplastics in laboratory-based studies. *Environmental Science and Pollution Research, 28*(4), 4209–4215. https://doi.org/10.1007/s11356-020-10801-4
20. Kalaronis, D., Ainali, N. M., Evgenidou, E., Kyzas, G. Z., Yang, X., Bikiaris, D. N., & Lambropoulou, D. A. (2022). Microscopic techniques as means for the determination of microplastics and nanoplastics in the aquatic environment: A concise review. *Green Analytical Chemistry, 3*, 100036. https://doi.org/10.1016/j.greeac.2022.100036
21. Shruti, V. C., Pérez-Guevara, F., Roy, P. D., & Kutralam-Muniasamy, G. (2022). Analyzing microplastics with Nile Red: Emerging trends, challenges, and prospects. *Journal of Hazardous Materials, 423*, 127171. https://doi.org/10.1016/j.jhazmat.2021.127171
22. Mahidin, Gani, A., Saiful, Irham, M., Windari, W., & Erdiwansyah. (2022). An overview of the potential risks, sources, and analytical methods for microplastics in soil. *AIMS Environmental Science, 9*(2), 169–200. https://doi.org/10.3934/environsci.2022013
23. Bhat, M. A., & Gaga, E. O. (2024). Unveiling the invisible: First discovery of micro- and nanoplastic size segregation in indoor commercial markets using a cascade impactor. *Air Quality, Atmosphere & Health*. https://doi.org/10.1007/s11869-024-01624-0
24. Browne, M. A., Galloway, T. S., & Thompson, R. C. (2010). Spatial patterns of plastic debris along estuarine shorelines. *Environmental Science & Technology, 44*(9), 3404–3409. https://doi.org/10.1021/es903784e
25. Dris, R., Gasperi, J., Mirande, C., Mandin, C., Guerrouache, M., Langlois, V., & Tassin, B. (2017). A first overview of textile fibers, including microplastics, in indoor and outdoor environments. *Environmental Pollution, 221*, 453–458. https://doi.org/10.1016/j.envpol.2016.12.013
26. Hickey, S. M., Ung, B., Bader, C., Brooks, R., Lazniewska, J., Johnson, I. R. D., Sorvina, A., Logan, J., Martini, C., Moore, C. R., Karageorgos, L., Sweetman, M. J., & Brooks, D. A. (2021). Fluorescence microscopy—An outline of hardware, biological handling, and fluorophore considerations. *Cells, 11*(1), 35. https://doi.org/10.3390/cells11010035
27. Rossi, M., Minicozzi, G., Pascarella, G., & Capasso, A. (2020). ESG, competitive advantage and financial performances: Preliminary research. *Handle.net, 1*(1), 969–986.
28. Sturm, M. T., Myers, E., Schober, D., Korzin, A., & Schuhen, K. (2023). Development of an inexpensive and comparable microplastic detection method using fluorescent staining with novel Nile Red derivatives. *Analytica, 4*(1), 27–44. https://doi.org/10.3390/analytica4010004
29. Igburi, J., Ighalo, J., & Pande, C. (2024). Microplastics in African and Asian environments. In *Emerging contaminants and associated treatment technologies*. Springer. https://doi.org/10.1007/978-3-031-64253-1
30. Priyanka, & Chaudhari, O. P. (2018). Uses of transmission electron microscope in microscopy and its advantages and disadvantages. *International Journal of Current Microbiology and Applied Sciences, 7*, 743–747. https://doi.org/10.20546/ijcmas.2018.705.090
31. Cassano, D., Bogni, A., La Spina, R., Gilliland, D., & Ponti, J. (2023). Investigating the cellular uptake of model nanoplastics by single-cell ICP-MS. *Nanomaterials (Basel), 13*(3), 594. https://doi.org/10.3390/nano13030594
32. Man Thaiba, B., Sedai, T., Bastakoti, S., Karki, A., Anuradha K.C., Khadka, G., Acharya, S., Kandel, B., Giri, B., & Bhakta Neupane, B. (2023). A review on analytical performance of micro- and nanoplastics analysis methods. *Arabian Journal of Chemistry, 16*(5), 104686. https://doi.org/10.1016/j.arabjc.2023.104686
33. Sun, X., Chen, B., Li, Q., Liu, N., Xia, B., Zhu, L., & Qu, K. (2018). Toxicities of polystyrene nano- and microplastics toward marine bacterium Halomonas alkaliphila. *Science of The Total Environment, 642*, 1378–1385. https://doi.org/10.1016/j.scitotenv.2018.06.141

34. Pradel, A., Ferreres, S., Veclin, C., El Hadri, H., Gautier, M., Grassl, B., & Gigault, J. (2021). Stabilization of fragmental polystyrene nanoplastic by natural organic matter: Insight into mechanisms. *ACS ES&T Water*. https://doi.org/10.1021/acsestwater.0c00283
35. Caputo, F., Vogel, R., Savage, J., Vella, G., Law, A., Della Camera, G., Hannon, G., Peacock, B., Mehn, D., Ponti, J., Geiss, O., Aubert, D., Prina-Mello, A., & Calzolai, L. (2021). Measuring particle size distribution and mass concentration of nanoplastics and microplastics: Addressing some analytical challenges in the sub-micron size range. *Journal of Colloid and Interface Science, 588*, 401–417. https://doi.org/10.1016/j.jcis.2020.12.039
36. Lionetto, F., Corcione, C. E., Rizzo, A., & Maffezzoli, A. (2021). Production and characterization of polyethylene terephthalate nanoparticles. *Polymers, 13*(21), 3745. https://doi.org/10.3390/polym13213745
37. Akhatova, F., Ishmukhametov, I., Fakhrullina, G., & Fakhrullin, R. (2022). Nanomechanical atomic force microscopy to probe cellular microplastics uptake and distribution. *International Journal of Molecular Sciences, 23*(2), 806. https://doi.org/10.3390/ijms23020806
38. Girão, A. V., Caputo, G., & Ferro, M. C. (2017, January 1). Chapter 6—Application of scanning electron microscopy–energy dispersive X-ray spectroscopy (SEM-EDS) (T. A. P. Rocha-Santos & A. C. Duarte, Eds.). ScienceDirect; Elsevier. https://www.sciencedirect.com/science/article/pii/S0166526X16301519
39. Hossain, M., & Steinmann, P. (2013). Degree of cure-dependent modelling for polymer curing processes at small-strain. Part I: Consistent reformulation. *Computational Mechanics, 53*(4), 777–787. https://doi.org/10.1007/s00466-013-0929-5
40. Joy, D. C. (1984). Beam interactions, contrast and resolution in the SEM. *Journal of Microscopy, 136*(2), 241–258. https://doi.org/10.1111/j.1365-2818.1984.tb00532.x
41. Kwon, Y. E., Kim, J. K., Kim, Y. J., & Kim, J. G. (2019). Development of SEM and STEM-in-SEM grid holders for EDS analysis and their applications to apatite phases. *Journal of Analytical Science and Technology, 10*(27). https://doi.org/10.1186/s40543-019-0186-0
42. Chernoff, D., & Mujallid, M. S. (1995). Environmental SEMs: A new way to look at samples. *Microscopy Today, 3*(8), 14–15. https://doi.org/10.1017/s155192950006291x
43. Mitrano, D. M., Beltzung, A., Frehland, S., Schmiedgruber, M., Cingolani, A., & Schmidt, F. (2019b). Synthesis of metal-doped nanoplastics and their utility to investigate fate and behaviour in complex environmental systems. *Nature Nanotechnology, 14*(4), 362–368. https://doi.org/10.1038/s41565-018-0360-3
44. Ranjani, M., Veerasingam, S., Venkatachalapathy, R., Mugilarasan, M., Bagaev, A., Mukhanov, V., & Vethamony, P. (2021). Assessment of potential ecological risk of microplastics in the coastal sediments of India: A meta-analysis. *Marine Pollution Bulletin, 163*, 111969. https://doi.org/10.1016/j.marpolbul.2021.111969
45. Takahisa Ishimura, Iwai, I., Matsui, K., Mattonai, M., Watanabe, A., Robberson, W., Cook, A.-M., Allen, H., Pipkin, W., Norio Teramae, Hajime Ohtani, & Watanabe, C. (2021). Qualitative and quantitative analysis of mixtures of microplastics in the presence of calcium carbonate by pyrolysis-GC/MS. *Journal of Analytical and Applied Pyrolysis, 157*, 105188. https://doi.org/10.1016/j.jaap.2021.105188
46. Khatoon, N., Mallah, M. A., Yu, Z., Qu, Z., Ali, M., & Liu, N. (2024). Recognition and detection technology for microplastic, its source and health effects. *Environmental Science and Pollution Research International, 31*(8), 11428–11452. https://doi.org/10.1007/s11356-023-31655-6
47. Kaur, J., Kaur, K., Kaur, K., Matharu, A. S., & Mehta, S. K. (2023). Greener aspects of nanoparticle synthesis for water remediation: Challenges and future perspective. *Advances in Environmental and Engineering Research, 4*(2), 1–56. https://doi.org/10.21926/aeer.2302027
48. Allison, D. P., Mortensen, N. P., Sullivan, C. J., & Doktycz, M. J. (2010). Atomic force microscopy of biological samples. *WIREs Nanomedicine and Nanobiotechnology, 2*(6), 618–634. https://doi.org/10.1002/wnan.104
49. Reinstädtler, M., Rabe, U., Scherer, V., Hartmann, U., Goldade, A., Bhushan, B., & Arnold, W. (2003). On the nanoscale measurement of friction using atomic-force microscope

cantilever torsional resonances. *Applied Physics Letters, 82*(16), 2604–2606. https://doi.org/10.1063/1.1565179

50. Haggerty, L., & Lenhoff, A. M. (1993). STM and AFM in biotechnology. *Biotechnology Progress, 9*(1), 1–11. https://doi.org/10.1021/bp00019a001
51. Alemansour, H.(2022). *Development of probe-based devices and methods for applications in micro/nano characterization and fabrication*. ProQuest. Proquest.com. https://search.proquest.com/openview/94834b4e684faf4a176193e1807bf078/1?pq-origsite=gscholar&cbl=18750&diss=y
52. Aoki, J., Gao, W., Kiyono, S., & Ono, T. (2005). A high precision AFM for nanometrology of large area micro-structured surfaces. *Key Engineering Materials, 295–296*, 65–70. https://doi.org/10.4028/www.scientific.net/kem.295-296.65
53. Wan, Q. (2017). *Scanning electron microscopy investigation of bio-polymer composites morphology*. https://etheses.whiterose.ac.uk/id/eprint/19125/1/Thesis.pdf
54. Kalousek, R., Lopour, F., Dub, P., & ikola, T. (2000). Vibrational analysis of the cantilever in non-contact scanning force microscopy. *Surface and Interface Analysis, 30*(1), 292–296. https://doi.org/10.1002/1096-9918(200008)30:1<292::aid-sia795>3.0.co;2-9
55. Galluzzi, M., Lancia, M., Zheng, C., Re, V., Castelvetro, V., Guo, S., & Viaroli, S. (2025). Atomic force microscopy (AFM) nanomechanical characterization of micro- and nanoplastics to support environmental investigations in groundwater. *Emerging Contaminants, 11*(2), 100478. https://doi.org/10.1016/j.emcon.2025.100478
56. Butt, H.-J., Cappella, B., & Kappl, M. (2005). Force measurements with the atomic force microscope: Technique, interpretation and applications. *Surface Science Reports, 59*(1–6), 1–152. https://doi.org/10.1016/j.surfrep.2005.08.003
57. Shan, X., Luo, L., Yu, Z., & You, J. (2022). Recent advances in versatile inverse lyotropic liquid crystals. *Journal of Controlled Release, 348*, 1–21. https://doi.org/10.1016/j.jconrel.2022.05.036
58. Mohamed, M. A., Jaafar, J., Ismail, A. F., Othman, M. H. D., & Rahman, M. A. (2017, January 1). Chapter 1—Fourier transform infrared (FTIR) spectroscopy (N. Hilal, A. F. Ismail, T. Matsuura, & D. Oatley-Radcliffe, Eds.). ScienceDirect; Elsevier. https://www.sciencedirect.com/science/article/pii/B9780444637765000012
59. Ţucureanu, V., Matei, A., & Avram, A. M. (2016). FTIR spectroscopy for carbon family study. *Critical Reviews in Analytical Chemistry, 46*(6), 502–520. https://doi.org/10.1080/10408347.2016.1157013
60. Guo, C., Luo, X., Zhou, X., Shi, B., Wang, J., Zhao, J., & Zhang, X. (2017). Quantitative analysis of binary polymorphs mixtures of fusidic acid by diffuse reflectance FTIR spectroscopy, diffuse reflectance FT-NIR spectroscopy, Raman spectroscopy and multivariate calibration. *Journal of Pharmaceutical and Biomedical Analysis, 140*, 130–136. https://doi.org/10.1016/j.jpba.2017.02.053
61. Song, Y. K., Hong, S. H., Jang, M., Kang, J.-H., Kwon, O. Y., Han, G. M., & Shim, W. J. (2014). Large accumulation of micro-sized synthetic polymer particles in the sea surface microlayer. *Environmental Science & Technology, 48*(16), 9014–9021. https://doi.org/10.1021/es501757s
62. Turner, A., & Holmes, L. (2011). Occurrence, distribution and characteristics of beached plastic production pellets on the island of Malta (Central Mediterranean). *Marine Pollution Bulletin, 62*(2), 377–381. https://doi.org/10.1016/j.marpolbul.2010.09.027
63. Dazzi, A., Saunier, J., Kjoller, K., & Yagoubi, N. (2015). Resonance enhanced AFM-IR: A new powerful way to characterize blooming on polymers used in medical devices. *International Journal of Pharmaceutics, 484*(1–2), 109–114. https://doi.org/10.1016/j.ijpharm.2015.02.046
64. Bunaciu, A., Fleschin, S., & Aboul-Enein, H. (2013). Infrared microspectroscopy applications—Review. *Current Analytical Chemistry, 10*(1), 132–139. https://doi.org/10.2174/1573411011410010011

65. Harville, D. G., & Freeman, D. L. (1988). The benefits and application of rapid mineral analysis provided by Fourier transform infrared spectroscopy. *SPE Annual Technical Conference and Exhibition*. https://doi.org/10.2118/18120-ms
66. Campanale, C., Savino, I., Massarelli, C., & Uricchio, V. F. (2023). Fourier transform infrared spectroscopy to assess the degree of alteration of artificially aged and environmentally weathered microplastics. *Polymers, 15*(4), 911. https://doi.org/10.3390/polym15040911
67. Morgado, V., Gomes, L., Bettencourt da Silva, R. J. N., & Palma, C. (2021). Validated spreadsheet for the identification of PE, PET, PP and PS microplastics by micro-ATR-FTIR spectra with known uncertainty. *Talanta, 234*, 122624. https://doi.org/10.1016/j.talanta.2021.122624
68. Lapworth, D. J., & Shockley, D. J. (2022). *Microplastics in UK groundwater and stygobites: Protocols for sampling, analysis and pilot study results*. NERC Open Research Archive. Nerc.ac.uk. https://nora.nerc.ac.uk/id/eprint/532670/1/OR22015.pdf
69. Pagter, E., Frias, J., Kavanagh, F., & Nash, R. (2020). Varying levels of microplastics in benthic sediments within a shallow coastal embayment. *Estuarine, Coastal and Shelf Science, 243*, 106915. https://doi.org/10.1016/j.ecss.2020.106915
70. Kauffmann, T. H., Kokanyan, N., & Fontana, M. D. (2018). Use of Stokes and anti-Stokes Raman scattering for new applications. *Journal of Raman Spectroscopy, 50*(3), 418–424. https://doi.org/10.1002/jrs.5523
71. Ivleva, N. P. (2021). Chemical analysis of microplastics and nanoplastics: Challenges, advanced methods, and perspectives. *Chemical Reviews, 121*(19), 11886–11936. https://doi.org/10.1021/acs.chemrev.1c00178
72. Laurens, & Meirer, F. (2022). Spectro-microscopic techniques for studying nanoplastics in the environment and in organisms. *Angewandte Chemie, 62*(2). https://doi.org/10.1002/anie.202210494
73. Schymanski, D., Oßmann, B. E., Benismail, N., Boukerma, K., Dallmann, G., von der Esch, E., Fischer, D., Fischer, F., Gilliland, D., Glas, K., Hofmann, T., Käppler, A., Lacorte, S., Marco, J., Rakwe, M. E., Weisser, J., Witzig, C., Zumbülte, N., & Ivleva, N. P. (2021). Analysis of microplastics in drinking water and other clean water samples with micro-Raman and micro-infrared spectroscopy: Minimum requirements and best practice guidelines. *Analytical and Bioanalytical Chemistry, 413*(24), 5969–5994. https://doi.org/10.1007/s00216-021-03498-y
74. Finnie, P., Ouyang, J., & Lefebvre, J. (2020). Full spectrum Raman excitation mapping spectroscopy. *Scientific Reports, 10*(1). https://doi.org/10.1038/s41598-020-65757-9
75. Schmidt, R., Nachtnebel, M., Dienstleder, M., Mertschnigg, S., Schroettner, H., Zankel, A., Poteser, M., Hutter, H.-P., Eppel, W., & Fitzek, H. (2021). Correlative SEM-Raman microscopy to reveal nanoplastics in complex environments. *Micron (1993), 144*, 103034. https://doi.org/10.1016/j.micron.2021.103034
76. Puhakka, E., & Santala, V. (2022). Method for acrylic acid monomer detection with recombinant biosensor cells for enhanced plastic degradation monitoring from water environments. *Marine Pollution Bulletin, 178*, 113568. https://doi.org/10.1016/j.marpolbul.2022.113568
77. Bouhroum, A. (2011). *Characterisation of the physico-chemical and solid-state behaviour of drugs in HFA-134 and 227*. Nottingham ePrints. Nottingham.ac.uk. https://eprints.nottingham.ac.uk/12004/1/A_Bouhroum_PhD_thesis%203.pdf
78. Schröder, K., Kossel, E., & Lenz, M. (2021). Microplastic abundance in beach sediments of the Kiel Fjord, Western Baltic Sea. *Environmental Science and Pollution Research*. https://doi.org/10.1007/s11356-020-12220-x
79. Anger, P. M., von der Esch, E., Baumann, T., Elsner, M., Niessner, R., & Ivleva, N. P. (2018). Raman microspectroscopy as a tool for microplastic particle analysis. *TrAC Trends in Analytical Chemistry, 109*, 214–226. https://doi.org/10.1016/j.trac.2018.10.010
80. Hou, Y. (2015). *Nano handling and measurement of biological cells in culture*. Openrepository.com. http://hdl.handle.net/10547/618562

81. Jiang, N. (2021). *Tip-enhanced Raman spectroscopy: Chemical analysis with nanoscale to Angstrom scale resolution*. APS March Meeting Abstracts, 2021, F26.003. https://ui.adsabs.harvard.edu/abs/2021APS..MARF26003J/abstract
82. Buzzini, P., & Massonnet, G. (2015). The analysis of colored acrylic, cotton, and wool textile fibers using micro-Raman spectroscopy. Part 2: Comparison with the traditional methods of fiber examination. *Journal of Forensic Sciences, 60*(3), 712–720. https://doi.org/10.1111/1556-4029.12654
83. Rani, M. (2022). *Analysis and characterization of microplastics through vibrational spectroscopic techniques for environmental monitoring*. Unibs.it. http://hdl.handle.net/11379/563081
84. Nava, V., Frezzotti, M. L., & Leoni, B. (2021). Raman spectroscopy for the analysis of microplastics in aquatic systems. *Applied Spectroscopy, 75*(11), 1341–1357. https://doi.org/10.1177/00037028211043119
85. Yakovenko, N., Carvalho, A., & ter Halle, A. (2020). Emerging use thermo-analytical method coupled with mass spectrometry for the quantification of micro(nano)plastics in environmental samples. *TrAC Trends in Analytical Chemistry, 131*, 115979. https://doi.org/10.1016/j.trac.2020.115979
86. Peñalver, R., Arroyo-Manzanares, N., López-García, I., & Hernández-Córdoba, M. (2020). An overview of microplastics characterization by thermal analysis. *Chemosphere, 242*, 125170. https://doi.org/10.1016/j.chemosphere.2019.125170
87. Yu, J., Wang, P., Ni, F., Cizdziel, J., Wu, D., Zhao, Q., & Zhou, Y. (2019). Characterization of microplastics in environment by thermal gravimetric analysis coupled with Fourier transform infrared spectroscopy. *Marine Pollution Bulletin, 145*, 153–160. https://doi.org/10.1016/j.marpolbul.2019.05.037
88. Sommer, C. (2024). *Laser-induced breakdown spectroscopy for the characterization and investigation of (micro)plastics and their weathering-induced oxidation behavior*. Uni-Marburg.de. https://archiv.ub.uni-marburg.de/diss/z2024/0110
89. Lion, A., & Yagimli, B. (2008). Differential scanning calorimetry—Continuum mechanical considerations with focus to the polymerisation of adhesives. *Zeitschrift für Angewandte Mathematik und Mechanik, 88*(5), 388–402. https://doi.org/10.1002/zamm.200800006
90. Golebiewski, J., & Galeski, A. (2007). Thermal stability of nanoclay polypropylene composites by simultaneous DSC and TGA. *Composites Science and Technology, 67*(15–16), 3442–3447. https://doi.org/10.1016/j.compscitech.2007.03.007
91. Reichel, J., Graßmann, J., Letzel, T., & Drewes, J. E. (2020). Systematic development of a simultaneous determination of plastic particle identity and adsorbed organic compounds by thermodesorption–pyrolysis GC/MS (TD-Pyr-GC/MS). *Molecules, 25*(21), 4985. https://doi.org/10.3390/molecules25214985
92. Majewsky, M., Bitter, H., Eiche, E., & Horn, H. (2016). Determination of microplastic polyethylene (PE) and polypropylene (PP) in environmental samples using thermal analysis (TGA-DSC). *Science of the Total Environment, 568*, 507–511. https://doi.org/10.1016/j.scitotenv.2016.06.017
93. Nene, A., Sadeghzade, S., Viaroli, S., Yang, W., Uchenna, U. P., Kandwal, A., Liu, X., Somani, P., & Galluzzi, M. (2025). Recent advances and future technologies in nano-microplastics detection. *Environmental Sciences Europe, 37*(1). https://doi.org/10.1186/s12302-024-01044-y
94. Wang, F. C.-Y. (1999). Polymer analysis by pyrolysis gas chromatography. *Journal of Chromatography A, 843*(1–2), 413–423. https://doi.org/10.1016/s0021-9673(98)01051-6
95. Tang, Y., Clark, T. J., & Yoon, J. (2023). Receptor-based detection of microplastics and nanoplastics: Current and future. *Biosensors and Bioelectronics, 234*, 115361. https://doi.org/10.1016/j.bios.2023.115361
96. Boodoo, C., Dester, E., David, J., Patel, V., Kc, R., & Alocilja, E. C. (2023). Multi-probe nano-genomic biosensor to detect S. aureus from magnetically-extracted food samples. *Biosensors, 13*(6), 608–608. https://doi.org/10.3390/bios13060608

97. Furst, A. L., & Francis, M. B. (2018). Impedance-based detection of bacteria. *Chemical Reviews, 119*(1), 700–726. https://doi.org/10.1021/acs.chemrev.8b00381
98. Naresh, V., & Lee, N. (2021). A review on biosensors and recent development of nanostructured materials-enabled biosensors. *Sensors, 21*(4), 1109. https://doi.org/10.3390/s21041109
99. Raghavan, S., Chee, S., Li, J., Poschmann, J., Nagarajan, N., Wei, S. J., Verma, C., & Ghadessy, F. J. (2019). Development and application of a transcriptional sensor for detection of heterologous acrylic acid production in E. coli. *Microbial Cell Factories, 18*(1). https://doi.org/10.1186/s12934-019-1185-y
100. Disciplinare, S., Dainelli, D., Prof, M., Coppi, A., Co-Supervisori, Gonnelli, C., Colzi, I., Coordinatore, C., & Duccio. (n.d.). *Dottorato di ricerca in biologia evoluzionistica ed ecologia. Unraveling the impacts of micro-and nano-plastics in the plant world.* Retrieved August 2, 2025, from https://tesidottorato.depositolegale.it/bitstream/20.500.14242/210765/1/Tesi%20DAINELLI_Marco_%20DT31439.pdf
101. Shukla, N. B., Daraboina, N., & Madras, G. (2009). Oxidative and photooxidative degradation of poly (acrylic acid). *Polymer Degradation and Stability, 94*(8), 1238–1244. https://doi.org/10.1016/j.polymdegradstab.2009.04.020
102. Scales, B. S., Cable, R. N., Duhaime, M. B., Gerdts, G., Fischer, F., Fischer, D., Mothes, S., Hintzki, L., Moldaenke, L., Ruwe, M., Kalinowski, J., Kreikemeyer, B., Pedrotti, M.-L., Gorsky, G., Elineau, A., Labrenz, M., & Oberbeckmann, S. (2021). Cross-hemisphere study reveals geographically ubiquitous, plastic-specific bacteria emerging from the rare and unexplored biosphere. *MSphere, 6*(3). https://doi.org/10.1128/msphere.00851-20
103. Denaro, R., Aulenta, F., Crisafi, F., Di Pippo, F., Cruz Viggi, C., Matturro, B., Tomei, P., Smedile, F., Martinelli, A., Di Lisio, V., Venezia, C., & Rossetti, S. (2020). Marine hydrocarbon-degrading bacteria breakdown poly(ethylene terephthalate) (PET). *Science of the Total Environment, 749*, 141608. https://doi.org/10.1016/j.scitotenv.2020.141608
104. Xie, M., Gao, M., Yun, Y., Malmsten, M., Rotello, V. M., Zbořil, R., Akhavan, O., Kraskouski, A., Amalraj, J., Cai, X., Lu, J., Zheng, H., & Li, R. (2023). Antibacterial nanomaterials: Mechanisms, impacts on antimicrobial resistance and design principles. *Angewandte Chemie International Edition, 62*(17). https://doi.org/10.1002/anie.202217345
105. Teuten, E. L., Saquing, J. M., Knappe, D. R. U., Barlaz, M. A., Jonsson, S., Björn, A., Rowland, S. J., Thompson, R. C., Galloway, T. S., Yamashita, R., Ochi, D., Watanuki, Y., Moore, C., Viet, P. H., Tana, T. S., Prudente, M., Boonyatumanond, R., Zakaria, M. P., Akkhavong, K., & Ogata, Y. (2009). Transport and release of chemicals from plastics to the environment and to wildlife. *Philosophical Transactions of the Royal Society B: Biological Sciences, 364*(1526), 2027–2045. https://doi.org/10.1098/rstb.2008.0284
106. Saunders, D. (2023, December 6). *Concentration and characterization of microplastics in Blue mussels (Mytilus edulis) and Eastern oysters (Crassostrea virginica) from Nova Scotia.* Library.dal.ca. https://dalspace.library.dal.ca/items/2b581901-665b-4588-92ea-5eb1a803308e
107. Shimomura, M., Nomura, Y., Zhang, W., Sakino, M., Lee, K.-H., Ikebukuro, K., & Karube, I. (2001). Simple and rapid detection method using surface plasmon resonance for dioxins, polychlorinated biphenylx and atrazine. *Analytica Chimica Acta, 434*(2), 223–230. https://doi.org/10.1016/S0003-2670(01)00809-1
108. Scharnweber, T., Fisher, M., Suchànek, M., Knopp, D., & Niessner, R. (2001). Monoclonal antibody to polycyclic aromatic hydrocarbons based on a new benzo[a]pyrene immunogen. *Fresenius' Journal of Analytical Chemistry, 371*(5), 578–585. https://doi.org/10.1007/s002160101012
109. Ragavan, K. V., Rastogi, N. K., & Thakur, M. S. (2013). Sensors and biosensors for analysis of bisphenol-A. *TrAC Trends in Analytical Chemistry, 52*, 248–260. https://doi.org/10.1016/j.trac.2013.09.006
110. Kharitonov, A. B., Zayats, M., Lichtenstein, A., Katz, E., & Willner, I. (2000). Enzyme monolayer-functionalized field-effect transistors for biosensor applications. *Sensors and Actuators B: Chemical, 70*(1–3), 222–231. https://doi.org/10.1016/s0925-4005(00)00573-6

111. Csáki, A., Stranik, O., & Fritzsche, W. (2018). Localized surface plasmon resonance based biosensing. *Expert Review of Molecular Diagnostics, 18*(3), 279–296. https://doi.org/10.1080/14737159.2018.1440208
112. Adey, N. B., Mataragnon, A. H., Rider, J. E., Carter, J. M., & Kay, B. K. (1995). Characterization of phage that bind plastic from phage-displayed random peptide libraries. *Gene, 156*(1), 27–31. https://doi.org/10.1016/0378-1119(95)00058-e
113. Apitius, L., Rübsam, K., Jakesch, C., Jakob, F., & Schwaneberg, U. (2019). Ultrahigh-throughput screening system for directed polymer binding peptide evolution. *Biotechnology and Bioengineering, 116*(8), 1856–1867. https://doi.org/10.1002/bit.26990
114. Rübsam, K., Weber, L., Jakob, F., & Schwaneberg, U. (2017). Directed evolution of polypropylene and polystyrene binding peptides. *Biotechnology and Bioengineering, 115*(2), 321–330. https://doi.org/10.1002/bit.26481
115. Li, N., Kang, J., Jiang, L., He, B., Lin, H., & Huang, J. (2017). PSBinder: A web service for predicting polystyrene surface-binding peptides. *BioMed Research International, 2017*, 1–5. https://doi.org/10.1155/2017/5761517
116. Oh, S., Hur, H., Kim, Y., Shin, S., Woo, H., Choi, J., & Lee, H. H. (2021). Peptide specific nanoplastic detection based on sandwich typed localized surface plasmon resonance. *Nanomaterials, 11*(11), 2887. https://doi.org/10.3390/nano11112887
117. Cao, L. (2023). *Electron-transfer mediated photo-switching in nanoparticles*. https://www.proquest.com/openview/7914e08f6eff4e46eaf2a3b5df7406ba/1?pq-origsite=gscholar&cbl=18750&diss=y
118. Vodnik, M., Štrukelj, B., & Lunder, M. (2012). HWGMWSY, an unanticipated polystyrene binding peptide from random phage display libraries. *Analytical Biochemistry, 424*(2), 83–86. https://doi.org/10.1016/j.ab.2012.02.013
119. Motalebizadeh, A., Fardindoost, S., & Hoorfar, M. (2024). Selective on-site detection and quantification of polystyrene microplastics in water using fluorescence-tagged peptides and electrochemical impedance spectroscopy. *Journal of Hazardous Materials, 480*, 136004. https://doi.org/10.1016/j.jhazmat.2024.136004
120. Seggio, M., Arcadio, F., Cennamo, N., Zeni, L., & Bossi, A. M. (2023). A plasmonic gold nano-surface functionalized with the estrogen receptor for fast and highly sensitive detection of nanoplastics. *Talanta, 267*, 125211. https://doi.org/10.1016/j.talanta.2023.125211
121. Ribeiro, F., Duarte, A. C., & da Costa, J. P. (2024). Staining methodologies for microplastics screening. *TrAC Trends in Analytical Chemistry, 172*, 117555. https://doi.org/10.1016/j.trac.2024.117555
122. Maes, T., Jessop, R., Wellner, N., Haupt, K., & Mayes, A. G. (2017). A rapid-screening approach to detect and quantify microplastics based on fluorescent tagging with Nile Red. *Scientific Reports, 7*(1). https://doi.org/10.1038/srep44501
123. Maxwell S, H., Melinda K, F., & Matthew, G. (2020). Counterstaining to separate Nile Red-stained microplastic particles from terrestrial invertebrate biomass. *Environmental Science & Technology, 54*(9), 5580–5588. https://doi.org/10.1021/acs.est.0c00711
124. Park, S. J., & Ahn, Y. H. (2022). Detection of polystyrene microplastic particles in water using surface-functionalized terahertz microfluidic metamaterials. *Applied Sciences, 12*(14), 7102. https://doi.org/10.3390/app12147102
125. Lin, J.-F., Kukkola, J., Sipola, T., Raut, D., Samikannu, A., Mikkola, J.-P., Mohl, M., Toth, G., Su, W.-F., Laurila, T., & Kordas, K. (2015). Trifluoroacetylazobenzene for optical and electrochemical detection of amines. *Journal of Materials Chemistry A, 3*(8), 4687–4694. https://doi.org/10.1039/c4ta05358c

# Chapter 4
# Microplastics in Water Management

**Abdulhalim Musa Abubakar**

## Introduction

Venghaus & Barjenbruch [1], Fältström & Anderberg [2], Kim et al. [3] and Ou et al. [4] all wrote on microplastic occurrence in urban water management. Microplastics in water come from different sources, which are generally divided into primary and secondary, as well as industrial and domestic. Primary microplastics are tiny plastic particles that are manufactured intentionally in small sizes, including microbeads used in cosmetics, scrubbing agents in cleaning products, and plastic pellets used as raw materials in the plastics industry. Commonly, the particles enter water bodies directly through wastewater discharge or accidental spills. Secondary microplastics are formed when larger plastic items break down into smaller pieces due to sunlight, waves, or physical wear. In line with that, plastic bags, bottles, fishing nets, and packaging materials degrade over time and release fragments into rivers, lakes, and oceans. Thus, secondary sources are basically the most widespread and contribute significantly to microplastic pollution. Industrial sources of microplastics include wastewater from textile factories, plastic manufacturing plants, and tire wear from vehicles that gets washed into drains and rivers. Certain tablets, capsules, and controlled-release drugs use polymeric excipients or coatings (e.g., polyethylene glycol, polymethacrylates, or other synthetic polymers) that could degrade into microplastics in the environment [5]. Domestic sources include synthetic fibers shed from clothes during washing, microbeads from personal care products, and fragments from household plastics. When wastewater treatment plants cannot fully remove these particles, they end up in natural water

A. M. Abubakar (✉)
Department of Chemical Engineering, Faculty of Engineering, Modibbo Adama University, Yola, Adamawa, Nigeria
e-mail: abdulhalim@mau.edu.ng


S. Kulkarni, S. Das (eds.), *Microplastics and Nanoplastics*, Lecture Notes in Nanoscale Science and Technology 49,
https://doi.org/10.1007/978-3-032-17963-0_4

systems. In the UK, the INCA microplastics model was applied to the River Thames catchment to simulate microplastic transport, loads, and concentrations, showcasing serious accumulation and potential ecological impacts [6]. This study highlights the specific role of wastewater treatment plants (WWTPs) as both pathways and partial barriers to microplastic pollution, and pointed out the lack of standardized methodologies for their detection and monitoring. It introduces recent models and innovative approaches, such as hydrogeoplastic modeling for groundwater transport, new dispersion–drag force models, and advanced analytical techniques. Adsorption can remove microplastics from water. Many natural and synthetic adsorbents, such as activated carbon, biochar, clays, and modified polymers, have been tested for this purpose. Similar materials traps microplastic particles on their surfaces and reduce the amount of plastic in the water. The efficiency of adsorption depends on the size and type of microplastics, the surface properties of the adsorbent, and the pH and temperature conditions of the water Although adsorption is not a complete solution, it shows promise as part of a combined treatment method for removing microplastics from water. We further present numerous studies stating the amount of microplastics discovered in the surface waters of known locations around the world. The review is unique and extensive as it combines occurrence, detection, modeling, treatment, and policy measures into a single framework. The study calls for integration of microplastic monitoring into water management frameworks and linking it to sustainability goals.

## Occurrence and Distribution of Microplastics

### *Surface Water*

Microplastics comes from larger plastic items applied in cosmetics and cleaning agents, eventually finding itself in rivers, lakes, oceans and other surface waters. Not just that, they enter surface water through many sources. Agricultural mulch film (AMF) once contribute microplastics to Chinese surface waters [7]. Wastewater from homes and industries, plastic litter, and even synthetic clothing fibers washed during laundry can all contribute [8]. But stormwater is a major and under-researched pathway for transporting microplastics to surface waters, with contributions greater than wastewater effluents [9]. WWTPs are significant point sources of microplastics in rivers [10, 11], contributing high concentrations of fibers, pellets, and fragments downstream, and facilitating the transport of unique microbial communities on their surfaces [12]. Raritan River is a river of the US state of New Jersey investigated for possible microplastic content discharge from 4 WWTP effluent [13]. Not all the source of microplastic to water bodies are known. For example, the sources and transport routes of microplastics reaching the Arctic are still not well understood [14]. Over time, larger plastic waste breaks down into smaller particles due to sunlight and water movement. Once in the water, microplastics are difficult to remove.

Although microplastics are traditionally considered non-degradable, various physical, chemical, and biological degradation pathways—especially involving diverse microbes—are increasingly observed in surface waters [15]. Sunlight-driven photodegradation of polypropylene (PP) microplastics in calm surface waters primarily causes carbon loss and shrinkage through dissolution, rather than fragmentation [16]. The same author suggests that plastic lifespan should be estimated using linear, not exponential, degradation models. Saipolbahri et al. [17], dos Santos et al. [18] and Suteja et al. [19] revealed that polyethylene (PE) and PP were the predominant polymers found in surface waters. Specifically, the Fuyang River and Zhang River in North China are majorly polluted with PE and PP [20]. Yin et al. [21] said over 89.5% of the PP measured were smaller than 2 mm in size for urban lakes in Changsha, China. Aquatic animals often mistake them for food, which later harm their health and move up the food chain to humans [22, 23]. For instance, 36% of shrimp found in Victoria freshwater in Australia contains microplastics [24]; whereas all the 7 ponds of Nakhon Pathom Province of Central Thailand contains $66.17 \pm 29.19$ items per individual shrimp [25]. Higher concentrations of microplastics in fish is found downstream of the sewage treatment plant (STP) in the Tanchon Stream, in South Korea [26]. Nylon were detected in fish from the Escravos Estuary, where Onyena et al. [27] found that crabs is one of the contaminated organisms (at 29 particles/kg/DW). Microplastics also carry harmful chemicals that may pollute water and affect both animals and people. A pioneering study by Mehboob et al. [28] documents microplastic contamination in Red-wattled Lapwings from the Ravi River in Pakistan, where they revealed the high prevalence and tissue (kidney & liver) damage linked to ingested microplastics by the bird. Microplastic pollution in China's inland waters is widespread and influenced by urbanization [29], as observed in Table 4.1, with fibers and fragments being the most common types [30].

Scientists are studying the full effects, but it is clear that microplastics are a growing environmental problem. Reduction in plastic use and the improvement waste management, has the potential to limit microplastic pollution in surface water. Several methods for the estimation the content of microplastics in surface water has been reported. In Zhao et al. [183], high-speed cameras were employed to visually identify and characterize microplastics in surface water and sediment samples from urban rivers in Shanghai. Dacewicz et al. [184] demonstrates that digital image analysis, combined with microscopic and multivariate methods, will effectively characterize the 2D and 3D morphology of microplastics in surface waters. A stereomicroscope is basically used to count and visually identify microplastics trapped on filters after extraction from water and sediment samples of the Zayandeh-rud River [117]. Again, a stereomicroscope can be used to record the size, color, and type of microplastics in both surface water and fish samples [64, 185]. Composition of microplastics was identified using infrared spectrometer analysis by Han et al. [45] and Diez-Pérez et al. [134], while scanning electron microscope (SEM) and microscope photo observation were used to determine their type and origin [75, 186]. Laser direct infrared imaging (LDIR) imaging was used previously by Hansen et al. [173] as a fast and efficient method to analyze microplastics in the 20–300 μm size range following microwave-assisted digestion. Near-infrared hyperspectral

**Table 4.1** Notable water bodies and their level of microplastic contamination

| Country | Surface water | Amount discovered | Source |
|---|---|---|---|
| Australia | Freshwater in Victoria | 0.40 ± 0.27 items/L | [24] |
| Bangladesh | Saint Martin Island in the Bay of Bengal | ~2–19 particles/30 min trawl | [31] |
| Bangladesh | Patenga and Parki beaches (Northern Bay of Bengal) | 305.56–116.67 MPs/$m^3$ | [32] |
| Bangladesh | Moheshkhali channel (Bay of Bengal) | 0–0.1 particles/$m^3$ | [33] |
| Bangladesh | Buriganga River | 4.33–43.67 (± 0.58) items $L^{-1}$ | [34] |
| Bangladesh | Meghna estuary (Bengal Delta Coast) | 128.89 ± 67.94 item/$m^3$ | [35] |
| Bangladesh | Jamuna River and Turag River | 16.11 ± 2.22 particles/100 L & 29.00 ± 10.68 particles/100 L | [36] |
| Bangladesh | Buriganga River | 350–660 items/$m^3$ | [37] |
| Bangladesh | Surma River | 5–20 items/L | [38] |
| Bangladesh | Karnafully River | 0.57 ± 0.07–6.63 ± 0.52 items/L | [39] |
| Bangladesh | Hatiya Island | 139 ± 44 items/$m^3$ | [40] |
| Brazil | Todos dos Santos Bay (TDB) (state of Bahia) | 33,000 items $m^{-3}$ | [41] |
| Brazil | Jurujuba (Guanabara Bay) & Itaipu embayments | 138.41 items $kg^{-1}$ | [42] |
| Canada | St. Lawrence Estuary (SLE) | 120 ± 42 SD particles $m^{-3}$ | [43] |
| Canada | Lake Superior | 9000–40,000 particles/$km^2$ | [44] |
| China | Estuary of Yellow River (Eastern Shandong Province) | <200 μm (87.94%) | [45] |
| China | Lanzhou section of the Yellow River | 0.98 particles $L^{-1}$ | [46] |
| China | Pearl River Delta | 0–400 items $m^{-3}$ | [47] |
| China | Pearl River (Guangzhou City) | 379–7924 items·$m^{-3}$ | [48] |
| China | Three Gorges Reservoir, Yangtze River | 3407.7 × $10^3$–13,617.5 × $10^3$ items/$km^2$ | [49] |

**Table 4.1** (continued)

| Country | Surface water | Amount discovered | Source |
|---|---|---|---|
| China | Three Gorges Reservoir (Wanzhou section) | $5.27 \pm 3.47 \times 10^7$ items $km^{-2}$ | [50] |
| China | Three Gorges Reservoir (Chongqing municipality and Hubei Province) | 1597–12,611 n/$m^3$ | [51] |
| China | Yangtze River (Chongqing City) | 46.7–204 n $L^{-1}$ | [52] |
| China | Three Gorges Reservoir, Yangtze River | $2.80 \pm 1.86 \times 10^6$ items/$km^2$ | [53] |
| China | West River downstream | 2.99–9.87 items/L | [54] |
| China | Wei River | 3.67–10.7 items /L | [55] |
| China | Dagu River (Jiaodong Peninsula) | 0.68–3.96 items $L^{-1}$ | [56] |
| China | Small-scale Estuaries in Shanghai | $27.84 \pm 11.81$ particles $L^{-1}$ | [57] |
| China | Chongming Island in the Yangtze Estuary | 0–259 items $m^{-3}$ | [58] |
| China | Yangtze River Basin | 3441 particles/$m^3$ | [59] |
| China | Yangtze River Basin | $8797 \pm 12{,}281$ items/$m^3$ | [60] |
| China | Hanjiang River and Yangtze River (Wuhan) & Bei Lake | $1660.0 \pm 639.1$–$8925 \pm 1591$ n/$m^3$ | [61] |
| China | Yangtze River Estuary Region | $0.17 \pm 0.14$ items/$m^3$ | [62] |
| China | Yangtze River Delta | 4.4–10.8 items/L | [63] |
| China | Yangtze Estuary System (East China Sea) | $4137.3 \pm 2461.5$ & $0.167 \pm 0.138$ n/$m^3$ | [64] |
| China | Dongting Lake and Hong Lake | 900–2800 & 1250–4650 n/$m^3$ | [65] |
| China | West Dongting Lake & South Dongting Lake | 616.67–2216.67 & 716.67–2316.67 items/$m^3$ | [66] |
| China | Qin River in Beibu Gulf (Guangxi Province) | 67.2% PE & 72.3% PP | [67] |
| China | Beibu Gulf | $0.56 \pm 1.02$ pieces/$m^3$ | [68] |
| China | Manas River Basin (Northeast China) | 21–49 (±3) items/L | [69] |

(continued)

**Table 4.1** (continued)

| Country | Surface water | Amount discovered | Source |
|---|---|---|---|
| China | Manas River Basin (Northeast China) | 22 ± 5–14 ± 3 (in April) and 19 ± 2–10 ± 1 items/L (in July) | [70] |
| China | Guangdong Coastal Areas | 850–3500 items/L | [71] |
| China | Maozhou River (Guangdong-Hong Kong-Macao Greater Bay Area) | 4.0 ± 1.0–25.5 ± 3.5 (dry season) & 3.5 ± 1.0–10.5 ± 2.5 items·$L^{-1}$ (wet season) | [72] |
| China | Guangdong Coastal Areas | 8895 items/$m^3$ | [73] |
| China | Haizhou Bay, Lianyungang | 2.60 ± 1.40 n/g | [74] |
| China | Changsha Urban Lake | 2425 ± 247.5–7050 ± 1060.66 items/$m^3$ | [21] |
| China | Changsha City (CS) Section of Xiangjiang (XJ) River (Hunan Province) | 0.72–18.6 (7.32 ± 2.36) items $L^{-1}$ | [75] |
| China | Bohai Sea | 0.33 ± 0.34 particles/$m^3$ | [76] |
| China | Bohai and Huanghai Seas | 3–162 particles per 100 $m^3$ | [77] |
| China | Haihe Estuary (HHE) & Yondingxinhe Estuary (YDXE) of Bohai Bay | 1485.7 ± 819.9 (HHE) & 788.0 ± 464.2 items/$m^3$ (YDXE) | [78] |
| China | Changjiang Estuary | 23.1 ± 18.2 n/100 L | [79] |
| China | Seas around Shenzhen | 3.8–7.8 items per liter | [80] |
| China | Middle Route of the South-to-North Water Diversion Project (SNWDP) | 516 items $m^{-3}$ | [81] |
| China | Minjiang River Watershed | 0.12–2.72 n/L | [82] |
| China | Xinyi River & Ashe River (Harbin) | 13 & 12 counts | [83] |
| China | South China Sea (SCS) | 0.03 particles/$m^3$ | [84] |
| China | Northern SCS, Western SCS & Western Pacific Ocean | 797 ± 512, 744 ± 330 & 201 ± 134 particles/$m^3$ | [85] |
| China | Wuliangsuhai Lake in Inner Mongolia | 3.12–11.25 n/L | [86] |

**Table 4.1** (continued)

| Country | Surface water | Amount discovered | Source |
|---|---|---|---|
| China | Meishe River, Haikou | 3–10 items/L | [87] |
| China | Huangshui River Basin (Qinghai Province) | 665–8780 n $m^{-3}$ | [88] |
| China | Feilaixia Reservoir (Beijiang River, Guangdong Province) | 0.56 ± 0.45 items/$m^3$ | [89] |
| China | Huangbai River | 7.33 ± 4.66 n $L^{-1}$ | [90] |
| China | Hong Kong Tap Water | 2.181 ± 0.165 n $L^{-1}$ | [91] |
| China | Xi'an City Park Water Body | 2900–6970 items/$m^3$ | [92] |
| Colombia | Luruaco Lake, Low Basin Magdalena River | 0–3.83 MPs·$L^{-1}$ | [93] |
| Colombia | Caribbean and Pacific Coast | 0.01–8.96 items $m^{-3}$ | [94] |
| Denmark | Kattegat/Skagerrak area | 11–87 MP $m^{-3}$ | [95] |
| France | Bay of Marseille & Rhône River | 0.07–118 & 0.20–21.32 kg | [96] |
| India | Haryana (Rohtak district) | 16–28 particles/L | [97] |
| India | Mula river in Pune | 1808 ± 697 particles/L | [98] |
| India | Subarnarekha and Kharkai Rivers (Jamshedpur) | 9.55 ± 4.87 items/L | [99] |
| India | Ganges River (Bay of Bengal) | 91% fibers & 9% fragments | [100] |
| India | Ganges River | 0.05 MP $L^{-1}$ | [101] |
| India | Bay of Bengal & Eastern Indian Ocean (EIO) | 2.04 ± 2.26 items $m^{-2}$ & 0.16 ± 0.17 items $m^{-2}$ | [102] |
| India | Northern Indian Ocean | 15,200 ± 7999 no./$km^2$ (pre-monsoon), 18,223 ± 14,725 no./$km^2$ (monsoon) & 72,381 ± 77,692 no./$km^2$ (post-monsoon) | [103] |
| India | Chennai coastal aquifers | 22 particles/5 g DS | [104] |
| India | Ennore to Kovalam water | 6–30 items/Location | [105] |
| India | Drinking Water Supply Reservoirs (Ranchi) | 3699–6873 counts | [106] |

(continued)

**Table 4.1** (continued)

| Country | Surface water | Amount discovered | Source |
|---|---|---|---|
| India | Periyar River, Kerala | 120.50 items/L (pre-monsoon), 124.95 items/L (monsoon) & 123.21 items/L (post-monsoon) | [107] |
| India | Shahpura Lake & Kerwa Dam (Bhopal city) | 1640 ± 162.69 & 510 ± 43.20 particles/$m^3$ | [108] |
| India | Chalakudy River, Kerala | 55.81 ± 34 particles $L^{-1}$ | [109] |
| India | Thoothukudi Region, south Tamil Nadu | 54 ± 41–619 ± 377 items/L | [110] |
| Indonesia | Ciwalengke River, Majalaya district | 5.85 ± 3.28 particles/L | [8] |
| Indonesia | Maninjau Lake, Agam | 180–335 $L^{-1}$ particles | [111] |
| Indonesia | Cisadane River | 13.33–113.33 particles $m^{-3}$ | [112] |
| Indonesia | Jatiluhur Reservoir | $0.71\times10^4$–$4.59\times10^5$ particles/$km^2$ | [113] |
| Indonesia | Benoa Bay, Bali | Mostly PS, PE & PP | [19] |
| Indonesia | Jakarta Bay | 70.9 particles/L | [114] |
| Indonesia | Megacity of Jakarta Bay | 30,000 particles/$m^3$ | [115] |
| Indonesia | Lake Singkarak (Tanah Datar, West Sumatra) | 117.5–202.5 particles/L | [116] |
| Iran | Zayandeh-Rud River | 0–51 ± 16.5 particles/$m^3$ | [117] |
| Iran | Caspian Sea (Iranian Rivers) | 1.406 ± 0.1380 (snowmelts) & 0.4070 ± 0.01500 microplastics/$m^3$ (dry season) | [118] |
| Italy | Subalpine Lakes (Lake Maggiore, Iseo & Garda) | 39,000, 40,000 & 25,000 particles/$km^2$ | [119] |
| Jamaica | Kingston Harbour | 0–5.73 particles/$m^3$ | [120] |
| Japan | Japanese River | 300–1240 particles/$m^3$ | [121] |
| Japan | Island of Okinawa | 75% PE | [122] |
| Japan | Tokai region and Tokyo Bay | 1000–5900 pieces $m^{-3}$ | [123] |
| Kenya | Lake Naivasha | 0.407 ± 0.135 particles/$m^2$ | [124] |
| Latvia | Daugavpils & Liepaja Cities | 16.63 ± 20.29 particles/L | [125] |
| Lebanon | Sidon water (Eastern Mediterranean Basin) | 6.7 MPs/$m^3$ | [126] |
| Lithuania | Klaipeda & Siauliai Cities | 16.63 ± 20.29 particles/L | [125] |
| Malaysia | Kelantan Bay | Mostly PP | [17] |

**Table 4.1** (continued)

| Country | Surface water | Amount discovered | Source |
|---|---|---|---|
| Mexico | WWTP into Todos Santos Bay (TSB) | 0.01–0.70 plastic particles/m$^3$ | [127] |
| Morocco | Al Hoceima Bay (Southwestern Mediterranean) | 4.70 ± 4.50 items/m$^3$ | [128] |
| Morocco | Sebou Estuary and Atlantic Coast | 10–168 particles/m$^3$ & 10–300 particles/kg | [129] |
| Nepal | Lake Phewa | 2.96 ± 1.83 (winter) & 1.51 ± 0.62 particles/L (rainy season) | [130] |
| New Zealand | Vava'u archipelago, Tonga | 1.05 ± 0.13 pcs m$^{-3}$ | [131] |
| Nigeria | Escravos Estuary (Niger Delta) | 50 particles/L | [27] |
| Nigeria | Lagos Lagoon | 65.2 particles/L | [132] |
| Pakistan | Ravi River | 5150 ± 7.5 particles/m$^3$ | [28] |
| Pakistan | Rawal Lake | 6.4 ± 0.5–8.8 ± 0.5 particles/m$^3$ | [133] |
| Paraguay | Bay of Asuncion | 13.2 ± 13.4 items.m$^{-3}$ | [134] |
| Philippines | Laguna de Bay | 14.29 items/m$^3$ | [135] |
| Philippines | Cañas, Meycauayan, Parañaque, Pasig & Tullahan River (Manila Bay) | 1580–57,665 particles/m$^3$ | [136] |
| Poland | Vistula River | 1.6–2.55 items L$^{-1}$ | [137] |
| Portugal | Algarve Coast | 1.36 ± 4.03 particles/m$^3$ | [138] |
| Russia | Ob and Tom Rivers (Western Siberia) | 44.2–51.2 items/m$^3$ | [139] |
| Scotland | Scottish Marine Regions (SMR) & Offshore Marine Regions (OMR) | 0–91,128 microplastics km$^{-2}$ | [140] |
| South Korea | WWTP or STP of Tanchon stream (tributary of the Han River) | 37.4 ± 37.0 particles/m$^3$ (rainy season) & 28.2 ± 22.2 particles/m$^3$ (dry season) | [26] |
| South Korea | Geoje Eastern Bay and Jinhae Bay | NR | [141] |
| South Korea | Incheon/Kyeonggi Coastal Region | 1602 ± 1274 particles/m$^3$ | [142] |
| South Africa | Vaal River in Johannesburg | 0.13–2.52 particles/m$^3$ | [143] |
| South Africa | Vaal River in Johannesburg | 0.61 ± 0.57 particles/m$^3$ & $4.6\times10^2 \pm 2.8\times10^2$ particles/kg DW | [144] |
| Spain | Llobregat River Basin | 544 μg/L | [145] |
| Sweden | West Coast | 0.04 & 0.1 particles/m$^{-3}$ | [146] |

(continued)

**Table 4.1** (continued)

| Country | Surface water | Amount discovered | Source |
|---|---|---|---|
| Thailand | Songkhla Lagoon (Southeast Asia's coastline) | 0.43 ± 0.18 (wet season) & 0.34 ± 0.08 items/L (dry season) | [147] |
| Thailand | Nakhon Pathom Province Ponds | NR | [25] |
| Thailand | Bueng Boraphet Wetland | 0–19.57 particles/$m^3$ | [148] |
| Thailand | Inner Gulf of Thailand | 34.59 ± 46.02 (wet season) & 8.70 ± 15.34 pieces/L (dry season) | [149] |
| Thailand | Inner Gulf of Thailand | 9.97 pieces/L | [150] |
| Thailand | Chao Phraya River Estuary | 48 ± 8 items/$m^3$ | [151] |
| Tunisia | Gulf of Gabes (southern Mediterranean Sea) | 63,739 items/$km^2$ | [152] |
| Turkey | Sea of Marmara | 1.263 item/$m^2$ | [153] |
| Turkey | Konya Closed Basin | 1139–23,444 particles/$m^3$ | [154] |
| Turkey | Ergene River | 4.65 ± 2.06 & 6.90 ± 5.16 items $L^{-1}$ | [155] |
| Turkey | Sürgü Dam Reservoir (Malatya) | 106.63–200 particle $m^{-3}$ | [156] |
| United Kingdom | Surface trawls around Plymouth | 0.26–0.68 n $m^{-3}$ | [157] |
| United Kingdom | Southern North Sea | 0.1–245.4 particles $m^{-3}$ | [158] |
| United Kingdom | Southern North Sea (North West Europe) | 0–1.5 microplastic/$m^3$ | [159] |
| United States | Lower Hudson River Estuary | ~70% microfibers | [160] |
| United States | Lakes & Playa Wetlands (Lubbock, Texas) | 0.79 ± 0.88–1.56 ± 1.64 & 0.31 ± 0.72–1.25 ± 1.98 mg/L | [161] |
| United States | Chesapeake Bay | 32,000 particles/$km^2$ | [162] |
| United States | Laurentian Great Lakes | 43,000 particles/$km^2$ | [163] |
| Vietnam | Da Nang City Lakes | 293.3 ± 23.1–6510.0 ± 380.4 (dry season) & 366.7 ± 23.1–1143.3 ± 110.2 items $m^{-3}$ (wet season) | [164] |
| Vietnam | Mekong River Delta | 53.8 ± 140.7 items $m^{-3}$ | [165] |
| Vietnam | Day River downstream (Cau Que, Cau Do & Do Thong) | 269,693 ± 60,624 to 863,005 ± 131,925 items/$m^3$ | [166] |
| Vietnam | West Lake, Yen So Lake, and Bay Mau Lake (in Ha Noi) | 589.46 items $m^{-3}$, 139.86 items $m^{-3}$ & 154.92 items $m^{-3}$ | [167] |

**Table 4.1** (continued)

| Country | Surface water | Amount discovered | Source |
|---|---|---|---|
| Vietnam | Dau Tieng & Tri An Reservoirs | 3.96 ± 1.38 & 4.04 ± 1.35 items $m^{-3}$ | [168] |
| Vietnam | Hai Tien, Sam Son & Hai Hoa Beaches (Thanh Hoa province) | 15.5–44.1 items/$m^3$ | [169] |
| Bordered by Myanmar, India, Indonesia, Thailand & Malaysia | Andaman Sea | 0.02–0.74 particles/$m^3$ | [170] |
| Bordered by Kenya, Tanzania & Uganda | Lake Victoria (East Africa) | 0.02–2.19 particles/$m^3$ | [171] |
| Bordered by Russia, Kazakhstan, Turkmenistan, Iran & Azerbaijan | Caspian Sea | 34,490 particles per $km^2$ | [172] |
| Bordered by United Kingdom, France, Ireland, Spain, Portugal & Norway | Northeast Atlantic Ocean | 47–2154 particles/$m^3$ | [173] |
| Panama, Colombia, and Central America (Nicaragua, Honduras, Costa Rica, Guatemala & Belize) | Southwestern Caribbean Sea | 1.72 (rainy season) & 0.22 MP/$m^3$ (dry season) | [174] |
| Antarctic Seawater | Ross Sea & Dumont d'Urville Sea | 0.10 ± 0.14 items/$m^3$ & 1.66 ± 1.20 items/$m^3$ | [175] |
| Arctic Ocean | Arctic Central Basin | 0.7 particles $m^{-3}$ | [176] |
| Arctic and Sub-Arctic Region | Kongsfjorden, Svalbard Seawater | 0–10.0 items $m^{-3}$ | [177] |
| Arctic & Sub-Arctic Ocean | East Asian Seas to the Arctic Central Basin | 2.91 ± 1.93 items/$m^3$ | [178] |
| Arctic Region | Northeast Greenland | 2.4 items/$m^3$ ± 0.8 SD | [179] |
| Arctic Ecosystem | Arctic fjords of Western Svalbard | 3.5 particles/$m^3$ | [180] |
| Baltic Region | Northern Europe | 5.71 (summer) & 0.75 particles/$m^3$ (winter) | [181] |
| Baltic Sea | Gulf of Riga & Eastern Gotland Basin | 0.09–4.43 particles/$m^{-3}$ | [182] |

*NR* not reported, *DW* dry weight, *SD* standard deviation, *DS* dry sediment imaging is also capable of identifying most microplastic particles [146]. Di & Wang [51] show how micro-Raman spectroscopy was used to identify the types of microplastics and analyze contaminants adsorbed on their surfaces. Both Huang et al. [54] and Ding et al. [55] state that Fourier Transform Infrared (FTIR) spectroscopy can be used to identify the polymer types of microplastics found in freshwater environments. Alternatively, Focal Plane Array (FPA) FTIR imaging is used to analyze and identify small microplastics (11–500 μm) in environmental samples [158]. Pyrolysis-Gas Chromatography Mass Spectrometry (Py-GC-MS) is a validated method to identify and quantify the six major microplastic polymers in surface and drinking water [44, 145]. Sampling of microplastic in the Arctic Central Basin was conducted using the bow water system at a fixed depth (8.5 m) and a Conductivity, Temperature, and Depth (CTD) rosette sampler across multiple depths (8–4369 m) to assess abundance and vertical distribution [176]. According to Karlsson et al. [187], the differences in microplastic sampling methods (specifically manta trawls and *in situ* pumps) affect accuracy, concentration, and composition data, and stated that the need for standardized, high-volume sampling and improved identification protocols is essential to ensure reliable, quantitative comparisons. But microplastic sampling in surface waters using Manta nets lacks standardized methods and consistent protocols, which makes data comparison across studies difficult [188]. Nile red fluorescence tagging method is effectively used as a rapid screening tool to detect and map microplastics across various environmental compartments [189]. Saipolbahri et al. [17] states that microplastics in surface water and sediment from Kelantan Bay were isolated using wet peroxide oxidation (WPO) and density separation, with samples collected using Water and Ekman Grab Samplers. To date, no single assessment method can be fully relied upon for the efficiention detection of microplastics in water bodies [190]. Because, Koelmans et al. [191] discovered that the overall data quality in studies on microplastics in drinking water is low, with only 4 out of 50 studies meeting all quality criteria.

## *Groundwater*

When microplastics are released into the environment, they travel through soil and enter groundwater systems, which serve as a major source of drinking water [192–194]. The presence of microplastics in groundwater has become a growing environmental concern [195–197]. Because groundwater is a major drinking water source, it is crucial to understand and monitor how microplastics enter it from surface water, sewage, agricultural runoff, seawater, and soil [198–201]. Open groundwater contains higher microplastic abundance and diversity than closed groundwater due to differing entry pathways [202]. Particles of plastics 'microplastic' originate from industrial waste, tire abrasion, water processing systems, agricultural runoff, municipal waste disposal, landfill leachate, old plastic distribution systems and household products sources [203, 204]. Wei et al. [205] presented a new dispersion–drag force coupled model that more accurately simulates microplastic migration in soil–groundwater systems by accounting for particle size, plastic type, irrigation factors, and aquifer heterogeneity. Viaroli et al. [206] introduces a new "Hydrogeoplastic Model" to better understand their movement and risks in aquifers. By using a cost-effective, hydrogeology-based method to detect microplastics in an alluvial aquifer, Severini et al. [207] found that groundwater microplastic transport is strongly shaped by aquifer hydrogeological conditions. Priya et al. [208] and Li et al. [209] states that groundwater near dumping sites faces high microplastic pollution risks from leachate, heavy metal and seawater intrusion, as shown by the Microplastic Pollution Index. Accurate measurement of microplastics in groundwater requires sound field techniques and a strong thought on hydrogeology [210, 211]. Perraki et al. [212] identifies microplastics in water samples by first visually classifying particles under a microscope based on their physical appearance and then confirming their polymer types using Raman microspectroscopy. Mintenig et al. [213] and Kim et al. [214] states how micro-FTIR (μ-FTIR) imaging was used to detect and identify microplastics larger than 20 μm in groundwater and drinking water samples along the supply chain. Groundwater microplastic research faces challenges due to the lack of standardized sampling protocols, and the common use of Polyvinyl chloride (PVC) well casings contribute to contamination [215]. Once in the soil, microplastics migrate downward through cracks and pores, eventually contaminating underground water reserves [216–219]. Their small size allows them to persist and escape conventional filtration processes used in water treatment. Though advanced membrane filtration technologies can remove up to 99% of them [220]. This contamination poses potential risks to human health and aquatic ecosystems, as microplastics carry toxic chemicals and harmful microorganisms [221, 222]. PVC microplastics hinder the effectiveness of nano zerovalent iron in removing heavy metals from groundwater by promoting its aging [223]. According to Khant & Kim [224], PE and polyethylene terephthalate (PET) fragments and fibers are increasingly infiltrating groundwater through soil, yet remained largely overlooked compared to surface waters. Research on microplastics in groundwater is growing, but global understanding of their distribution, characteristics, transport factors, and health risks remains limited [225] (Fig. 4.1).

Some countries or regions have managed to report the level of pollution of their groundwaters by microplastics [226, 227]. Small fibers of polystyrene (PS), PE, and PET were detected in groundwater wells in Iran at concentrations up to 1.3 MP $L^{-1}$, due to varied distribution linked to land use and hydrogeology [228]. Small PET and polyurethane (PU) particles were detected at high levels in all sampled groundwater sites in China [229]. China's Yulong River karst area contains PP, PS, and PET fibers present in surface water, sediments, and especially groundwater [230]. Zhang et al. [231] reveals that groundwater in Shanghai contains high levels of PE and PP microplastics from agricultural activities, atmospheric deposition, landfills, and construction. On a relatively undeveloped bedrock island in the South China Sea, groundwater contained substantial amounts of microplastics [232]. In China, microplastics—mainly PS, PE, PET, and fiber-shaped particles—have been found in both karst and northern drinking-water groundwater systems at low to medium ecological risk levels, with their abundance influenced by rainfall, erosion, water

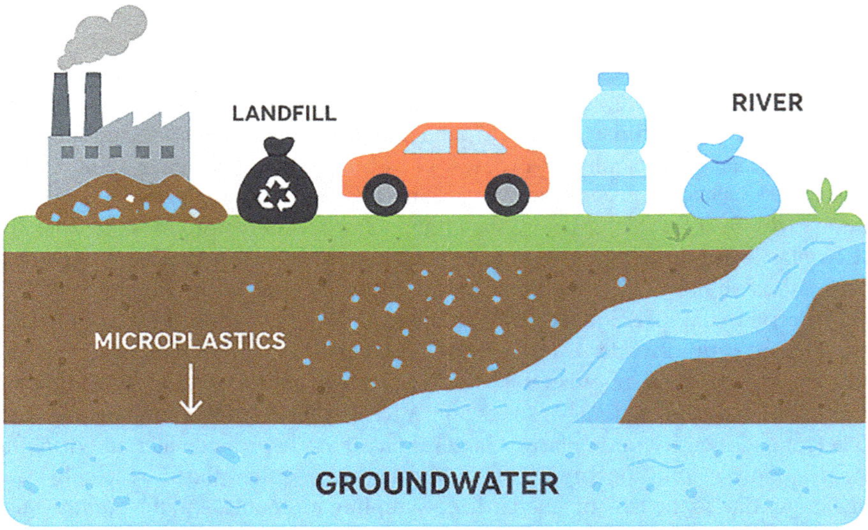


**Fig. 4.1** Microplastic transport from soil surface to groundwater

flow, and intensified by human activities in densely populated areas [233, 234]. Other studies from China shows that groundwater in a drinking-water source area is contaminated with both antibiotics and microplastics [235]. Limited research on groundwater pollution by microplastic in Korea, in which according to Lee et al. [236] is limited, is now getting the right attention. Even agricultural groundwater without obvious pollution sources in the country contains microplastics [237, 238]. Cha et al. [239] discusses how sampling groundwater in Korean agricultural areas showed microplastic concentrations varying by season and sample volume. In Korea, microplastics from agricultural activities are present even in deep groundwater and cave water, with higher concentrations in the dry season [240]. Also, in South Korea, greenhouses are identified as the primary source of microplastics entering soils and groundwater [214, 241]. The Government of India officially prohibited plastic bags thinner than 50 μm starting in 2016 [242]. Manikanda et al. [243] and Valsan et al. [244] reports that groundwater near waste dumpsites in South India is contaminated with various colored microplastics (primarily nylon and PP-derived particles from buried plastic debris) ranging from 2–80 items per liter; as well as heavy metals [245]. For the first time, Samandra et al. [246] and Cha et al. [247] mentioned that even capped alluvial aquifers in Australia contain PE and PVC microplastics (up to 97 particles/L), likely permeating through soil. Panno et al. [248] and Shockley & Lapworth [249] reveals that karst aquifers in Illinois are vulnerable to septic effluent pollution, with detected microplastic fibers (up to 15.2 particles/L). Similar research was reported in France [250], Greece [212], Mexico [251], Canary Island [252], Switzerland [253], United Kingdom [254], Vietnam [255] and Italy [256].

## Impacts of Microplastics on Health and Ecosystem

Little is known as to the impact of microplastics on human health [257, 258]. But VOSviewer software was used by Mittal et al. [259] to perform a bibliographic analysis of research on microplastics and their effects on gut health, where a sharp increase in studies since 2016 is noted. According to Yee et al. [260] and Yang et al. [261]), once microplastics are released into the environment, they harm human health by entering the body through food, water, and air, potentially causing inflammation, chemical exposure, physical damage and other adverse biological effects (from the cellular to the organ system level). Both nano- and microplastics cause cell apoptosis and possibly exert genotoxic and cytotoxic effects [262]. Sources of microplastics include water, fish, shellfish, poultry meat, fruits, vegetables, canned goods, milk, salt, sugar, honey, tea, beer, wine, soft drinks, and energy drinks [263–266]. Non-edible sources are agricultural and industrial products, and packaging materials [267]. Microplastics enter the human body through ingestion, fish consumption, inhalation, and skin/dermal contact [268, 269], thereby causing digestive issues, organ dysfunction, immune disruption, hormonal disruption, neurotoxicity, respiratory and cardiovascular problems, cancer risk, obesity, sleep disruption, oxidative stress, gut homeostasis disruption, digestive problem, metabolic disorder, DNA damage, infertility, diabetes, and even carrying harmful bacteria [268, 270–279]. Prathipa et al. [265], Karganovic et al. [280] and Barceló et al. [281] examines the accumulation of microplastics in the human body, and also highlighted their detection in lung, placentas, liver, various bodily fluids (sputum, stool, blood & breast milk), and joints of the lower limbs. To gain clearer insights into the effects of microplastics on the human body, their impacts on each organ or disease condition should be classified based on the type, size, and shape of the particles [282]. Karbalaei et al. [283] notes that microplastics have also been detected in air samples, indicating that inhalation is a potential exposure route that may pose health risks to humans. The immune system's failure to eliminate synthetic particles can result in persistent inflammation and elevate the risk of developing neoplasia [284]. In aquatic ecosystems reviewed by Liu & Li [285], microplastics are often mistaken for food by fish, plankton, and other marine animals [286]. This can block their digestive systems, reduce nutrient absorption, and lead to death in severe cases. *Siganus rivulatus* are herbivorous fish in the MENA region that have been found to contain over 1000 microplastic particles per individual from ingesting contaminated seaweed [287]. Microplastics can also absorb harmful chemicals, which then enter the food chain when animals eat them [288]. On land, microplastics change soil structure and reduce its fertility, affecting plant growth. Potentially, it stood a chance of altering soil permeability, which then disrupt essential ecosystem functions and indirectly affect human health [289]. In agricultural soils, microplastic affect soil health, microbial communities, and crop production [290]. Specifically, they harm earthworms and other soil organisms that are essential for nutrient cycling. Microplastics affect crop quality, lower the leaf number in wheat, and shorten the root length in garden cress, maize, black gram and water spinach [291]. Over time, these effects disrupt food webs, reduce biodiversity, and threaten ecosystem stability. Food security is notably threatened by microplastic contamination in seafood, which makes human exposure inevitable and poses risks to human health [292]. Microplastics even travel long distances, making their impact widespread and difficult to reverse. Despite being considered markers of a new geological epoch, microplastics act as carriers and sources of hazardous chemicals, and their health impacts remain poorly understood [293]. Current in vitro models for studying microplastic and nanoplastic toxicity have limitations; as such, Sadique et al. [294] emphasized the need for more advanced in vivo models to better simulate real-life human exposures. Because Hampton et al. [295] states that the concentrations at which microplastics cause harm to humans and aquatic life are still uncertain due to significant data gaps. Also, Weis et al. [296] highlights that microplastics and nanoplastics are a potential environmental and health concern for New Jersey in the United States, but significant data gaps exist. Hence, more localized research and monitoring to determine these critical thresholds is needed [297]. Currently, Canada has responded to concerns about microplastics by banning the manufacture and import of personal care products containing microbeads [298].

## Advanced Treatment Technologies

### *Membrane Filtration*

Landfill leachate treatment releases microplastics into surface waters, and while several filtration and membrane-based methods show potential for removal, more research is urgently needed to develop effective, sustainable mitigation strategies [299]. Poerio et al. [300] highlights that plastic pollution, especially micro- and nanoplastics 'particles' from wastewater treatment plants, remains a serious issue. While wastewater treatment plants play a key role in pollution control, their effectiveness against microplastics is limited without proper tertiary treatment [301]. Acarer [302] also said that conventional treatment plants cannot fully remove microplastics, but membrane technologies (Fig. 4.2) such as ultrafiltration (UF), microfiltration (MF), nanofiltration (NF), reverse osmosis (RO) and membrane bioreactors (MBRs) are the most effective options. Aging degree of PP microplastics impacts membrane fouling during ultrafiltration, with different chemical aging conditions altering cake formation behavior and filtration performance [303]. Enfrin et al. [304] shows that nano- and microplastics increase fouling during ultrafiltration of water, thereby reducing membrane performance. Even though rapid gravity filtration and ultrafiltration reduce microplastics and microlitter from wastewater effluents, González-Camejo et al. [305]'s ultrafiltration achieves higher removal despite higher energy use, indicating post-treatment is needed beyond conventional WWTP processes. Bayo et al. [306] found that both MBRs and rapid sand filtration (RSF) in wastewater treatment plants reduce microplastics (around 75–79%

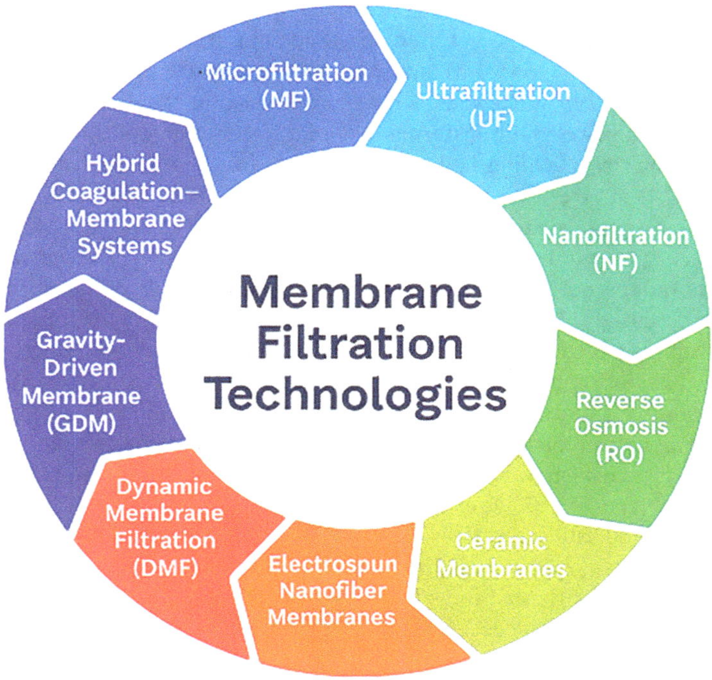


**Fig. 4.2** List of membrane filtration technologies

removal), with better elimination of particulate types than fibers. Membrane filtration is a method used to remove microplastics from water. It works by passing the water through a thin filter called a membrane. Many of such membrane are characterized with very small holes that can trap tiny plastic particles [307]. Flow cytometry is a rapid and effective way to quantify microplastics in water and evaluate membrane removal performance, which reveals how smaller particles can still pass through larger-pore membranes depending on conditions [308]. After comparing polymeric vs ceramic membrane performance, Kook & Park [309] reveals that ceramic membranes remove more particles by turbidity but their larger pores miss some smaller microplastics. Commercial laundry microplastic filters perform poorly, but smaller-pore polymeric and ceramic membranes are more effective alternatives for retaining microplastic fibers and organic contaminants in washing machine effluent [310]. Aluminium-coated polycarbonate membrane filters provide an ideal surface for micro-Raman spectroscopy by minimizing background interference and enabling automatic detection of microplastics down to 1 μm in food samples [311].

Carter et al. [312] compares five different analytical filter membranes and shows that gold-coated microslit silicon nitride (MSSN-Au) performs best in terms of filtration rate and optical imaging for microplastic monitoring applications. So et al.

[313] shows that current wastewater treatments do not effectively remove microplastics from urban sources, but high-speed ceramic membranes, along with fluid imaging microscopy (FlowCam) analysis can successfully detect and reduce even small microplastics in wastewater. FlowCam is a useful rapid tool for identifying and tracking shape-related transformations of different microplastics during membrane filtration, by revealing how their geometry influences retention and fouling behavior [314]. As the water flows through, the microplastics are left behind while clean water comes out. This process is simple, effective, and does not require harmful chemicals. Low-cost membranes, including polycarbonate, cellulose acetate, and polytetrafluoroethylene are 94% efficient in removing microplastics from water in domestic settings [315]. Though performance varies based on membrane properties, microplastic types, and operational factors such as abrasion, fouling, and pressure, with cellulose acetate proving most suitable for long-term use. Even low levels of microplastics in wastewater increase membrane fouling during filtration, but backwashing helps reduce this effect and restore membrane performance [316]. Li et al. [317] presents a new two-step sample-pretreatment method (membrane filtration followed by cloud-point extraction) that enables separate detection of microplastics and nanoplastics in water with high efficiency and sensitivity. Electrospinning is a highly adaptable technique for producing filter membranes [318]. Membrane filtration is now widely used in water treatment to help keep water safe and clean. Miclea et al. [319] introduces a new cascade filtration system using silicon and alumina filters to selectively capture and analyze nano- and microplastics. Sometimes, the membranes themselves may release these particles due to wear and aging, raising concerns about contamination and emphasizing the need for improved membrane design and standardized testing [320]. Adding smaller or higher doses of microplastics to gravity-driven membrane systems reduces flux by promoting thicker bio-cake formation and alters the biofilm microbial community [321]. A gravity-driven dynamic membrane filtration reactor using stainless-steel mesh supports can remove >99% of microplastics from plastic-recycling wastewater [322]. Using octadecyl-quaternium hybrid coagulant (OQHC) in pre-coagulated gravity-driven membrane systems significantly improves microplastic removal from surface water [323]. Combining electrocoagulation-electroflotation and membrane filtration processes can achieve up to 100% removal of microplastics from wastewater [324].

## *Coagulation-Flocculation*

In coagulation, chemicals called coagulants are added to water to neutralize the charges on tiny suspended particles, including microplastics, which prevents them from repelling each other. Once neutralized, the particles come together [325]. Coagulation aids added sometimes, function by enhancing floc formation, strengthening floc structure, and improving settling or filtration efficiency during the coagulation process [326]. Flocculation follows coagulation. In this step, gentle mixing helps the small neutralized particles stick together to form larger clumps called flocs, which are heavier and can either settle at the bottom or be filtered out more easily. Microplastics, which are usually very small and difficult to remove by normal filtration, attach to the flocs and are carried down with them. Research by He et al. [327] show that chemical pretreatment of PP microplastics alters their surface properties, which in turn affects floc size, structure, and strength during flocculation. Ballasted flocculation is an enhanced water treatment process where a dense material, such as sand or microspheres, is added during coagulation-flocculation to increase floc weight, to make them settle faster [328]. Ballasted flocculation is far more effective than conventional flocculation, as it achieves 99.6% removal of PVC in Godoy et al. [329], with much larger floc sizes and significantly faster settling velocities. A particle size analyzer (PSA) is normally employed to determine floc sizes under varying pH conditions [330]. Electro-coagulation, on the other hand, generates coagulant agents in situ by applying an electric current to sacrificial metal electrodes. In a study by Pu [331], the removal efficiency of microplastics using conventional coagulation ranges from 8.28% to 98.2%, while electro-coagulation achieves between 8.24% and 99.4%. Ultrafiltration completely reject PE microplastics, and causes slight membrane fouling, which is less severe when combined with coagulation due to the heterogeneous cake layer formation [332]. Many studies have shown that coagulation-flocculation dwindles the amount of microplastics in water [333]. Again, it is simple, cost-effective, and widely applied in drinking water and wastewater treatment plants [334]. Afterwards, sedimentation does the remaining separation [335]. Table 4.2 gave us practical case studies. Additional recycling of microplastics from water would create valuable products such as hydrogen, liquid fuels, carbon nanocomposites, open-cell foams, and materials for flame conservation [336].

Gravity-driven membrane filtration is promising for drinking water treatment, but Chen et al. [359] mentioned that adding polyaluminum chloride (PACl) pre-coagulation helps control fouling and improves flux in the presence of microplastics by forming a more stable and uniform cake layer. Dissolved air flotation (DAF), combined with aluminum (Al)- or iron (Fe)-based coagulants, is a highly effective advanced treatment method for realizing 96.10% microplastic removal from greywater [345]. Greywater is wastewater generated from household activities such as bathing, laundry, and dishwashing, excluding toilet waste [360]. Coagulant selection is important. Azizi et al. [361] concludes that since the five different coagulants examined showed no significant difference in microplastic removal efficiency, the best coagulant is aluminum hydroxide ($Al(OH)_3$) because it achieves effective removal at the lowest dosage. Badawi et al. [362] emphasizes that natural coagulants from plant, animal, and microbial sources offer a sustainable and highly effective alternative for microplastic removal in water, with hybrid systems achieving near-complete efficiency. As observed in Table 4.2 and stated by Khan et al. [363], optimizing reaction conditions—such as coagulant type, dosage, pH, and microplastic characteristics—is essential for maximizing coagulation efficiency in microplastic removal. At the moment, machine learning models, especially the Radial Basis Function (RBF) model, can accurately predict microplastic removal efficiency

**Table 4.2** List of coagulants employed in previous experiments

| Coagulant | Dosage | Condition | Microplastic | Removal efficiency | Author |
|---|---|---|---|---|---|
| MMHC | 200 mg/L | pH = 5–9 | PE | 92% | [337] |
| Ferrate (dissolved HA) | 2.7 & 3.6 mg/L | PE particle size = 40–48 μm & PET powder particle size = 300 μm | PE & PET | >99 & 75% | [338] |
| Chitosan & Alum | 1 & 20 mg/L | pH 5 | LDPE | 92.13% | [339] |
| Chitosan-based & $AlCl_3 \cdot 6H_2O$ (plus HA) | 5 mg/L | pH = 4 & 7 | PEST fibre | 78.3 & 88.46% | [335] |
| Cationic PAM & Alum | 100 μL/min | 15 μm PE, 140 μm PS & 90 μm PEST; 22°C | PS, PE & PEST fibre | 97 & 99% | [340] |
| Alum | 30 mg/L | 6 μm PS | PS | 75.6 & 85.2% | [341] |
| Al-based | 3.4 mg Al/L | pH = 5, contact time = 7 min, bead size = 100 μm | Low-density PS bead | 98.9% | [342] |
| $FeCl_3$ and PACl (dissolved HA) | 62.5 mg/L | 300 rpm stirring & 20 min contact | PS | 98.2 & 97.9% | [343] |
| $FeCl_3$, PACl & Polyamine | 31–34.8 mg/L | pH = 7.3–6.5 & PS size = 1 and 6.3 μm | PS spheres | 99.4% | [344] |
| Al-based & Fe-based | 14.46 mg/L | 680.5 mL/min water flow rate, 5.95 bar saturation pressure & pH 6 | PE | 96.10 & 70.56% | [345] |
| $Al(OH)_3$ | 5 & 10 mg/L Al | 16 NTU initial turbidity & 20 mg/L of surfactant | PE fibres | 0.5–1.0 NTU final turbidity | [346] |
| PACl & APAM | NR | Promoter = HA | PE | > 99% | [347] |
| Fe(II) & Al(III) sulfates | 10–80 mg/L | 7.6–7.8 pH, 300 rpm & 5 min. | PS & PVC | 5% rise in efficiency | [348] |
| $AlCl_3 \cdot 6H_2O$ & $FeCl_3 \cdot 6H_2O$ | 0.05 g/L | pH = 7 | PS & PVC | 95 & 80% | [349] |
| PACl & PAM | 0.4 mmol/L & 3 mg/L | pH = 8, 240 rpm mixing (1 min), 35 rpm flocculation (13 min) & 25 min sedimentation | PS beads | 97.1% | [350–352] |
| $Al_2(SO_4)_3$ & *Moringa oleifera* | 40–60 mg/L | Coupling with APAM | PA, PE & PS | 92.99, 28.94 & 80.48% | [353] |

**Table 4.2** (continued)

| Coagulant | Dosage | Condition | Microplastic | Removal efficiency | Author |
|---|---|---|---|---|---|
| *Moringa oleifera* seeds | NR | Coupling with APAM and SA & 0.54 cm/min settling velocity | PVC | 69% | [328] |
| Alum & PAM | NR | pH = 3–5 | LDPE, HDPE & PP | 92, 84 & 96% | [354] |
| OMA & PAFC | 15 & 30 mg/L | pH = 9 | PS | 94.8% | [355] |
| PAM & PACl | 21 & 200 ppm | pH = 9 | PP | 18 & 19.69% | [330] |
| PAFC | 60 mg/L | Coagulation aids: APAM, SA, ASA | PS | 94.28% | [326] |
| Fe- or Al-based | 40–60 & 20–60 mg L | pH 5–8 & pH 7–8 | PVC | 80% | [356] |
| $FeCl_3$ & chitosan | 2 & 7 mg/L | pH = 6.3, stirring speed = 100 rpm & settling time = 30 min | PS | 99.8% | [357] |
| PAS | 7.5 g/L | Aging treatment with $O_3$, NaOH, $CH_3OH$, $H_2SO_4$ & NaClO | PS | 90.4% | [358] |

*MMHC* magnetic magnesium hydroxide coagulant, *HA* humic acid, *PEST* polyester, *PAM* polyacrylamide, *APAM* anionic polyacrylamide, *SA* sodium alginate, *PAFC* polyaluminum ferric chloride, *OMA* Opuntia Milpa Alta, *ASA* active silicic acid, *PAS* polyaluminum sulfate, *LDPE* low density polyethylene

by coagulation under varying conditions [364]. Specialized reactor types in coagulation-flocculation-sedimentation provide controlled mixing, settling, and contact conditions that optimize coagulant performance, enhance floc formation, and improve microplastic and pollutant removal efficiency. For instance, applying the $FeCl_3$–chitosan coagulation system to effluents from moving bed biofilm reactors (MBBR) and sequencing batch reactors (SBR) previously achieved over 98% microplastic removal under optimized conditions [357]. Not just coagulants, microplastic morphology and density influence removal efficiency in coagulation–flocculation–sedimentation (CFS) system, where PET and PP fibrous microplastic demonstrate a higher removal rates than particulate PA [365]. Yao et al. [358] and Zhang et al. [366] also affirmed how density, size, morphology and other surface characteristics of microplastics adsorbed substances, and functional groups, strongly influence coagulation efficiency. There is need to experiment CFS on different water sources, as examined in Zhang et al. [367].

## *Advanced Oxidation Processes (AOP)*

At the Landau wastewater treatment plant (WWTP) pilot plant, Germany, AOP combined with granular activated carbon (GAC) effectively decreased microplastics by roughly 61% [368]. However, most existing WWTPs are not designed to effectively remove organic micropollutants (OMPs), including microplastics [369, 370]. AOPs are treatment methods to remove microplastics from water by breaking them down into smaller, less harmful substances [371]. Traditional methods for degrading or removing microplastics are challenging because their structural complexity and hydrophobic nature lead to extended treatment durations [372]. AOPs work by generating highly reactive hydroxyl radicals, which then attacks and degrade the plastic's chemical structure [373]. The processes are mainly carried out using plasma, persulfate oxidation, sonolysis, ozone treatment, Fenton reaction, UV/solar photolysis, UV light with hydrogen peroxide, photocatalysis, electrochemical oxidation, heat-activated persulphate and peroxymonosulphate [374–376]. Over time, the strong oxidation reactions break the microplastics into carbon dioxide, water, and other harmless compounds. Simply because AOPs utilize highly reactive oxygen species (ROS), including superoxide radicals ($O_2^-$), hydrogen peroxide ($H_2O_2$), sulfate radicals ($SO_4^-$) and hydroxyl radicals (OH) [377, 378]. Ultraviolet C (UVC)/$H_2O_2$ increase the leaching of dissolved organic matter from microplastics, resulting in higher trihalomethane formation during chlorination, with some thalomethanes (THMs) re-adsorbing onto the microplastics and potentially raising their toxicity [379]. Again, UVC/$H_2O_2$ treatment notably degrade plastic microfibres from laundry wastewater, with over 50% mass loss in synthetic samples, complete chemical oxygen demand (COD) removal and partial microplastic degradation [380]. Thomas et al. [381] formerly reported how UVC/ $H_2O_2$ treatment degrades polyester microfibres from laundry water into $CO_2$ without harmful by-products, with measurable physical and chemical breakdown observed through spectroscopy and imaging techniques. Research by Dos Santos et al. [382] show that conventional Fenton processes degrade microplastics less effectively than other AOPs. Previously, Fenton-like process using nano zero valent iron effectively remove cyan dye from wastewater containing PE microplastics, achieving up to 92% removal under optimized conditions [383]. Fenton process combined with post-oxidation coagulation removes PET microplastics effectively, but UV-weathered and non-weathered PET behave differently due to surface changes from weathering [384]. Raksaman et al. [385] shows that hydrogen peroxide–based AOPs (particularly UV/$H_2O_2$), can effectively decolorize dye wastewater, but the presence and size of microplastics influence the process efficiency and kinetics. Ozonation, with or without hydrogen peroxide, alters the physical and chemical properties of common microplastics in water treatment, causing surface degradation, increased dissolved organic carbon, and leaching of harmful phthalic acid esters additives [386]. Still, coupling ozonation with hydrogen peroxide accelerates the aging and fragmentation of polyethylene microplastics by generating hydroxyl radicals, increasing surface oxidation and adsorption capacity, and achieving significant mass loss [387]. Adeel et al. [388] states that both UV/$H_2O_2$ and ozonation disinfection efficiencies decrease in the presence of polyethylene microplastics. Jihye [389] reveals that UV photolysis and ozonation in water treatment alters the physical and chemical properties of microplastics, with PVC showing the highest resistance and PS the lowest. Some studies looked at ozonation alone. For instance, Lee et al. [390] assessed how varying ozone concentrations influence trace organic compound

(TOrC) removal in the presence of microplastics, also microplastic type and adsorption behavior. Findings show that wet oxidation removed more microplastics than ozonation from wastewater, with wet oxidation mainly achieving removal through chemical degradation and ozonation largely through physical entrainment [391]. Table 4.3 mention other techniques and their key importance.

Generally, AOPs stood effective because they degrade even very small and stubborn particles. For instance, tire-based microplastics (60% of total microplastic aquatic pollution) can potentially be mitigated using AOPs that degrade both the particles and their leached contaminants [401]. Although AOPs are green approaches for degrading microplastics in water, their use remains largely confined to laboratory settings because of prolonged reaction times, low degradation efficiency, and the risk of generating nanoplastics [402]. That is to say, they require careful control to ensure efficiency and to avoid producing unwanted by-products. To address this, Yang et al. [403] proposed the development of efficient catalytic materials, in line with the recommendation of Zoppas et al. [404]. Combining conducting polymers with nanoparticles creates low-cost, reusable catalytic materials with unique properties that enhance AOP for efficient pollutant removal in water treatment [405]. The efficiency of photocatalytic oxidation is influenced by the choice of photocatalyst, the physicochemical properties of microplastics, and the prevailing reaction conditions [396]. Ma et al. [406] also provides a critical analysis of mechanistic pathways, degradation efficiencies and reaction kinetics. Because detecting transformed metabolites, understanding degradation mechanisms, and conducting toxicity bioassays are necessary to validate the feasibility of AOPs as remediation alternatives [407]. It was researched that electrooxidation could degrades PS microplastics in water by up to 58% within an hour, and converts them directly into gaseous products without generating smaller particles or harmful by-products [397]. In addition, using a boron-doped diamond (BDD) anode with sodium sulfate as the supporting electrolyte has demonstrated excellent performance in microplastic removal [408]. Cui et al. [409] and Zahmatkesh et al. [410] compared the strength and limitations of the three treatment methods discussed herein.

## *Emerging Solutions and Innovations*

Moraczewska-Majkut et al. [411] shows that nanofiltration and ultrafiltration membranes remained capacitated to remove pollutants leached from microplastics, but adsorption on activated carbon achieves even higher removal efficiency. Microplastics and antibiotics in water do physically and chemically interact—mainly through adsorption—forming more toxic, harder-to-remove complexes whose interactions are shaped by both pollutant properties and environmental conditions [412, 413]. It is also known that some plant's root does adsorb this size of plastics. Senavirathna et al. [414] states that microplastics were not absorbed into *Egeria densa* tissues but adhered to the roots, with 0.25 mg $L^{-1}$ identified as the threshold concentration affecting the plant. Machine learning models (random

**Table 4.3** Comparison of AOPs for microplastic removal from water

| AOP method | Mechanism of action | Key reactive species | Advantages in microplastic removal | Limitations or challenges | Emphasis |
|---|---|---|---|---|---|
| Persulfate oxidation | Activation of persulfate to generate sulfate radicals for oxidative degradation | $SO_4^{\bullet -}$ & $^{\bullet}OH$ | Strong oxidizing power; effective under wide pH range; can break down persistent polymers | Requires activation (heat, UV, metals); may form secondary pollutants | [392, 393] |
| Sonolysis | Use of ultrasonic waves to create cavitation bubbles that collapse, generating heat and radicals | $^{\bullet}OH$ & $H^{\bullet}$ | Mechanical fragmentation plus oxidation; no chemical additives needed | Low energy efficiency; limited scalability | [394] |
| Ozone treatment | Direct oxidation by ozone and indirect oxidation via hydroxyl radicals in water | $O_3$ & $^{\bullet}OH$ | Strong oxidizer; effective for both particles and organic leachates | Ozone generation cost; less effective in turbid water | [386, 389] |
| Fenton reaction | Reaction of hydrogen peroxide with ferrous ions to produce hydroxyl radicals | $^{\bullet}OH$ | Simple, low-cost, effective in acidic conditions | Requires acidic pH (~3); iron sludge generation | [383] |
| UV/solar photolysis | Direct photodegradation of polymers via high-energy photons | $^{\bullet}OH$ (indirect) & direct photolysis | Uses renewable solar energy; simple setup | Limited efficiency for certain plastics; weather-dependent for solar | [395] |
| UV light with hydrogen peroxide | UV activates $H_2O_2$ to produce hydroxyl radicals | $^{\bullet}OH$ | High oxidation potential; effective for both fragmentation and leachate degradation | Requires UV penetration; higher operational cost | [381, 385] |
| Photocatalysis | Semiconductor catalyst (e.g., $TiO_2$) activated by light generates reactive radicals | $^{\bullet}OH$ & $O_2^{\bullet -}$ | Works under UV or visible light (with doped catalysts); reusable catalyst | Catalyst recovery; reduced efficiency in colored/ turbid water | [396] |

**Table 4.3** (continued)

| AOP method | Mechanism of action | Key reactive species | Advantages in microplastic removal | Limitations or challenges | Emphasis |
|---|---|---|---|---|---|
| Electrochemical oxidation | Electric current generates ROS at electrode surfaces | $^{\bullet}OH$ & $O_2^{\bullet-}$ | No need for chemical oxidants; precise control over reaction | High energy cost; electrode fouling | [397] |
| Heat-activated persulfate | Thermal activation of persulfate releases sulfate radicals | $SO_4^{\bullet-}$ & $^{\bullet}OH$ | Effective at high temperatures; strong degradation potential | High heat requirement; not energy-efficient for large volumes | [398, 399] |
| Heat-activated Peroxymonosulfate (PMS) | Heating PMS generates sulfate and hydroxyl radicals | $SO_4^{\bullet-}$ & $^{\bullet}OH$ | Works well for persistent microplastics; dual radical generation | Similar heat energy limitations as persulfate; chemical handling required | [400] |

forest, support vector machine, and artificial neural network) can accurately predict microplastic–water partition coefficients associated with adsorption [415, 416]. Random forest regression once accurately predicted methylene blue removal by PE microplastics [417]. Five QSPR models were developed by Li et al. [418] to predict microplastic–water partition coefficients for various polymer–water systems, to rapidly estimate the adsorption capacity of organic pollutants under different water conditions. Innovatively, a pilot-scale RSF removed 98% of microplastics smaller than 10 μm from tap water through interception, trapping, entanglement [419]. As stated by Verma et al. [420] and Mulindwa et al. [421], numerous adsorbent exist for the purpose of attenuating microplastic from water. Table 4.4 enumerated few among those that has been experimented. Literature findings show that microplastics can act both as adsorbent and adsorbate.

Microplastics have large specific surface areas, which give them high adsorption capacities for toxic pollutants in water and soil [417, 439]. From 2000 to 2025, research shows common microplastics (PE, PP, PS, PVC, PET) can adsorb dyes, heavy metals [440, 441], pesticides [442], fungicide [443], hydrophobic organics [444, 445] and some pharmaceuticals [446, 447], with adsorption strengthened by aging and controlled by water chemistry [422]. Chromium can be adsorbed by PE [448] and lead and copper by PE, PP, PVC and PES [449]. Across dozens of studies, microplastics consistently act as weak-to-moderate adsorbents [450]: hydrophobic interactions and π–π stacking dominate for organic pollutants [418, 451], while electrostatic attraction and surface complexation explain much of the metal uptake [425, 452]. Environmental "aging" (UV/oxidation, biofouling) roughens surfaces

**Table 4.4** Notable studies on microplastic adsorption

| Adsorbent | Microplastic type | Wastewater type | Adsorption capacity/ efficiency | Study |
|---|---|---|---|---|
| Magnetic corncob biochar | Polyamide (PA) | Natural water | 97% | [422] |
| Graphene oxide | PS | Tap & Lake water | 617.28 mg/g | [423] |
| Quaternary ammonium salt | Modified microplastic | Natural water | 146.38 mg/g | [424] |
| Fe-modified lignin-based magnetic biochar | PS | Natural water | 68.57 mg/g | [425] |
| Magnetic sponge carbon | PS | Natural water | 369 mg/g | [426] |
| Mg/Zn modified magnetic biochars | PS | Real water | 94.81–99.46% | [427] |
| Magnetic Janus microparticles | PS & PE | Natural water | 2.72 & 2.42 mg/mg | [428] |
| Fibrous framework of chitin and cellulose | PET, PS, PP & polymethyl methacrylate | Real water | 98–99.9% | [429] |
| PEG-modified $Fe_3O_4$ magnetic nanoparticles | PE | Natural water | 2163 mg/g | [430] |
| RSF | <10 $\mu$m microplastic | Tap water | 98% | [419] |
| Polydopamine-modified sodium alginate hydrogel | Unspecified | Drinking water | 99.6% | [431] |
| Modified maifanite | PS | Natural water | 98.46% | [432] |
| Dialdehyde modified and cationic aerogel | Unspecified | Environmental waters | 90.01–97.61% | [433] |
| Graphene oxide layer | PET | Natural water | NR | [434] |
| Cellulose benzoate CNT | PS | Natural water | 97% | [435] |
| Loofah plant—Derived sponge | PS | Aqueous suspensions | 381–569 mg/g (99%) | [436] |
| Hydrophobic cattail (*Typha Latifolia*) fibres | PVC, PP, LDPE, HDPE & Nylon 6 | Industrial wastewater & Type II distilled water | 89–100% | [437] |
| Hardwood vessel-inspired chitosan-based sponge | PS | Natural water | 0.26 mg/g | [438] |

*NR* not reported, *CNT* carbon nanotube, *HDPE* high density polyethylene

and adds oxygen-containing groups, often raising capacities by an order of magnitude [453, 454]—for example, aged PE's methylene-blue uptake increased from ~0.6 to ~8 mg $g^{-1}$ in You et al. [455]. Kinetics commonly fit pseudo-second-order and isotherms fit Freundlich, indicating multilayer, heterogeneous adsorption [449, 456, 457]; salinity, competing ions, and dissolved organics can suppress or reshape uptake [445], and desorption is frequently substantial [458, 459], meaning microplastics can serve as mobile vectors rather than permanent sinks [460, 461]. Beyond unmodified microplastics, several papers convert waste plastics to carbonized/activated materials that remove dyes more effectively, hinting at upcycling routes for treatment media [462–464]. Evidence since 2000 supports that microplastics do adsorb diverse contaminants, but with highly variable, condition-dependent capacities and significant reversibility [465, 466].

## Sampling, Monitoring and Detection Techniques

By definition, sampling of microplastics in water is the process of collecting water samples to check for the presence of tiny plastic particles [467, 468]. Different methods are used depending on the location and purpose of the study [469–472]. In the laboratory, density separation, filtration, sieving, visual sorting, and digestion methods are used to separate microplastics from seawater [473]. In surface water sampling, nets (e.g., neuston, plankton, bongo or manta trawls) are dragged across the water surface to collect floating microplastics. The manta trawl and plankton net shown in Fig. 4.3 are among the most employed sampling methods [474–476]. MuMi sampler comes to actually improve the manta trawl, according to Montoto-Martínez et al. [477]. Grab sampling is an alternative method where bottles or containers are used to directly collect water from water bodies. Bottle grab sampling method once revealed the presence of microplastic in Ascension Island and the Falkland Islands coastal waters [478]. For deeper waters, specialized pumps [479] or sediment traps may be used to capture microplastics suspended below the surface. In Shi et al. [480], shallow-water plankton pump (SPP) and deep-water plankton pump (DPP) yielded similar microplastic abundances to the manta trawl, but fibers dominated (>70%), whereas the submersible pump with on-site filtration (50 μm mesh) captured far higher abundances. Zheng et al. [481] found that the efficiency of membrane filtration with a 0.45 μm pore size is very high compared to sieve pre-concentration and trawling. In-line filtration for microplastic sampling in drinking water is a method where water passes directly through fine filters installed within the supply line to capture microplastics without needing separate collection containers [482–484].

After collection, the water samples are usually filtered to separate microplastics from other materials. Pittroff et al. [485] presents a novel, validated workflow for microplastic analysis in water using a portable filter cascade unit (100–20–5 μm) that enables sampling of several cubic meters down to 5 μm. Covernton et al. [486], Kamil et al. [487] and Becucci et al. [488] reported that with greater mesh size, microplastic concentrations decline exponentially. Some separation methods include sodium chloride solution, elutriation column and sodium tungstate dihydrate solution [489]. The particles are then dried, weighed, and examined under a microscope or analyzed using instruments to identify their size, shape, and type of plastic [490]. Pretreatment is also discussed by Sun et al. [491]. Care must be taken to prevent contamination during sampling and analysis. Continuous flow centrifugation is an efficient and reliable sampling configuration for capturing the full load of suspended microplastics without causing contamination, filter clogging, or particle size discrimination [492]. After isolating microplastic from riverine areas, qualitative analyses approaches are necessary to determine the chemical composition of the inherent particles [493]. Erhunmwunse et al. [494] discusses several analytical techniques—such as FTIR, SEM, Raman spectroscopy, and near-infrared spectroscopy—as effective methods for identifying and quantifying microplastics in water. Other techniques are pyrolysis, microscopies and thermal desorption gas chromatography, imaging [495]. UV irradiation was once applied by Kumar [496], where dialysis tubing emerged as the most effective wastewater filtration method, cutting microplastic levels by around 50%. So far, no standardized methods exist for detecting microplastics in environmental samples; hence, it is difficult to compare reported data across studies [483, 497–499]. In view of that, Pasquier et al. [500] foretold how Jellyfishbot® (an aquatic drone) could be a new standardized way to sample microplastic from waterbodies. Bruge et al. [501] shows that uncertainty in microplastic estimation decreases with more replicate samples, and collecting triplicates provides a reasonable balance between accuracy and time efficiency for river monitoring. Standardizing sampling, sample preparation, and detection methods is essential to generate comparable data on microplastics in the environment [502]. As such, Briggs et al. [503] documents for the first time the sampling and extraction of

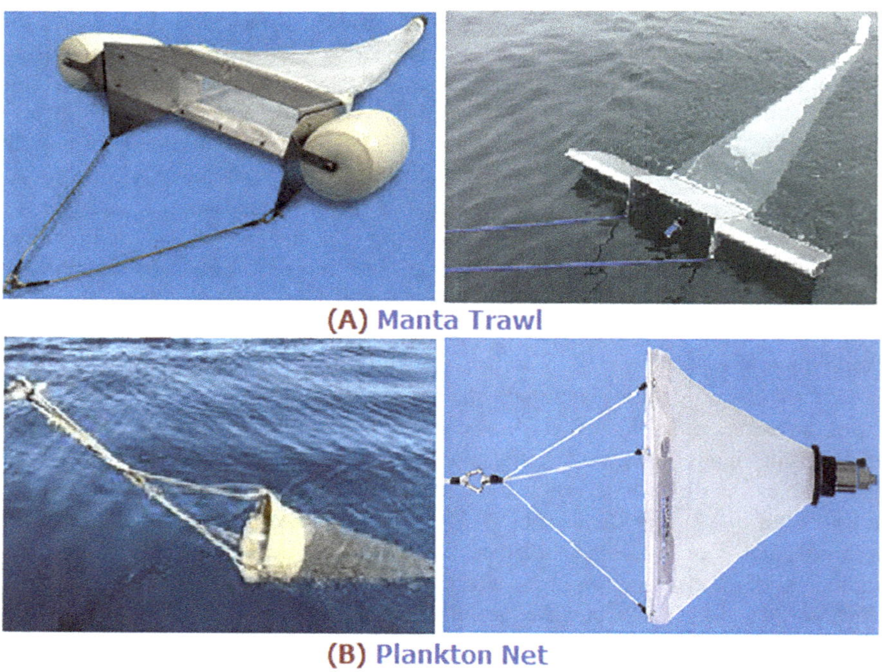


**Fig. 4.3** Manta trawl and small plankton net in practical use

LDPE microplastics in Nigeria, where they present an optimized protocol for surface water. Similarly, Zobkov et al. [504] describe a novel sampling method that uses a new tool capable of collecting several cubic meters of water from specific depths down to 100 m.

## Policy, Regulation, and Public Awareness

Synthetic textiles are a major source of persistent microplastic fibre pollution, and addressing it requires coordinated policies and technological innovation [505]. Strong policies and clear regulations usually set limits on microplastic discharge, require industries to adopt cleaner production methods, and mandate the use of effective removal technologies in WWTPs [506]. Microbeads can escape WWTPs [507]. Formulating laws will encourage innovation in recycling and product/material design to reduce plastic pollution at its source [508]. Public awareness is equally important, as it helps people understand the dangers of microplastics to human health and the environment [509]. Through education campaigns, community programs, and media outreach, individuals are encouraged to reduce plastic use, properly dispose of waste, and support policies aimed at cleaner water [510]. Public perceptions and preferences influence marine microplastic governance, with most people in the Jiaozhou Bay region favoring monitoring and recycling over plastic-use restrictions [511]. Amanda et al. [512] indicates that fewer than 25% of respondents are willing to pay full price for high-efficiency washing machine filters that could preemptively prevent a category of microplastics from entering waterways. Convincingly, effective governance and informed citizens create a powerful approach to tackling microplastic pollution. Garcia-Vazquez & Garcia-Ael [513] and Deme et al. [514] calls for more research in Africa, and recommends better communication and education to drive pro-environmental action. A survey around Lagos Lagoon found over half of respondents were aware of microplastics but had limited knowledge and poor perception of the issue, though they showed willingness to learn and support clean-up efforts [515]. Between 2015 and 2017, several developed nations enacted bans on microbeads in cosmetics or personal care products (PCP) to curb waterway pollution, though differences in restrictions, timelines, and loopholes existed [496, 516–519]. Malaysians are generally aware of plastic waste and support microplastic reduction policies, but limited knowledge, poor waste management, and unclear government actions highlight the need for stronger public education and awareness efforts [520]. Jani et al. [521] reviews microplastic removal methods in wastewater and drinking water treatment plants and outlines existing U.S. regulations, aiming to inform future research and policy for better microplastic management. One of such is the United States Microbead Free Waters Act (MFWA) of 2015 [519]. Other federal statutes formulated in the country includes the Clean Water Act, Safe Drinking Water Act, Toxic Substances Control Act, Resource Conservation and Recovery Act, Clean Air Act, and the Federal Food, Drug, and Cosmetic Act, alongside various state-level initiatives [522]. In

Kerala, India, rapid urbanization, poor waste management, and monsoon runoff have intensified microplastic pollution in water, soil, and marine life [523]. The country later introduced the 2016 Plastic Waste Management Rules and the 2022 single-use plastic ban policies to curb microplastic pollution, but gaps in infrastructure, enforcement, and specific legislation hinder full effectiveness [524]. In China, rapid urbanization has intensified microplastic pollution in freshwater systems, but limited research, non-standardized sampling, and underdeveloped legislation hinder effective management and regulation [525]. Policy implications of microplastic accumulation in the River Thames in UK, embodied some strategies to address their ecological impacts [6]. The MICROPLASMA (MIcro and maCRO PLAStic pollution Monitoring with Advanced technologies) research project, funded by the Apulia region, aims to develop an integrated, innovative monitoring system for an urban river in southern Italy [526]. Munhoz et al. [527] compiles and analyzes global policies and strategies for tackling microplastic pollution, and emphasize that prevention and multifaceted approaches are key, while downstream measures serve as supportive solutions. Namely, the Global Partnership on Marine Litter, United Nations Environment Programme's (UNEP's), International Initiative on Water Quality of UNESCO's Intergovernmental Hydrological Programme (IHP) and the United Nations Environment Assembly have been established to promote coordinated efforts, capacity building, and knowledge sharing in tackling microplastic pollution [528–530]. This is because, plastic pollution transcends national and regional boundaries, making it a challenge no single country can tackle independently [531, 532]. Despite numerous policies to combat plastic pollution, gaps in regulation, implementation, and research—especially on microplastics in food and water—demand stronger, coordinated global and national action [533].

## Circular Economy

It is believed that material flow analysis is a tool to track microplastic movement and identify leakage points—but data gaps and degradation challenges limit its effectiveness [534]. Thus, circular economy encourages the development of technologies and strategies that not only extract microplastics but also find ways to repurpose or recycle them into useful products [535]. Instead of simply disposing of the collected microplastics, it is processed into raw materials for manufacturing, to eliminate the need for virgin plastics. In view of that, Sadia et al. [536] explains how WWTP release microplastics and reviewed sustainable removal methods that are capable of recovering and converting them into renewable energy. Moreso, Rai et al. [537] describes how agricultural waste, specifically rice straws, gets transformed into biodegradable, microplastic-free drinking straws. Ornamental plant-based wetlands can genuinely treat domestic wastewater by removing contaminants and retaining microplastics, while enabling safe water reuse for crop irrigation [538]. Removing microplastics from wastewater is technically feasible and cost-effective, with membrane bioreactors plus sludge incineration being the most efficient option

[539]. There is even a pilot-scale, two-stage treatment that removes over 98% of microplastics and 94% of COD from industrial wastewater, proposed by Sturm et al. [540]. Life cycle assessment (LCA) is another key tool for managing microplastics and reducing their environmental impact within a circular economy [541]. Those approaches will prevent further plastic pollution, conserves resources, and promotes sustainable water management, and lastly ensures a cleaner water systems and a healthier environment. As recommended solutions to microplastic menace, Sadeghi et al. [542] proposed few complementary mechanisms that include financial incentives, equipment and product modification, on-site recycling, economic justification, long-term policies, and the reuse of plastic waste. Lately, various laws have been enacted to limit plastic production, usage, and disposal, but their potential to meaningfully reduce environmental microplastic pollution remains uncertain [543, 544]. Pursuant to that, Antohi et al. [545] develops a new policy-focused approach, supported by statistical and econometric analysis, to address microplastic water pollution within the framework of the circular economy in Europe. Circular economy models, including AI-based applications, is embodied with the potential to boost efficiency, create revenue, and reduce the carbon footprint in water and wastewater sectors [546].

## Conclusion

This review looked at up to 546 published works on microplastics. Evidently, research in this area is plenty and increasing. Microplastics have become an emerging pollutant of concern in water management, with widespread occurrence in both surface and groundwater. They originate from multiple sources; industrial, domestic, primary, and secondary, and persist in the environment due to their durability and small size. WWTPs reduce but do not completely remove microplastics. In fact, the effluents act as pathways into rivers and lakes. Microplastic impacts extend from aquatic ecosystems to human health, where ingestion and bioaccumulation pose risks. Advanced treatment methods such as membrane filtration, coagulation-flocculation, and advanced oxidation processes show promise for effective removal, though challenges of cost, energy demand, fouling, and standardization remain. This work highlights the urgent need for harmonized protocols in sampling, preparation, and detection to improve comparability of data and to guide management practices. All removal techniques should be examined for a particular waterbody in a location to find out the most suitable, due to the type and sizes present in them. Because, the current literature is dominated by concentrations found in Chinese waters. Old and new WWTPs design must address their inability to trap microplastics. It is possible to see that waterbodies that have been cleaned from microplastic in the past, record more or additional concentrations in subsequent analysis. Hence, preventing microplastics from getting into surface and groundwaters could be an effective measure to take. Perhaps, adsorption now opens the door for laboratory scale removal assessment of microplastics, by also looking at the kinetics, isotherm and thermodynamic studies involved.

## References


1. Venghaus, D., & Barjenbruch, M. (2017). Microplastics in urban water management. *Technical Transactions, 1*(114), 137–146. https://doi.org/10.4467/2353737XCT.17.011.6108
2. Fältström, E., & Anderberg, S. (2020). Towards control strategies for microplastics in urban water. *Environmental Science and Pollution Research, 27*, 40421–40433. https://doi.org/10.1007/s11356-020-10064-z
3. Kim, J., Wager, Y. Z., Miller, C., & Norton, J. (2022). Occurrence and removal of microplastics in drinking water systems (ch. 2). In B.-J. Ni, Q. Xu, & W. Wei (Eds.), *Microplastics in urban water management* (1st ed., pp. 53–89). Wiley. https://doi.org/10.1002/9781119759379
4. Ou, H., Liu, R., Liao, Z., & Zeng, E. Y. (2024). Occurrence and fate of microplastics in urban water management systems (ch. 7). In E. Y. Zeng (Ed.), *Microplastic contamination in aquatic environments—An emerging matter of environmental urgency* (2nd ed., pp. 181–228). Elsevier. https://doi.org/10.1016/B978-0-443-15332-7.00006-5
5. Mansouri, F., Chouchene, K., Roche, N., & Ksibi, M. (2021). Removal of pharmaceuticals from water by adsorption and advanced oxidation processes: State of the art and trends. *Applied Sciences, 11*(14). https://doi.org/10.3390/app11146659
6. Whitehead, P. G., et al. (2021). Modelling microplastics in the River Thames: Sources, sinks and policy implications. *Water (Basel), 13*(6). https://doi.org/10.3390/w13060861
7. Ren, S.-Y., Kong, S.-F., & Ni, H.-G. (2021). Contribution of mulch film to microplastics in agricultural soil and surface water in China. *Environmental Pollution, 291*, 118227. https://doi.org/10.1016/j.envpol.2021.118227
8. Alam, F. C., Sembiring, E., Muntalif, B. S., & Suendo, V. (2019). Microplastic distribution in surface water and sediment river around slum and industrial area (case study: Ciwalengke River, Majalaya district, Indonesia). *Chemosphere, 224*, 637–645. https://doi.org/10.1016/j.chemosphere.2019.02.188
9. Shafi, M., et al. (2024). Are we underestimating stormwater? Stormwater as a significant source of microplastics in surface waters. *Journal of Hazardous Materials, 465*, 133445. https://doi.org/10.1016/j.jhazmat.2024.133445
10. Wei, H., et al. (2023). Are wastewater treatment plants as the source of microplastics in surface water and soil? *Journal of Hazardous Materials, 459*, 132154. https://doi.org/10.1016/j.jhazmat.2023.132154
11. Nocoń, W., Moraczewska-Majkut, K., & Wiśniowska, E. (2018). Microplastics in surface water under strong anthropopression. *Desalination and Water Treatment, 134*, 174–181. https://doi.org/10.5004/dwt.2018.22833
12. McCormick, A. R., Hoellein, T. J., London, M. G., Hittie, J., Scott, J. W., & Kelly, J. J. (2016). Microplastic in surface waters of urban rivers: Concentration, sources, and associated bacterial assemblages. *Ecosphere, 7*(11). https://doi.org/10.1002/ecs2.1556
13. Estahbanati, S., & Fahrenfeld, N. L. (2016). Influence of wastewater treatment plant discharges on microplastic concentrations in surface water. *Chemosphere, 162*, 277–284. https://doi.org/10.1016/j.chemosphere.2016.07.083
14. Lusher, A. L., Tirelli, V., O'Connor, I., & Officer, R. (2015). Microplastics in Arctic polar waters: The first reported values of particles in surface and sub-surface samples. *Scientific Reports, 5*, 14947. https://doi.org/10.1038/srep14947
15. Niu, L., Wang, Y., Li, Y., Lin, L., Chen, Y., & Shen, J. (2023). Occurrence, degradation pathways, and potential synergistic degradation mechanism of microplastics in surface water: A review. *Current Pollution Reports, 9*, 312–326. https://doi.org/10.1007/s40726-023-00262-x

16. Tuttle, E., Wiman, C., Muñoz, S., Law, K. L., & Stubbins, A. (2024). Sunlight-driven photochemical removal of polypropylene microplastics from surface waters follows linear kinetics and does not result in fragmentation. *Environmental Science & Technology, 58*(12), 5461–5471. https://doi.org/10.1021/acs.est.3c07161
17. Saipolbahri, N., Bitlus, M. L. A., Ismail, N. A., Fauzi, N. M., & Subki, N. S. (2020). Determination of microplastics in surface water and sediment of Kelantan Bay. In *2nd International Conference on Tropical Resources and Sustainable Sciences 10–11 August 2020, Universiti Malaysia Kelantan, City Campus, Malaysia* (pp. 1–6). IOP Publishing Ltd. https://doi.org/10.1088/1755-1315/549/1/012059
18. dos Santos, C. R., D. A, G. P., Moreira, V. R., de S. Santos, L. V., & Amaral, M. C. S. (2023). Microplastics in surface water: Occurrence, ecological implications, quantification methods and remediation technologies. *Chemical Engineering Journal, 474*, 144936. https://doi.org/10.1016/j.cej.2023.144936
19. Suteja, Y., Atmadipoera, A. S., Riani, E., Nurjaya, W., Nugroho, D., & Cordova, M. R. (2021). Spatial and temporal distribution of microplastic in surface water of tropical estuary: Case study in Benoa Bay, Bali, Indonesia. *Marine Pollution Bulletin, 163*, 111979. https://doi.org/10.1016/j.marpolbul.2021.111979
20. Li, K., Zhao, R., & Meng, X. (2025). Spatio-temporal distribution of microplastics in surface water of typical urban rivers in North China, risk assessment and influencing factors. *Journal of Contaminant Hydrology, 273*, 104626. https://doi.org/10.1016/j.jconhyd.2025.104626
21. Yin, L., et al. (2019). Microplastic pollution in surface water of urban lakes in Changsha, China. *International Journal of Environmental Research and Public Health, 16*(9). https://doi.org/10.3390/ijerph16091650
22. Felicitas, G., et al. (2025). Microplastics in surface water and gastrointestinal tracts of demersal fishes (Oreochromis niloticus and Cyprinus carpio) in the largest wetland of the Philippines. *International Journal of Environmental Research, 19*, 96. https://doi.org/10.1007/s41742-025-00768-w
23. Sánchez-Paz, A., Encinas-García, T., Ávila-Félix, L., & Mendoza-Cano, F. (2025). Microplastics in surface waters surrounding a touristic beach at the Gulf of California, Mexico. *International Journal of Environmental Health Research*. https://doi.org/10.1080/09603123.2025.2483356
24. Nan, B., Su, L., Kellar, C., Craig, N. J., Keough, M. J., & Pettigrove, V. (2020). Identification of microplastics in surface water and Australian freshwater shrimp Paratya australiensis in Victoria, Australia. *Environmental Pollution, 259*, 113865. https://doi.org/10.1016/j.envpol.2019.113865
25. Vitheepradit, A., & Prommi, T.-O. (2023). Microplastics in surface water and tissue of white leg shrimp, Litopenaeus vannamei, in a cultured pond in Nakhon Pathom Province, Central Thailand. *AIMS Environmental Science, 10*(4), 478–503. https://doi.org/10.3934/environsci.2023027
26. Park, T.-J., Lee, S.-H., Lee, M.-S., Lee, J.-K., Park, J.-H., & Zoh, K.-D. (2020). Distributions of microplastics in surface water, fish, and sediment in the vicinity of a sewage treatment plant. *Water (Basel), 12*(12). https://doi.org/10.3390/w12123333
27. Onyena, A. P., Tekeme, M. E., Uwakwe, J. C., Aderibigbe, D., & Sam, K. (2025). Baseline characterisation of microplastics in surface water, sediment, and seafood from the Escravos Estuary, Nigeria. *Scientific African, 27*, e02591, 1–17. https://doi.org/10.1016/j.sciaf.2025.e02591
28. Mehboob, S., Anjum, K. M., Azmat, H., & Imran, M. (2025). The measurement of microplastics in surface water and their impact on histopathological structures in wading birds of district Lahore. *Frontiers in Toxicology, 6*, 1–15. https://doi.org/10.3389/ftox.2024.1484724
29. Li, Y., et al. (2024). Occurrence and distribution characteristics of microplastics in surface water of China: A review. *Water, Air, and Soil Pollution, 235*, 21. https://doi.org/10.1007/s11270-023-06850-3

30. Fan, J., Zou, L., Duan, T., Qin, L., Qi, Z., & Sun, J. (2022). Occurrence and distribution of microplastics in surface water and sediments in China's inland water systems: A critical review. *Journal of Cleaner Production, 331*, 129968. https://doi.org/10.1016/j.jclepro.2021.129968
31. Al Nahian, S., et al. (2022). Occurrence, spatial distribution, and risk assessment of microplastics in surface water and sediments of Saint Martin Island in the Bay of Bengal. *Marine Pollution Bulletin, 179*, 113720. https://doi.org/10.1016/j.marpolbul.2022.113720
32. Bappy, M. M., et al. (2025). Contamination, potential sources, and risk assessment of microplastics in surface waters of two public bathing beaches along the Northern Bay of Bengal. *Regional Studies in Marine Science, 83*, 104067, 1–11. https://doi.org/10.1016/j.rsma.2025.104067
33. Al Nahian, S., Rakib, R. J., Kumar, R., Haider, S. M. B., Sharma, P., & Idris, A. M. (2023). Distribution, characteristics, and risk assessments analysis of microplastics in shore sediments and surface water of Moheshkhali channel of Bay of Bengal, Bangladesh. *Science of the Total Environment, 855*, 158892. https://doi.org/10.1016/j.scitotenv.2022.158892
34. Islam, M. S., Islam, Z., & Hasan, R. (2022). Pervasiveness and characteristics of microplastics in surface water and sediment of the Buriganga River, Bangladesh. *Chemosphere, 307*, 135945. https://doi.org/10.1016/j.chemosphere.2022.135945
35. Hossain, M. B., et al. (2023). Microplastics in surface water from a mighty subtropical estuary: First observations on occurrence, characterization, and contamination assessment. *Environmental Research Letters, 226*, 115594, 1–9. https://doi.org/10.1016/j.envres.2023.115594
36. Hossain, S., Siddika, A., Saifullah, A. S. M., Sheikh, S., & Uddin, M. J. (2022). Abundance and distribution of microplastics in surface water and sediment of two selected rivers in Bangladesh. *Environmental Engineering and Management Journal, 21*(6), 1047–1058. http://www.eemj.icpm.tuiasi.ro/
37. Riya, K. K., et al. (2024). Characteristics, contamination levels, and ecosystem risk assessment of microplastics in surface water of a highly urbanized river from a developing country. *ACS Omega, 9*(52), 50922–50932. https://doi.org/10.1021/acsomega.4c01528
38. Rahman, H., et al. (2025). Occurrence and risk assessment of microplastics in surface water, sediment, and biota of Surma River, Bangladesh. *Journal of Contaminant Hydrology, 273*, 104620. https://doi.org/10.1016/j.jconhyd.2025.104620
39. Hossain, J., et al. (2022). Surface water, sediment, and biota: The first multi-compartment analysis of microplastics in the Karnafully river, Bangladesh. *Marine Pollution Bulletin, 180*, 113820. https://doi.org/10.1016/j.marpolbul.2022.113820
40. Fardullah, M., et al. (2024). Occurrence and spatial distribution of microplastics in water and sediments of Hatiya Island, Bangladesh and their risk assessment. *Journal of Environmental Management, 370*, 122697. https://doi.org/10.1016/j.jenvman.2024.122697
41. Trindade, L. d. S., Gloaguen, T. V., Benevides, T. d. S. F., Valentim, A. C. S., Bomfim, M. R., & Santos, J. A. G. (2023). Microplastics in surface waters of tropical estuaries around a densely populated Brazilian bay. *Environmental Pollution, 323*, 121224. https://doi.org/10.1016/j.envpol.2023.121224
42. Castro, R. O., da Silva, M. L., Marques, M. R. C., & de Araújo, F. V. (2020). Spatio-temporal evaluation of macro, meso and microplastics in surface waters, bottom and beach sediments of two embayments in Niterói, RJ, Brazil. *Marine Pollution Bulletin, 160*, 111537. https://doi.org/10.1016/j.marpolbul.2020.111537
43. Kelly, N. E. (2024). Spatial distribution and risk assessment of microplastics in surface waters of the St. Lawrence estuary. *Science of the Total Environment, 946*, 174324, 1–14. https://doi.org/10.1016/j.scitotenv.2024.174324
44. Minor, E. C., Lin, R., Burrows, A., Cooney, E. M., Grosshuesch, S., & Lafrancois, B. (2020). An analysis of microlitter and microplastics from Lake Superior beach sand and surface-water. *Science of the Total Environment, 744*, 140824. https://doi.org/10.1016/j.scitotenv.2020.140824

45. Han, M., et al. (2020). Distribution of microplastics in surface water of the lower Yellow River near estuary. *Science of the Total Environment, 707*, 135601. https://doi.org/10.1016/j.scitotenv.2019.135601
46. Liu, Z., et al. (2024). Contributions from typical sources to microplastics in surface water of a semiarid urban river. *Journal of Hazardous Materials, 478*, 135570. https://doi.org/10.1016/j.jhazmat.2024.135570
47. Gao, Y., et al. (2022). Abundance, composition, and potential ecological risks of microplastics in surface water at different seasons in the Pearl River Delta, China. *Water, 14*(16), 1–17. https://doi.org/10.3390/w14162545
48. Lin, L., et al. (2015). Occurrence and distribution of microplastics in an urban river: A case study in the Pearl River along Guangzhou City, China. *Archives of Environmental Contamination and Toxicology, 69*, 269–278. https://doi.org/10.1016/j.scitotenv.2018.06.327
49. Zhang, K., Gong, W., Lv, J., Xiong, X., & Wu, C. (2015). Accumulation of floating microplastics behind the Three Gorges Dam. *Environmental Pollution, 204*, 117–123. https://doi.org/10.1016/j.envpol.2015.04.023
50. He, Y., et al. (2023). Seasonal pollution and surface characteristics of microplastics in surface water in the Wanzhou section of the Three Gorges Reservoir, China. *Environmental Science and Pollution Research, 30*, 67854–67864. https://doi.org/10.1007/s11356-023-27185-w
51. Di, M., & Wang, J. (2018). Microplastics in surface waters and sediments of the Three Gorges Reservoir, China. *Science of the Total Environment, 616–617*, 1620–1627. https://doi.org/10.1016/j.scitotenv.2017.10.150
52. Fan, J., Zou, L., & Zhao, G. (2021). Microplastic abundance, distribution, and composition in the surface water and sediments of the Yangtze River along Chongqing City, China. *Journal of Soils and Sediments, 21*, 1840–1851. https://doi.org/10.1007/s11368-021-02902-5
53. He, D., et al. (2021). Microplastics contamination in the surface water of the Yangtze River from upstream to estuary based on different sampling methods. *Environmental Research Letters, 196*, 110908. https://doi.org/10.1016/j.envres.2021.110908
54. Huang, D., et al. (2021). The occurrence and abundance of microplastics in surface water and sediment of the West River downstream, in the south of China. *Science of the Total Environment, 756*, 143857. https://doi.org/10.1016/j.scitotenv.2020.143857
55. Ding, L., Mao, R. F., Guo, X., Yang, X., Zhang, Q., & Yang, C. (2019). Microplastics in surface waters and sediments of the Wei River, in the northwest of China. *Science of the Total Environment, 667*, 427–434. https://doi.org/10.1016/j.scitotenv.2019.02.332
56. Jiao, J., Chen, G., Yang, Z., Li, Z., & Hu, H. (2022). Microplastics in surface waters and floodplain sediments of the Dagu River in the Jiaodong Peninsula, China. *Journal of Ocean University of China, 21*, 1538–1548. https://doi.org/10.1007/s11802-022-5211-z
57. Zhang, J., et al. (2019). Microplastics in the surface water of small-scale estuaries in Shanghai. *Marine Pollution Bulletin, 149*, 110569. https://doi.org/10.1016/j.marpolbul.2019.110569
58. Li, Y., Lu, Z., Zheng, H., Wang, J., & Chen, C. (2020). Microplastics in surface water and sediments of Chongming Island in the Yangtze Estuary, China. *Environmental Sciences Europe, 32*, 15. https://doi.org/10.1186/s12302-020-0297-7
59. Yang, H., et al. (2023). Distribution, abundance, and risks posed by microplastics in surface waters of the Yangtze River Basin, China. *Environmental Pollution, 333*, 122086. https://doi.org/10.1016/j.envpol.2023.122086
60. Luo, Q., Gao, J., & Gao, B. (2025). Microplastics in surface water in the Yangtze River, China: Basin-Wide observation, multiple ecological risk assessment and sustainability. *Sustainability, 17*(9). https://doi.org/10.3390/su17094162
61. Wang, W., Ndungu, A. W., Li, Z., & Wang, J. (2017). Microplastics pollution in inland freshwaters of China: A case study in urban surface waters of Wuhan, China. *Science of the Total Environment, 575*, 1369–1374. https://doi.org/10.1016/j.scitotenv.2016.09.213
62. Ji, X., Yan, S., He, Y., He, H., & Liu, H. (2023). Distribution characteristics of microplastics in surface seawater off the Yangtze River Estuary section and analysis of ecological risk assessment. *Toxics, 11*(11). https://doi.org/10.3390/toxics11110889

63. Yu, F., Pei, Y., Zhang, X., Wu, X., Zhang, G., & Ma, J. (2023). Occurrence and distribution characteristics of aged microplastics in the surface water, sediment, and crabs of the aquaculture pond in the Yangtze River Delta of China. *Science of the Total Environment, 871*, 162039. https://doi.org/10.1016/j.scitotenv.2023.162039
64. Zhao, S., Zhu, L., Wang, T., & Li, D. (2014). Suspended microplastics in the surface water of the Yangtze Estuary System, China: First observations on occurrence, distribution. *Marine Pollution Bulletin, 86*(1–2), 562–568. https://doi.org/10.1016/j.marpolbul.2014.06.032
65. Wang, W., Yuan, W., Chen, Y., & Wang, J. (2018). Microplastics in surface waters of Dongting Lake and Hong Lake, China. *Science of the Total Environment, 633*, 539–545. https://doi.org/10.1016/j.scitotenv.2018.03.211
66. Jiang, C., et al. (2018). Microplastics in sediment and surface water of West Dongting Lake and South Dongting Lake: Abundance, source and composition. *International Journal of Environmental Research and Public Health, 15*(10). https://doi.org/10.3390/ijerph15102164
67. Zhang, L., et al. (2020). Distribution of microplastics in surface water and sediments of Qin river in Beibu Gulf, China. *Science of the Total Environment, 708*, 135176. https://doi.org/10.1016/j.scitotenv.2019.135176
68. Wu, L.-S., et al. (2021). Distribution characteristics of microplastics in surface water of Beibu Gulf. *Research of Environmental Sciences, 34*(11), 2556–2562. https://doi.org/10.13198/j.issn.1001-6929.2021.08.13
69. Wang, G., Lu, J., Tong, Y., Liu, Z., Zhou, H., & Xiayihazi, N. (2020). Occurrence and pollution characteristics of microplastics in surface water of the Manas River Basin, China. *Science of the Total Environment, 710*, 136099. https://doi.org/10.1016/j.scitotenv.2019.136099
70. Wang, G., et al. (2021). Seasonal variation and risk assessment of microplastics in surface water of the Manas River Basin, China. *Ecotoxicology and Environmental Safety, 208*, 111477, 1–9. https://doi.org/10.1016/j.ecoenv.2020.111477
71. Li, Y., et al. (2021). Microplastics in surface waters and sediments from Guangdong Coastal Areas, South China. *Sustainability, 13*(5). https://doi.org/10.3390/su13052691
72. Wu, P., et al. (2020). Spatial-temporal distribution of microplastics in surface water and sediments of Maozhou River within Guangdong-Hong Kong-Macao Greater Bay Area. *Science of the Total Environment, 717*, 135187. https://doi.org/10.1016/j.scitotenv.2019.135187
73. Zhang, C., et al. (2020). Microplastic pollution in surface water from east coastal areas of Guangdong, South China and preliminary study on microplastics biomonitoring using two marine fish. *Chemosphere, 256*, 127202. https://doi.org/10.1016/j.chemosphere.2020.127202
74. Li, Z., et al. (2020). Distribution characteristics of microplastics in surface water and sediments of Haizhou Bay, Lianyungang. *Huan Jing Ke Xue, 41*(7), 3212–3221. https://doi.org/10.13227/j.hjkx.201910005
75. Zhao, W., et al. (2024). Seasonal variations of microplastics in surface water and sediment in an inland river drinking water source in southern China. *Science of the Total Environment, 908*, 168241. https://doi.org/10.1016/j.scitotenv.2023.168241
76. Zhang, W., et al. (2017). Microplastic pollution in the surface waters of the Bohai Sea, China. *Environmental Pollution, 231*, 541–548. https://doi.org/10.1016/j.envpol.2017.08.058
77. Mai, L., Bao, L.-J., Shi, L., Liu, L.-Y., & Zeng, E. Y. (2018). Polycyclic aromatic hydrocarbons affiliated with microplastics in surface waters of Bohai and Huanghai Seas, China. *Environmental Pollution, 241*, 834–840. https://doi.org/10.1016/j.envpol.2018.06.012
78. Wu, N., et al. (2019). Occurrence and distribution of microplastics in the surface water and sediment of two typical estuaries in Bohai Bay, China. *Environmental Science. Processes & Impacts, 21*(7), 1143–1152. https://doi.org/10.1039/C9EM00148D
79. Xu, P., Peng, G., Su, L., Gao, Y., Gao, L., & Li, D. (2018). Microplastic risk assessment in surface waters: A case study in the Changjiang Estuary, China. *Marine Pollution Bulletin, 133*, 647–654. https://doi.org/10.1016/j.marpolbul.2018.06.020
80. Ding, R., et al. (2022). A case study of distribution and characteristics of microplastics in surface water and sediments of the seas around Shenzhen, southern coastal area of China. *Science of the Total Environment, 838*, 156063. https://doi.org/10.1016/j.scitotenv.2022.156063

81. Huang, S., et al. (2021). Spatiotemporal distribution of microplastics in surface water, biofilms, and sediments in the world's largest drinking water diversion project. *Science of the Total Environment, 789*, 148001. https://doi.org/10.1016/j.scitotenv.2021.148001
82. Huang, Y., et al. (2020). Coupled effects of urbanization level and dam on microplastics in surface waters in a coastal watershed of Southeast China. *Marine Pollution Bulletin, 154*, 111089. https://doi.org/10.1016/j.marpolbul.2020.111089
83. Liu, S., et al. (2024). Temporal and spatial variation of microplastics in the urban rivers of Harbin. *Science of the Total Environment, 910*, 168373. https://doi.org/10.1016/j.scitotenv.2023.168373
84. Huo, C., et al. (2025). Distribution characteristics of microplastics in surface waters in the western Pacific Ocean. *Journal of Oceanology and Limnology, 43*, 492–501. https://doi.org/10.1007/s00343-025-3288-0
85. Cui, Y., et al. (2022). Microplastics in the surface waters of the South China sea and the western Pacific Ocean: Different size classes reflecting various sources and transport. *Chemosphere, 299*, 134456. https://doi.org/10.1016/j.chemosphere.2022.134456
86. Mao, R., Hu, Y., Zhang, S., Wu, R., & Guo, X. (2020). Microplastics in the surface water of Wuliangsuhai Lake, northern China. *Science of the Total Environment, 723*, 137820. https://doi.org/10.1016/j.scitotenv.2020.137820
87. Wen, S., Yu, C., Lin, F., & Diao, X. (2022). Comparative assessment of microplastics in surface water and sediments of Meishe River, Haikou, China. *Sustainability, 14*(20). https://doi.org/10.3390/su142013099
88. Fan, M. Y., Huang, Y. M., Zhang, H. X., Li, H. H., & Huang, Q. (2022). Distribution, risk, and influencing factors of microplastics in surface water of Huangshui River Basin. *Huan Jing Ke Xue, 43*(10), 4430–4439. https://doi.org/10.13227/j.hjkx.202201087
89. Tan, X., Yu, X., Cai, L., Wang, J., & Peng, J. (2019). Microplastics and associated PAHs in surface water from the Feilaixia Reservoir in the Beijiang River, China. *Chemosphere, 221*, 834–840. https://doi.org/10.1016/j.chemosphere.2019.01.022
90. Fang, Z.-J., et al. (2025). Occurrence characteristics and influencing factors of microplastics in surface water, sediments, and benthic macroinvertebrates in Huangbai River. *Huan Jing Ke Xue, 46*(2), 1181–1192. https://doi.org/10.13227/j.hjkx.202401168
91. Lam, T. W. L., Ho, H. T., Ma, A. T. H., & Fok, L. (2020). Microplastic contamination of surface water-sourced tap water in Hong Kong—A preliminary study. *Applied Sciences, 10*(10). https://doi.org/10.3390/app10103463
92. Yu, X., et al. (2022). Abundance and characteristics of microplastics in the surface water and sediment of parks in Xi'an city, Northwest China. *Science of the Total Environment, 806*, 150953. https://doi.org/10.1016/j.scitotenv.2021.150953
93. Rojas-Luna, R. A., Oquendo-Ruiz, L., García-Alzate, C. A., Arana, V. A., García-Alzate, R., & Trilleras, J. (2023). Identification, abundance, and distribution of microplastics in surface water collected from Luruaco Lake, Low Basin Magdalena River, Colombia. *Water (Basel), 15*(2). https://doi.org/10.3390/w15020344
94. Garcés-Ordóñez, O., Espinosa, L. F., Muniz, M. C., Pereira, L. B. S., & dos Anjos, R. M. (2021). Abundance, distribution, and characteristics of microplastics in coastal surface waters of the Colombian Caribbean and Pacific. *Environmental Science and Pollution Research, 28*, 43431–43442. https://doi.org/10.1007/s11356-021-13723-x
95. Gunaalan, K., et al. (2023). Abundance and distribution of microplastics in surface waters of the Kattegat/ Skagerrak (Denmark). *Environmental Pollution, 318*, 120853, 1–11. https://doi.org/10.1016/j.envpol.2022.120853
96. Schmidt, N., Thibault, D., Galgani, F., Paluselli, A., & Sempéré, R. (2018). Occurrence of microplastics in surface waters of the Gulf of Lion (NW Mediterranean Sea). *Progress in Oceanography, 163*, 214–220. https://doi.org/10.1016/j.pocean.2017.11.010
97. Narwal, N., & Katyal, D. (2025). The abundance and analytical characterization of microplastics in the surface water of Haryana, India. *Microscopy Research and Technique, 88*(1), 139–153. https://doi.org/10.1002/jemt.24657

98. Verma, M., Singh, P., Pradhan, V., & Dhanorkar, M. (2025). Spatial and seasonal variations in abundance, distribution characteristics, and sources of microplastics in surface water of Mula river in Pune, India. *Environmental Pollution, 373*, 126091. https://doi.org/10.1016/j.envpol.2025.126091
99. Patidar, K., Ambade, B., Younis, A. M., & Alluhayb, A. H. (2024). Characteristics, fate, and sources of microplastics contaminant in surface water and sediments of river water. *Physics and Chemistry of the Earth Parts A/B/C, 134*, 103596. https://doi.org/10.1016/j.pce.2024.103596
100. Napper, I. E., et al. (2021). The abundance and characteristics of microplastics in surface water in the transboundary Ganges River. *Environmental Pollution, 274*, 116348. https://doi.org/10.1016/j.envpol.2020.116348
101. Napper, I. E., et al. (2023). The distribution and characterisation of microplastics in air, surface water and sediment within a major river system. *Science of the Total Environment, 901*, 166640. https://doi.org/10.1016/j.scitotenv.2023.166640
102. Li, C., et al. (2021). Pelagic microplastics in surface water of the Eastern Indian Ocean during monsoon transition period: Abundance, distribution, and characteristics. *Science of the Total Environment, 755*, 142629. https://doi.org/10.1016/j.scitotenv.2020.142629
103. Janakiram, R., et al. (2023). Seasonal distribution of microplastics in surface waters of the Northern Indian Ocean. *Marine Pollution Bulletin, 190*, 114838. https://doi.org/10.1016/j.marpolbul.2023.114838
104. Srihari, S., Subramani, T., Prapanchan, V. N., & Li, P. (2022). Human health risk perspective study on characterization, quantification and spatial distribution of microplastics in surface water, groundwater and coastal sediments of thickly populated Chennai coast of South India. *Human and Ecological Risk Assessment: An International Journal*, 1–24. https://doi.org/10.1080/10807039.2022.2154635
105. Velmurugan, P. M., Vijayaprabhakaran, K., & Devika, P. T. (2023). Baseline study on identification, characterization, distribution and abundance of microplastics in surface water from Ennore to Kovalam along the east coast of India. *Physics and Chemistry of the Earth, Parts A/B/C, 130*, 103391. https://doi.org/10.1016/j.pce.2023.103391
106. Bhola, S., & Chakraborty, S. (2024). Morphology and polymeric composition-based source apportionment of microplastics in surface water and sediment of drinking water supply reservoirs in Ranchi, India. *Environmental Processes, 11*, 32. https://doi.org/10.1007/s40710-024-00708-4
107. Mukhopadhyay, P., & Valsalan, S. A. (2024). Seasonal variation, spatial distribution and risk assessment of microplastics in surface waters of Periyar River, Kerala, India. *Environmental Monitoring and Assessment, 196*, 679. https://doi.org/10.1007/s10661-024-12820-x
108. Gupta, D. K., et al. (2025). First evidence of microplastics in surface water of urban waterbodies in Bhopal city, India- abundance and their characteristics. *Journal of Contaminant Hydrology, 273*, 104575. https://doi.org/10.1016/j.jconhyd.2025.104575
109. Kumar, S. K. M., et al. (2023). Characterization of suspended microplastics in surface waters of Chalakudy River, Kerala, India. *Chemistry and Ecology, 39*(3), 268–287. https://doi.org/10.1080/02757540.2023.2178648
110. Keerthika, K., Padmavathy, P., Rani, V., Jeyashakila, R., Aanand, S., & Kutty, R. (2022). Spatial, seasonal and ecological risk assessment of microplastics in sediment and surface water along the Thoothukudi, south Tamil Nadu, south east India. *Environmental Monitoring and Assessment, 194*, 820. https://doi.org/10.1007/s10661-022-10468-z
111. Deswati, D., et al. (2023). Preliminary detection of microplastics in surface water of Maninjau Lake in Agam, Indonesia. *Aquaculture, Aquarium, Conservation & Legislation, 16*(5), 2601–2614. http://www.bioflux.com.ro/aacl
112. Sulistyowati, L., Nurhasanah, Riani, E., & Cordova, M. R. (2022). The occurrence and abundance of microplastics in surface water of the midstream and downstream of the Cisadane River, Indonesia. *Chemosphere, 291*, 133071. https://doi.org/10.1016/j.chemosphere.2021.133071

113. Ramadan, A. H., & Sembiring, E. (2020). Occurrence of nicroplastic in surface water of Jatiluhur Reservoir. In *The 6th Environmental Technology and Management Conference (ETMC) in Conjunction with The 12th AUN/SEED-Net Regional Conference on Environmental Engineering (RC EnvE 2019)* (pp. 1–4). https://doi.org/10.1051/e3sconf/202014807004
114. Takarina, N. D., Purwiyanto, A. I. S., Rasud, A. A., Arifin, A. A., & Suteja, Y. (2022). Microplastic abundance and distribution in surface water and sediment collected from the coastal area. *Global Journal of Environmental Science and Management, 8*(2), 183–196. https://doi.org/10.22034/gjesm.2022.02.03
115. Henny, C., Suryono, T., Rohaningsih, D., Yoga, G. P., Sudarso, J., & Waluyo, A. (2023). The occurrence of microplastics in the surface water of several urban lakes in the Megacity of Jakarta. *6th International Symposium on Green Technology for Value Chains 2022, 1201*, 012023, 1–8. https://doi.org/10.1088/1755-1315/1201/1/012023
116. Deswati, D., Yusuf, Y., Putra, Z. A., & Putra, A. (2025). Abundance and characteristics of microplastics in surface water of Lake Singkarak in Tanah Datar, West Sumatra, Indonesia. *Pollution, 11*(2), 464–476. https://doi.org/10.22059/poll.2024.381179.2516
117. Rami, Y., Shoshtari-Yeganeh, B., Ebrahimi, A., & Ebrahimpour, K. (2023). Occurrence and characteristics of microplastics in surface water and sediment of Zayandeh-rud river, Iran. *Environmental Health Engineering and Management Journal, 10*(2), 207–216. https://doi.org/10.34172/EHEM.2023.23
118. Manbohi, A., Mehdinia, A., Rahnama, R., Hamzehpour, A., & Dehbandi, R. (2023). Distribution of microplastics in upstream and downstream surface waters of the Iranian rivers discharging to the southern Caspian Sea. *Environmental Science and Pollution Research, 30*, 92359–92370. https://doi.org/10.1007/s11356-023-28898-8
119. Sighicelli, M., et al. (2018). Microplastic pollution in the surface waters of Italian Subalpine Lakes. *Environmental Pollution, 236*, 645–651. https://doi.org/10.1016/j.envpol.2018.02.008
120. Rose, D., & Webber, M. (2019). Characterization of microplastics in the surface waters of Kingston Harbour. *Science of the Total Environment, 664*, 753–760. https://doi.org/10.1016/j.scitotenv.2019.01.319
121. Kameda, Y., Yamada, N., & Fujita, E. (2021). Source- and polymer-specific size distributions of fine microplastics in surface water in an urban river. *Environmental Pollution, 284*, 117516. https://doi.org/10.1016/j.envpol.2021.117516
122. Ripken, C., Kotsifaki, D. G., & Chormaic, S. N. (2021). Analysis of small microplastics in coastal surface water samples of the subtropical island of Okinawa, Japan. *Science of the Total Environment, 760*, 143927. https://doi.org/10.1016/j.scitotenv.2020.143927
123. Xu, H., Nakano, H., Tokai, T., Miyazaki, T., Hamada, H., & Arakawa, H. (2022). Contamination of sea surface water offshore the Tokai region and Tokyo Bay in Japan by small microplastics. *Marine Pollution Bulletin, 185*, 114245. https://doi.org/10.1016/j.marpolbul.2022.114245
124. Migwi, F. K., Ogunah, J. A., & Kiratu, J. M. (2020). Occurrence and spatial distribution of microplastics in the surface waters of Lake Naivasha, Kenya. *Environmental Toxicology and Chemistry, 39*(4), 765–774. https://doi.org/10.1002/etc.4677
125. Pashaei, R., et al. (2023). Assessing the occurrence and distribution of microplastics in surface freshwater and wastewaters of Latvia and Lithuania. *Toxics, 11*(4). https://doi.org/10.3390/toxics11040292
126. Kazour, M., Jemaa, S., Issa, C., Khalaf, G., & Amara, R. (2019). Microplastics pollution along the Lebanese coast (Eastern Mediterranean Basin): Occurrence in surface water, sediments and biota samples. *Science of the Total Environment, 696*, 133933. https://doi.org/10.1016/j.scitotenv.2019.133933
127. Ramírez-Álvarez, N., et al. (2020). Microplastics: Sources and distribution in surface waters and sediments of Todos Santos Bay, Mexico. *Science of the Total Environment, 703*, 134838. https://doi.org/10.1016/j.scitotenv.2019.134838
128. Bouadil, O., Benomar, M., El Ouarghi, H., Aboulhassan, M. A., & Benbrahim, S. (2024). Identification and quantification of microplastics in surface water of a southwestern

Mediterranean Bay (Al Hoceima, Morocco). *Waste Management Bulletin, 2*(1), 142–151. https://doi.org/10.1016/j.wmb.2024.01.003
129. Haddout, S., Gimiliani, G. T., Priya, K. L., Hoguane, A. M., Casila, J. C. C., & Ljubenkov, I. (2022). Microplastics in surface waters and sediments in the Sebou Estuary and Atlantic Coast, Morocco. *Analytical Letters, 55*(2), 256–268. https://doi.org/10.1080/00032719.2021.1924767
130. Malla-Pradhan, R., Suwunwong, T., Phoungthong, K., Joshi, T. P., & Pradhan, B. L. (2022). Microplastic pollution in urban Lake Phewa, Nepal: The first report on abundance and composition in surface water of lake in different seasons. *Environmental Science and Pollution Research, 29*, 39928–39936. https://doi.org/10.1007/s11356-021-18301-9
131. Markic, A., Bridson, J. H., Morton, P., Hersey, L., Maes, T., & Bowen, M. (2022). Microplastic pollution in the surface waters of Vava'u, Tonga. *Marine Pollution Bulletin, 185*, 114243. https://doi.org/10.1016/j.marpolbul.2022.114243
132. Akinhanmi, F. O., Ayanda, O. I., & Dedeke, G. A. (2023). Occurrence and characteristics of microplastics in the surface water and sediment of Lagos Lagoon, Nigeria. In P. O. Isibor, P. Akinduti, S. U. Oranusi, & J. O. Popoola (Eds.), *Biotechnological approaches to sustainable development goals* (pp. 103–118). Springer. https://doi.org/10.1007/978-3-031-33370-5_7
133. Bashir, A., & Hashmi, I. (2022). Detection in influx sources and estimation of microplastics abundance in surface waters of Rawal Lake, Pakistan. *Heliyon, 8*(3). https://doi.org/10.1016/j.heliyon.2022.e09166
134. Diez-Pérez, D. B., Arenas, I., Maidana, E., López-Rosales, A., Andrade, J. M., & Muniategui-Lorenzo, S. (2023). Microplastics in surface water of the Bay of Asunción, Paraguay. *Marine Pollution Bulletin, 192*, 115075, 1–11. https://doi.org/10.1016/j.marpolbul.2023.115075
135. Arcadio, C. G. L. A., et al. (2023). Microplastics in surface water of Laguna de Bay: First documented evidence on the largest lake in the Philippines. *Environmental Science and Pollution Research, 30*, 29824–29833. https://doi.org/10.1007/s11356-022-24261-5
136. Osorio, E. D., Tanchuling, M. A. N., & Diola, M. B. L. D. (2021). Microplastics occurrence in surface waters and sediments in five river mouths of Manila Bay. *Frontiers in Environmental Science, 9*. https://doi.org/10.3389/fenvs.2021.719274
137. Sekudewicz, I., Dąbrowska, A. M., & Syczewski, M. D. (2021). Microplastic pollution in surface water and sediments in the urban section of the Vistula River (Poland). *Science of the Total Environment, 762*, 143111. https://doi.org/10.1016/j.scitotenv.2020.143111
138. Lechthaler, S., Schwarzbauer, J., Reicherter, K., Stauch, G., & Schüttrumpf, H. (2020). Regional study of microplastics in surface waters and deep sea sediments south of the Algarve Coast. *Regional Studies in Marine Science, 40*, 101488. https://doi.org/10.1016/j.rsma.2020.101488
139. Frank, Y. A., et al. (2021). Preliminary screening for microplastic concentrations in the surface water of the Ob and Tom Rivers in Siberia, Russia. *Sustainability, 13*(1). https://doi.org/10.3390/su13010080
140. Russell, M., & Webster, L. (2021). Microplastics in sea surface waters around Scotland. *Marine Pollution Bulletin, 166*, 112210, 1–9. https://doi.org/10.1016/j.marpolbul.2021.112210
141. Kang, J.-H., Kwon, O.-Y., & Shim, W. J. (2015). Potential threat of microplastics to Zooplanktivores in the surface waters of the southern sea of Korea. *Archives of Environmental Contamination and Toxicology, 69*, 340–351. https://doi.org/10.1007/s00244-015-0210-3
142. Chae, D.-H., Kim, I.-S., Kim, S.-K., Song, Y. K., & Shim, W. J. (2015). Abundance and distribution characteristics of microplastics in surface seawaters of the Incheon/Kyeonggi Coastal Region. *Archives of Environmental Contamination and Toxicology, 69*, 269–278. https://doi.org/10.1007/s00244-015-0173-4
143. Saad, D., Ramaremisa, G., Ndlovu, M., & Chimuka, L. (2024). Morphological and chemical characteristics of microplastics in surface water of the Vaal River, South Africa. *Environmental Processes, 11*, 16. https://doi.org/10.1007/s40710-024-00693-8
144. Ramaremisa, G., Ndlovu, M., & Saad, D. (2022). Comparative assessment of microplastics in surface waters and sediments of the Vaal River, South Africa: Abundance, composi-

tion, and sources. *Environmental Toxicology and Chemistry, 41*(12), 3029–3040. https://doi.org/10.1002/etc.5482

145. Dalmau-Soler, J., Boleda, M. R., & Lacorte, S. (2025). Implementation plan to monitor microplastics in surface and drinking water using Py-GC-MS according to Decision (EU) 2024/1441. *Environmental Pollution, 373*, 126174. https://doi.org/10.1016/j.envpol.2025.126174
146. Schönlau, C., et al. (2020). Microplastics in sea-surface waters surrounding Sweden sampled by manta trawl and in-situ pump. *Marine Pollution Bulletin, 153*, 111019. https://doi.org/10.1016/j.marpolbul.2020.111019
147. Pradit, S., Noppradit, P., Sornplang, K., Jitkaew, P., Jiwarungrueangkul, T., & Muenhor, D. (2024). Occurrence and abundance of microplastics in surface water of Songkhla Lagoon. *PeerJ, 12*, e17822. https://doi.org/10.7717/peerj.17822
148. Sarin, C., & Klomjek, P. (2022). Spatial and seasonal distribution of microplastic in surface water of Bueng Boraphet Wetland—A Ramsar wetland in Thailand. *Environmental Monitoring and Assessment, 194*, 904. https://doi.org/10.1007/s10661-022-10578-8
149. Vibhatabandhu, P., & Srithongouthai, S. (2022). Influence of seasonal variations on the distribution characteristics of microplastics in the surface water of the Inner Gulf of Thailand. *Marine Pollution Bulletin, 180*, 113747. https://doi.org/10.1016/j.marpolbul.2022.113747
150. Vibhatabandhu, P., & Srithongouthai, S. (2022). Abundance and characteristics of microplastics contaminating the surface water of the Inner Gulf of Thailand. *Water, Air, and Soil Pollution, 233*, 50. https://doi.org/10.1007/s11270-022-05531-x
151. Ta, A. T., & Babel, S. (2020). Microplastics pollution with heavy metals in the aquaculture zone of the Chao Phraya River Estuary, Thailand. *Marine Pollution Bulletin, 161*, 111747. https://doi.org/10.1016/j.marpolbul.2020.111747
152. Zayen, A., Sayadi, S., Chevalier, C., Boukthir, M., Ben Ismail, S., & Tedetti, M. (2020). Microplastics in surface waters of the Gulf of Gabes, southern Mediterranean Sea: Distribution, composition and influence of hydrodynamics. *Estuarine, Coastal and Shelf Science, 242*, 106832. https://doi.org/10.1016/j.ecss.2020.106832
153. Tunçer, S., Artüz, O. B., Demirkol, M., & Artüz, M. L. (2018). First report of occurrence, distribution, and composition of microplastics in surface waters of the Sea of Marmara, Turkey. *Marine Pollution Bulletin, 135*, 283–289. https://doi.org/10.1016/j.marpolbul.2018.06.054
154. Aydin, S., Ulvi, A., & Aydin, M. E. (2024). Occurrence, characteristics, and risk assessment of microplastics and polycyclic aromatic hydrocarbons associated with microplastics in surface water and sediments of the Konya Closed Basin, Turkey. *Environmental Science and Pollution Research, 31*, 57989–58009. https://doi.org/10.1007/s11356-024-35029-4
155. Akdogan, Z., Guven, B., & Kideys, A. E. (2023). Microplastic distribution in the surface water and sediment of the Ergene River. *Environmental Research Letters, 234*, 116500. https://doi.org/10.1016/j.envres.2023.116500
156. Turhan, D. Ö. (2022). Evaluation of microplastics in the surface water, sediment and fish of Sürgü Dam Reservoir (Malatya) in Turkey. *Turkish Journal of Fisheries and Aquatic Sciences, 22*(7). https://doi.org/10.4194/TRJFAS20157
157. Higgins, C., & Turner, A. (2023). Microplastics in surface coastal waters around Plymouth, UK, and the contribution of boating and shipping activities. *Science of the Total Environment, 893*, 164695. https://doi.org/10.1016/j.scitotenv.2023.164695
158. Lorenz, C., et al. (2019). Spatial distribution of microplastics in sediments and surface waters of the southern North Sea. *Environmental Pollution, 252*, 1719–1729. https://doi.org/10.1016/j.envpol.2019.06.093
159. Maes, T., et al. (2017). Microplastics baseline surveys at the water surface and in sediments of the North-East Atlantic. *Frontiers in Marine Science, 4*. https://doi.org/10.3389/fmars.2017.00135
160. Polanco, H., Hayes, S., Roble, C., Krupitsky, M., & Branco, B. (2020). The presence and significance of microplastics in surface water in the Lower Hudson River Estuary 2016–2019:

A research note. *Marine Pollution Bulletin, 161*, 111702. https://doi.org/10.1016/j.marpolbul.2020.111702
161. Lasee, S., et al. (2017). Microplastics in a freshwater environment receiving treated wastewater effluent. *Integrated Environmental Assessment and Management, 13*(3), 528–532. https://doi.org/10.1002/ieam.1915
162. Bikker, J., Lawson, J., Wilson, S., & Rochman, C. M. (2020). Microplastics and other anthropogenic particles in the surface waters of the Chesapeake Bay. *Marine Pollution Bulletin, 156*, 111257. https://doi.org/10.1016/j.marpolbul.2020.111257
163. Eriksen, M., et al. (2013). Microplastic pollution in the surface waters of the Laurentian Great Lakes. *Marine Pollution Bulletin, 77*(1–2), 177–182. https://doi.org/10.1016/j.marpolbul.2013.10.007
164. Tran-Nguyen, Q. A., Le, T. M., Nguyen, H. N. Y., Nguyen, Q. T., & Trinh-Dang, M. (2024). Microplastics in the surface water of urban lakes in central Vietnam: Pollution level, characteristics, and ecological risk assessment. *Case Studies in Chemical and Environmental Engineering, 9*, 100622. https://doi.org/10.1016/j.cscee.2024.100622
165. Kieu-Le, T.-C., Thuong, Q.-T., Truong, T.-N.-S., Le, T.-M.-T., Tran, Q.-V., & Strady, E. (2023). Baseline concentration of microplastics in surface water and sediment of the northern branches of the Mekong River Delta, Vietnam. *Marine Pollution Bulletin, 187*, 114605. https://doi.org/10.1016/j.marpolbul.2023.114605
166. Oanh, D. T., et al. (2021). Preliminary results on microplastics in surface water from the downstream of the Day River. *Vietnam Journal of Earth Sciences*, 1–11. https://doi.org/10.15625/2615-9783/16504
167. Mai, H., et al. (2023). Distribution of microplastics in surface water of tropical urban lakes: A case study in Ha Noi, Vietnam. *Vietnam Journal of Earth Sciences, 45*(3), 388–404. https://doi.org/10.15625/2615-9783/18595
168. Truong, T.-N.-S., Kieu-Le, T.-C., Ngo, T.-P., & Strady, E. (2024). Abundance of microplastics in surface water of tropical reservoirs during contrasted season, the case of Dau Tieng and Tri An, Vietnam. *International Journal of Environmental Science and Technology, 21*, 9391–9408. https://doi.org/10.1007/s13762-024-05543-6
169. Cham, D. D., et al. (2022). Distribution and characteristics of microplastics in surface water at some beaches in Thanh Hoa province, Viet Nam. *Vietnam Journal of Catalysis and Adsorption, 10*(1S), 193–200. https://doi.org/10.51316/jca.2021.120
170. Cherdsukjai, P., Sangmanee, C., Worachananant, S., & Phaksopa, J. (2022). Microplastics in the inshore and offshore surface water in the Andaman Sea. *Water, Air, and Soil Pollution, 233*, 500. https://doi.org/10.1007/s11270-022-05975-1
171. Egessa, R., Nankabirwa, A., Ocaya, H., & Pabire, W. G. (2020). Microplastic pollution in surface water of Lake Victoria. *Science of the Total Environment, 741*, 140201. https://doi.org/10.1016/j.scitotenv.2020.140201
172. Mataji, A., Taleshi, M. S., & Balimoghaddas, E. (2020). Distribution and characterization of microplastics in surface waters and the Southern Caspian Sea Coasts sediments. *Archives of Environmental Contamination and Toxicology, 78*, 86–93. https://doi.org/10.1007/s00244-019-00700-2
173. Hansen, J., Hildebrandt, L., Zimmermann, T., El Gareb, F., Fischer, E. K., & Pröfrock, D. (2023). Quantification and characterization of microplastics in surface water samples from the Northeast Atlantic Ocean using laser direct infrared imaging. *Marine Pollution Bulletin, 190*, 114880, 1–10. https://doi.org/10.1016/j.marpolbul.2023.114880
174. Tigreros-Benavides, P., et al. (2024). Microplastics and plastisphere at surface waters in the Southwestern Caribbean Sea. *Journal of Environmental Management, 358*, 120745. https://doi.org/10.1016/j.jenvman.2024.120745
175. Zhang, S., et al. (2022). Distribution characteristics of microplastics in surface and subsurface Antarctic seawater. *Science of the Total Environment, 838*, 156051. https://doi.org/10.1016/j.scitotenv.2022.156051

176. Kanhai, L. D. K., Gårdfeldt, K., Lyashevska, O., Hassellöv, M., Thompson, R. C., & O'Connor, I. (2018). Microplastics in sub-surface waters of the Arctic Central Basin. *Marine Pollution Bulletin, 130*, 8–18. https://doi.org/10.1016/j.marpolbul.2018.03.011
177. Li, Y., et al. (2024). Microplastics in surface seawater of Kongsfjorden, Svalbard, Arctic. *Environmental Technology & Innovation, 36*, 103819. https://doi.org/10.1016/j.eti.2024.103819
178. Huang, Y., et al. (2022). Systematical insights into distribution and characteristics of microplastics in near-surface waters from the East Asian Seas to the Arctic Central Basin. *Science of the Total Environment, 814*, 151923. https://doi.org/10.1016/j.scitotenv.2021.151923
179. Morgana, S., et al. (2018). Microplastics in the Arctic: A case study with sub-surface water and fish samples off Northeast Greenland. *Environmental Pollution, 242*, 1078–1086. https://doi.org/10.1016/j.envpol.2018.08.001
180. Carlsson, P., Singdahl-Larsen, C., & Lusher, A. L. (2021). Understanding the occurrence and fate of microplastics in coastal Arctic ecosystems: The case of surface waters, sediments and walrus (Odobenus rosmarus). *Science of the Total Environment, 792*, 148308. https://doi.org/10.1016/j.scitotenv.2021.148308
181. Barone, M., et al. (2024). What to monitor? Microplastics in a freshwater lake—From seasonal surface water to bottom sediments. *Environmental Advances, 17*, 100577. https://doi.org/10.1016/j.envadv.2024.100577
182. Aigars, J., Barone, M., Suhareva, N., Putna-Nimane, I., & Dimante-Deimantovica, I. (2021). Occurrence and spatial distribution of microplastics in the surface waters of the Baltic Sea and the Gulf of Riga. *Marine Pollution Bulletin, 172*, 112860. https://doi.org/10.1016/j.marpolbul.2021.112860
183. Zhao, X., et al. (2020). Pollution status and pollution behavior of microplastic in surface water and sediment of urban rivers. *Huanjing Kexue/Chinese Journal of Environmental Science, 41*(8), 3612–3620. https://doi.org/10.13227/j.hjkx.201912236
184. Dacewicz, E., Łobos-Moysa, E., & Łobos-Moysa, E. (2024). Identification tools of microplastics from surface water integrating digital image processing and statistical techniques. *Materials (Basel), 17*(15). https://doi.org/10.3390/ma17153701
185. Zhang, C., Wang, S., Sun, D., Pan, Z., & Zou, J. (2023). Investigation of microplastics in surface water and Estuarine Mullet Mugil cephalus from 23 Estuary Areas, South China. *Sustainability, 15*(5). https://doi.org/10.3390/su15054193
186. Sagawa, N., Kawaai, K., & Hinata, H. (2018). Abundance and size of microplastics in a coastal sea: Comparison among bottom sediment, beach sediment, and surface water. *Marine Pollution Bulletin, 133*, 532–542. https://doi.org/10.1016/j.marpolbul.2018.05.036
187. Karlsson, T. M., Kärrman, A., Rotander, A., & Hassellöv, M. (2020). Comparison between manta trawl and in situ pump filtration methods, and guidance for visual identification of microplastics in surface waters. *Environmental Science and Pollution Research, 27*, 5559–5571. https://doi.org/10.1007/s11356-019-07274-5
188. Pasquier, G., et al. (2022). Manta Net: The golden method for sampling surface water microplastics in aquatic environments. *Frontiers in Environmental Science, 10*. https://doi.org/10.3389/fenvs.2022.811112
189. Bakir, A., et al. (2020). Occurrence and abundance of meso and microplastics in sediment, surface waters, and marine biota from the South Pacific region. *Marine Pollution Bulletin, 160*, 111572. https://doi.org/10.1016/j.marpolbul.2020.111572
190. Sonbhadra, S., & Pandey, L. M. (2023). Assessment of microplastics from surface water bodies: Challenges and future scopes. *Water, Air, and Soil Pollution, 234*, 80. https://doi.org/10.1007/s11270-023-06094-1
191. Koelmans, A. A., Nor, N. H. M., Hermsen, E., Kooi, M., & Mintenig, S. M. (2019). Microplastics in freshwaters and drinking water: Critical review and assessment of data quality. *Water Research, 155*, 410–422. https://doi.org/10.1016/j.watres.2019.02.054
192. Ren, Z., Gui, X., Xu, X., Zhao, L., Qiu, H., & Cao, X. (2021). Microplastics in the soil-groundwater environment: Aging, migration, and co-transport of contaminants—A criti-

cal review. *Journal of Hazardous Materials, 419*, 126455. https://doi.org/10.1016/j.jhazmat.2021.126455
193. Cheng, D., Liu, H., Qian, W., Yao, R., & Wang, X. (2024). Migration characteristics of microplastics in riparian soils and groundwater. *Environmental Monitoring and Assessment, 196*, 796. https://doi.org/10.1007/s10661-024-12962-y
194. Belkhiri, A. H., Carre, F., & Quiot, F. (2022). State of knowledge and future research needs on microplastics in groundwater. *Journal of Water and Health, 20*(10), 1479–1496. https://doi.org/10.2166/wh.2022.048
195. Jin, R., Li, H., Li, X., & Shen, M. (2025). Research advances of micro/nanoplastics in groundwater: Occurrence, environmental impacts and control strategies. *Environmental Pollution, 383*, 126899. https://doi.org/10.1016/j.envpol.2025.126899
196. Lee, J.-Y., Cha, J., Ha, K., & Viaroli, S. (2024). Microplastic pollution in groundwater: A systematic review. *Environmental Pollution and Bioavailability, 36*(1). https://doi.org/10.1080/26395940.2023.2299545
197. Singh, S., & Bhagwat, A. (2022). Microplastics: A potential threat to groundwater resources. *Groundwater for Sustainable Development, 19*, 100852. https://doi.org/10.1016/j.gsd.2022.100852
198. Re, V. (2019). Shedding light on the invisible: Addressing the potential for groundwater contamination by plastic microfibers. *Hydrogeology Journal, 27*, 2719–2727. https://doi.org/10.1007/s10040-019-01998-x
199. Sforzi, L., Sarti, C., Santini, S., Martellini, T., & Cincinelli, A. (2024). Global status, risk assessment, and knowledge gaps of microplastics in groundwater: A bibliometric analysis. *Groundwater for Sustainable Development, 27*, 101375. https://doi.org/10.1016/j.gsd.2024.101375
200. Dey, U., et al. (2023). Microplastics in groundwater: An overview of source, distribution, mobility constraints and potential health impacts during the anthropocene. *Groundwater for Sustainable Development, 23*, 101036. https://doi.org/10.1016/j.gsd.2023.101036
201. Calero, E. C., Viršek, M. K., & Mali, N. (2024). Microplastics in groundwater: Pathways, occurrence, and monitoring challenges. *Water (Basel), 16*(9). https://doi.org/10.3390/w16091228
202. He, Y.-Q., McDonough, L. K., Zainab, S. M., Guo, Z.-F., Chen, C., & Xu, Y.-Y. (2024). Microplastic accumulation in groundwater: Data-scaled insights and future research. *Water Research, 258*, 121808. https://doi.org/10.1016/j.watres.2024.121808
203. Gündoğdu, S., et al. (2023). Micro and nano plastics in groundwater systems: A review of current knowledge and future perspectives. *TrAC Trends in Analytical Chemistry, 165*, 117119. https://doi.org/10.1016/j.trac.2023.117119
204. Sumam, S. B., Gautam, S., & Arulraj, P. (2025). Understanding microplastic pollution in groundwater: Pathways, health implications and solutions. *Water Environment Research, 97*(7). https://doi.org/10.1002/wer.70132
205. Wei, Y., et al. (2024). Modeling of microplastics migration in soil and groundwater: Insights into dispersion and particle property effects. *Environmental Science & Technology, 58*(34), 15224–15235. https://doi.org/10.1021/acs.est.4c05202
206. Viaroli, S., Lancia, M., & Re, V. (2022). Microplastics contamination of groundwater: Current evidence and future perspectives. A review. *Science of the Total Environment, 824*, 153851. https://doi.org/10.1016/j.scitotenv.2022.153851
207. Severini, E., Ducci, L., Sutti, A., Robottom, S., Sutti, S., & Celico, F. (2022). River–groundwater interaction and recharge effects on microplastics contamination of groundwater in confined alluvial aquifers. *Water (Basel), 14*(12). https://doi.org/10.3390/w14121913
208. Priya, K. L., et al. (2023). Implications of solid waste dumps on the microplastic abundance in groundwater in Kollam, India. *Journal of Environmental Management, 348*, 119224. https://doi.org/10.1016/j.jenvman.2023.119224

209. Li, X., et al. (2025). Risk assessment of potentially toxic elements, microplastics, and microorganisms in groundwater around municipal solid waste landfill. *Journal of Hazardous Materials, 487*, 137240. https://doi.org/10.1016/j.jhazmat.2025.137240
210. Jeon, C., & Kim, H. (2024). Microplastics and nanoplastics in groundwater: Occurrence, analysis, and identification. *Trends in Environmental Analytical Chemistry, 44*, e00246. https://doi.org/10.1016/j.teac.2024.e00246
211. Lee, J.-Y., Jung, J., & Raza, M. (2022). Good field practice and hydrogeological knowledge are essential to determine reliable concentrations of microplastics in groundwater. *Environmental Pollution, 308*, 119617. https://doi.org/10.1016/j.envpol.2022.119617
212. Perraki, M., et al. (2024). Identification of microplastics using μ-Raman spectroscopy in surface and groundwater bodies of SE Attica, Greece. *Water, 16*(6). https://doi.org/10.3390/w16060843
213. Mintenig, S. M., Löder, M. G. J., Primpke, S., & Gerdts, G. (2019). Low numbers of microplastics detected in drinking water from ground water sources. *Science of the Total Environment, 648*, 631–635. https://doi.org/10.1016/j.scitotenv.2018.08.178
214. Kim, Y.-I., Jeong, E., Lee, J.-Y., Chia, R. W., & Raza, M. (2023). Microplastic contamination in groundwater on a volcanic Jeju Island of Korea. *Environmental Research, Engineering and Management, 226*, 115682. https://doi.org/10.1016/j.envres.2023.115682
215. Viaroli, S., et al. (2024). Limits, challenges, and opportunities of sampling groundwater wells with plastic casings for microplastic investigations. *Science of the Total Environment, 946*, 174259, 1–8. https://doi.org/10.1016/j.scitotenv.2024.174259
216. Wei, Y., & Chen, Y. (2023). The urgent need to investigate microplastic contamination in groundwater: Soil and groundwater interactions as key drivers. *ACS ES&T Water, 3*(12), 3736–3740. https://doi.org/10.1021/acsestwater.3c00645
217. Su, S., Zhou, S., & Lin, G. (2021). Existence of microplastics in soil and groundwater in Jiaodong Peninsula. *International Conference on Tourism, Economy, and Environmental Sustainability (TEES 2021), 251*, 02045, 1–5. https://doi.org/10.1051/e3sconf/202125102045
218. Dong, S., Yu, Z., Huang, J., & Gao, B. (2022). Fate and transport of microplastics in soils and groundwater (ch. 9). In B. Gao (Ed.), *Emerging contaminants in soil and groundwater systems-occurrence, impact, fate and transport* (pp. 301–329). Elsevier. https://doi.org/10.1016/B978-0-12-824088-5.00001-X
219. Moeck, C., Davies, G., Krause, S., & Schneidewind, U. (2023). Microplastics and nanoplastics in agriculture—A potential source of soil and groundwater contamination? *Grundwasser, 28*, 23–35. https://doi.org/10.1007/s00767-022-00533-2
220. Hoang, V.-H., et al. (2025). Microplastic characteristics, transport, risks, and remediation in groundwater: A review. *Environmental Chemistry Letters, 23*, 817–837. https://doi.org/10.1007/s10311-025-01825-8
221. Samayamanthula, D. R., Alsabti, B., Al-Rashidi, A., & Chidambaram, S. (2024). Review of microplastics in groundwater: Sources, identification, and health effects (ch. 15). In S. Sekar, V. Senapathi, C. Sabarathinam, & P. M. Viswanathan (Eds.), *Meso and microplastic risk assessment in marine environments-new threats and challenges* (pp. 293–307). Elsevier. https://doi.org/10.1016/B978-0-323-90980-8.00002-9
222. Refat, J. R., et al. (2025). Exploring the impact of microplastics in groundwater on human health: A comprehensive review. *Current Opinion in Environmental Science & Health, 44*, 100605. https://doi.org/10.1016/j.coesh.2025.100605
223. Luo, Z., Zhu, J., Yu, L., & Yin, K. (2021). Heavy metal remediation by nano zero-valent iron in the presence of microplastics in groundwater: Inhibition and induced promotion on aging effects. *Environmental Pollution, 287*, 117628. https://doi.org/10.1016/j.envpol.2021.117628
224. Khant, N. A., & Kim, H. (2022). Review of current issues and management strategies of microplastics in groundwater environments. *Water (Basel), 14*(7). https://doi.org/10.3390/w14071020

225. Xu, J., et al. (2024). Global distribution, drivers, and potential hazards of microplastics in groundwater: A review. *Science of the Total Environment, 954*, 176194. https://doi.org/10.1016/j.scitotenv.2024.176194
226. Liu, S., et al. (2025). Microplastics in groundwater: Environmental fate and possible interactions with coexisting contaminants. *Environmental Pollution, 372*, 126026. https://doi.org/10.1016/j.envpol.2025.126026
227. Sangkham, S., et al. (2023). Evidence of microplastics in groundwater: A growing risk for human health. *Groundwater for Sustainable Development, 23*, 100981. https://doi.org/10.1016/j.gsd.2023.100981
228. Esfandiari, A., et al. (2020). Distribution and transport of microplastics in groundwater (Shiraz aquifer, Southwest Iran). *Water Research, 220*, 118622. https://doi.org/10.1016/j.watres.2022.118622
229. Mu, H., et al. (2022). High abundance of microplastics in groundwater in Jiaodong Peninsula, China. *Science of the Total Environment, 839*, 156318. https://doi.org/10.1016/j.scitotenv.2022.156318
230. Shu, X., Xu, L., Yang, M., Qin, Z., Zhang, Q., & Zhang, L. (2023). Spatial distribution characteristics and migration of microplastics in surface water, groundwater and sediment in karst areas: The case of Yulong River in Guilin, Southwest China. *Science of the Total Environment, 868*, 161578. https://doi.org/10.1016/j.scitotenv.2023.161578
231. Zhang, L., Zhao, W., Yan, R., Lyu, S., & Sui, Q. (2025). Occurrence, sources, and ecological risk of microplastics in groundwater: Impacts by agricultural activities and atmospheric deposition. *Water Research, 281*, 123585. https://doi.org/10.1016/j.watres.2025.123585
232. Gong, X., Tian, L., Wang, P., Wang, Z., Zeng, L., & Hu, J. (2023). Microplastic pollution in the groundwater under a bedrock Island in the South China Sea. *Environmental Research, 239*, 117277. https://doi.org/10.1016/j.envres.2023.117277
233. An, X., Li, W., Lan, J., & Adnan, M. (2022). Preliminary study on the distribution, source, and ecological risk of typical microplastics in Karst groundwater in Guizhou Province, China. *International Journal of Environmental Research and Public Health, 19*(22). https://doi.org/10.3390/ijerph192214751
234. Wu, B., et al. (2022). Microplastics contamination in groundwater of a drinking-water source area, northern China. *Environmental Research, Engineering, and Management, 214*, 114048. https://doi.org/10.1016/j.envres.2022.114048
235. Shi, J., et al. (2022). Groundwater antibiotics and microplastics in a drinking-water source area, northern China: Occurrence, spatial distribution, risk assessment, and correlation. *Environmental Research, 210*, 112855. https://doi.org/10.1016/j.envres.2022.112855
236. Lee, J.-Y., Cha, J., & Chia, R. W. (2022). Current status of researches on microplastics in groundwater and perspectives. *Journal of the Geological Society of Korea, 58*(2), 233–241. https://doi.org/10.14770/jgsk.2022.58.2.233
237. Lee, J.-Y., Cha, J., Jeong, E., & Chia, R. W. (2024). Microplastics in soil and groundwater of Korea: Occurrence, characteristics and risk assessment. *Episodes Journal of International Geoscience, 47*(3), 685–696. https://doi.org/10.18814/epiiugs/2023/02403s01
238. Cha, J., Lee, J.-Y., & Chia, R. W. (2023). Microplastics contamination and characteristics of agricultural groundwater in Haean Basin of Korea. *Science of the Total Environment, 864*, 161027. https://doi.org/10.1016/j.scitotenv.2022.161027
239. Cha, J., Lee, J.-Y., & Lee, J. (2024). Effects of groundwater sample volume on identified microplastics in groundwater of an agricultural area in Korea. *Science of the Total Environment, 911*, 168650. https://doi.org/10.1016/j.scitotenv.2023.168650
240. Jeong, E., Kim, Y.-I., Lee, J.-Y., & Raza, M. (2023). Microplastic contamination in groundwater of rural area, eastern part of Korea. *Science of the Total Environment, 895*, 165006. https://doi.org/10.1016/j.scitotenv.2023.165006
241. Chia, R. W., Lee, J.-Y., Kim, H., & Jang, J. (2021). Microplastic pollution in soil and groundwater: A review. *Environmental Chemistry Letters, 19*, 4211–4224. https://doi.org/10.1007/s10311-021-01297-6

242. Kumar, R., & Sharma, P. (2021). Microplastics pollution pathways to groundwater in India. *Current Science, 120*(2), 249–251. https://d1wqtxts1xzle7.cloudfront.net/86314341/0249-libre.pdf
243. Manikanda, B. K., Natesan, U., Vaikunth, R., Praveen, K. R., Ruthra, R., & Srinivasalu, S. (2021). Spatial distribution of microplastic concentration around landfill sites and its potential risk on groundwater. *Chemosphere, 277*, 130263. https://doi.org/10.1016/j.chemosphere.2021.130263
244. Valsan, G., Warrier, A. K., Amrutha, K., Anusree, S., & Rangel-Buitrago, N. (2023). Exploring the presence and distribution of microplastics in subterranean estuaries from Southwest India. *Marine Pollution Bulletin, 190*, 114820. https://doi.org/10.1016/j.marpolbul.2023.114820
245. Selvam, S., Jesuraja, K., Venkatramanan, S., Roy, P. D., & Kumari, V. J. (2021). Hazardous microplastic characteristics and its role as a vector of heavy metal in groundwater and surface water of coastal South India. *Journal of Hazardous Materials, 402*, 123786. https://doi.org/10.1016/j.jhazmat.2020.123786
246. Samandra, S., et al. (2022). Microplastic contamination of an unconfined groundwater aquifer in Victoria, Australia. *Science of the Total Environment, 802*, 149727. https://doi.org/10.1016/j.scitotenv.2021.149727
247. Cha, J., Lee, J.-Y., & Chia, R. W. (2022). Comment on the paper 'Microplastic contamination of an unconfined groundwater aquifer in Victoria, Australia'. *Science of the Total Environment, 820*, 153121. https://doi.org/10.1016/j.scitotenv.2022.153121
248. Panno, S. V., et al. (2019). Microplastic contamination in Karst groundwater systems. *Groundwater, 57*(2), 189–196. https://doi.org/10.1111/gwat.12862
249. Shockley, D. J., & Lapworth, D. J. (2022). *Microplastics in groundwater: A literature review*. National Environment Research Council (NERC) Open Research Archive, Keyworth, Nottingham. https://nora.nerc.ac.uk/id/eprint/532669/1/OR22014.pdf
250. Ledieu, L., et al. (2023). May a former municipal landfill contaminate groundwater in microplastics? First investigations from the 'Prairie de Mauves Site' (Nantes, France). *Microplastics, 2*(1), 93–106. https://doi.org/10.3390/microplastics2010007
251. Alvarado-Zambrano, D., Rivera-Hernández, J. R., & Green-Ruiz, C. (2023). First insight into microplastic groundwater pollution in Latin America: The case of a coastal aquifer in Northwest Mexico. *Environmental Science and Pollution Research, 30*, 73600–73611. https://doi.org/10.1007/s11356-023-27461-9
252. Rodríguez-Alcántara, J. S., et al. (2024). Presence of microplastics in the groundwater of volcanic islands, El Hierro and La Palma (Canary Islands). *Journal of Contaminant Hydrology, 263*, 104340, 1–9. https://doi.org/10.1016/j.jconhyd.2024.104340
253. Velasco, A. N., Ellero, A., Gentile, S. R., Zimmermann, S., Ramaciotti, P., & Stoll, S. (2024). Impact of a nanofiltration system on microplastic contamination in Geneva groundwater (Switzerland). *Environmental Science and Pollution Research, 31*, 13512–13522. https://doi.org/10.1007/s11356-024-31940-y
254. Lapworth, D. J., & Shockley, D. J. (2022). *Microplastics in UK groundwater and stygobites: Protocols for sampling, analysis and pilot study results*. NERC Open Research Archives & British Geological Survey (BGS). https://nora.nerc.ac.uk/id/eprint/532670/
255. Thao, L. X. T., et al. (2022). Microplastics occurrence in groundwater of Da Nang City, Viet Nam. *Vietnam Journal of Science and Technology, 60*(5B), 39–49. https://doi.org/10.15625/2525-2518/17352
256. Sforzi, L., et al. (2024). (Micro-)plastics in saturated and unsaturated groundwater bodies: First evidence of presence in groundwater fauna and habitats. *Sustainability, 16*(6). https://doi.org/10.3390/su16062532
257. Blackburn, K., & Green, D. (2022). The potential effects of microplastics on human health: What is known and what is unknown. *Ambio, 51*, 518–530. https://doi.org/10.1007/s13280-021-01589-9
258. Shi, Q., Tang, J., Liu, R., & Wang, L. (2022). Toxicity in vitro reveals potential impacts of microplastics and nanoplastics on human health: A review. *Critical Reviews in Environmental*

*Science and Technology, 52*(21), 3863–3895. https://doi.org/10.1080/10643389.2021.1951528

259. Mittal, N., Tiwari, N., Singh, D., Tripathi, P., & Sharma, S. (2024). Toxicological impacts of microplastics on human health: A bibliometric analysis. *Environmental Science and Pollution Research, 31*, 57417–57429. https://doi.org/10.1007/s11356-023-30801-4
260. Yee, M. S.-L., et al. (2021). Impact of microplastics and nanoplastics on human health. *Nanomaterials, 11*(2). https://doi.org/10.3390/nano11020496
261. Yang, X., Man, Y. B., Wong, M. H., Owen, R. B., & Chow, K. L. (2022). Environmental health impacts of microplastics exposure on structural organization levels in the human body. *Science of the Total Environment, 825*, 154025. https://doi.org/10.1016/j.scitotenv.2022.154025
262. Winiarska, E., Jutel, M., & Zemelka-Wiacek, M. (2024). The potential impact of nano- and microplastics on human health: Understanding human health risks. *Environmental Research, Engineering, And Management, 251*, 118535. https://doi.org/10.1016/j.envres.2024.118535
263. Bocker, R., & Silva, E. K. (2025). Microplastics in our diet: A growing concern for human health. *Science of the Total Environment, 968*, 178882. https://doi.org/10.1016/j.scitotenv.2025.178882
264. Sharma, P., & Vidyarthi, V. K. (2024). Impact of microplastic intake via poultry products: Environmental toxicity and human health. *Journal of Hazardous Materials Advances, 14*, 100426. https://doi.org/10.1016/j.hazadv.2024.100426
265. Prathipa, R., Gopalakrishnan, N. K., Chaudhary, M., & Pappuswamy, A. (2024). Impact of microplastics on human health: A comprehensive review. *Journal of Preventive, Diagnostic and Treatment Strategies in Medicine, 3*(2), 76–84. https://doi.org/10.4103/jpdtsm.jpdtsm_24_24
266. Kadac-Czapska, K., Knez, E., & Grembecka, M. (2024). Food and human safety: The impact of microplastics. *Critical Reviews in Food Science and Nutrition, 64*(11), 3502–3521. https://doi.org/10.1080/10408398.2022.2132212
267. Eze, C. G., Nwankwo, C. E., Dey, S., Sundaramurthy, S., & Okeke, E. S. (2024). Food chain microplastics contamination and impact on human health: A review. *Environmental Chemistry Letters, 22*, 1889–1927. https://doi.org/10.1007/s10311-024-01734-2
268. Li, Y., Tao, L., Wang, Q., Wang, F., Li, G., & Song, M. (2023). Potential health impact of microplastics: A review of environmental distribution, human exposure, and toxic effects. *Environmental Health, 1*(4), 249–257. https://doi.org/10.1021/envhealth.3c00052
269. Kumar, S., Rajesh, M., Rajesh, K. M., Suyani, N. K., Ahamed Rasheeq, K. S., & Pratiksha, A. (2020). Impact of microplastics on aquatic organisms and human health: A review. *International Journal of Environmental Sciences and Natural Resources, 26*(2), 59–64. https://juniperpublishers.com/ijesnr/pdf/IJESNR.MS...
270. Inam, Ö. (2025). Impact of microplastics on female reproductive health: Insights from animal and human experimental studies: A systematic review. *Archives of Gynecology and Obstetrics, 312*, 77–92. https://doi.org/10.1007/s00404-024-07929-w
271. Temel, H. (2025). The impact of microplastics on the environment and human (living) health. In M. P. Shah & G. Y. Töre (Eds.), *Microplastics (MPs) in wastewater: Determination-treatment methods and effects on climate change. environmental science and engineering* (pp. 305–322). Springer. https://doi.org/10.1007/978-3-031-76949-8_12
272. Yuldasheva, Z. Z. (2025). Impact of microplastics on human health. *International Journal of Scientific Research, 10*(1), 77–80. https://inlibrary.uz/index.php/ijsr/article/view/69216
273. Moiniafshari, K., Zanut, A., Tapparo, A., Pastore, P., Bogialli, S., & Monikh, F. A. (2025). A perspective on the potential impact of microplastics and nanoplastics on the human central nervous system. *Perspective, 12*, 1809–1820. https://doi.org/10.1039/D4EN01017E
274. Fournier, E., Etienne-Mesmin, L., Blanquet-Diot, S., & Mercier-Bonin, M. (2021). Impact of microplastics in human health—A focus on the gastrointestinal tract. In *Handbook of microplastics in the environment* (pp. 1–25). https://doi.org/10.1007/978-3-030-10618-8_48-1

275. Abbas, G., Ahmed, U., & Ahmad, M. A. (2025). Impact of microplastics on human health: Risks, diseases, and affected body systems. *Microplastics, 4*(2). https://doi.org/10.3390/microplastics4020023
276. Yang, W., et al. (2022). Impacts of microplastics on immunity. *Frontiers in Toxicology, 4*. https://doi.org/10.3389/ftox.2022.956885
277. Van Tran, T., Jalil, A. A., Nguyen, T. M., Nguyen, T. T. T., Nabgan, W., & Nguyen, D. T. C. (2023). A review on the occurrence, analytical methods, and impact of microplastics in the environment. *Environmental Toxicology and Pharmacology, 102*, 104248. https://doi.org/10.1016/j.etap.2023.104248
278. Ghosh, S., Sinha, J. K., Ghosh, S., Vashisth, K., Han, S., & Bhaskar, R. (2023). Microplastics as an emerging threat to the global environment and human health. *Sustainability, 15*(14). https://doi.org/10.3390/su151410821
279. Emenike, E. C., et al. (2023). From oceans to dinner plates: The impact of microplastics on human health. *Heliyon, 9*(10). https://doi.org/10.1016/j.heliyon.2023.e20440
280. Karganovic, Z., Vracaric, A., & Djuricic, I. (2025). Impact of microplastics on health: Scientific review. In *14th International Scientific Conference. Science and Higher Education in Function of Sustainable Development [SED 2025]* (Vol. 5, pp. 90–97) https://sed.akademijazs.edu.rs/fajlovi/papers25/proceedings/5-16.pdf
281. Barceló, D., Picó, Y., & Alfarhan, A. H. (2023). Microplastics: Detection in human samples, cell line studies, and health impacts. *Environmental Toxicology and Pharmacology, 101*, 104204. https://doi.org/10.1016/j.etap.2023.104204
282. Kim, O.-H., Chang, E. S., Kang, H., & Lee, H. J. (2025). Unveiling the impact of microplastics: A perspective on size, shape, and composition in human health. *Molecular and Cellular Toxicology*. https://doi.org/10.1007/s13273-025-00549-8
283. Karbalaei, S., Hanachi, P., Walker, T. R., & Cole, M. (2018). Occurrence, sources, human health impacts and mitigation of microplastic pollution. *Environmental Science and Pollution Research, 25*, 36046–36063. https://doi.org/10.1007/s11356-018-3508-7
284. Prata, J. C., da Costa, J. P., Lopes, I., Duarte, A. C., & Rocha-Santos, T. (2020). Environmental exposure to microplastics: An overview on possible human health effects. *Science of the Total Environment, 702*, 134455. https://doi.org/10.1016/j.scitotenv.2019.134455
285. Liu, J., & Li, C. (2025). Impact of microplastics on aquatic ecosystems. *Water, 17*(14). https://doi.org/10.3390/w17142124
286. Smith, M., Love, D. C., Rochman, C. M., & Neff, R. A. (2018). Microplastics in seafood and the implications for human health. *Current Environmental Health Reports, 5*, 375–386. https://doi.org/10.1007/s40572-018-0206-z
287. Alziny, N., et al. (2025). The impact of microplastics in food and drugs on human health: A review of the MENA region. *International Journal of Environmental Research and Public Health, 22*(3). https://doi.org/10.3390/ijerph22030380
288. Sana, S. S., Dogiparthi, L. K., Gangadhar, L., Chakravorty, A., & Abhishek, N. (2020). Effects of microplastics and nanoplastics on marine environment and human health. *Environmental Science and Pollution Research, 27*, 44743–44756. https://doi.org/10.1007/s11356-020-10573-x
289. Prata, J. C., da Costa, J. P., Lopes, I., Andrady, A. L., Duarte, A. C., & Rocha-Santos, T. (2021). A One Health perspective of the impacts of microplastics on animal, human and environmental health. *Science of the Total Environment, 777*, 146094. https://doi.org/10.1016/j.scitotenv.2021.146094
290. Hasan, M., & Tarannum, M. N. (2025). Adverse impacts of microplastics on soil physicochemical properties and crop health in agricultural systems. *Journal of Hazardous Materials Advances, 17*, 100528. https://doi.org/10.1016/j.hazadv.2024.100528
291. Bello, F. A., Folorunsho, A. B., Chia, R. W., Lee, J.-Y., & Fasusi, S. A. (2025). Microplastics in agricultural soils: Sources, impacts on soil organisms, plants, and humans. *Environmental Monitoring and Assessment, 197*, 448. https://doi.org/10.1007/s10661-025-13874-1

292. De-la-Torre, G. E. (2020). Microplastics: An emerging threat to food security and human health. *Journal of Food Science and Technology, 57*, 1601–1608. https://doi.org/10.1007/s13197-019-04138-1
293. Campanale, C., Massarelli, C., Savino, I., Locaputo, V., & Uricchio, V. F. (2020). A detailed review study on potential effects of microplastics and additives of concern on human health. *International Journal of Environmental Research and Public Health, 17*(4). https://doi.org/10.3390/ijerph17041212
294. Sadique, S. A., Konarova, M., Niu, X., Szilagyi, I., Nirmal, N., & Li, L. (2025). Impact of microplastics and nanoplastics on human health: Emerging evidence and future directions. *Emerging Contaminants, 11*(3). https://doi.org/10.1016/j.emcon.2025.100545
295. Hampton, L. M. T., et al. (2022). Research recommendations to better understand the potential health impacts of microplastics to humans and aquatic ecosystems. *Microplastics and Nanoplastics, 2*, 18. https://doi.org/10.1186/s43591-022-00038-y
296. Weis, J., et al. (2015). *Human health impacts of microplastics and nanoplastics*. Public Health Standing Committee, NJDEP Science Advisory Board. https://superfoodvegan.co.il/wp-content/uploads/2023/11/njdep-sab-phsc-final-2016.pdf
297. Megha, K. B., Anvitha, D., Parvathi, S., Neeraj, A., Sonia, J., & Mohanan, P. V. (2025). Environmental impact of microplastics and potential health hazards. *Critical Reviews in Biotechnology, 45*(1), 97–127. https://doi.org/10.1080/07388551.2024.2344572
298. Katyal, D., Kong, E., & Villanueva, J. (2020). Microplastics in the environment: Impact on human health and future mitigation strategies. *Environmental Health Review, 63*(1). https://doi.org/10.5864/d2020-005
299. Sangkham, S., Pitakpong, A., & Kumar, R. (2025). Advanced approaches to microplastic removal in landfill leachate: Advanced oxidation processes (AOPs), biodegradation, and membrane filtration. *Case Studies in Chemical and Environmental Engineering, 11*, 101056, 1–8. https://doi.org/10.1016/j.cscee.2024.101056
300. Poerio, T., Piacentini, E., & Mazzei, R. (2019). Membrane processes for microplastic removal. *Molecules, 24*(22). https://doi.org/10.3390/molecules24224148
301. Le, L.-T., Bui, X.-B., Tran, C.-S., Chiemchaisri, C., & Pandey, A. (2023). Membrane and filtration processes for microplastic removal (ch. 9). In X.-T. Bui, W. Guo, C. Chiemchaisri, & A. Pandey (Eds.), *Current developments in biotechnology and bioengineering-membrane technology for sustainable water and energy management* (pp. 203–220). Elsevier. https://doi.org/10.1016/B978-0-443-19180-0.00019-5
302. Acarer, S. (2023). A review of microplastic removal from water and wastewater by membrane technologies. *Water Science and Technology, 88*(1), 199–219. https://doi.org/10.2166/wst.2023.186
303. He, W., Chen, X., Xu, C., Zhou, C., & Luo, J. (2023). Effect of microplastic aging degree on filter cake formation and membrane fouling characteristics in ultrafiltration process with pre-coagulation. *Separation and Purification Technology, 310*, 123221. https://doi.org/10.1016/j.seppur.2023.123221
304. Enfrin, M., Lee, J., Le-Clech, P., & Dumée, L. F. (2020). Kinetic and mechanistic aspects of ultrafiltration membrane fouling by nano- and microplastics. *Journal of Membrane Science, 601*, 117890. https://doi.org/10.1016/j.memsci.2020.117890
305. González-Camejo, J., et al. (2023). Feasibility of rapid gravity filtration and membrane ultrafiltration for the removal of microplastics and microlitter in sewage and wastewater from plastic industry. *Journal of Water Process Engineering, 51*, 103452, 1–10. https://doi.org/10.1016/j.jwpe.2022.103452
306. Bayo, J., López-Castellanos, J., & Olmos, S. (2020). Membrane bioreactor and rapid sand filtration for the removal of microplastics in an urban wastewater treatment plant. *Marine Pollution Bulletin, 156*, 111211. https://doi.org/10.1016/j.marpolbul.2020.111211
307. Shen, M., et al. (2023). Recent advances on micro/nanoplastic pollution and membrane fouling during water treatment: A review. *Science of the Total Environment, 881*, 163467. https://doi.org/10.1016/j.scitotenv.2023.163467

308. LaRue, R. J., Koo, S., Warren, A., McKay, Y. G., & Latulippe, D. R. (2024). A strategy for quantifying microplastic particles in membrane filtration processes using flow cytometry. *Chemosphere, 368*, 143613. https://doi.org/10.1016/j.chemosphere.2024.143613
309. Kook, H., & Park, C. (2022). Engineered approaches to facile identification of tiny microplastics in polymeric and ceramic membrane filtrations for wastewater treatment. *Membranes (Basel), 12*(6). https://doi.org/10.3390/membranes12060565
310. Kim, S., Hyeon, Y., Rho, H., & Park, C. (2024). Ceramic membranes as a potential high-performance alternative to microplastic filters for household washing machines. *Separation and Purification Technology, 344*, 127278. https://doi.org/10.1016/j.seppur.2024.127278
311. Oßmann, B. E., Sarau, G., Schmitt, S. W., Holtmannspötter, H., Christiansen, S. H., & Dicke, W. (2017). Development of an optimal filter substrate for the identification of small microplastic particles in food by micro-Raman spectroscopy. *Analytical and Bioanalytical Chemistry, 409*, 4099–4109. https://doi.org/10.1007/s00216-017-0358-y
312. Carter, J., et al. (2023). Comparative evaluation of filtration and imaging properties of analytical filters for microplastic capture and analysis. *Chemosphere, 332*, 138811. https://doi.org/10.1016/j.chemosphere.2023.138811
313. So, Y., Kim, S. Y., Kim, S., & Par, C. (2025). Innovative approaches to high-speed ceramic membrane filtration for microplastic mitigation in urban wastewater treatment facilities. *Separation and Purification Technology, 363*, 132013. https://doi.org/10.1016/j.seppur.2025.132013
314. Hyeon, Y., Kim, S., & Park, C. (2024). Exploring the transformation of polyethylene and polyamide microplastics during membrane filtration through FlowCam analysis. *Separation and Purification Technology, 334*, 126036. https://doi.org/10.1016/j.seppur.2023.126036
315. Pizzichetti, A. R. P., Pablos, C., Álvarez-Fernández, C., Reynolds, K., Stanley, S., & Marugán, J. (2021). Evaluation of membranes performance for microplastic removal in a simple and low-cost filtration system. *Case Studies in Chemical and Environmental Engineering, 3*, 100075, 1–5. https://doi.org/10.1016/j.cscee.2020.100075
316. LaRue, R. J., Patterson, B., O'Brien, S., & Latulippe, D. R. (2022). Evaluation of membrane fouling by microplastic particles in tertiary wastewater treatment processes. *ACS ES&T Water, 2*(6), 955–966. https://doi.org/10.1021/acsestwater.1c00430
317. Li, Q.-C., et al. (2021). Sequential isolation of microplastics and nanoplastics in environmental waters by membrane filtration, followed by cloud-point extraction. *Analytical Chemistry, 93*(10), 4559–4566. https://doi.org/10.1021/acs.analchem.0c04996
318. Rist, M., & Greiner, A. (2024). Bio-based electrospun polyamide membrane—Sustainable multipurpose filter membranes for microplastic filtration. *RSC Applied Polymers, 2*, 642–655. https://doi.org/10.1039/D3LP00201B
319. Miclea, P.-T., Richter, S., & Hagendorf, C. (2023). Nanoporous membrane filter cascade for size-selective analysis of nano- and microplastic particles. *Applied Research, 2*(5). https://doi.org/10.1002/appl.202200094
320. Ding, H., et al. (2021). Do membrane filtration systems in drinking water treatment plants release nano/microplastics? *Science of the Total Environment, 755*, 142658. https://doi.org/10.1016/j.scitotenv.2020.142658
321. Hube, S., Veronelli, S., Li, T., Burkhardt, M., Brynjólfsson, S., & Wu, B. (2024). Microplastics affect membrane biofouling and microbial communities during gravity-driven membrane filtration of primary wastewater. *Chemosphere, 353*, 141650. https://doi.org/10.1016/j.chemosphere.2024.141650
322. Colakoglu, E. B., Uyanik, I., Elbir, H., Sahinkaya, E., & Yurtsever, A. (2025). A novel gravity-driven dynamic membrane filtration reactor for microplastic removal from plastic recycling facility wastewater. *Journal of Environmental Chemical Engineering, 13*(2). https://doi.org/10.1016/j.jece.2025.115793
323. Chen, M., et al. (2024). Superior microplastic removal and gravity-driven membrane filtration optimization: The role of octadecyl-quaternium hybrid coagulant and molecular

dynamics insights. *Chemical Engineering Journal, 497*, 154817. https://doi.org/10.1016/j.cej.2024.154817

324. Akarsu, C., Kumbur, H., & Kideys, A. E. (2021). Removal of microplastics from wastewater through electrocoagulation-electroflotation and membrane filtration processes. *Water Science and Technology, 84*(7), 1648–1662. https://doi.org/10.2166/wst.2021.356
325. Girish, N., Parashar, N., & Hait, S. (2023). Coagulative removal of microplastics from aqueous matrices: Recent progresses and future perspectives. *Science of the Total Environment, 899*, 165723. https://doi.org/10.1016/j.scitotenv.2023.165723
326. Yu, K., Zhang, Y., Yang, S., & Chen, L. (2024). Enhanced removal of polystyrene microplastics from water through coagulation using polyaluminum ferric chloride with coagulant aids. *Environmental Engineering Science, 41*(11). https://doi.org/10.1089/ees.2023.0332
327. He, W., Chen, X., Xu, C., Zhou, C., & Wang, C. (2023). Internal interaction between chemically-pretreated polypropylene microplastics and floc growth during flocculation: Critical effect on floc properties and flocculation mechanisms. *Separation and Purification Technology, 306*, 122710. https://doi.org/10.1016/j.seppur.2022.122710
328. Godoy, L. G. R., de Oliveira, M. G. C., Moruzzi, R. B., Sharifi, S., & dos Reis, A. G. (2025). Improving microplastics aggregate settling velocity through ballasted flocculation via natural-based coagulants and nonintrusive image monitoring. *Journal of Water Process Engineering, 74*, 107761. https://doi.org/10.1016/j.jwpe.2025.107761
329. Godoy, L. G. R., Moruzzi, R. B., Odjegba, E. E., Sharifi, S., & dos Reis, A. G. (2025). Enhanced removal of microplastics from drinking water using ballasted flocculation: A comparative study with conventional methods using non-intrusive image analysis. *Brazilian Journal of Chemical Engineering*. https://doi.org/10.1007/s43153-024-00530-3
330. Adib, D., Mafigholami, R., Tabeshkia, H., & Walker, T. R. (2022). Optimization of polypropylene microplastics removal using conventional coagulants in drinking water treatment plants via response surface methodology. *Journal of Environmental Health Science and Engineering, 20*, 565–577. https://doi.org/10.1007/s40201-022-00803-4
331. Pu, K. (2023). Coagulation technologies for separation of microplastics in water: Current status. In *The 3rd International Conference on Materials Chemistry and Environmental Engineering (CONFMCEE 2023) 18/03/2023–18/03/2023 Stanford, United States of America*. IOP Publishing Ltd. https://doi.org/10.1088/1742-6596/2608/1/012056
332. Ma, B., Xue, W., Ding, Y., Hu, C., Liu, H., & Qu, J. (2019). Removal characteristics of microplastics by Fe-based coagulants during drinking water treatment. *Journal of Environmental Sciences, 78*, 267–275. https://doi.org/10.1016/j.jes.2018.10.006
333. Putranto, P. A., Khoironi, A., & Baihaqi, R. A. (2023). Optimisation of chitosan as a natural flocculant for microplastic remediation. *Journal of emergency science and engineering, 1*(2), 44–50. https://doi.org/10.61435/jese.2023.7
334. Reza, T., Riza, Z. H. M., Abdullah, S. R. S., Hasan, H. A., 'Izzati Ismail, N., & Othman, A. R. (2024). Microplastic removal in wastewater treatment plants (WWTPs) by natural coagulation: A literature review. *Toxics, 12*(1). https://doi.org/10.3390/toxics12010012
335. Thathsarani, N., Khiadani, M., Azhar, M. R., & Zargar, M. (2025). Enhanced removal of microplastic fibres using aluminium and chitosan-based coagulants assisted with microbubble technology. *Journal of Environmental Chemical Engineering, 13*(3). https://doi.org/10.1016/j.jece.2025.116780
336. Gao, Y., & Liu, Y. (2022). Removal of microplastics by coagulation treatment in waters and prospect of recycling of separated microplastics: A mini-review. *Journal of Environmental Chemical Engineering, 10*(5). https://doi.org/10.1016/j.jece.2022.108197
337. Zhang, Y., et al. (2021). Coagulation removal of microplastics from wastewater by magnetic magnesium hydroxide and PAM. *Journal of Water Process Engineering, 43*, 102250. https://doi.org/10.1016/j.jwpe.2021.102250
338. Lee, J., Wang, J., Oh, Y., & Jeong, S. (2023). Highly efficient microplastics removal from water using in-situ ferrate coagulation: Performance evaluation by micro-Fourier-transformed

infrared spectroscopy and coagulation mechanism. *Chemical Engineering Journal, 451*, 138556. https://doi.org/10.1016/j.cej.2022.138556
339. Eamrat, R., Rujakom, S., Pussayanavin, T., Taweesan, A., Witthayaphirom, C., & Kamei, T. (2024). Optimizing biocoagulant aid from shrimp shells (Litopenaeus vannamei) for enhancing microplastics removal from aqueous solutions. *Environmental Technology and Innovation, 33*, 103457. https://doi.org/10.1016/j.eti.2023.103457
340. Lapointe, M., Farner, J. M., Hernandez, L. M., & Tufenkji, N. (2020). Understanding and improving microplastic removal during water treatment: Impact of coagulation and flocculation. *Environmental Science & Technology, 54*(14), 8719–8727. https://doi.org/10.1021/acs.est.0c00712
341. Xue, J., Peldszus, S., Van Dyke, M. I., & Huck, P. M. (2021). Removal of polystyrene microplastic spheres by alum-based coagulation-flocculation-sedimentation (CFS) treatment of surface waters. *Chemical Engineering Journal, 422*, 130023. https://doi.org/10.1016/j.cej.2021.130023
342. Li, C., Busquets, R., Moruzzi, R. B., & Campos, L. C. (2021). Preliminary study on low-density polystyrene microplastics bead removal from drinking water by coagulation-flocculation and sedimentation. *Journal of Water Process Engineering, 44*, 102346. https://doi.org/10.1016/j.jwpe.2021.102346
343. Peydayesh, M., et al. (2021). Sustainable removal of microplastics and natural organic matter from water by coagulation–flocculation with protein amyloid fibrils. *Environmental Science & Technology, 55*(13), 8848–8858. https://doi.org/10.1021/acs.est.1c01918
344. Rajala, K., Grönfors, O., Hesampour, M., & Mikola, A. (2020). Removal of microplastics from secondary wastewater treatment plant effluent by coagulation/flocculation with iron, aluminum and polyamine-based chemicals. *Water Research, 183*, 116045. https://doi.org/10.1016/j.watres.2020.116045
345. Esfandiari, A., & Mowla, D. (2021). Investigation of microplastic removal from greywater by coagulation and dissolved air flotation. *Process Safety and Environmental Protection, 151*, 341–354. https://doi.org/10.1016/j.psep.2021.05.027
346. Skaf, D. W., Punzi, V. L., Rolle, J. T., & Kleinberg, K. A. (2020). Removal of micron-sized microplastic particles from simulated drinking water via alum coagulation. *Chemical Engineering Journal, 386*, 123807. https://doi.org/10.1016/j.cej.2019.123807
347. He, J., et al. (2022). Understanding and characteristics of coagulation removal of composite pollution of microplastic and norfloxacin during water treatment. *Science of the Total Environment, 831*, 154826. https://doi.org/10.1016/j.scitotenv.2022.154826
348. Wiśniowska, E., Moraczewska-Majkut, K., Nocoń, W., & Popenda, A. (2024). Microplastics removal from natural surface water by coagulation process. *Desalination and Water Treatment, 319*, 100462. https://doi.org/10.1016/j.dwt.2024.100462
349. Ziembowicz, S., Kida, M., & Koszelnik, P. (2023). Elimination of a mixture of microplastics using conventional and detergent-assisted coagulation. *Materials (Basel), 16*(11). https://doi.org/10.3390/ma16114070
350. Li, C. (2023). *Bench-scale study into the effectiveness of microplastic removal from surface water by coagulation flocculation and sedimentation.* Degree of Doctor of Philosophy in Environmental Engineering, Department of Civil, Environmental and Geomatic Engineering, Faculty of Engineering, University College London (UCL). https://discovery.ucl.ac.uk/id/eprint/10175050/
351. Li, C., Busquets, R., & Campos, L. C. (2024). Enhancing microplastic removal from natural water using coagulant aids. *Chemosphere, 364*, 143145. https://doi.org/10.1016/j.chemosphere.2024.143145
352. Li, C. (2023). *Coagulation-flocculation-sedimentation for the removal of microplastics from surface water*. Degree of Doctor of Philosophy (PhD) in Environmental Engineering, Department of Civil, Environmental and Geomatic Engineering, Faculty of Engineering, University College London (UCL). https://discovery.ucl.ac.uk/id/eprint/10175052/

353. Avazpour, S., & Noshadi, M. (2024). Enhancing the coagulation process for the removal of microplastics from water by anionic polyacrylamide and natural-based Moringa oleifera. *Chemosphere, 358*, 142215. https://doi.org/10.1016/j.chemosphere.2024.142215
354. Monira, S., Haque, M. A., Nawshad, B., & Pramanik, B. K. (2021). Assess the performance of chemical coagulation process for microplastics removal from stormwater. *Process Safety and Environmental Protection, 155*, 11–16. https://doi.org/10.1016/j.psep.2021.09.002
355. Zhang, Y., Fu, S., & Chen, L. (2024). Enhanced removal of polystyrene microplastics through coagulation using polyaluminum ferric chloride with Opuntia Milpa Alta particles. *Journal of Environmental Chemical Engineering, 12*(5). https://doi.org/10.1016/j.jece.2024.113802
356. Prokopova, M., Novotna, K., Pivokonska, L., Cermakova, L., Cajthaml, T., & Pivokonsky, M. (2021). Coagulation of polyvinyl chloride microplastics by ferric and aluminium sulphate: Optimisation of reaction conditions and removal mechanisms. *Journal of Environmental Chemical Engineering, 9*(6). https://doi.org/10.1016/j.jece.2021.106465
357. Raj, S., Mahanty, B., & Hait, S. (2024). Coagulative removal of polystyrene microplastics from aqueous matrices using $FeCl_3$-chitosan system: Experimental and artificial neural network modeling. *Journal of Hazardous Materials Letters, 468*, 133818. https://doi.org/10.1016/j.jhazmat.2024.133818
358. Yao, J., et al. (2023). Surface characteristics of polystyrene microplastics mainly determine their coagulation performances. *Marine Pollution Bulletin, 186*, 114347. https://doi.org/10.1016/j.marpolbul.2022.114347
359. Chen, M., Nan, J., Xu, Y., Yao, J., Wang, H., & Zu, X. (2022). Effect of microplastics on the physical structure of cake layer for pre-coagulated gravity-driven membrane filtration. *Separation and Purification Technology, 288*, 120632. https://doi.org/10.1016/j.seppur.2022.120632
360. Hassanzad, M. (2024). *A comprehensive study on the investigation of microplastic removal in greywater treatment using adsorbent materials and coagulants*. Masters Thesis, Department of Environmental Engineering. https://dea.lib.unideb.hu/items/b7bbed9c-4192-41ef-bc7f-53e55136aa10
361. Azizi, N., Pirsaheb, M., Jaafarzadeh, N., & Nodehi, R. N. (2023). Microplastics removal from aquatic environment by coagulation: Selecting the best coagulant based on variables determined from a systematic review. *Heliyon, 9*(5). https://doi.org/10.1016/j.heliyon.2023.e15664
362. Badawi, A. K., Hasan, R., & Ismail, B. (2025). Sustainable coagulative removal of microplastic from aquatic systems: Recent progress and outlook. *RSC Advances, 15*, 25256–25273. https://doi.org/10.1039/D5RA04074D
363. Khan, M. T., et al. (2023). Microplastic removal by coagulation: A review of optimizing the reaction conditions and mechanisms. *Water Emerging Contaminants & Nanoplastics, 2*(4). https://doi.org/10.20517/wecn.2023.39
364. Hosseinzadeh, A., Amirkhani, F., Azizi, N., Dashti, A., Zhou, J. L., & Altaee, A. (2025). Machine learning modeling of microplastics removal by coagulation in water and wastewater treatment. *Journal of Water Process Engineering, 76*, 108108, 1–9. https://doi.org/10.1016/j.jwpe.2025.108108
365. Kumar, A. (2022). *Removal characteristics of fibrous and particulate microplastics: A study of the effectiveness of coagulation-flocculation-sedimentation*. Master Programme in Earth Science, Department of Earth Sciences, Uppsala University. https://www.diva-portal.org/smash/get/diva2:1687510/FULLTEXT01.pdf
366. Zhang, Y., Zhao, J., Li, W., & Yuan, H. (2024). Coagulation properties of magnetic magnesium hydroxide for removal of microplastics in the presence of kaolin and humic acid. *Environmental Technology, 45*(8), 1459–1470. https://doi.org/10.1080/09593330.2022.2144766
367. Zhang, X., et al. (2024). Effect of microplastic ingredient on the removal of microplastics by calcium alginate flocculation. *Chemical Engineering Journal, 498*, 155701. https://doi.org/10.1016/j.cej.2024.155701

368. Sturm, M. T., Myers, E., Schober, D., Korzin, A., Thege, C., & Schuhen, K. (2023). Comparison of AOP, GAC, and novel organosilane-based process for the removal of microplastics at a municipal wastewater treatment plant. *Water (Basel), 15*(6). https://doi.org/10.3390/w15061164
369. Wang, D., Junker, A. L., Sillanpää, M., Jiang, Y., & Wei, Z. (2023). Photo-based advanced oxidation processes for zero pollution: Where are we now? *Engineering Applications of Technology, 23*, 19–23. https://doi.org/10.1016/j.eng.2022.08.005
370. Wypart-Pawul, A., Neczaj, E., & Grobelak, A. (2023). Advanced oxidation processes for removal of organic micropollutants from wastewater. *Desalination and Water Treatment, 305*, 114–128. https://doi.org/10.5004/dwt.2023.29665
371. John, K. I., Omorogie, M. O., Bayode, A. A., Adeleye, A. T., & Helmreich, B. (2023). Environmental microplastics and their additives—A critical review on advanced oxidative techniques for their removal. *Chemical Papers, 77*, 657–676. https://doi.org/10.1007/s11696-022-02505-5
372. Quilumbaquin, W., Vaca-Escobar, K., & Espinoza-Montero, P. J. (2025). Advanced oxidation processes for degrading microplastics in aqueous media. In I. Haq & M. P. Shah (Eds.), *Advanced treatment technologies for the removal of microplastics in wastewater* (1st ed., pp. 1–28). CRC Press. https://doi.org/10.1201/9781003379386
373. Hu, K., et al. (2021). Microplastics remediation in aqueous systems: Strategies and technologies. *Water Research, 198*, 117144. https://doi.org/10.1016/j.watres.2021.117144
374. Wang, H., Wang, Y., & Dionysiou, D. D. (2023). Advanced oxidation processes for removal of emerging contaminants in water. *Water (Basel), 15*(3). https://doi.org/10.3390/w15030398
375. Gayathri, P. V., Joseph, S., Mohan, M., & Pillai, D. (2023). Advanced oxidation processes for the degradation of microplastics from the environment: A review. *Water Environment Journal, 37*(4), 686–701. https://doi.org/10.1111/wej.12883
376. Kim, S., Sin, A., Nam, H., Park, Y., Lee, H., & Han, C. (2022). Advanced oxidation processes for microplastics degradation: A recent trend. *Chemical Engineering Journal Advances, 9*, 100213. https://doi.org/10.1016/j.ceja.2021.100213
377. Surana, M., Pattanayak, D. S., Yadav, V., Singh, V. K., & Pal, D. (2024). An insight decipher on photocatalytic degradation of microplastics: Mechanism, limitations, and future outlook. *Environmental Research, 247*, 118268. https://doi.org/10.1016/j.envres.2024.118268
378. Jadoun, S., Fuentes, J. P., Yepsen, O., & Yáñez, J. (2023). Removal of environmental microplastics by advanced oxidation processes. In C. Wang, S. Babel, & E. Lichtfouse (Eds.), *Microplastic occurrence, fate, impact, and remediation* (Vol. 73, pp. 109–125). Springer. https://doi.org/10.1007/978-3-031-36351-1_5
379. Yao, J., Qian, H., Yan, Z., Zhao, X., Gao, N., & Zhang, Z. (2024). Insight into the effect of UVC-based advanced oxidation processes on the interaction of typical microplastics and their derived disinfection byproducts during disinfection. *Journal of Hazardous Materials, 472*, 134597. https://doi.org/10.1016/j.jhazmat.2024.134597
380. Easton, T., Koutsos, V., & Chatzisymeon, E. (2023). Removal of polyester fibre microplastics from wastewater using a UV/$H_2O_2$ oxidation process. *Journal of Environmental Chemical Engineering, 11*(1). https://doi.org/10.1016/j.jece.2022.109057
381. Thomas, E., Koutsos, V., & Efthalia, C. (2021). Tackling microplastic pollution with advanced oxidation processes: Development of a wastewater treatment method for polyester fibre removal. In *SPEA11* (pp. 1–2) https://www.researchgate.net/profile/Thomas_Easton3/publication/380185866
382. Dos Santos, N. d. O., Busquets, R., & Campos, L. C. (2023). Insights into the removal of microplastics and microfibres by advanced oxidation processes. *Science of the Total Environment, 861*, 160665. https://doi.org/10.1016/j.scitotenv.2022.160665
383. Prica, M., et al. (2020). Application of advanced oxidation process for the removal of synthetic water-based printing dye and microplastics from aqueous solution. In *GRID* (pp. 243–249). https://doi.org/10.24867/GRID-2020-p26

384. Kovačić, M., et al. (2024). Pristine and UV-weathered PET microplastics as water contaminants: Appraising the potential of the Fenton process for effective remediation. *Processes, 12*(4). https://doi.org/10.3390/pr12040844
385. Raksaman, S., Kloysuwan, C., Khrueakham, A., & Sairiam, S. (2024). Effect of spunbond nonwoven microplastics on dye wastewater treatment via hydrogen peroxide–based catalyst–assisted advanced oxidation processes. *Environmental Advances, 17*, 100567. https://doi.org/10.1016/j.envadv.2024.100567
386. Ziembowicz, S., & Kida, M. (2024). The effect of water ozonation in the presence of microplastics on water quality and microplastics degradation. *Science of the Total Environment, 929*, 172595. https://doi.org/10.1016/j.scitotenv.2024.172595
387. Hu, J., Lim, F. Y., & Hu, J. (2023). Ozonation facilitates the aging and mineralization of polyethylene microplastics from water: Behavior, mechanisms, and pathways. *Science of the Total Environment, 866*, 161290. https://doi.org/10.1016/j.scitotenv.2022.161290
388. Adeel, M., Maniakova, G., & Rizzo, L. (2024). Tertiary/quaternary treatment of urban wastewater by UV/$H_2O_2$ or ozonation: Microplastics may affect removal of E. coli and contaminants of emerging concern. *Science of the Total Environment, 907*, 167940. https://doi.org/10.1016/j.scitotenv.2023.167940
389. Jihye, S. (2021). *Oxidation behaviors of microplastics in water during UV photolysis and ozonation* (Vol. 2). https://s-space.snu.ac.kr/handle/10371/175498
390. Lee, H., Im, S.-J., Kim, Y., Lee, G., & Jang, A. (2021). Effects of microplastics on the removal of trace organic compounds during ozonation: Oxidation and adsorption of trace organic compounds and byproducts. *Environmental Pollution, 280*, 116878. https://doi.org/10.1016/j.envpol.2021.116878
391. Solís-Balbín, C., Sol, D., Laca, A., Laca, A., & Díaz, M. (2023). Destruction and entrainment of microplastics in ozonation and wet oxidation processes. *Journal of Water Process Engineering, 51*, 103456. https://doi.org/10.1016/j.jwpe.2022.103456
392. Wang, L., et al. (2022). Transfer of microplastics in sludge upon Fe(II)-persulfate conditioning and mechanical dewatering. *Science of the Total Environment, 838*, 156316. https://doi.org/10.1016/j.scitotenv.2022.156316
393. Lee, Y., et al. (2025). Activated persulfate oxidation as pretreatment for quantification of microplastics in natural water. *ACS ES&T Engineering, 5*(5), 1131–1139. https://doi.org/10.1021/acsestengg.4c00821
394. Lopez-Arago, N., et al. (2025). Sonolytic degradation of benzophenone-3 in water matrices: Reaction mechanism, transformation products, ecotoxicological implications & microplastic interaction. *Science of the Total Environment, 996*, 180087. https://doi.org/10.1016/j.scitotenv.2025.180087
395. Zawadzki, P. (2022). Visible light–driven advanced oxidation processes to remove emerging contaminants from water and wastewater: A review. *Water, Air, and Soil Pollution, 233*, 374. https://doi.org/10.1007/s11270-022-05831-2
396. Sacco, N. A., Zoppas, F. M., Devard, A., del P. G. Muñoz, M., García, G., & Marchesini, F. A. (2023). Recent advances in microplastics removal from water with special attention given to photocatalytic degradation: Review of scientific research. *Microplastics, 2*(3), 278–303. https://doi.org/10.3390/microplastics2030023
397. Kiendrebeogo, M., Estahbanati, M. R. K., Mostafazadeh, A. K., Drogui, P., & Tyagi, R. D. (2021). Treatment of microplastics in water by anodic oxidation: A case study for polystyrene. *Environmental Pollution, 269*, 116168. https://doi.org/10.1016/j.envpol.2020.116168
398. Chokejaroenrat, C., et al. (2024). Impacts of microplastic decomposition using heat-activated persulfate on antibiotic adsorption and environmental toxicity. *Marine Pollution Bulletin, 205*, 116576. https://doi.org/10.1016/j.marpolbul.2024.116576
399. Zhang, X., et al. (2023). UV-photoaging behavior of polystyrene microplastics enhanced by thermally-activated persulfate. *Journal of Environmental Chemical Engineering, 11*(5). https://doi.org/10.1016/j.jece.2023.110508

400. Kim, Y.-E., et al. (2023). Role of inorganic anions in improving the oxidizing capacity of heat-activated peroxymonosulfate: Identification of primary degradative pathways. *Chemical Engineering Journal, 478*, 147472. https://doi.org/10.1016/j.cej.2023.147472
401. Surendran, D., Sakai, H., Takagi, S., & Dimapilis, D. A. (2025). Tire-based microplastics: Composition, detection, and impacts of advanced oxidation processes in drinking water treatment. *Science of the Total Environment, 972*, 179114. https://doi.org/10.1016/j.scitotenv.2025.179114
402. Shen, M., Song, B., Zhou, C., Hu, T., Zeng, G., & Zhang, Y. (2022). Advanced oxidation processes for the elimination of microplastics from aqueous systems: Assessment of efficiency, perspectives and limitations. *Science of the Total Environment, 842*, 156723. https://doi.org/10.1016/j.scitotenv.2022.156723
403. Yang, Z., Li, Y., & Zhang, G. (2024). Degradation of microplastic in water by advanced oxidation processes. *Chemosphere, 357*, 141939. https://doi.org/10.1016/j.chemosphere.2024.141939
404. Zoppas, F. M., Sacco, N., Soffietti, J., Devard, A., Akhter, F., & Marchesini, F. A. (2023). Catalytic approaches for the removal of microplastics from water: Recent advances and future opportunities. *Chemical Engineering Journal Advances, 16*, 100529. https://doi.org/10.1016/j.ceja.2023.100529
405. Ergun, C. (2023). A current review on conducting polymer-based catalysts: Advanced oxidation processes for the removal of aquatic pollutants. *Water, Air, and Soil Pollution, 234*, 524. https://doi.org/10.1007/s11270-023-06526-y
406. Ma, Y., et al. (2025). Advanced oxidation in the treatment of microplastics in water: A review. *Desalination and Water Treatment, 322*, 101135. https://doi.org/10.1016/j.dwt.2025.101135
407. Rizwan, K., & Bilal, M. (2022). Developments in advanced oxidation processes for removal of microplastics from aqueous matrices. *Environmental Science and Pollution Research, 29*, 86933–86953. https://doi.org/10.1007/s11356-022-23545-0
408. Roy, H., et al. (2025). Degradation of microplastics in the aquatic environment by advanced oxidation process. In T. A. Kurniawan & A. Anouzla (Eds.), *Handbook of microplastic pollution in the environment* (1st ed., pp. 1–31). CRC Press. https://doi.org/10.1201/9781003487555
409. Cui, B., et al. (2024). Chemical methods to remove microplastics from wastewater: A review. *Environmental Research Letters, 249*, 118416. https://doi.org/10.1016/j.envres.2024.118416
410. Zahmatkesh, S., et al. (2023). Various advanced wastewater treatment methods to remove microplastics and prevent transmission of SARS-CoV-2 to airborne microplastics. *International Journal of Environmental Science and Technology, 20*, 2229–2246. https://doi.org/10.1007/s13762-022-04654-2
411. Moraczewska-Majkut, K., et al. (2023). Application of membrane filtration and adsorption processes in the removal of micropollutants released from microplastics. *Desalination and Water Treatment, 315*, 458–468. https://doi.org/10.5004/dwt.2023.30178
412. Li, J., Zhang, K., & Zhang, H. (2018). Adsorption of antibiotics on microplastics. *Environmental Pollution, 237*, 460–467. https://doi.org/10.1016/j.envpol.2018.02.050
413. Li, M., & Sui, M. (2025). Interaction between microplastics and antibiotics in aquatic environments and the characteristics of composite pollution removal by coagulation and advanced oxidation method. *Progress in Chemistry, 37*(1), 124–132. https://doi.org/10.7536/PC240617
414. Senavirathna, M. D. H. J., Zhaozhi, L., & Fujino, T. (2022). Root adsorption of microplastic particles affects the submerged freshwater Macrophyte Egeria densa. *Water, Air, and Soil Pollution, 233*, 80. https://doi.org/10.1007/s11270-022-05556-2
415. Godasiaei, S. H. (2025). Predictive modeling of microplastic adsorption in aquatic environments using advanced machine learning models. *Science of the Total Environment, 958*, 178015. https://doi.org/10.1016/j.scitotenv.2024.178015
416. Astray, G., Soria-Lopez, A., Barreiro, E., Mejuto, J. C., & Cid-Samamed, A. (2023). Machine learning to predict the adsorption capacity of microplastics. *Nanomaterials, 13*(6). https://doi.org/10.3390/nano13061061

417. Bahrami, M., Amiri, M. J., Rajabi, S., & Mahmoudi, M. (2024). The removal of methylene blue from aqueous solutions by polyethylene microplastics: Modeling batch adsorption using random forest regression. *Alexandria Engineering Journal, 95*, 101–113. https://doi.org/10.1016/j.aej.2024.03.100
418. Li, M., Yu, H., Wang, Y., Li, J., Ma, G., & Wei, X. (2020). QSPR models for predicting the adsorption capacity for microplastics of polyethylene, polypropylene and polystyrene. *Scientific Reports, 10*, 14597. https://doi.org/10.1038/s41598-020-71390-3
419. Chabi, K., et al. (2024). Rapid sand filtration for <10 μm-sized microplastic removal in tap water treatment: Efficiency and adsorption mechanisms. *Science of the Total Environment, 912*, 169074. https://doi.org/10.1016/j.scitotenv.2023.169074
420. Verma, A., et al. (2024). Microplastic pollutants in water: A comprehensive review on their remediation by adsorption using various adsorbents. *Chemosphere, 352*, 141365. https://doi.org/10.1016/j.chemosphere.2024.141365
421. Mulindwa, P., Kasule, J. S., Nantaba, F., Wasswa, J., & Expósito, A. J. (2024). Bioadsorbents for removal of microplastics from water ecosystems: A review. *International Journal of Sustainable Engineering, 17*(1), 582–599. https://doi.org/10.1080/19397038.2024.2374003
422. Li, J., Chen, X., Yu, S., & Cui, M. (2023). Removal of pristine and aged microplastics from water by magnetic biochar: Adsorption and magnetization. *Science of the Total Environment, 875*, 162647. https://doi.org/10.1016/j.scitotenv.2023.162647
423. Yuan, F., Yue, L., Zhao, H., & Wu, H. (2020). Study on the adsorption of polystyrene microplastics by three-dimensional reduced graphene oxide. *Water Science and Technology, 81*(10), 2163–2175. https://doi.org/10.2166/wst.2020.269
424. Zhuang, J., Rong, N., Wang, X., Chuchu, C., & Xu, Z. (2022). Adsorption of small size microplastics based on cellulose nanofiber aerogel modified by quaternary ammonium salt in water. *Separation and Purification Technology, 293*, 121133. https://doi.org/10.1016/j.seppur.2022.121133
425. Jiao, Y., et al. (2025). Adsorption efficiency and in-situ catalytic thermal degradation behaviour of microplastics from water over Fe-modified lignin-based magnetic biochar. *Separation and Purification Technology, 353*, 128468. https://doi.org/10.1016/j.seppur.2024.128468
426. Li, H., et al. (2023). Theoretical and experimental investigation on rapid and efficient adsorption characteristics of microplastics by magnetic sponge carbon. *Science of the Total Environment, 897*, 165404. https://doi.org/10.1016/j.scitotenv.2023.165404
427. Wang, J., Sun, C., Huang, Q.-X., Chi, Y., & Yan, J.-H. (2021). Adsorption and thermal degradation of microplastics from aqueous solutions by Mg/Zn modified magnetic biochars. *Journal of Hazardous Materials, 419*, 126486. https://doi.org/10.1016/j.jhazmat.2021.126486
428. Li, W., Liu, S., Huang, K., Qin, S., Liang, B., & Wang, J. (2023). Preparation of magnetic Janus microparticles for the rapid removal of microplastics from water. *Science of the Total Environment, 903*, 166627. https://doi.org/10.1016/j.scitotenv.2023.166627
429. Wu, Y., et al. (2024). Revivable self-assembled supramolecular biomass fibrous framework for efficient microplastic removal. *Science Advances, 10*(48). https://doi.org/10.1126/sciadv.adn8662
430. Zhang, Y., et al. (2025). Efficient magnetic capture of PE microplastic from water by PEG modified $Fe_3O_4$ nanoparticles: Performance, kinetics, isotherms and influence factors. *Journal of Environmental Sciences, 147*, 677–687. https://doi.org/10.1016/j.jes.2023.07.025
431. Han, M., Wang, Z., Xie, Z., Hou, M., & Gao, Z. (2025). Polydopamine-modified sodium alginate hydrogel for microplastics removal: Adsorption performance, characteristics, and kinetics. *International Journal of Biological Macromolecules, 297*, 139947. https://doi.org/10.1016/j.ijbiomac.2025.139947
432. Shi, C., Wu, H., Wang, W., Zhao, J., Niu, F., & Geng, J. (2024). Microplastic removal from water using modified maifanite with rotating magnetic field affected. *Journal of Cleaner Production, 434*, 140111. https://doi.org/10.1016/j.jclepro.2023.140111
433. Liu, F., Zhang, Y., Zhong, L., Feng, Q., Dong, Z., & Xu, Z. (2024). Dialdehyde modified and cationic aerogel for efficient microplastics adsorption from environmental waters.

*International Journal of Biological Macromolecules, 256*, 128326. https://doi.org/10.1016/j.ijbiomac.2023.128326
434. Sumaryada, T., Nabilah, F., Handayasari, F., Kartono, A., & Hardhienata, H. (2025). Microplastic contaminant adsorption by graphene oxide layer. *Journal of Biological Physics, 51*, 12. https://doi.org/10.1007/s10867-025-09677-7
435. Liu, S., & Wang, J. (2023). Exploring the potential of cellulose benzoate adsorbents modified with carbon nanotubes and magnetic carbon nanotubes for microplastic removal from water. *Chemical Engineering Journal, 469*, 143910. https://doi.org/10.1016/j.cej.2023.143910
436. Ha, T. T. V., Viet, N. M., Thanh, P. T., & Quan, V. T. (2023). Loofah plant—Derived biodegradable superhydrophobic sponge for effective removal of oil and microplastic from water. *Environmental Technology & Innovation, 32*, 103265. https://doi.org/10.1016/j.eti.2023.103265
437. Bhagwat, K. P., Rodrigue, D., & Romero-Zerón, L. (2024). Effective removal of microplastic particles from wastewater using hydrophobic bio-substrates. *Pollutants, 4*(2), 231–250. https://doi.org/10.3390/pollutants4020015
438. Xu, J., et al. (2023). Hardwood vessel-inspired chitosan-based sponge with superior compressibility, superfast adsorption and remarkable recyclability for microplastics removal in water. *Chemical Engineering Journal, 475*, 146130. https://doi.org/10.1016/j.cej.2023.146130
439. Zhao, M., et al. (2022). Adsorption of different pollutants by using microplastic with different influencing factors and mechanisms in wastewater: A review. *Nanomaterials, 12*(13). https://doi.org/10.3390/nano12132256
440. Wang, Y., et al. (2021). Effects of exposure of polyethylene microplastics to air, water and soil on their adsorption behaviors for copper and tetracycline. *Chemical Engineering Journal, 404*, 126412. https://doi.org/10.1016/j.cej.2020.126412
441. Turner, A., & Holmes, L. A. (2015). Adsorption of trace metals by microplastic pellets in fresh water. *Environment and Chemistry, 12*(5), 600–610. https://doi.org/10.1071/EN14143
442. Li, H., Wang, F., Li, J., Deng, S., & Zhang, S. (2021). Adsorption of three pesticides on polyethylene microplastics in aqueous solutions: Kinetics, isotherms, thermodynamics, and molecular dynamics simulation. *Chemosphere, 264*, 128556. https://doi.org/10.1016/j.chemosphere.2020.128556
443. Fang, S., Yu, W., Li, C., Liu, Y., Qiu, J., & Kong, F. (2019). Adsorption behavior of three triazole fungicides on polystyrene microplastics. *Science of the Total Environment, 691*, 1119–1126. https://doi.org/10.1016/j.scitotenv.2019.07.176
444. Lee, Y. K., & Hur, J. (2020). Adsorption of microplastic-derived organic matter onto minerals. *Water Research, 187*, 116426. https://doi.org/10.1016/j.watres.2020.116426
445. Costigan, E., et al. (2022). Adsorption of organic pollutants by microplastics: Overview of a dissonant literature. *Journal of Hazardous Materials Advances, 6*, 100091. https://doi.org/10.1016/j.hazadv.2022.100091
446. Puckowski, A., Cwięk, W., Mioduszewska, K., Stepnowski, P., & Białk-Bielińska, A. (2021). Sorption of pharmaceuticals on the surface of microplastics. *Chemosphere, 263*, 127976. https://doi.org/10.1016/j.chemosphere.2020.127976
447. Wang, L., Yang, H., Guo, M., Wang, Z., & Zheng, X. (2023). Adsorption of antibiotics on different microplastics (MPs): Behavior and mechanism. *Science of the Total Environment, 863*, 161022. https://doi.org/10.1016/j.scitotenv.2022.161022
448. Zhang, W., et al. (2020). The mechanism for adsorption of Cr(VI) ions by PE microplastics in ternary system of natural water environment. *Environmental Pollution, 257*, 113440. https://doi.org/10.1016/j.envpol.2019.113440
449. Purwiyanto, A. I. S., et al. (2020). Concentration and adsorption of Pb and Cu in microplastics: Case study in aquatic environment. *Marine Pollution Bulletin, 158*, 111380. https://doi.org/10.1016/j.marpolbul.2020.111380
450. Munoz, M., Ortiz, D., Nieto-Sandoval, J., de Pedro, Z. M., & Casas, J. A. (2021). Adsorption of micropollutants onto realistic microplastics: Role of microplastic nature, size, age, and NOM fouling. *Chemosphere, 283*, 131085. https://doi.org/10.1016/j.chemosphere.2021.131085

451. Fu, L., Li, J., Wang, G., Luan, Y., & Dai, W. (2021). Adsorption behavior of organic pollutants on microplastics. *Ecotoxicology and Environmental Safety, 217*, 112207. https://doi.org/10.1016/j.ecoenv.2021.112207
452. He, W., Wang, X., Zhang, Y., Zhu, B., & Wu, H. (2022). Adsorption behavior of aged polystyrene microplastics (PSMPs) for manganese in water: Critical role of hydrated functional zone surrounding the microplastic surface. *Journal of Environmental Chemical Engineering, 10*(6). https://doi.org/10.1016/j.jece.2022.109040
453. Guo, C., et al. (2023). UV and chemical aging alter the adsorption behavior of microplastics for tetracycline. *Environmental Pollution, 318*, 120859. https://doi.org/10.1016/j.envpol.2022.120859
454. Bhagat, K., et al. (2022). Aging of microplastics increases their adsorption affinity towards organic contaminants. *Chemosphere, 298*, 134238. https://doi.org/10.1016/j.chemosphere.2022.134238
455. You, H., et al. (2021). Adsorption-desorption behavior of methylene blue onto aged polyethylene microplastics in aqueous environments. *Marine Pollution Bulletin, 167*, 112287. https://doi.org/10.1016/j.marpolbul.2021.112287
456. Li, S., Yang, M., Wang, H., & Jiang, Y. (2022). Adsorption of microplastics on aquifer media: Effects of the action time, initial concentration, ionic strength, ionic types and dissolved organic matter. *Environmental Pollution, 308*, 119482. https://doi.org/10.1016/j.envpol.2022.119482
457. Chakraborty, T. K., et al. (2023). Adsorption of acid and basic dye from the simulated wastewater using carbonized microplastic particles synthesized from recycled polyethylene terephthalate plastic waste bottles: An integrated approach for experimental and practical applications. *AQUA—Water Infrastructure, Ecosystems and Society, 72*(4), 491–506. https://doi.org/10.2166/aqua.2023.211
458. McDougall, L., Thomson, L., Brand, S., Wagstaff, A., Lawton, L. A., & Petrie, B. (2022). Adsorption of a diverse range of pharmaceuticals to polyethylene microplastics in wastewater and their desorption in environmental matrices. *Science of the Total Environment, 808*, 152071. https://doi.org/10.1016/j.scitotenv.2021.152071
459. Zhou, Y., Yang, Y., Liu, G., He, G., & Liu, W. (2020). Adsorption mechanism of cadmium on microplastics and their desorption behavior in sediment and gut environments: The roles of water pH, lead ions, natural organic matter and phenanthrene. *Water Research, 184*, 116209. https://doi.org/10.1016/j.watres.2020.116209
460. You, H., Cao, C., Sun, X., Huang, B., Qian, Q., & Chen, Q. (2023). Microplastics as an emerging vector of Cr(VI) in water: Correlation of aging properties and adsorption behavior. *Science of the Total Environment, 904*, 166480. https://doi.org/10.1016/j.scitotenv.2023.166480
461. Atugoda, T., Wijesekara, H., Werellagama, D. R. I. B., Jinadasa, K. B. S. N., Bolan, N. S., & Vithanage, M. (2020). Adsorptive interaction of antibiotic ciprofloxacin on polyethylene microplastics: Implications for vector transport in water. *Environmental Technology & Innovation, 19*, 100971. https://doi.org/10.1016/j.eti.2020.100971
462. Abuwatfa, W. H., Al-Muqbel, D., Al-Othman, A., Halalsheh, N., & Tawalbeh, M. (2021). Insights into the removal of microplastics from water using biochar in the era of COVID-19: A mini review. *Case Studies in Chemical and Environmental Engineering, 4*, 100151. https://doi.org/10.1016/j.cscee.2021.100151
463. Anuwa-Amarh, N. A., Dizbay-Onat, M., Venkiteshwaran, K., & Wu, S. (2024). Carbon-based adsorbents for microplastic removal from wastewater. *Materials (Basel), 17*(22). https://doi.org/10.3390/ma17225428
464. Siipola, V., Pflugmacher, S., Romar, H., Wendling, L., & Koukkari, P. (2020). Low-cost biochar adsorbents for water purification including microplastics removal. *Applied Sciences, 10*(3). https://doi.org/10.3390/app10030788
465. Titov, I., Semerád, J., Boháčková, J., Beneš, H., & Cajthaml, T. (2024). Microplastics meet micropollutants in a central European river stream: Adsorption of pollutants to microplastics

under environmentally relevant conditions. *Environmental Pollution, 360*, 124616. https://doi.org/10.1016/j.envpol.2024.124616

466. Lim, C., Kim, N., Lee, J., & Yoon, Y. (2022). Potential of adsorption of diverse environmental contaminants onto microplastics. *Water (Basel), 14*(24). https://doi.org/10.3390/w14244086
467. Ziajahromi, S., Neale, P. A., Rintoul, L., & Leusch, F. D. L. (2017). Wastewater treatment plants as a pathway for microplastics: Development of a new approach to sample wastewater-based microplastics. *Water Research, 112*, 93–99. https://doi.org/10.1016/j.watres.2017.01.042
468. Hung, C., Klasios, N., Zhu, X., Sedlak, M., Sutton, R., & Rochman, C. M. (2021). Methods matter: Methods for sampling microplastic and other anthropogenic particles and their implications for monitoring and ecological risk assessment. *Integrated Environmental Assessment and Management, 17*(1), 282–291. https://doi.org/10.1002/ieam.4325
469. Zong, C., et al. (2024). Vertical distribution of microplastics in the Western Pacific Warm Pool: In situ results comparison of different sampling method. *Journal of Hazardous Materials, 479*, 135722. https://doi.org/10.1016/j.jhazmat.2024.135722
470. Pasquier, G., Doyen, P., Dehaut, A., Veillet, G., Duflos, G., & Amara, R. (2023). Vertical distribution of microplastics in a river water column using an innovative sampling method. *Environmental Monitoring and Assessment, 195*, 1302. https://doi.org/10.1007/s10661-023-11915-1
471. Nousheen, R., Hashmi, I., Rittschof, D., & Capper, A. (2022). Comprehensive analysis of spatial distribution of microplastics in Rawal Lake, Pakistan using trawl net and sieve sampling methods. *Chemosphere, 308*, 136111. https://doi.org/10.1016/j.chemosphere.2022.136111
472. Stock, F., Kochleus, C., Bänsch-Baltruschat, B., Brennholt, N., & Reifferscheid, G. (2019). Sampling techniques and preparation methods for microplastic analyses in the aquatic environment—A review. *TrAC Trends in Analytical Chemistry, 113*, 84–92. https://doi.org/10.1016/j.trac.2019.01.014
473. Cutroneo, L., et al. (2020). Microplastics in seawater: Sampling strategies, laboratory methodologies, and identification techniques applied to port environment. *Environmental Science and Pollution Research, 27*, 8938–8952. https://doi.org/10.1007/s11356-020-07783-8
474. De-la-Torre, G. E., et al. (2022). Are we underestimating floating microplastic pollution? A quantitative analysis of two sampling methodologies. *Marine Pollution Bulletin, 178*, 113592. https://doi.org/10.1016/j.marpolbul.2022.113592
475. Lenz, R., & Labrenz, M. (2018). Small microplastic sampling in water: Development of an encapsulated filtration device. *Water (Basel), 10*(8). https://doi.org/10.3390/w10081055
476. Razeghi, N., Hamidian, A. H., Wu, C., Zhang, Y., & Yang, M. (2021). Microplastic sampling techniques in freshwaters and sediments: A review. *Environmental Chemistry Letters, 19*, 4225–4252. https://doi.org/10.1007/s10311-021-01227-6
477. Montoto-Martínez, T., Meléndez-Díez, C., Melián-Ramírez, A., Hernández-Brito, J. J., & Gelado-Caballero, M. D. (2022). Comparison between the traditional Manta net and an innovative device for microplastic sampling in surface marine waters. *Marine Pollution Bulletin, 185*, 114237. https://doi.org/10.1016/j.marpolbul.2022.114237
478. Green, D. S., et al. (2018). A comparison of sampling methods for seawater microplastics and a first report of the microplastic litter in coastal waters of Ascension and Falkland Islands. *Marine Pollution Bulletin, 137*, 695–701. https://doi.org/10.1016/j.marpolbul.2018.11.004
479. Tamminga, M., Stoewer, S.-C., & Fischer, E. K. (2019). On the representativeness of pump water samples versus manta sampling in microplastic analysis. *Environmental Pollution, 254*, 112970. https://doi.org/10.1016/j.envpol.2019.112970
480. Shi, H., Wang, X., Zhu, L., & Li, D. (2023). Comprehensive comparison of various microplastic sampling methods in sea water: Implications for data compilation. *Water (Basel), 15*(6). https://doi.org/10.3390/w15061035
481. Zheng, Y., et al. (2021). Comparative study of three sampling methods for microplastics analysis in seawater. *Science of the Total Environment, 765*, 144495. https://doi.org/10.1016/j.scitotenv.2020.144495

482. D'Ascanio, N. A., Glienke, J., Almuhtaram, H., & Andrews, R. C. (2025). Comparison of ASTM and in-line microplastic sampling methods for drinking water. *Microplastics and Nanoplastics, 5*, 17. https://doi.org/10.1186/s43591-025-00124-x
483. Almuhtaram, H., & Andrews, R. C. (2022). Sampling microplastics in water matrices: A need for standardization. *ACS ES&T Water, 2*(8), 1276–1278. https://doi.org/10.1021/acsestwater.2c00236
484. Yuan, C., Almuhtaram, H., McKie, M. J., & Andrews, R. C. (2022). Assessment of microplastic sampling and extraction methods for drinking waters. *Chemosphere, 286*, 131881. https://doi.org/10.1016/j.chemosphere.2021.131881
485. Pittroff, M., Müller, Y. K., Witzig, C. S., Scheurer, M., Storck, F. R., & Zumbülte, N. (2021). Microplastic analysis in drinking water based on fractionated filtration sampling and Raman microspectroscopy. *Environmental Science and Pollution Research, 28*, 59439–59451. https://doi.org/10.1007/s11356-021-12467-y
486. Covernton, G. A., et al. (2019). Size and shape matter: A preliminary analysis of microplastic sampling technique in seawater studies with implications for ecological risk assessment. *Science of the Total Environment, 667*, 124–132. https://doi.org/10.1016/j.scitotenv.2019.02.346
487. Kamil, K., Agnieszka, P., Anita, K., Michał, W., Krassimira, I.-M., & Jerzy, R. (2023). The determination of microplastic contamination in freshwater environments using sampling methods: A case study. *Journal of Water Law & Development, 57*(IV–VI), 140–146. https://doi.org/10.24425/jwld.2023.145344
488. Becucci, M., Mancini, M., Campo, R., & Paris, E. (2022). Microplastics in the Florence wastewater treatment plant studied by a continuous sampling method and Raman spectroscopy: A preliminary investigation. *Science of the Total Environment, 808*, 152025. https://doi.org/10.1016/j.scitotenv.2021.152025
489. Pagter, E., Frias, J., & Nash, R. (2018). Microplastics in Galway Bay: A comparison of sampling and separation methods. *Marine Pollution Bulletin, 135*, 932–940. https://doi.org/10.1016/j.marpolbul.2018.08.013
490. Prata, J. C., da Costa, J. P., Duarte, A. C., & Rocha-Santos, T. (2019). Methods for sampling and detection of microplastics in water and sediment: A critical review. *TrAC Trends in Analytical Chemistry, 110*, 150–159. https://doi.org/10.1016/j.trac.2018.10.029
491. Sun, C., Ding, J., & Gao, F. (2021). Methods for microplastic sampling and analysis in the seawater and fresh water environment. *Methods in Enzymology, 648*, 27–45. https://doi.org/10.1016/bs.mie.2020.12.009
492. Hildebrandt, L., Voigt, N., Zimmermann, T., Reese, A., & Proefrock, D. (2019). Evaluation of continuous flow centrifugation as an alternative technique to sample microplastic from water bodies. *Marine Environmental Research, 151*, 104768. https://doi.org/10.1016/j.marenvres.2019.104768
493. Campanale, C., Savino, I., Pojar, I., Massarelli, C., & Uricchio, V. F. (2020). A practical overview of methodologies for sampling and analysis of microplastics in riverine environments. *Sustainability, 12*(17). https://doi.org/10.3390/su12176755
494. Erhunmwunse, N. O., Pajiah, T. J., & Ogwu, M. C. (2024). Microplastics as water pollutants and sustainable management strategies. In S. C. Izah, M. C. Ogwu, A. Loukas, & H. Hamidifar (Eds.), *Water crises and sustainable management in the global south* (pp. 253–278). Springer. https://doi.org/10.1007/978-981-97-4966-9_8
495. Picó, Y., & Barceló, D. (2019). Analysis and prevention of microplastics pollution in water: Current perspectives and future directions. *ACS Omega, 4*(4), 6709–6719. https://doi.org/10.1021/acsomega.9b00222
496. Kumar, H. (2024). Novel methods for microplastic remediation from cosmetic wastewater as a response to poor consumer awareness of microplastic contamination. *SSRN-Elsevier*. https://doi.org/10.2139/ssrn.4942230

497. Al-Azzawi, M. S. M., et al. (2020). Validation of sample preparation methods for microplastic analysis in wastewater matrices—Reproducibility and standardization. *Water (Basel), 12*(9). https://doi.org/10.3390/w12092445
498. Poli, V., Litti, L., & Lavagnolo, M. C. (2024). Microplastic pollution in the North-east Atlantic Ocean surface water: How the sampling approach influences the extent of the issue. *Science of the Total Environment, 947*, 174561. https://doi.org/10.1016/j.scitotenv.2024.174561
499. Wang, W., & Wang, J. (2018). Investigation of microplastics in aquatic environments: An overview of the methods used, from field sampling to laboratory analysis. *TrAC Trends in Analytical Chemistry, 108*, 195–202. https://doi.org/10.1016/j.trac.2018.08.026
500. Pasquier, G., Doyenb, P., Carlesic, N., & Amara, R. (2022). An innovative approach for microplastic sampling in all surface water bodies using an aquatic drone. *Heliyon, 8*(1). https://doi.org/10.1016/j.heliyon.2022.e11662
501. Bruge, A., Dhamelincourt, M., Lanceleur, L., Monperrus, M., Gasperi, J., & Tassin, B. (2020). A first estimation of uncertainties related to microplastic sampling in rivers. *Science of the Total Environment, 718*, 137319. https://doi.org/10.1016/j.scitotenv.2020.137319
502. Bannick, C. G., et al. (2019). Development and testing of a fractionated filtration for sampling of microplastics in water. *Water Research, 149*, 650–658. https://doi.org/10.1016/j.watres.2018.10.045
503. Briggs, E., de Moura, E. A. B. H. A. F., Cotrim, M. E. B., Oguzie, E. E., & Lugao, A. B. (2019). Microplastics: A novel method for surface water sampling and sample extraction in Elechi Creek, Rivers State, Nigeria. In B. Li, J. Li, S. Ikhmayies, M. Zhang, Y. E. Kalay, J. S. Carpenter, J.-Y. Hwang, S. N. Monteiro, C. Bai, J. P. Escobedo-Diaz, P. R. Spena, & R. Goswami (Eds.), *Characterization of minerals, metals, and materials 2019* (pp. 269–281). Springer. https://doi.org/10.1007/978-3-030-05749-7_27
504. Zobkov, M. B., Esiukova, E. E., Zyubin, A. Y., & Samusev, I. G. (2019). Microplastic content variation in water column: The observations employing a novel sampling tool in stratified Baltic Sea. *Marine Pollution Bulletin, 138*, 193–205. https://doi.org/10.1016/j.marpolbul.2018.11.047
505. Kentin, E., & Battaglia, G. (2022). Policies and perspectives on regulating microplastic fibre pollution. In J. S. Wei, F. De Falco, & M. Cocca (Eds.), *Polluting textiles* (1st ed., pp. 1–25). Routledge. https://doi.org/10.4324/9781003165385
506. Hettiarachchi, H., & Meegoda, J. N. (2023). Microplastic pollution prevention: The need for robust policy interventions to close the loopholes in current waste management practices. *International Journal of Environmental Research and Public Health, 20*(14). https://doi.org/10.3390/ijerph20146434
507. Graney, G. (2016). Slipping through the cracks: How tiny plastic microbeads are currently escaping water treatment plants and international pollution regulation. *Fordham International Law Journal, 39*(4), 1023–1028. https://ir.lawnet.fordham.edu/ilj/vol39/iss4/5/
508. Mitrano, D. M., & Wohlleben, W. (2020). Microplastic regulation should be more precise to incentivize both innovation and environmental safety. *Nature Communications, 11*, 5324. https://doi.org/10.1038/s41467-020-19069-1
509. Deng, L., Cai, L., Sun, F., Li, G., & Che, Y. (2020). Public attitudes towards microplastics: Perceptions, behaviors and policy implications. *Resources, Conservation and Recycling, 163*, 105096. https://doi.org/10.1016/j.resconrec.2020.105096
510. Onyena, A. P., Aniche, D. C., Ogbolu, B. O., Rakib, R. J., Uddin, J., & Walker, T. R. (2022). Governance strategies for mitigating microplastic pollution in the marine environment: A review. *Microplastics, 1*(1), 15–46. https://doi.org/10.3390/microplastics1010003
511. Zhao, Y.-N., & Li, J.-M. (2025). Public risk perception and treatment preferences for marine microplastic pollution: Choice experiments and the latent class model. *Journal for Nature Conservation, 86*, 126924. https://doi.org/10.1016/j.jnc.2025.126924
512. Amanda, G., Granek, E. F., Nielsen-Pincus, M., & Harkins, L. (2025). Perceptions about potential microplastic interventions: A study on knowledge, concerns, and willingness to pay. *Microplastics and Nanoplastics, 5*, 11. https://doi.org/10.1186/s43591-025-00119-8

513. Garcia-Vazquez, E., & Garcia-Ael, C. (2021). The invisible enemy. Public knowledge of microplastics is needed to face the current microplastics crisis. *Sustainable Production and Consumption, 28*, 1076–1089. https://doi.org/10.1016/j.spc.2021.07.032
514. Deme, G. G., et al. (2022). Macro problems from microplastics: Toward a sustainable policy framework for managing microplastic waste in Africa. *Science of the Total Environment, 804*, 150170. https://doi.org/10.1016/j.scitotenv.2021.150170
515. Omoyajowo, K., Raimi, M., Waleola, T., Odipe, O., & Ogunyebi, A. (2021). Public awareness, knowledge, attitude and perception on microplastics pollution around Lagos Lagoon. *Науково-технічний журнал, 2*(24), 35–46. https://doi.org/10.31471/2415-3184-2021-2(24)-35-46
516. Doughty, R., & Eriksen, M. (2014). The case for a ban on microplastics in personal care products. *Tulane Environmental Law Journal, 27*(2), 277–298. https://www.jstor.org/stable/43294166
517. Nelson, D., Sellers, K., Mackenzie, S., & Weinberg, N. (2019). Microbeads—A case study in how public outrage fueled the emergence of new regulations. *Current Pollution Reports, 5*, 172–179. https://doi.org/10.1007/s40726-019-00114-7
518. Li, Y. (2022). Legislation and policy on pollution prevention and the control of marine microplastics. *Water (Basel), 14*(18). https://doi.org/10.3390/w14182790
519. Venus, O. (2020). *National marine microbead policy in developed nations: How microbead bans have influenced microplastic pollution in waterways and begun the trend towards international collaboration*. Degree of Bachelor of Arts Thesis in Environment and Political Science, University of Michigan. https://deepblue.lib.umich.edu/bitstream/handle/2027.42/162658
520. Praveena, S. M. (2024). Exploring public awareness, influencing factors and policy implications towards microplastic pollution: Perspectives from Malaysia. *Marine Policy, 161*, 106042. https://doi.org/10.1016/j.marpol.2024.106042
521. Jani, V., Wu, S., & Venkiteshwaran, K. (2024). Advancements and regulatory situation in microplastics removal from wastewater and drinking water: A comprehensive review. *Microplastics, 3*(1), 98–123. https://doi.org/10.3390/microplastics3010007
522. Sorensen, R. M., Kanwar, R. S., & Jovanovi, B. (2023). Past, present, and possible future policies on plastic use in the United States, particularly microplastics and nanoplastics: A review. *Integrated Environmental Assessment and Management, 19*(2), 474–488. https://doi.org/10.1002/ieam.4678
523. Shini, S., Adukkadan, N., Pillai, S. U., Balan, B. M., Paili, T. A., & Mohan, M. (2025). Mitigating microplastic pollution in urban water systems: A global overview with governance perspectives from Kerala State, India. *Environmental Quality Management, 34*(4). https://doi.org/10.1002/tqem.70082
524. Mahajan, M., Singh, A. K., Singh, R. P., Gupta, P. K., Singh, S., & Pratap, M. (2025). Preventing microplastic pollution in surface waters: Legal frameworks and strategic actions. *Environmental Quality Management, 35*(1). https://doi.org/10.1002/tqem.70116
525. Xu, Y., et al. (2021). A critical review of microplastic pollution in urban freshwater environments and legislative progress in China: Recommendations and insights. *Critical Reviews in Environmental Science and Technology, 51*(22), 2637–2680. https://doi.org/10.1080/10643389.2020.1801308
526. Campanale, C., Massarelli, C., Bagnuolo, G., Savino, I., & Uricchio, V. F. (2019). The problem of microplastics and regulatory strategies in Italy. In F. Stock, G. Reifferscheid, N. Brennholt, & E. Kostianaia (Eds.), *Plastics in the aquatic environment—Part II* (pp. 255–276). Springer. https://doi.org/10.1007/698_2019_419
527. Munhoz, D. R., Harkes, P., Beriot, N., Larreta, J., & Basurko, O. C. (2023). Microplastics: A review of policies and responses. *Microplastics, 2*(1), 1–26. https://doi.org/10.3390/microplastics2010001

528. Zandaryaa, S. (2021). Freshwater microplastic pollution: The state of knowledge and research. In F. Stock, G. Reifferscheid, N. Brennholt, & E. Kostianaia (Eds.), *Plastics in the aquatic environment—Part I* (pp. 255–272). Springer. https://doi.org/10.1007/698_2021_748
529. Kamble, S. B., & Pawara, C. B. (2025). Review of current policies on plastic discharge and policy implications on microplastic reduction. In T. Kurniawan & A. Anouzla (Eds.), *Handbook of microplastic pollution in the environment* (1st ed., pp. 1–30). CRC Press. https://doi.org/10.1201/9781003487593
530. Alam, J., & Rahman, M. (2025). Various conventional and advanced management techniques and policies adopted at the global level for microplastics (ch. 18). In B. Singh & S. K. Upadhyay (Eds.), *Microplastics-origins, risks, and mitigation* (pp. 429–440). Elsevier. https://doi.org/10.1016/B978-0-443-29804-2.00018-4
531. Estahbanati, S., Upadhyaya, G., Wells, M. J. M., & Bell, K. Y. (2024). Primary and secondary microplastic and nanoplastic regulations: Perspectives on water industry impacts. *Journal of Environmental Protection (Irvine, Calif.), 15*(6), 697–715. https://doi.org/10.4236/jep.2024.156039
532. Li, J.-M., Yu, M.-F., Kao, C.-M., Surampalli, R. Y., & Zhang, T. C. (2025). Global strategies/policies and citizen science for microplastic management (ch. 20). In R. Y. Surampalli, T. C. Zhang, B. M. Al-Hashimi, C.-M. Kao, M. M. Ghangrekar, P. Bhunia, & S. Das (Eds.), *Microplastics in the environment: Fate, impacts, removal, and management*. Wiley Online Library. https://doi.org/10.1002/9781394251100.ch20
533. Usman, S., et al. (2022). The burden of microplastics pollution and contending policies and regulations. *International Journal of Environmental Research and Public Health, 19*(11). https://doi.org/10.3390/ijerph19116773
534. Cholewinski, A., et al. (2022). A critical review of microplastic degradation and material flow analysis towards a circular economy. *Environmental Pollution, 315*, 120334. https://doi.org/10.1016/j.envpol.2022.120334
535. Dhiman, S., Sharma, C., Kumar, A., Pathak, P., & Pathak, P. (2023). Microplastics in aquatic and food ecosystems: Remediation coupled with circular economy solutions to create resource from waste. *Sustainability, 15*(19). https://doi.org/10.3390/su151914184
536. Sadia, M., et al. (2022). Microplastics pollution from wastewater treatment plants: A critical review on challenges, detection, sustainable removal techniques and circular economy. *Environmental Technology & Innovation, 28*, 102946. https://doi.org/10.1016/j.eti.2022.102946
537. Rai, R., Ranjan, R., Kant, C., & Dhar, P. (2024). Microplastic and adhesive free, multifunctional, circular economy approach-based biomass-derived drinking straws. *iScience, 27*(5). https://doi.org/10.1016/j.isci.2024.109630
538. Aguilar, F. I. T., et al. (2025). Microplastic identification in domestic wastewater-treating constructed wetlands and its potential usage in a circular economy. *Processes, 13*(5). https://doi.org/10.3390/pr13051499
539. Vuori, L., & Ollikainen, M. (2022). How to remove microplastics in wastewater? A cost-effectiveness analysis. *Ecological Economics, 192*, 107246. https://doi.org/10.1016/j.ecolecon.2021.107246
540. Sturm, M. T., Myers, E., Schober, D., Korzin, A., & Schuhen, K. (2024). Beyond microplastics: Implementation of a two-stage removal process for Microplastics and chemical oxygen demand in industrial wastewater streams. *Water (Basel), 16*(2). https://doi.org/10.3390/w16020268
541. Elsamahy, T., et al. (2023). Strategies for efficient management of microplastics to achieve life cycle assessment and circular economy. *Environmental Monitoring and Assessment, 195*, 1361. https://doi.org/10.1007/s10661-023-11955-7
542. Sadeghi, P., Sadeghi, B., Marfavi, Y., Kowsari, E., Ramakrishna, S., & Chinnappan, A. (2021). Addressing the challenge of microfiber plastics as the marine pollution crisis using circular economy methods: A review. *Materials Circular Economy, 3*, 27. https://doi.org/10.1007/s42824-021-00038-y

543. Wiśniowska, E., & Kowalczyk, M. (2022). Recovery of cellulose, extracellular polymeric substances and microplastics from sewage sludge: A review. *Energies, 15*(20). https://doi.org/10.3390/en15207744
544. Capodaglio, A. G. (2024). Microplastics in the urban water cycle: A critical analysis of issues and of possible (needed?) solutions. *Science of the Total Environment, 954*, 176580. https://doi.org/10.1016/j.scitotenv.2024.176580
545. Antohi, V. M., Ionescu, R. V., Zlati, M. L., Iticescu, C., & Calmuc, P. L. G. M. (2023). Regional regression correlation model of microplastic water pollution control using circular economy tools. *International Journal of Environmental Research and Public Health, 20*(5). https://doi.org/10.3390/ijerph20054014
546. Mahmoud, A. E. D., & Mahmoud, E. (2023). Circular economy models in water and wastewater sectors. In A. E. D. Mahmoud, M. Fawzy, & N. A. Khan (Eds.), *Artificial intelligence and modeling for water sustainability* (1st ed., pp. 1–20). CRC Press. https://doi.org/10.1201/9781003260455-11

# Chapter 5
# Microplastics Pollution: Natural and Anthropogenic Sources, Cause and Impact on Ecosystem

**Jeba Sweetly Dharmadhas, R. Beaulah Mary, K. V. Shalini, Priyanka Jayachandran, and Aminu Abdullahi**

## Introduction

Microplastics, which range in size from 1 to 5000 μm, are simply plastic waste. Pollution from microplastics is already occurring all over the world through everyday items like makeup essential items for personal hygiene, in addition to synthetic clothing, bags, and water bottles. These substances can readily infiltrate the environment by a number of methods, including as human activity, rural activities, hospital waste discharge, and apartment building effluent. The waste materials are found in rivers, oceans, the environment, and the chemosphere all over the world, despite the fact that we use treatment facilities to treat microplastics [1]. Between 60 and 80 percent of marine debris and 90% of waste floating on the ocean and seas is made of plastic. Air circulation upstream is hampered when plastic waste begins to break down into microplastics that float on the water's top or sink to the bottom. Plastic pollution damages the maritime environment and impacts marine creatures [2]. Microplastics gradually disintegrated into tiny particles including pellets, microbeads, fibers, and pieces as a result of the degradation process. In contrast, researchers discovered that they were widely dispersed throughout 21 sites in Vietnam, including bays, lakes, rivers, and beaches [3].

J. S. Dharmadhas (✉) · K. V. Shalini · P. Jayachandran
Department of Biochemistry and Biotechnology, Avinashilingam Institute for Home Science and Higher Education for Women, Coimbatore, Tamil Nadu, India
e-mail: jeba_bc@avinuty.ac.in

R. B. Mary
Department of Allied Health Science, Ganga College of Nursing, Coimbatore, India

A. Abdullahi
Department of Biotechnology, Modibbo Adama University, Gombi, Nigeria



S. Kulkarni, S. Das (eds.), *Microplastics and Nanoplastics*, Lecture Notes in Nanoscale Science and Technology 49,
https://doi.org/10.1007/978-3-032-17963-0_5

Nowadays, people are well aware of the negative consequences of the tiny plastic particles found in seafood, which can be dangerous to one's health [4]. While fibers, which accounted for 12% of the microplastics in the industrial division, were primarily from textile laundering, flakes, which accounted for 31% of the microplastics, were produced from containers like bottles, cutlery, pots, folding tubes, and toilet cans. Significantly, the bulk of microplastics discovered in aquatic environments are composed of polyester and polypropylene. Most microplastics are classified as primary or secondary depending on their size. Given that larger plastic objects break up and produce more microplastics, this section looks at the sources and environmental paths of macroplastics. Of the plastic debris found in the ocean, terrestrial sources account for about 75–90%, while aquatic sources account for 10–25%. Littering, inappropriate plastic waste disposal, inefficient garbage collection, and poor landfill management are the primary causes of this pollution [5].

## Sources of Microplastics

According to [6], microplastics originate from a range of man-made sources, such as fishing and aqua tourism. By producing plastic pellets and air-blasting resins, the plastic production sector contributes significantly to the contamination of plastics in marine environments. The polymers can be separated into two groups according to their size, specifically [7]. Microplastics that are primary are synthetic polymers that are smaller than 2 mm. While these are frequently used in the production of cosmetics and personal hygiene products like plastic pellets and microbeads are mostly utilized in the transportation and industrial manufacturing sectors as well as the synthetic textiles sector.

However, there are a number of ways that plastic pollution might reach the environment directly, including product washing, spills during industry manufacture, and scratches during textile washing [8]. The second kind of microplastic is secondary microplastic, which is made up of broken pieces of larger macro or mesoplastic. Environmental events such as thermo-oxidative degradation, hydrolysis, and thermal degradation, as well as weathering, UV exposure, wind exposure, and contact with water waves, are the main causes of secondary microplastic [9] (Fig. 5.1). There are nanoplastics smaller than 1 μm, and some studies have shown that they can encourage the bioaccumulation and bioamplification of a range of contaminants because of their large surface area [6]. The primary plastic monomers that comprise microplastics are polyvinyl chloride (PVC), polyethylene (PE), polystyrene (PS), polyurethane (PUR), polychlorinated bipenyls (PCBs), polyamides (PA), polylactic acid (PLA), polypropylene (PP), polytetrafluoroethylene (PTFE), polymethyl methacrylate (PMMA), and polyethylene terephthalate (PET) [10]. Some of the primary polymer components found in microplastics and microbeads are shown in Fig. 5.2.

Each type of plastic in the table has a different combination of materials. Each plastic's unique environmental effects are displayed in the environmental impact. The concerns connected to each type of plastic are indicated by the effects on health.

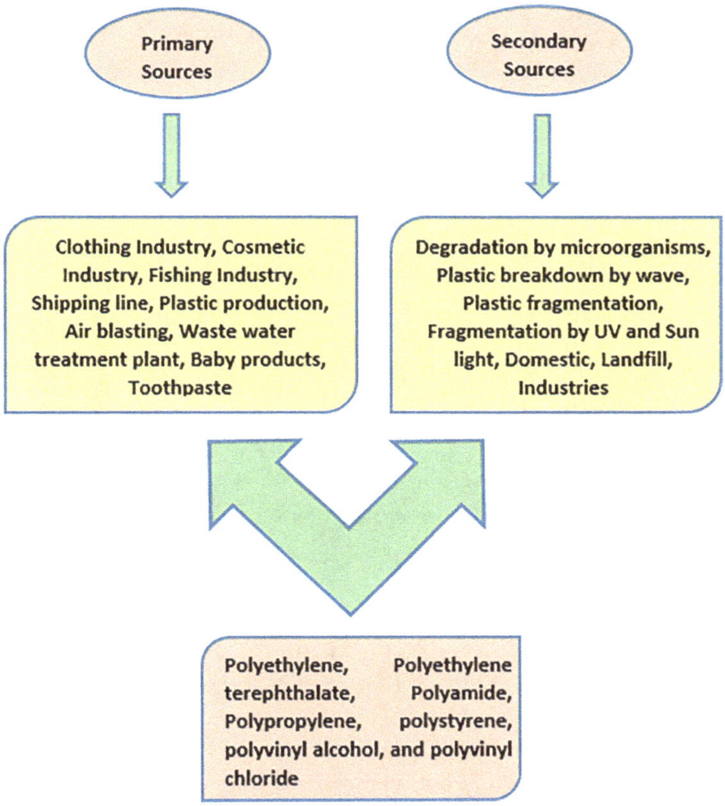


**Fig. 5.1** Primary and secondary MP types, sources, and generation processes

Making educated judgments about the use of plastic, promoting environmental preservation initiatives, and raising public knowledge of the health dangers associated with plastic are all crucial. By considering the size ranges, environmental impacts, and health consequences of different plastic kinds, individuals and organizations can make more sustainable decisions and contribute to reducing the harmful effects of plastic materials (Table 5.1).

## Effect on Ecosystem

### *Effects of Microplastics on the Environment*

Ecosystems and marine life are significantly impacted by microplastics. Because oceanic animals mistook microplastics for food, they ingested them, which caused them to accumulate in their bodies. Ultimately results in actual harm, such as

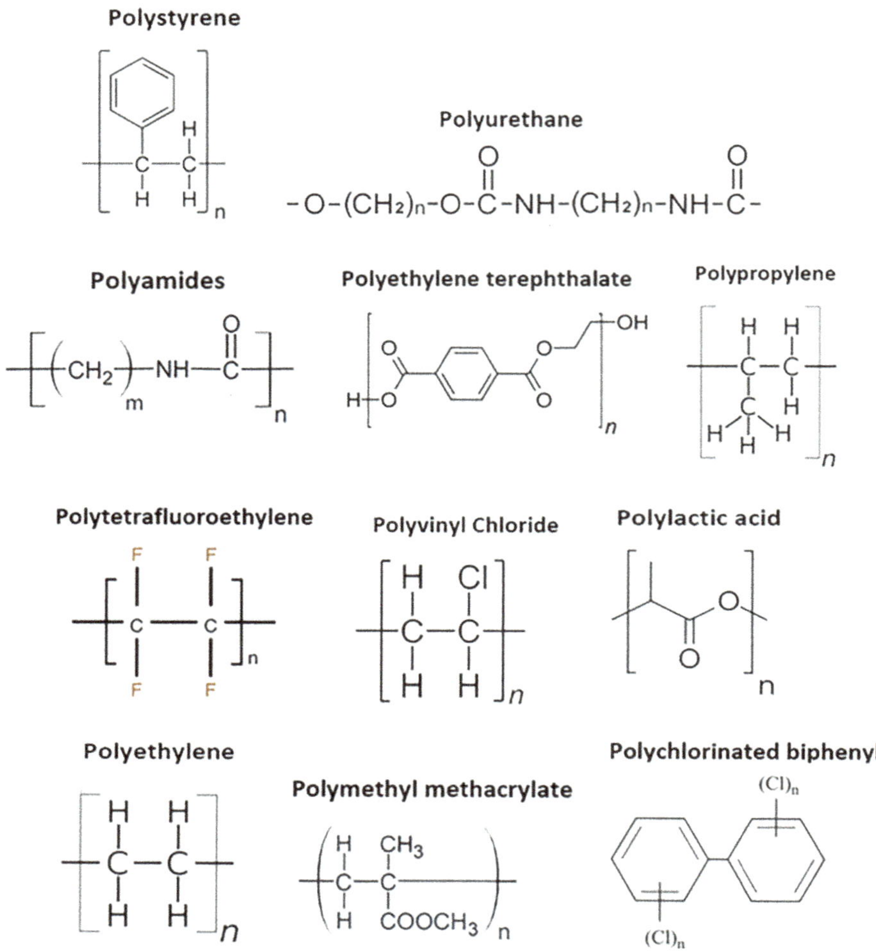


**Fig. 5.2** Several significant polymer products present in microplastics

damage to internal organs and obstructions in the digestive tract. Furthermore, by filling stomachs without offering any nutritional benefit, eating microplastics also results in malnutrition in marine species. Additionally, these stored microplastics disrupt marine creatures' reproductive systems, making them less able to swim and more susceptible to killers [38–40]. However, microplastics have a negative impact on a number of other ecosystems in addition to marine life (Table 5.2).

**Table 5.1** Various plastic types, their dimensions, environmental impacts, and potential health hazards

| S. No. | Plastic category | Size choice | Environmental effect | Health effects | References |
|---|---|---|---|---|---|
| 1. | Polyethylene | 0.06 to 11.06 μm | Damages marine organisms | Respiratory problems and potential consumption of contaminated seafood. | [11] |
| 2. | Terephthalate | Range < 4 μm and > 100 μm | Impacts aquatic life and contaminates water bodies | Breathing issues and distraction of hormone balance. | [12] |
| 3. | Polystyrene | 33 μm to 190 μm | Remains in the atmosphere | Breathing problem, hormone imbalance, and neurological effects. | [13–15] |
| 4. | Polypropylene | Range between 120 to 220 μm | Wildlife at risk | Respirational difficulties, gastrointestinal disorders, problem on the immune system | [16–19] |
| 5. | Polyvinyl chloride (PVC) | As much as 20 μm and less than 10 nm | Releases harmful substances that endanger aquatic environments | Potential reproductive, developmental, and respiratory disorders | [20–22] |
| 6. | Polycarbonate | 5–200 nm | Affects marine life | Hormone imbalance and harm to neurological development | [23] |
| 7. | Polyethylene Terephthalate (pet) | Microfiber measurements: Thickness: 12–18 μm length: Under 1 millimeter | Damages marine life and contaminates bodies of water. | Consumption of contaminated seafood leads to respiratory problems | [24, 25] |
| 8. | Polyurethane | < 5 mm | Presents hazards to ecosystems on land | Issues with breathing and skin irritation | [26–28] |

(continued)

**Table 5.1** (continued)

| S. No. | Plastic category | Size choice | Environmental effect | Health effects | References |
|---|---|---|---|---|---|
| 9. | Polyvinylidene Fluoride (PVDF) | Under 5 mm | Accumulates in marine organisms and endures in the environment | Some people may become irritated or sensitized after extended skin contact or inhalation of PVDF particles | [29] |
| 10. | Acrylic Copolymer | 80 to 110 nm | Adversely affecting marine environments as well as possibly making their way into the food chain | Acquire sensitivity to acrylic or allergic responses, which could result in to respiratory problems or rashes after exposure | [30–32] |
| 11. | Polyamides | 10–50 microns | Derived mainly from petroleum-based resources | Depending on exposure, may result in long-term consequences, skin irritation, or breathing problems | [33] |
| 12. | Polytetrafluoroethylene | ~100 nm | Does not readily break down in the environment, contributing to microplastic and nanoplastic contamination. | Associated with immune system impacts, liver damage, thyroid disruption, and cancer (e.g., kidney/testicular). | [34] |
| 13. | Polylactic acid | Typically 1–20 microns | Can compete with food crops for land use, leading to concerns about agricultural sustainability. | Can irritate eyes, throat, or lungs in poorly ventilated areas. | [35] |

**Table 5.1** (continued)

| S. No. | Plastic category | Size choice | Environmental effect | Health effects | References |
|---|---|---|---|---|---|
| 14. | Polymethyl methacrylate | Sizes from 100 nm to 10 μm | Used in cosmetics and industrial applications can enter waterways and contribute to microplastic pollution | Utilized frequently in medical implants (such as bone cement and dental applications) and in solid form (such as in lenses and displays) where it is non-toxic, inert, and generally safe | [36] |
| 15. | Polychlorinated bipenyls (PCBs) | 188 to 499 g/Mol | Accumulate in animal tissues, fish, and humans, moving up the food chain. | Linked to liver, skin, and possibly breast cancer. | [37] |

## *Effects of Microplastics on Soil*

It is well recognized that soil microplastics change the nutrient cycle and decrease water retention, hence altering the soil's characteristics [55, 56]. It is possible for microplastics to disrupt the food chain and jeopardize aquatic life in freshwater habitats. Furthermore, because microplastics have the ability to penetrate the food chain and endanger humans who consume contaminated seafood or water, they can be harmful to human health [57, 58]. Although more study is needed, microplastics might also harm the quality of the air [59].

## *Effects of Microplastics on Human Health*

Numerous trails, including consumption, breathing, and skin communication, might allow microplastics to enter our bodies (Fig. 5.2). The most frequent way that people are exposed to microplastics is through ingestion [57, 60]. Microplastic pollution can be found in adulterated marine food, water, packaged foods, and plastic utensils. After entering humans, microplastics may be stored in the digestive system and other organs [57, 60]. Inflammation, oxidative stress, and alterations in gut microbiota can result from microplastics damaging the intestinal wall during ingestion [61–64]. Neurodegenerative illnesses, cancer, and autoimmune disorders are among the excess health impacts that result from this. Another route is inhalation,

**Table 5.2** Effects of Microplastics on the Environment

| Affected area | Impact type | Details | References |
|---|---|---|---|
| Marine ecosystem | Ingestion by organisms | Fish, sea birds, marine mammals, zooplankton, mussels, whales consume microplastics leading to digestive tract obstruction, starvation, energy depletion and even death | [41–43] |
| | Habitat alteration | Microplastics in sediments alter microbial activity, sediment composition and water chemistry | |
| | Trophic level transfer | Microplastics can move up the food chain, affecting all predators | |
| | Altered behaviour and physiology | Fishes show altered feeding and predator avoidance behaviour leads to reduced survival and reproduction | |
| Freshwater ecosystems | Bioaccumulation | Microplastics store in freshwater organisms and disrupt aquatic ecosystems | [44–46] |
| | Water quality ruin | Microplatics can leach out toxic pollutants into the freshwater ecosystem | |
| | Filter feeding interference | Affects species like mussels, clams and fishes this will reduce water filtration | |
| Terrestrial ecosystem | Reduced soil quality | Usage of microplastic contaminated sewage sludge affects soil structure, porosity, water retention and microbial communities | [5, 45] |
| | Affects soil fauna | Ingestion of plastics disrupts digestion and affects reproduction and survival. It affects soil fertility | |
| Atmosphere | Microplastics found in air | Inhalation of microplastics found in air, causes potential health risks for humans and animals | [47] |
| Human health | Toxicity | Endocrine disruptors, carcinogens, hormonal imbalance and reproductive issues | [48–50] |
| | Contaminated foods | Consumption of seafood may cause long-term health risks | |
| Chemical pollution | Vectors for toxic pollutants | Acts as vectors for pesticides, heavy metals and pathogens | [51] |
| Climate change | Emission of greenhouse gases | During degradation microplastics release methane and ethylene | [52] |
| Waste management | Landfills | Microplastics leach into surrounding soil and groundwater during plastics decomposing | [53, 54] |
| | Incineration | Burning plastics releases dioxins and furans, it causes toxic air pollution | |

which primarily affects those who work in sectors that produce and handle plastic materials [65]. Microplastics have the potential to pollute the air during manufacturing processes. Chronic-obstructive pulmonary disease (COPD) and asthma are examples of respiratory conditions that can be brought on by inhalation, which also irritates and damages the respiratory system [59, 66].

On the other hand, utilizing personal care items like face cleansers and body washes that include microbeads could expose the skin to microplastics in a third way [67, 68]. In addition to causing skin irritation and inflammation, dermal exposure may also result in reproductive damage [69] and endocrine disruption [60]. In addition to personal care items, apparel and textiles composed of synthetic materials may also include microplastics, which can shed microfibers after being cleaned and put on [70]. Researchers found that exposure to microplastics has been linked to a number of detrimental effects on human health. Numerous studies indicate that microplastics actually cause cellular damage [71], genotoxicity in cells of humans [72], and inflammation [73]. Additionally, ambient microplastics can interfere with circadian cycles and melatonin production [75, 76], which can make it harder to fall asleep and raise the risk of sleep disorders [74]. (Fig. 5.2). Furthermore, circadian cycles and melatonin production can be disrupted by ambient microplastics [75, 76], which may lead to difficulty falling asleep and an increased risk of sleep disorders [60, 74]. Microplastic exposure can cause oxidative stress, disrupt hormone regulation, and inflammation, which can result in weight gain, obesity, and diabetes in cases of obesity [77–79]. Therefore, it is clear from past studies that microplastics could be detrimental to human health and the environment (Table 5.3).

### *Microplastics Effects on Marine Life*

Above all, microplastic pollution's consequences on marine life and other systems are becoming worse. The financial effects of businesses like fishing and tourism are greatly impacted by the presence of microplastics in marine habitats. Marine species tissues acquire microplastics, which most likely reduces their market value and quality [112, 113]. The economic viability of fishing is impacted by microplastic pollution in seafood, which may raise consumer worries about food safety [114]. Moreover, microplastics may affect tourism in addition to fishing. Microplastic-contaminated beaches have the potential to deter tourists, which lowers the amount of money generated in the impacted locations. In order to assist in stopping the release of microplastic contamination into the environment, efforts are being made to decrease plastic waste and enhance waste management procedures (Table 5.4).

## Properties of Microplastics

### *Physical Properties*

Microscopes can be used to measure the physical features of microplastics, such as their dimension, shape, surface area, crystallinity, and density. In this analysis, size is the most frequently used criterion. Depending on the sample collection and

**Table 5.3** Effects of Microplastics on Human Health

| Affected system | Potential diseases | References |
|---|---|---|
| Skin | Contact dermatitis | [80, 81] |
| Gastrointestinal tract | Pancreatic cancer, colorectal cancer, metabolic diseases, disturbance of the intestinal barrier and non-alcoholic fatty liver disease | [82–85] |
| Respiratory system | Lung cancer, occupational hypersensitivity pneumonitis, respiratory problems (coughing, dyspnea, wheeze), pleural effusion, and obstructive bronchiolitis | [86–89] |
| Blood neoplasms | Leukemia, lymphohematopoietic neoplasms, and bladder cancer | [90, 91] |
| Neurological effects | Degenerative neurological illnesses, encephalopathy, unidentified dementia, and neurotoxic consequences (headache, exhaustion, lightheadedness, etc.) | [92] |
| Osteolysis around bone prosthesis | Periprosthetic osteolysis | [82, 93] |
| Reproductive | Reduction in fertility, gonadal damage, reduced offspring weight | [94–96] |
| Genotoxicity | Damage to DNA and a higher incidence of sister chromatid exchange | [97] |
| Skin | Induces allergy | [98] |
| Inflammation | IL-8 expression is elevated | [99–101] |
| Oxidative stress and apoptosis | Apoptosis was induced in all intestinal epithelial cells | [102, 103] |
| Metabolic homeostasis | Changes in ion channel function and ionic homeostasis | [104–106] |
| Physical harm | Gastrointestinal discomfort, inflammation | [107] |
| Chemical toxicity | Acute toxicity, oxidative stress | [108] |
| Immune response | Acute inflammation, allergic reactions | [47] |
| Cellular damage | Cellular inflammation, oxidative stress | [109] |
| Respiratory issues | Respiratory irritation, coughing, shortness of breath | [110] |
| Neurological effects | Acute neuroinflammation | [49] |
| Metabolic disorders | Altered lipid metabolism, acute stress response | [111] |

separation techniques we employed, the size might change. Furthermore, the filter's pore size has an impact on the particle's size as well [154]. Size determination is more important in research because based on the size the features can be distinguished from old types of microplastics with new types [155]. Previous studies reported that size of microplastics affects the correctness of counting. Large size microplastics show small error when compared to large size particles [155]. It is heterogeneous in nature, and shows individual behavior, based on their shape, size and density [156–159]. During fragmentation and degradation of microplastics researchers observed that the physical properties such as density, shape and diameter can be changed [160, 161]. Microplastics degradation occurs through various processes like oxidation, bio-film growth, solar exposure, and also thermal aging [154], which changes the properties of microplastics [162–164]. During

**Table 5.4** Microplastics effects on marine environments

| Species | Impact | References |
|---|---|---|
| Zooplankton, fish, seabirds, marine turtles | Reduced vitality and viability | [115–118] |
| Worms, clams, oysters, mussels, and corals | Development and growth are impacted | [119, 120] |
| Fish, crabs, lobsters, sea stars, sea cucumbers | Decreased feeding behavior and food | [121–123] |
| Fish, mollusks, echinoderms, cnidarians, | Toxicology and health effects | [124–126] |
| Fish, seabirds, marine mammals, humans | The buildup of microplastics in higher food chains | [127] |
| Phytoplankton, zooplankton, benthic | Ecological knock-on effect | [128–131] |
| Coral reefs, seagrass beds, mangroves, salt | Impaired ecosystem function | [132–134] |
| Marine flora and fauna | Species diversity decline | [135, 136] |
| Humans | Human health risk | [137–139] |
| Microalgae | Decreased activity of photosynthetic processes | [140] |
| Copepods | Decreased capacity for nutrition, survival, hatching success, and reproduction | [141] |
| Artemia | Altered nauplii hatching sequence, cyst production, and survival rate | [137] |
| Fishes | Reduced eating rate and obstructions in the gut | [142, 143] |
| Crustaceans | Decreased food intake, delayed offspring development, and decreased fecundity | [144, 145] |
| Benthos | It was discovered that the sedimentary and sinking microplastics were consumed by benthic suspension and deposit feeders | [146] |
| Molluscs | Decreased mussel filtration and feeding efficiency, altered oyster eating, and disrupted oyster reproduction, human consumption of microplastics may come from contaminated molluscan species | [147, 148] |
| Coral reefs | Decreased the rates of coral skeleton mineralization and skeletal growth | [149, 150] |
| Marine plants | Polyethylene, polypropylene, and polystyrene were the most prevalent MP types found in some mangrove sediments | [151] |
| Humans | Our body excretes about 90% of the micro and nanoplastics that are consumed through feces | [152] |
| Marine plants | The most common MP types identified in some mangrove sediments were polyethylene, polypropylene, and polystyrene | [153] |

degradation the microplastic particles changed from micro-sized to nano-sized particles and accelerated their entry into the biotic ecosystem [165]. Microplastics in riverine environments can float, float, or sink [142]. Low-density particles can stay suspended, however, at low flow rates, high-density plastics can roll, slide, and occasionally bounce at the riverbed. Large microplastics were consequently deposited in the riverbed. Colonization between microplastics and microorganisms are quite common in the aquatic environment to form bio-films [166, 167].

## *Chemical Properties*

The chemical makeup and surface groups that microplastics include make up the majority of their chemical characteristics. It is composed of polymers, contaminants adsorbed on the surface of microplastics, colors, and additives like plasticizers and antioxidants. As plastics are made, used, and weathered, these compounds are readily discharged from the microplastics into the environment [168, 169]. PVC, which is thought to be the most dangerous because of its higher chlorine concentration and dioxin emission, is where microplastics leak from [170]. The physical characteristics of polymers, including their molecular size, porosity, and degree of degradation, are combined with chemical leaching [171]. Polyethylene (25%), polyamide (12%), polypropylene (14%), polyvinyl alcohol (6%), polyethylene terephthalate (16.5%), polystyrene (8.5%), and polyvinyl chloride (2%) are the most prevalent forms of microplastic polymers found in aquatic environments [172]. Among these polymers, polyvinyl chloride, polystyrene, and polyethylene terephthalate are amorphous, whereas polyethylene and polypropylene are semi-crystalline [173]. Researchers have found 17 microplastic polymers in urban wastewater, including acrylate, biopolymer, low-density polyethylene (LDPE), high-density polyethylene (HDPE), melamine, PP, PS, polyurethane (PUR), PVA, teflon, rubber, and a few other polymers that are unknown [174].

## *Biological Properties*

Due to their elevated level of crystallinity, microplastics are typically resistant to bacteria; as a result, biodegradation of microplastics is difficult [175]. Scientists have recently discovered that a range of microorganisms, including fungi and bacteria, aid in the breakdown of plastic [176]. The main factors prompting the breakdown of microplastics is the state of the soil [177]. Furthermore, the conversion process is influenced by biotic and abiotic conditions, organism types, the method of pretreatment applied, as well as properties of microplastics (such as mobility, molecular weight, harmfulness, crystallinity, the presence of surface functional groups and substitutes, the addition of scents or plasticizers), and more [178].

## Bioaccumulation and Biomagnification of Microplastics

Two key ideas that were utilized in ecological risk assessments to determine the amount of pollution carried within food webs are bioaccumulation and biomagnification. The dissolved content of chemical contamination is typically referred to when discussing bioaccumulation and biomagnification [179]. Here, we will talk about bioaccumulation, which is the process by which a pollutant from the environment is absorbed by prey, sediment, water, and other sources [180–182]. However, when intake exceeds an organism's capacity to expel a contaminant, bioaccumulation occurs [182]. These contaminants are biomagnified at higher trophic levels as a result of trophic level transfer in bioaccumulation [40]. When compared to the concentration in its prey, biomagnification across the food web can undoubtedly raise the contamination level in one organism [181].

## Detection Techniques of Microplastics

Identifying and classifying primary and secondary microplastics from aquatic and terrestrial environments is the main objective of this research. Several studies have been conducted to detect and measure microplastics using a range of analytical tools in addition to visual techniques [191]. Physical characterisation by microscopic microscopy and chemical characterization by spectroscopy are the two general techniques for identifying microplastics, according to [192]. Sample identification requires completion of the four most fundamental processes: density separation, filtering, sieving, and visual sorting of microplastics. According to [193], these four first identification methods make it easy to determine the size, shape, and color of bigger microplastic particles. Additionally, the combination of density separation with the fluorescence approach makes it simple and sensitive to identify the most common plastic polymer particles in marine sediments.

## Visual Techniques

The unaided eye or an optical microscope with objectives ranging from 10 to 50 times higher magnification can be used for visual identification. The microscope was occasionally used with image-analysis software like Histolab and Olympus Stream [194]. 79% of microplastics characterization studies use visual identification because it is straightforward and simple to apply [195]. However, because it can be impeded by a number of non-plastic particles, such as paint chips, cellulose, keratin, viscose rayon, and coal or fly ash, it is an unreliable method for identifying microplastic. This can have detrimental effects, as it can be difficult to detect translucent particles and microparticles smaller than 100 μm [195].

To improve a number of techniques used for the digestion of cotton fibers, cigarette filters, and human hair, wet peroxide oxidation (WPO) and the rapid staining process with Nile Red dye can be applied [193]. A digestion step is another excellent method in which organic molecules are broken down in approximately 1 h using a solution of nitric acid ($HNO_3$) and sodium hydroxide (NaOH). We also employ sodium iodide (NaI) in a different separation step if required to lessen the quantity of mineral remains in the samples. Researchers will also find it much easier if they use a scanning electron microscope (SEM) to view the particle surface characteristics. This method creates high-resolution images by using a highly concentrated electron beam to scan the surface material [196].

## Spectroscopic Techniques

While microscopic particles are extremely challenging to identify, a considerable number of microplastics can be readily observed using visual approaches, even when using chemical methods. Therefore, spectroscopic techniques make it easy to identify the small particles [197]. The composition of microplastics like PE, PP, PS, or PVC can be ascertained using two renowned and frequently applied analytical methods: Raman spectroscopy and Fourier transform infrared spectroscopy (FTIR) [193]. Spectroscopic methods are also used to evaluate the molecules in the samples [196]. FTIR, Raman, and near-infrared spectroscopy are also strongly advised for the detection of unknown plastic bits, particularly those that are smaller in size, according to [198]. (2020).

But according to experts, Raman spectroscopy is considerably more effective than FTIR spectroscopy for the measurement and interpretation of microplastic data [199]. We often use a range of methods, however we sometimes use other methods and get erroneous results from the sample. When analyzing microplastics, however, spectroscopic methods yield accurate results [200]. In the lab, this sample preparation approach also takes a lot of time and effort [201]. A novel method for the time-consuming task of spectroscopic data analysis was recently presented by a researcher by the name of Rennel et al. With the help of the library, a group of researchers quickly identified 98% of the different forms of plastic.

## Chromatographic Techniques

In conjunction with extraction procedures, chromatographic techniques are employed to identify both qualitative and quantitative polymers [202]. Serial improvements in chromatographic methods enable the rapid identification of microplastics. The series of developments includes pyrolysis GC–MS, liquid extraction with size-exclusion chromatography (SEC), and high temperature (HT) gel permeation chromatography. A sensitive and tried-and-true method for bulk characterisation and quantification is the GCMS methodology. This method works incredibly well for finding microplastics in environmental samples. According to [202, 203] Gas chromatography is used to separate the chemicals, and pyr-GC/MS materials are used to evaluate the degraded products by mass spectrometry once heat decomposition is finished. Many investigations have been conducted to detect and measure microplastics, but only a small number of these studies have yielded meaningful findings. Many detection techniques are now in use, although they are not very good at detecting microplastics.

It takes a few seconds to a few minutes to find the microplastics. Some of the current methods need to be replaced with more effective ones that produce results quickly. For example, castor oil is the most efficient substitute for the existing density separation method for the separation of microplastics [209]. Similarly, microplastics were extracted from wastewater that included a lot of organic compounds using Fenton's reagent, which is a mixture of $H_2O_2$ and the ferrous ion $Fe^{2+}$. This method decreased the estimated time needed for the identification and sample preparation procedures. According to [118], it offers a simple, rapid, and affordable method of identifying microplastics in environmental samples. Furthermore, the visual approaches are unable to identify the type of microplastic and other pertinent information.

The same is true for spectroscopic technologies, where time-consuming and effective sample preparation techniques prevented the problem from being solved. FTIR and Raman spectroscopy only are unable to enable the examination of microplastics from different samples in this type. To get over these issues, researchers here employ the effective Pyr-MS approach. Similar to FTIR and Raman spectroscopy, this instrument will measure microplastic particles in an extensive range and this technique is not limited by any parameters like shape, color, and size [171]. Furthermore, Raman Tweezers (RTs) created a new spectroscopy-related technique called optical tweezers in conjunction with Raman spectroscopy. This analytical tool is used to examine microplastics and nanoplastics in seawater and will have a significant influence on future studies on environmental pollution caused by microplastics and nanoplastics [196].

## Legislative Measures

Around the world, a number of administrative and legal actions have been taken to reduce the severity of microplastic contamination. The European Commission also identified microplastic pollution as a major concern; China issued the "Plastic Restriction Order" in 2008 and the "Opinions on Further Strengthening the Control of Plastic Pollution" in 2020; and the United Nations Environment Programme began the process of phase-out of the use of microplastics in cosmetics in 2015. However, laws and rules pertaining to the disposal of plastics must be strengthened and closely watched in order to avoid and control agricultural soil. Strengthening the legal framework for specific microplastics control and advancing legislation and amendments to address the issue of plastic waste, particularly plastic recycling, are imperative at the legal level. The oversight and accountability of businesses in the life cycle of plastic goods should be strengthened and defined at the governmental level. Public awareness needs to be raised, and we need to teach college and high school students how to properly dispose of plastic products. We also need to pay more attention to the pollution caused by microplastics.

## Prospects for the Future

Future research on MPs should be focused on the following factors in order to begin in-depth analysis:

1. We should carry out continuous investigations and assessments on different water bodies and habitats in order to progressively develop a model for point-line surface research in order to more accurately evaluate water environment contamination caused by MPs.
2. The majority of the current research on MP toxicity is conducted in laboratories; little is known about the hazardous consequences of MPs in complicated real-world environments. Since studies cannot demonstrate how intricate environmental elements and other chemicals interact to generate MP pollution, it is imperative to evaluate the detrimental impacts of MPs in the field. Furthermore, while examining the bioaccumulation of MPs in actual environmental settings, it is necessary to take into account the various eating patterns of organisms as well as the physical and chemical characteristics of MP particles (such as particle size, shape, and polymer type).
3. There is a need to create a long-term impact assessment mechanism for MPs in the environment due to their unique physicochemical properties and negative effects on aquatic life. Utilizing environmental exposure histology data and MP ecotoxicity data extensively is necessary to support research on the pathways, risks, and hazards of MPs in the human body as well as to systematically evaluate the health risks of MPs in different populations.

## Challenges of Controlling MPs Pollution

MPs' significant effects on freshwater and marine creatures have caused problems on a global scale [183, 184]. Characterizing microplastics in the water system is extremely challenging because they are often mistaken for biological materials. Removal of microplastics using engineering and biotechnology methods from environmental water sources [185]. Pollution from microplastics is difficult to regulate over a few months or years. Furthermore, the issue of unclear ownership persists since oceans and other bigger bodies of water are not contained within geographical borders, and various parties and governmental organizations bear duty both locally and internationally.

The 1972 London Convention on the Prevention of Marine Pollution by Dumping of Wastes and Other Matter, the International Convention for the Prevention of Pollution from Ships, and the Global Partnership on Marine Litter (GPML) are some of the initiatives that have been put into place through international cooperation to reduce MP pollution. Although significant progress has been made in many regions, marine MP contamination has not been effectively controlled by these techniques. The inability of global partnership approaches to create consistent policy approaches, monitor and enforce cross-border regulatory compliance, or fully manage the sources and entry ports of plastic pollution has been blamed for this failure [186, 187]. Given that the amount, composition, and distribution of plastic vary seasonally and geographically, MP pollution's dynamic nature is one of its most difficult characteristics [188]. Furthermore, new sources may be created by the ongoing waste stream, growing urban population, and plastic usage trends [155].

MPs are a common and slowly degrading contaminant that has the ability to take in additional pollutants, a long residence length, high stability, and a high propensity for fragmentation [189]. To eliminate such plastic materials, the entire range of MP types, shapes, and complete morphologies must be identified. Even with the growing amount of research, there are currently no appropriate and proven analytical techniques for measuring and detecting micro and nanoplastics [190]. Another difficulty is comprehending MPs' varied physicochemical characteristics, which make it difficult to perform suitable toxicological tests [155, 209]. The extraction, separation, identification, and quantification procedures in analytical methods for MPs are costly, time-consuming, and challenging. Furthermore, a lack of public awareness is a major barrier to reducing MP pollution. A long-term management approach will be necessary to address this pressing issue and reduce the detrimental impacts of MP contamination.

## References

1. Sun, N., Shi, H., Li, X., Gao, C., & Liu, R. (2023). Combined toxicity of micro/nanoplastics loaded with environmental pollutants to organisms and cells: Role, effects, and mechanism. *Environment International, 171*, 107711. [CrossRef].
2. Badylak, S., Phlips, E., Batich, C., Jackson, M., & Wachnicka, A. (2021). Polystyrene microplastic contamination versus microplankton abundances in two lagoons of the Florida keys. *Scientific Reports, 11*, 6029.
3. Barboza, L. G. A., Cunha, S. C., Monteiro, C., Fernandes, J. O., & Guilhermino, L. (2020). Bisphenol a and its analogs in muscle and liver of fish from the north East Atlantic Ocean in relation to microplastic contamination. Exposure and risk to human consumers. *Journal of Hazardous Materials, 393*, 122419.
4. Duis, K., & Coors, A. (2016). Microplastics in the aquatic and terrestrial environment: Sources (with a specific focus on personal care products), fate and effects. *Environmental Sciences Europe, 28*(2), 1–25.

5. da Costa, J. P., Santos, P. S. M., Duarte, A. C., & Rocha-Santos, T. (2016). (Nano)plastics in the environment—Sources, fates and efects. *Science of the Total Environment., 566*, 0048–9697. https://doi.org/10.1016/j.scitotenv.2016.05.041
6. Avio, C. G., Gorbi, S., & Regoli, F. (2016). Plastics and microplastics in the oceans: From emerging pollutants to emerged threat. *Marine Environmental Research, 128*, 2–11. https://doi.org/10.1016/j.marenvres.2016.05.012
7. Abarghouei, S., Hedayati, A., Raeisi, M., Hadavand, B. S., Rezaei, H., & AbedElmdoust, A. (2021). Size-dependent efects of microplastic on uptake, immune system, related gene expression and histopathology of goldfsh (Carassius auratus). *Chemosphere, 276*, 129977. https://doi.org/10.1016/j.chemosphere.2021.129977
8. Sharma, S., & Chatterjee, S. (2017). Microplastic pollution, a threat to marine ecosystem and human health: A short review. *Environmental Science and Pollution Research, 24*, 21530. https://doi.org/10.1007/s11356-017-9910-8
9. Bayo, J., López -Castellanos, J., & Olmos, S. (2020). Membrane bioreactor and rapid sand filtration for the removal of microplastics in an urban wastewater treatment plant. *Marine Pollution Bulletin, 156*, 111211. https://doi.org/10.1016/j.marpolbul.2020.111211
10. Bhattacharya, A., & Khare, S. K. (2020). Ecological and toxicological manifestations of microplastics: Current scenario, research gaps, and possible alleviation measures. *Journal of Environmental Science and Health, Part C, 38*, 1–20.
11. Ye, H., Ha, M., Yang, M., Yue, P., Xie, Z., & Liu, C. (2017). Di2-ethylhexyl phthalate disrupts thyroid hormone homeostasis through activating the Ras/Akt/TRHr pathway and inducing hepatic enzymes. *Scientific Reports, 7*, 40153.
12. Kik, K., Bukowska, B., & Sici ́nska, P. (2020). Polystyrene nanoparticles: Sources, occurrence in the environment, distribution in tissues, accumulation and toxicity to various organisms. *Environmental Pollution, 262*, 114297.
13. Li, X., Zhang, T., Lv, W., Wang, H., Chen, H., Xu, Q., Cai, H., & Dai, J. (2022). Intratracheal administration of polystyrene microplastics induces pulmonary fibrosis by activating oxidative stress and Wnt/β-catenin signaling pathway in mice. *Ecotoxicology and Environmental Safety, 232*, 113238.
14. Jin, H., Yang, C., Jiang, C., Li, L., Pan, M., Li, D., Han, X., & Ding, J. (2022). Evaluation of neurotoxicity in BALB/c mice following chronic exposure to polystyrene microplastics. *Environmental Health Perspectives, 130*, 107002.
15. Yedier, S., Yalçınkaya, S. K., & Bostancı, D. (2023). Exposure to polypropylene microplastics via diet and water induces oxidative stress in Cyprinus carpio. *Aquatic Toxicology, 259*, 106540.
16. Monclús, L., McCann Smith, E., Ciesielski, T. M., Wagner, M., & Jaspers, V. L. B. (2022). Microplastic ingestion induces size-specific effects in Japanese quail. *Environmental Science & Technology, 56*, 15902–15911.
17. Montero, D., Rimoldi, S., Torrecillas, S., Rapp, J., Moroni, F., Herrera, A., Gómez, M., Fernández-Montero, Á., & Terova, G. (2022). Impact of polypropylene microplastics and chemical pollutants on European sea bass (Dicentrarchus labrax) gut microbiota and health. *Science of the Total Environment, 805*, 150402.
18. Wang, H., Klosterhalfen, B., Klinge, U., Müllen, A., & Jockenhoevel, S. (2023). Influence of polypropylene mesh degradation on tissue inflammatory reaction. *Journal of Biomedial Materials Research Part A, 111*, 1110–1119.
19. Savrık, S. A., Balköse, D., Ulutan, S., & Ülkü, S. (2010). Characterization of poly(vinyl chloride) powder produced by emulsion polymeriza-tion. *Journal of Thermal Analysis and Calorimetry, 101*, 801–806.
20. Boyle, D., Catarino, A. I., Clark, N. J., & Henry, T. B. (2020). Polyvinyl chloride (PVC) plastic fragments release Pb additives that are bioavailable in zebrafish. *Environmental Pollution, 263*, 114422.
21. Liu, Y., Zhang, J., Zhao, H., Cai, J., Sultan, Y., Fang, H., Zhang, B., & Ma, J. (2022). Effects of polyvinyl chloride microplastics on reproduction, oxidative stress and reproduction and

detoxification-related genes in Daphnia magna. *Comparative Biochemistry and Physiology Part C: Toxicology & Pharmacology, 254*, 109269.

22. Anwar, Z., Anjum, F., & Ghayas, S. (2021). Polycarbonate plastics and neurological disorders: From exposure to preventive interventions. In M. S. H. Akash & K. Rehman (Eds.), *Environmental contaminants and neurological disorders* (pp. 147–183). Springer International Publishing.
23. Tian, L., Skoczynska, E., Siddhanti, D., van Putten, R.-J., Leslie, H. A., & Gruter, G.-J. M. (2022). Quantification of polyethylene terephthalate microplastics and nanoplastics in sands, indoor dust and sludge using a simplified in-matrix depolymerization method. *Marine Pollution Bulletin, 175*, 113403. [CrossRef].
24. Lionetto, F., & Esposito Corcione, C. (2021). An overview of the sorption studies of contaminants on poly(ethylene terephthalate) Microplastics in the Marine Environment. *Journal of Marine Science and Engineering, 9*, 445.
25. Albergamo, V., Wohlleben, W., & Plata, D. L. (2023). Photochemical weathering of polyurethane microplastics produced complex and dynamic mixtures of dissolved organic chemicals. *Environmental Science. Processes & Impacts, 25*, 432–444. [CrossRef] [PubMed].
26. ˇCerná, T., Pražanová, K., Beneš, H., Titov, I., Klubalová, K., Filipová, A., Kluso ˇn, P., & Cajthaml, T. (2021). Polycyclic aromatic hydrocarbon accumulation in aged and unaged polyurethane microplastics in contaminated soil. *Science of the Total Environment, 770*, 145254.
27. Segan, S., Jakobi, M., Khokhani, P., Klimosch, S., Billing, F., Schneider, M., Martin, D., Metzger, U., Biesemeier, A., Xiong, X., et al. (2020). Systematic investigation of polyurethane biomaterial surface roughness on human immune responses in vitro. *BioMed Research International, 2020*, 3481549.
28. Di Giannantonio, M., Gambardella, C., Miroglio, R., Costa, E., Sbrana, F., Smerieri, M., Carraro, G., Utzeri, R., Faimali, M., & Garaventa, F. (2022). Ecotoxicity of Polyvinylidene difluoride (PVDF) and Polylactic acid (PLA) Microplastics in Marine Zooplankton. *Toxics, 10*, 479.
29. Galloway, T. S., Dogra, Y., Garrett, N., Rowe, D., Tyler, C. R., Moger, J., Lammer, E., Landsiedel, R., Sauer, U. G., Scherer, G., et al. (2017). Ecotoxicological assessment of nanoparticle-containing acrylic copolymer dispersions in fairy shrimp and zebrafish embryos. *Environmental Science: Nano, 4*, 1981–1997.
30. Ivleva, N. P. (2021). Chemical analysis of microplastics and Nanoplastics: Challenges, advanced methods, and perspectives. *Chemical Reviews, 121*, 11886–11936. [CrossRef].
31. Kucharczyk, M., Słowik-Rylska, M., Cyran-Stemplewska, S., Gieron, M., Nowak-Starz, G., & Kr˛ecisz, B. (2021). Acrylates as a significant causes of allergic contact dermatitis—New sources of exposure. *Advances in Dermatology and Allergology, 38*, 555–560.
32. Ramakrishnan, D., & Sathiyamoorthy, M. (2025). Enhancing the remediation of polyamide microplastics: A comparative study of natural and synthetic coagulants. *Frontiers Environmental Science Sec. Water and Wastewater Management, 13*, 1. https://doi.org/10.3389/fenvs.2025.1620074
33. Karpenko, A. A., Odintsov, V. S., & Istomina, A. A. (2022). Micro-nano-sized polytetrafluoroethylene (teflon) particles as a model of plastic pollution detection in living organisms. *Environmental Science and Pollution Research, 29*, 11281–11290. https://doi.org/10.1007/s11356-021-16487-6
34. Ainali, N. M., Kalaronis, D., Evgenidou, E., Kyzas, G. Z., Bobori, D. C., Kaloyianni, M., Yang, X., Bikiaris, D. N., & Lambropoulou, D. A. (2022). Do poly(lactic acid) microplastics instigate a threat? A perception for their dynamic towards environmental pollution and toxicity. *Science of The Total Environment, 832*(1), 155014.
35. Mahadevan, G., & Val+iyaveettil, S. (2021). Understanding the interactions of poly(methyl methacrylate) and poly(vinyl chloride) nanoparticles with BHK-21 cell line. *Scientific Reports, 11*, 2089. https://doi.org/10.1038/s41598-020-80708-0

36. Guangxue, X., Qian, H., Lianzhen, L., Yan, X., Shaochong, L., & Xilin, S. (2024). Co-exposure of microplastics and polychlorinated biphenyls strongly influenced the cycling processes of typical biogenic elements in anoxic soil. *Journal of Hazardous Materials, 465*(5), 133277.
37. Romeo, T., Pietro, B., Peda, C., Consoli, P., Andaloro, F., & Fossi, M. C. (2015). First evidence of presence of plastic debris in stomach of large pelagic fish in the Mediterranean Sea. *Marine Pollution Bulletin, 95*, 358–361.
38. Mkuye, R., Gong, S., Zhao, L., Masanja, F., Ndandala, C., Bubelwa, E., Yang, C., & Deng, Y. (2022). Effects of microplastics on physiological performance of marine bivalves, potential impacts, and enlightening the future based on a comparative study. *Science of the Total Environment, 838*(Pt 1), 155933. [CrossRef].
39. Li, H. X., Shi, M., Tian, F., Lin, L., Liu, S., Hou, R., Peng, J. P., & Xu, X. R. (2022). Microplastics contamination in bivalves from the Daya Bay: Species variability and spatio-temporal distribution and human health risks. *Science of The Total Environment, 841*, 156749.
40. Wright, S. L., Thompson, R. C., & Galloway, T. S. (2013). The physical impacts of microplastics on marine organisms: A review. *Environmental Pollution, 178*, 483–492.
41. Troost, T. A., Desclaux, T., Leslie, H. A., et al. (2018). Do microplastics affect marine ecosystem productivity? *Marine Pollution Bulletin, 135*, 17–29.
42. Vroom, R. J., Koelmans, A. A., Besseling, E., et al. (2017). Aging of microplastics promotes their ingestion by marine zooplankton. *Environmental Pollution, 231*, 987–996.
43. Batel, A., Borchert, F., Reinwald, H., et al. (2018). Microplastic accumulation patterns and transfer of benzo [a] pyrene to adult zebrafish (*Danio rerio*) gills and zebrafish embryos. *Environmental Pollution, 235*, 918–930.
44. Horton, A. A., Walton, A., Spurgeon, D. J., et al. (2017). Microplastics in freshwater and terrestrial environments: Evaluating the current understanding to identify the knowledge gaps and future research priorities. *Science of the Total Environment, 586*, 127–141.
45. Kalčíková, G., Gotvajn, A. Ž., Kladnik, A., et al. (2017). Impact of polyethylene microbeads on the floating freshwater plant duckweed *Lemna minor*. *Environmental Pollution, 230*, 1108–1115.
46. Liu, Z., Huang, Q., Chen, L., Li, J., & Jia, H. (2022). Is the impact of atmospheric microplastics on human health underestimated? Uncertainty in risk assessment: A case study of urban atmosphere in Xi'an, Northwest China. *Science of The Total Environment, 851*(Pt 2), 158167. https://doi.org/10.1016/j.scitotenv.2022.158167. Epub 2022 Aug 20.
47. Yee, M. S., Hii, L. W., Looi, C. K., Lim, W. M., Wong, S. F., Kok, Y. Y., Tan, B. K., Wong, C. Y., & Leong, C. O. (2021). Impact of microplastics and Nanoplastics on human health. *Nanomaterials (Basel), 11*(2), 496. https://doi.org/10.3390/nano11020496. PMID: 33669327; PMCID: PMC7920297.
48. Verla, A. W., Enyoh, C. E., Verla, E. N., & Nwarnorh, K. O. (2019). Microplastic–toxic chemical interaction: A review study on quantified levels, mechanism and implication. *SN Applied Sciences, 1*, 1400.
49. Wu, P., Lin, S., Cao, G., Wu, J., Jin, H., Wang, C., Wong, M. H., Yang, Z., & Cai, Z. (2022). Absorption, distribution, metabolism, excretion and toxicity of microplastics in the human body and health implications. *Journal of Hazardous Materials, 437*, 129361. https://doi.org/10.1016/j.jhazmat.2022.129361. Epub 2022 Jun 16.
50. Li, J., Zhang, K., & Zhang, H. (2018). Adsorption of antibiotics on microplastics. *Environmental Pollution, 237*, 460–467. https://doi.org/10.1016/j.envpol.2018.02.050. Epub 2018 Mar 15.
51. Shen, M., Huang, W., Chen, M., Song, B., Zeng, G., & Zhang, Y. (2020). (Micro)plastic crisis: Un-ignorable contribution to global greenhouse gas emissions and climate change. *Journal of Cleaner Production, 254*, 120138. https://doi.org/10.1016/j.jclepro.2020.120138. January 13. https://www.sciencedirect.com/science/article/abs/pii/S0959652620301852.
52. Llorca, M., Schirinzi, G., Martínez, M., et al. (2018). Adsorption of perfluoroalkyl substances on microplastics under environmental conditions. *Environmental Pollution, 235*, 680–691.

53. Lamb, J. B., Willis, B. L., Fiorenza, E. A., et al. (2018). Plastic waste associated with disease on coral reefs. *Science, 359*(6374), 460–462.
54. Lozano, Y. M., Aguilar-Trigueros, C. A., Onandia, G., Maaß, S., Zhao, T., & Rillig, M. C. (2021). Effects of microplastics and drought on soil ecosystem functions and multi-functionality. *Journal Of Applied Ecology, 58*, 988–996.
55. de Souza Machado, A. A., Lau, C. W., Kloas, W., Bergmann, J., Bachelier, J. B., Faltin, E., Becker, R., Gorlich, A. S., & Rillig, M. C. (2019). Microplastics can change soil properties and affect plant performance. *Environmental Science & Technology, 53*, 6044–6052. [CrossRef] [PubMed].
56. Mamun, A. A., Prasetya, T. A. E., Dewi, I. R., & Ahmad, M. (2023). Microplastics in human food chains: Food becoming a threat to health safety. *Science of the Total Environment, 858*(Pt 1), 159834. [CrossRef].
57. Akoueson, F., Sheldon, L. M., Danopoulos, E., Morris, S., Hotten, J., Chapman, E., Li, J., & Rotchell, J. M. (2020). A preliminary analysis of microplastics in edible versus non-edible tissues from seafood samples. *Environmental Pollution, 263 Pt A*, 114452. [CrossRef].
58. Amato-Lourenco, L. F., Dos Santos Galvao, L., de Weger, L. A., Hiemstra, P. S., Vijver, M. G., & Mauad, T. (2020). An emerging class of air pollutants: Potential effects of microplastics to respiratory human health? *Science of the Total Environment, 749*, 141676. [CrossRef].
59. Campanale, C., Massarelli, C., Savino, I., Locaputo, V., & Uricchio, V. F. (2020). A detailed review study on potential effects of microplastics and additives of concern on human health. *International Journal of Environmental Research and Public Health, 17*, 1212. [CrossRef].
60. Osman, A. I., Hosny, M., Eltaweil, A. S., Omar, S., Elgarahy, A. M., Farghali, M., Yap, P.-S., Wu, Y.-S., Nagandran, S., & Batumalaie, K. (2023). Microplastic sources, formation, toxicity and remediation: A review. *Environmental Chemistry Letters, 21*, 2129–2169. [CrossRef].
61. Qiao, R., Sheng, C., Lu, Y., Zhang, Y., Ren, H., & Lemos, B. (2019). Microplastics induce intestinal inflammation, oxidative stress, and disorders of metabolome and microbiome in zebrafish. *Science of the Total Environment, 662*, 246–253. [CrossRef].
62. Hirt, N., & Body-Malapel, M. (2020). Immunotoxicity and intestinal effects of nano- and microplastics: A review of the literature. *Particle and Fibre Toxicology, 17*, 57. [CrossRef] [PubMed].
63. Niu, H., Liu, S., Jiang, Y., Hu, Y., Li, Y., He, L., Xing, M., Li, X., Wu, L., Chen, Z., et al. (2023). Are Microplastics toxic? A review from eco-toxicity to effects on the gut Microbiota. *Metabolites, 13*, 739. [CrossRef] [PubMed].
64. Kannan, K., & Vimalkumar, K. (2021). A review of human exposure to microplastics and insights into microplastics as Obesogens. *Frontiers in Endocrinology, 12*, 724989. [CrossRef] [PubMed].
65. Lu, K., Zhan, D., Fang, Y., Li, L., Chen, G., Chen, S., & Wang, L. (2022). Microplastics, potential threat to patients with lung diseases. *Frontiers in Toxicology, 4*, 958414. [CrossRef] [PubMed].
66. Wang, Y., Baynes, A., Renner, K. O., Zhang, M., Scrimshaw, M. D., & Routledge, E. J. (2022). Uptake, elimination and effects of cosmetic microbeads on the freshwater gastropod Biomphalaria glabrata. *Toxics, 10*, 87. [CrossRef] [PubMed].
67. Bashir, S. M., Kimiko, S., Mak, C.-W., Fang, J. K.-H., & Gonçalves, D. (2021). Personal care and cosmetic products as a potential source of environmental contamination by microplastics in a densely populated Asian City. *Frontiers in Marine Science, 8*, 683482. [CrossRef].
68. Dubey, I., Khan, S., & Kushwaha, S. (2022). Developmental and reproductive toxic effects of exposure to microplastics: A review of associated signaling pathways. *Frontiers in Toxicology, 4*, 901798. [CrossRef] [PubMed].
69. Enyoh, C. E., Shafea, L., Verla, A. W., Verla, E. N., Qingyue, W., Chowdhury, T., & Paredes, M. (2020). Microplastics exposure routes and toxicity studies to ecosystems: An overview. *Environmental Analysis Health and Toxicology, 35*, e2020004. [CrossRef].

70. Palaniappan, S., Sadacharan, C. M., & Rostama, B. (2022). Polystyrene and polyethylene microplastics decrease cell viability and dysregulate inflammatory and oxidative stress markers of MDCK and L929 cells in vitro. *Exposure and Health, 14*, 75–85. [CrossRef].
71. Roursgaard, M., Hezareh Rothmann, M., Schulte, J., Karadimou, I., Marinelli, E., & Moller, P. (2022). Genotoxicity of particles from grinded plastic items in Caco-2 and HepG2 cells. *Frontiers in Public Health, 10*, 906430. [CrossRef].
72. Caputi, S., Diomede, F., Lanuti, P., Marconi, G. D., Di Carlo, P., Sinjari, B., & Trubiani, O. (2022). Microplastics affect the inflammation pathway in human gingival fibroblasts: A study in the Adriatic Sea. *International Journal of Environmental Research and Public Health, 19*, 7782. [CrossRef].
73. Sinha, J. K., & Ghosh, S. (2010). Scoring more than ten plus century—Antiquity in gerontology? *The Indian Journal of Medical Research, 131*, 586–587.
74. Melendez-Fernandez, O. H., Liu, J. A., & Nelson, R. J. (2023). Circadian rhythms disrupted by light at night and mistimed food intake Alter hormonal rhythms and metabolism. *International Journal of Molecular Sciences, 24*, 3392. [CrossRef].
75. Moralia, M. A., Quignon, C., Simonneaux, M., & Simonneaux, V. (2022). Environmental disruption of reproductive rhythms. *Frontiers in Neuroendocrinology, 66*, 100990. [CrossRef].
76. Ghosh, S., Sinha, J. K., & Raghunath, M. (2019). 'Obesageing': Linking obesity & ageing. *The Indian Journal of Medical Research, 149*, 610–615. [PubMed].
77. Huang, D., Zhang, Y., Long, J., Yang, X., Bao, L., Yang, Z., Wu, B., Si, R., Zhao, W., Peng, C., et al. (2022). Polystyrene microplastic exposure induces insulin resistance in mice via dysbacteriosis and pro-inflammation. *Science of the Total Environment, 838*(Pt 1), 155937. [CrossRef].
78. Landrigan, P. J., Raps, H., Cropper, M., Bald, C., Brunner, M., Canonizado, E. M., Charles, D., Chiles, T. C., Donohue, M. J., Enck, J., et al. (2023). The Minderoo-Monaco commission on plastics and human health. *Annals of Global Health, 89*, 23. [CrossRef] [PubMed].
79. Heilig, S., Adams, D. R., & Zaenglein, A. L. (2011). Persistent allergic contact dermatitis to plastic toilet seats. *Pediatric Dermatology, 28*(5), 587–590. https://doi.org/10.1111/j.1525-1470.2010.01211.x. Epub 2011 Apr 26.
80. Yan, Z., Liu, Y., Zhang, T., Zhang, F., Ren, H., & Zhang, Y. (2022). Analysis of microplastics in human feces reveals a correlation between fecal microplastics and inflammatory bowel disease status. *Environmental Science & Technology, 56*(1), 414–421. https://doi.org/10.1021/acs.est.1c03924. Epub 2021 Dec 22. PMID: 34.
81. Oddone, E., Modonesi, C., & Gatta, G. (2014). Occupational exposures and colorectal cancers: A quantitative overview of epidemiological evidence. *World Journal of Gastroenterology, 20*(35), 12431–12444. https://doi.org/10.3748/wjg.v20.i35.12431
82. Chen, X., Zhuang, J., Chen, Q., Xu, L., Yue, X., & Qiao, D. (2022). Polyvinyl chloride microplastics induced gut barrier dysfunction, microbiota dysbiosis and metabolism disorder in adult mice. *Ecotoxicology and Environmental Safety, 241*, 113809. https://doi.org/10.1016/j.ecoenv.2022.113809. Epub 2022 Jul 4.
83. Auguet, T., Bertran, L., Barrientos-Riosalido, A., Fabregat, B., Villar, B., Aguilar, C., & Sabench, F. (2022). Are ingested or inhaled microplastics involved in nonalcoholic fatty liver disease? *International Journal of Environmental Research and Public Health, 19*(20), 13495. https://doi.org/10.3390/ijerph192013495. PMID: 36294076; PMCID: PMC9602632.
84. Molina, E., & Bened'e, S. (2022). Is there evidence of health risks from exposure to micro- and nanoplastics in foods? *Frontiers in Nutrition, 9*, 910094. https://doi.org/10.3389/fnut.2022.910094. PMID: 35836585; PMCID: PMC9274238.
85. Ruder, A. M., & Bertke, S. J. (2017). Cancer incidence among boat-building workers exposed to styrene. *American Journal of Industrial Medicine, 60*(7), 651–657. https://doi.org/10.1002/ajim.22735
86. Sartorelli, P., d'Hauw, G., Spina, D., Volterrani, L., & Mazzei, M. A. (2020). A case of hypersensitivity pneumonitis in a worker exposed to terephthalic acid in the production of polyethylene terephthalate. *International Journal of Occupational Medicine and*

*Environmental Health, 33*(1), 119–123. https://doi.org/10.13075/ijomeh.1896.01465. Epub 2019 Oct 30.

87. Nett, R. J., Cox-Ganser, J. M., Hubbs, A. F., Ruder, A. M., Cummings, K. J., Huang, Y. T., & Kreiss, K. (2017). Non-malignant respiratory disease among workers in industries using styrene-a review of the evidence. *American Journal of Industrial Medicine, 60*(2), 163–180. https://doi.org/10.1002/ajim.22655. PMID: 28079275; PMCID: PMC5379121.
88. Bertke, S. J., Yiin, J. H., & Daniels, R. D. (2018). Cancer mortality update with an exposure response analysis among styrene-exposed workers in the reinforced plastics boatbuilding industry. *American Journal of Industrial Medicine, 61*(7), 566–571. https://doi.org/10.1002/ajim.22853. Epub 2018 Apr 11.
89. Daniels, R. D., & Bertke, S. J. (2020). Exposure-response assessment of cancer mortality in styrene-exposed boatbuilders. *Occupational And Environmental Medicine, 77*(10), 706–712. Epub 2020 May 29. PMID: 32471836; PMCID: PMC7494518. https://doi.org/10.1136/oemed-2020-106445
90. Christensen, M. S., Vestergaard, J. M., d'Amore, F., Gørløv, J. S., Toft, G., Ramlau-Hansen, C. H., Stokholm, Z. A., Iversen, I. B., Nissen, M. S., & Kolstad, H. A. (2018). Styrene exposure and risk of lymphohematopoietic malignancies in 73,036 reinforced plastics workers. *Epidemiology, 29*(3), 342–351. https://doi.org/10.1097/EDE.0000000000000819
91. Werder, E. J., Engel, L. S., Richardson, D. B., Emch, M. E., Gerr, F. E., Kwok, R. K., & Sandler, D. P. (2018). Environmental styrene exposure and neurologic symptoms in U.S. Gulf coast residents. Pt 1. *Environment International, 121*, 480–490. https://doi.org/10.1016/j.envint.2018.09.025. Epub 2018 Oct 1. PMID: 30278311; PMCID: PMC6712572.
92. Bitar, D., & Parvizi, J. (2015). Biological response to prosthetic debris, 2. *world J. Orthoped., 6*, 172–189. https://doi.org/10.5312/wjo.v6.i2.172. PMID: 25793158; PMCID:PMC4363800.
93. Yu, C. W., Luk, T. C., & Liao, V. H. (2021). Long-term nanoplastics exposure results in multi and trans-generational reproduction decline associated with germline toxicity and epigenetic regulation in Caenorhabditis elegans. *Journal of Hazardous Materials, 412*, 125173. https://doi.org/10.1016/j.jhazmat.2021.125173. Epub 2021 Jan 19.
94. Sarasamma, S., Audira, G., Siregar, P., Malhotra, N., Lai, Y. H., Liang, S. T., Chen, J. R., Chen, K. H., & Hsiao, C. D. (2020). Nanoplastics cause neurobehavioral impairments, reproductive and oxidative damages, and biomarker responses in zebrafish: Throwing up alarms of wide spread health risk of exposure. *International Journal of Molecular Science, 21*, 1410. https://doi.org/10.3390/ijms21041410. PMID: 32093039; PMCID:PMC7073134.
95. Qiang, L., & Cheng, J. (2021). Exposure to polystyrene microplastics impairs gonads of zebrafish (Danio rerio). *Chemosphere, 263*, 128161. https://doi.org/10.1016/j.chemosphere.2020.128161. Epub 2020 Aug 31.
96. Brachner, A., Fragouli, D., Duarte, I. F., Farias, P. M. A., Dembski, S., Ghosh, M., Barisic, I., Zdzieblo, D., Vanoirbeek, J., Schwabl, P., & Neuhaus, W. (2020). Assessment of human health risks posed by nano-and microplastics is currently not feasible, 23. *International Journal of Environmental Research and Public Health, 17*, 8832. https://doi.org/10.3390/ijerph17238832. PMID: 33261100; PMCID: PMC7730001.
97. Rose, R. F., Lyons, P., Horne, H., & Mark Wilkinson, S. (2009). A review of the materials and allergens in protective gloves. *Contact Dermatitis, 61*(3), 129–137. https://doi.org/10.1111/j.1600-0536.2009.01580.x
98. Deng, Y., Zhang, Y., Lemos, B., & Ren, H. (2017). Tissue accumulation of microplastics in mice and biomarker responses suggest widespread health risks of exposure. *Scientific Reports, 7*(1), 46687–46710. https://doi.org/10.1038/srep46687
99. Forte, M., Iachetta, G., Tussellino, M., Carotenuto, R., Prisco, M., De Falco, M., et al. (2016). Polystyrene nanoparticles internalization in human gastric adenocarcinoma cells. *Toxicology In Vitro, 31*, 126–136. https://doi.org/10.1016/j.tiv.2015.11.006
100. Fuchs, A.-K., Syrovets, T., Haas, K. A., Loos, C., Musyanovych, A., Mailänder, V., et al. (2016). Carboxyl-and amino-functionalized polystyrene nanoparticles differentially affect

the polarization profile of M1 and M2 macrophage subsets. *Biomaterials, 85*, 78–87. https://doi.org/10.1016/j.biomaterials.2016.01.064
101. Mahadevan, G., & Valiyaveettil, S. (2021). Understanding the interactions of poly (methyl methacrylate) and poly (vinyl chloride) nanoparticles with BHK-21 cell line. *Scientific Reports, 11*(1), 1–15. https://doi.org/10.1038/s41598-020-80708-0
102. Inkielewicz-Stepniak, I., Tajber, L., Behan, G., Zhang, H., Radomski, M. W., Medina, C., et al. (2018). The role of mucin in the toxicological impact of polystyrene nanoparticles. *Materials, 11*(5), 724. https://doi.org/10.3390/ma11050724
103. Luo, T., Zhang, Y., Wang, C., Wang, X., Zhou, J., Shen, M., et al. (2019a). Maternal exposure to different sizes of polystyrene MPs during gestation causes metabolic disorders in their offspring. *Environmental Pollution, 255*, 113122. https://doi.org/10.1016/j.envpol.2019.113122
104. Jin, Y., Xia, J., Pan, Z., Yang, J., Wang, W., & Fu, Z. (2018). Polystyrene MPs induce microbiota Dysbiosis and inflammation in the gut of adult zebrafish. *Environmental Pollution, 235*, 322–329. https://doi.org/10.1016/j.envpol.2017.12.088
105. McCarthy, J., Gong, X., Nahirney, D., Duszyk, M., & Radomski, M. W. (2011). Polystyrene nanoparticles activate ion transport in human airway epithelial cells. *International Journal of Nanomedicine, 6*, 1343. https://doi.org/10.2147/ijn.s21145
106. Gouin, T. (2020). Toward an improved understanding of the ingestion and trophic transfer of microplastic particles: Critical review and implications for future research. *Environmental Toxicology and Chemistry, 39*(6), 1119–1137. https://doi.org/10.1002/etc.4718. PMID: 32223000; PMCID: PMC7383496.
107. Menéndez-Pedriza, A., & Jaumot, J. (2020). Interaction of environmental pollutants with microplastics: A critical review of sorption factors, bioaccumulation and Ecotoxicological effects. *Toxics, 8*(2), 40. https://doi.org/10.3390/toxics8020040. PMID: 32498316; PMCID: PMC7355763.
108. Huang, W., Song, B., Liang, J., Niu, Q., Zeng, G., Shen, M., Deng, J., Luo, Y., Wen, X., & Zhang, Y. (2021). Microplastics and associated contaminants in the aquatic environment: A review on their ecotoxicological effects, trophic transfer, and potential impacts to human health. *Journal of Hazardous Materials, 405*, 124187. https://doi.org/10.1016/j.jhazmat.2020.124187. Epub 2020 Oct 15.
109. Prata, J. C. (2018). Airborne microplastics: Consequences to human health? *Environmental Pollution, 234*, 115–126. https://doi.org/10.1016/j.envpol.2017.11.043. Epub 2017 Dec 21.
110. Sun, A., & Wang, W. X. (2023). Human exposure to microplastics and its associated health risks. *Environmental Health, 1*(3), 139–149.
111. Daniel, D. B., Ashraf, P. M., Thomas, S. N., & Thomson, K. T. (2021). Microplastics in the edible tissues of shellfishes sold for human consumption. *Chemosphere, 264*(Pt 2), 128554. [CrossRef].
112. Bhuyan, M. S. (2022). Effects of Microplastics on fish and in human health. *Frontiers in Environmental Science, 10*, 250.
113. Alberghini, L., Truant, A., Santonicola, S., Colavita, G., & Giaccone, V. (2022). Microplastics in fish and fishery products and risks for human health: A review. *International Journal of Environmental Research and Public Health, 20*, 789.
114. Bhuyan, M. S. (2022). Effects of Microplastics on fish and in human health. *Frontiers in Environmental Sciences, 10*, 250.
115. Yong, C. Q. Y., Valiyaveettil, S., & Tang, B. L. (2020). Toxicity of microplastics and Nanoplastics in mammalian systems. *International Journal of Environmental Research and Public Health, 17*, 1509.
116. Mason, V. G., Skov, M. W., Hiddink, J. G., & Walton, M. (2022). Microplastics alter multiple biological processes of marine benthic fauna. *Science of The Total Environment, 845*, 157362.
117. Morkunas, J., Biveinyt ˙e, V., Balˇci ¯unas, A., & Mork ¯un ˙e, R. (2021). The broader isotopic niche of Long-tailed duck Clangula hyemalis implies a higher risk of ingesting plastic and non-plastic debris than for other diving seabirds. *Marine Pollution Bulletin, 173*, 113065.

118. Hamilton, B. M., Bourdages, M. P. T., Geoffroy, C., Vermaire, J. C., Mallory, M. L., Rochman, C. M., & Provencher, J. F. (2021). Microplastics around an Arctic seabird colony: Particle community composition varies across environmental matrices. *Science of the Total Environment, 773*, 145536. [CrossRef].
119. Pantos, O. (2022). Microplastics: Impacts on corals and other reef organisms. *Emerging Topics In Life Sciences, 6*, 81–93. [CrossRef].
120. Iwalaye, O. A., Moodley, G. K., & Robertson-Andersson, D. V. (2020). The possible routes of microplastics uptake in sea cucumber Holothuria cinerascens (Brandt, 1835). *Environmental Pollution, 264*, 114644. [CrossRef].
121. Cau, A., Avio, C. G., Dessì, C., Moccia, D., Pusceddu, A., Regoli, F., Cannas, R., & Follesa, M. C. (2020). Benthic crustacean digestion can modulate the environmental fate of microplastics in the Deep Sea. *Environmental Science & Technology, 54*, 4886–4892. [CrossRef] [PubMed].
122. Zhang, T., Sun, Y., Song, K., Du, W., Huang, W., Gu, Z., & Feng, Z. (2021). Microplastics in different tissues of wild crabs at three important fishing grounds in China. *Chemosphere, 271*, 129479. [CrossRef] [PubMed].
123. Fackelmann, G., Pham, C. K., Rodríguez, Y., Mallory, M. L., Provencher, J. F., Baak, J. E., & Sommer, S. (2023). Current levels of microplastic pollution impact wild seabird gut microbiomes. *Nature Ecology & Evolution, 7*, 698–706. [CrossRef] [PubMed].
124. Wang, R., Mou, H., Lin, X., Zhu, H., Li, B., Wang, J., Junaid, M., & Wang, J. (2021). Microplastics in mollusks: Research Progress, current contamination status, analysis approaches, and future perspectives. *Frontiers in Marine Science, 8*, 759919.
125. Devereux, R., Hartl, M. G. J., Bell, M., & Capper, A. (2021). The abundance of microplastics in cnidaria and ctenophora in the North Sea. *Marine Pollution Bulletin, 173*, 112992.
126. Fallon, B. R., & Freeman, C. J. (2021). Plastics in Porifera: The occurrence of potential microplastics in marine sponges and seawater from Bocas del Toro, Panamá. *PeerJ, 9*, e11638. [CrossRef].
127. Hitchcock, J. N. (2022). Microplastics can alter phytoplankton community composition. *Science of the Total Environment, 819*, 153074. [CrossRef][PubMed].
128. Kvale, K., Prowe, A. E. F., Chien, C. T., Landolfi, A., & Oschlies, A. (2021). Zooplankton grazing of microplastic can accelerate global loss of ocean oxygen. *Nature Communications, 12*, 2358.
129. Pagter, E., Nash, R., Frias, J., & Kavanagh, F. (2021). Assessing microplastic distribution within infaunal benthic communities in a coastal embayment. *Science of the Total Environment, 791*, 148278. [CrossRef].
130. Xu, X., Wang, S., Gao, F., Li, J., Zheng, L., Sun, C., He, C., Wang, Z., & Qu, L. (2019). Marine microplastic-associated bacterial community succession in response to geography, exposure time, and plastic type in China's coastal seawaters. *Marine Pollution Bulletin, 145*, 278–286. [CrossRef].
131. Pinheiro, L. M., Britz, L. M. K., Agostini, V. O., Pérez-Parada, A., García-Rodríguez, F., Galloway, T. S., & Pinho, G. L. L. (2022). Salt marshes as the final watershed fate for meso- and microplastic contamination: A case study from southern Brazil. *Science of the Total Environment, 838*, 156077. [CrossRef].
132. Huang, W., Chen, M., Song, B., Deng, J., Shen, M., Chen, Q., Zeng, G., & Liang, J. (2021). Microplastics in the coral reefs and their potential impacts on corals: A mini-review. *Science of the Total Environment, 762*, 143112. [CrossRef] [PubMed].
133. Gerstenbacher, C. M., Finzi, A. C., Rotjan, R. D., & Novak, A. B. (2022). A review of microplastic impacts on seagrasses, epiphytes, and associated sediment communities. *Environmental Pollution, 303*, 119108. [CrossRef] [PubMed].
134. Corinaldesi, C., Canensi, S., Dell'Anno, A., Tangherlini, M., Di Capua, I., Varrella, S., Willis, T. J., Cerrano, C., & Danovaro, R. (2021). Multiple impacts of microplastics can threaten marine habitat-forming species. *Communications Biology, 4*, 431.

135. Dias, M. P., Martin, R. W., Pearmain, E. J., Burfield, I. J., Small, C., Phillips, R. A., Yates, O., Lascelles, B. G., Borboroglu, P. G., & Croxall, J. P. (2019). Threats to seabirds: A global assessment. *Biological Conservation, 237*, 525–537.
136. Meaza, I., Toyoda, J. H., & Wise, J. P., Sr. (2021). Microplastics in sea turtles, marine mammals and humans: A one environmental health perspective. *Frontiers in Environmental Science, 8*, 298.
137. Lundebye, A.-K., Lusher, A. L., & Bank, M. S. (2022). Marine microplastics and seafood: Implications for food security. In M. S. Bank (Ed.), *Microplastic in the environment: Pattern and process* (pp. 131–153). Springer International Publishing.
138. Mercogliano, R., Avio, C. G., Regoli, F., Anastasio, A., Colavita, G., & Santonicola, S. (2020). Occurrence of microplastics in commercial seafood under the perspective of the human food chain. A review. *Journal of Agricultural and Food Chemistry, 68*, 5296–5301. [CrossRef].
139. Bhattacharya, P., Lin, S., Turner, J. P., & Ke, P. C. (2010). Physical adsorption of charged plastic nanoparticles affects algal photosynthesis. *The Journal of Physical Chemistry C, 114*, 16556–16561.
140. Cole, M., Webb, H., Lindeque, P. K., Fileman, E. S., Halsband, C., & Galloway, T. S. (2014). Isolation of microplastics in biota-rich seawater samples and marine organisms. *Scientific Reports, 4*, 4528.
141. Ory, N., Chagnon, C., Felix, F., Fernández, C., Ferreira, J. L., Gallardo, C., Ordonez, O. G., Henostroza, A., Laaz, E., Mizraji, R., & Mojica, H. (2018). Low prevalence of microplastic contamination in planktivorous fish species from the Southeast Pacific Ocean. *Marine Pollution Bulletin, 127*, 211–216.
142. Cheung, L. T., Lui, C. Y., & Fok, L. (2018). Microplastic contamination of wild and captive flathead grey mullet (Mugil cephalus). *International Journal of Environmental Research and Public Health, 15*(4), 597.
143. Cole, M., Lindeque, P., Fileman, E., Halsband, C., & Galloway, T. S. (2015). The impact of polystyrene microplastics on feeding, function and fecundity in the marine copepod Calanus helgolandicus. *Environmental Science & Technology, 49*(2), 1130–1137.
144. Devriese, L. I., Van der Meulen, M. D., Maes, T., Bekaert, K., Paul-Pont, I., Frère, L., Robbens, J., & Vethaak, A. D. (2015). Microplastic contamination in brown shrimp (Crangon crangon, Linnaeus 1758) from coastal waters of the southern North Sea and channel area. *Marine Pollution Bulletin, 98*(1–2), 179–187.
145. Egbeocha, C. O., Malek, S., Emenike, C. U., & Milow, P. (2018). Feasting on microplastics: Ingestion by and effects on marine organisms. *Aquatic Biology, 27*, 93–106.
146. Sussarellu, R., Suquet, M., Thomas, Y., Lambert, C., Fabioux, C., Pernet, M. E., Le Goic, N., Quillien, V., Mingant, C., Epelboin, Y., & Corporeau, C. (2016). Oyster reproduction is affected by exposure to polystyrene microplastics. *Proceedings of the National Academy of Sciences of the United States of America, 113*(9), 2430–2435.
147. Naji, A., Nuri, M., Amiri, P., & Niyogi, S. (2019). Small microplastic particles (S-MPPs) in sediments of mangrove ecosystem on the northern coast of the Persian Gulf. *Marine Pollution Bulletin, 146*, 305–311.
148. Reichert, J., Schellenberg, J., Schubert, P., & Wilke, T. (2018). Responses of reef building corals to microplastic exposure. *Environmental Pollution, 237*, 955–960.
149. Chapron, L., Peru, E., Engler, A., Ghiglione, J. F., Meistertzheim, A. L., Pruski, A. M., Vetion, G., Galand, P. E., & Lartaud, F. (2018). Macro-and microplastics affect cold-water corals growth, feeding and behaviour. *Scientific Reports, 8*(1), 1–8.
150. Barasarathi, J., Agamuthu, P., Emenike, C. U., & Fauziah, S. H. (2014). Microplastic abundance in selected mangrove forest in Malaysia. In *Proceeding of the ASEAN conference on science and technology* (pp. 1–5).
151. da Costa, J. P., Santos, P. S., Duarte, A. C., & Rocha-Santos, T. (2016). Nano plastics in the environment-sources, fates and effects. *Science of The Total Environment, 566*, 5–26.
152. Li, R., Zhang, L., Xue, B., & Wang, Y. (2018). Abundance and characteristics of microplastics in the mangrove sediment of the semi-enclosed Maowei Sea of the South China Sea: New

implications for location, rhizosphere, and sediment compositions. *Environmental Pollution, 244*, 685–692.

153. Andrady, A. L. (2017). The plastic in microplastics: A review. *Marine Pollution Bulletin, 119*, 12–22.
154. Lambert, S., & Wagner, M. (2016). Characterisation of nanoplastics during the degradation of polystyrene. *Chemosphere, 145*, 265–268.
155. Zbyszewski, M., Corcoran, P. L., & Hockin, A. (2014). Comparison of the distribution and degradation of plastic debris along shorelines of the Great Lakes, North America. *Journal of Great Lakes Research, 40*, 288–299.
156. Wagner, M., Scherer, C., Alvarez-Muñoz, D., Brennholt, N., Bourrain, X., Buchinger, S., Fries, E., Grosbois, C., Klasmeier, J., & Marti, T. (2014). Microplastics in freshwater ecosystems: What we know and what we need to know. *Environmental Sciences Europe, 26*, 12.
157. Nizzetto, L., Bussi, G., Futter, M. N., Butterfield, D., & Whitehead, P. G. (2016). A theoretical assessment of microplastic transport in river catchments and their retention by soils and river sediments. *Environmental Science. Processes & Impacts, 18*, 1050–1059.
158. Wang, X., Bolan, N., Tsang, D. C., Sarkar, B., Bradney, L., & Li, Y. (2021). A review of microplastics aggregation in aquatic environment: Influence factors, analytical methods, and environmental implications. *Journal of Hazardous Materials, 402*, 123496.
159. Kaiser, D., Kowalski, N., & Waniek, J. J. (2017). Effects of biofouling on the sinking behavior of microplastics. *Environmental Research Letters, 12*, 124003. [CrossRef].
160. Waldschläger, K., & Schüttrumpf, H. (2019). Effects of particle properties on the settling and rise velocities of microplastics in freshwater under laboratory conditions. *Environmental Science & Technology, 53*, 1958–1966. [CrossRef] [PubMed].
161. Fazey, F. M. C., & Ryan, P. G. (2016). Biofouling on buoyant marine plastics: An experimental study into the effect of size on surface longevity. *Environmental Pollution, 210*, 354–360.
162. Rouillon, M., & Taylor, M. P. (2016). Can field portable X-ray fluorescence (pXRF) produce high quality data for application in environmental contamination research? *Environmental Pollution, 214*, 255–264.
163. Luo, H., Zhao, Y., Li, Y., Xiang, Y., He, D., & Pan, X. (2020). Aging of microplastics affects their surface properties, thermal decomposition, additives leaching and interactions in Simulated fluids. *Science of the Total Environment, 714*, 136862. https://doi.org/10.1016/j.scitotenv.2020.136862
164. El Hadri, H., Gigault, J., Maxit, B., Grassl, B., & Reynaud, S. (2020). Nanoplastic from mechan-ically degraded primary and secondary microplastics for environmental assessments. *NanoImpact, 17*, 100206. https://doi.org/10.1016/j.impact.2019.100206
165. Zettler, E. R., Mincer, T. J., & Amaral-Zettler, L. A. (2013). Life in the "Plastisphere ": Microbial communities on plastic marine debris. *Environmental Science & Technology, 47*, 7137–7146.
166. Johansen, M. P., Prentice, E., Cresswell, T., & Howell, N. (2018). Initial data on adsorption of Cs and Sr to the surfaces of microplastics with biofilm. *Journal of Environmental Radioactivity, 190–191*, 130–133.
167. Hammer, J., Kraak, M. H., & Parsons, J. R. (2012). Plastics in the marine environment: The dark side of a modern gift. In D. M. Whitacre (Ed.), *Reviews of environmental contamination and toxicology* (pp. 1–44). Springer.
168. Guerranti, C., Cannas, S., Scopetani, C., Fastelli, P., Cincinelli, A., & Renzi, M. (2017). Plastic litter in aquatic environments of Maremma Regional Park (Tyrrhenian Sea, Italy): Contribution by the Ombrone river and levels in marine sediments. *Marine Pollution Bulletin, 117*, 366–370. [CrossRef].
169. Schmidt, C., Kumar, R., Yang, S., & Büttner, O. (2020). Microplastic particle emission from wastewater treatment plant effluents into river networks in Germany: Loads, spatial patterns of concentrations and potential toxicity. *Science of the Total Environment, 737*, 139544. [CrossRef] [PubMed].

170. De Villiers, S. (2019). Microfibre pollution hotspots in river sediments adjacent to South Africa's coastline. *Water SA, 45*, 97–102.
171. Burns, E. E., & Boxall, A. B. A. (2018). Microplastics in the aquatic environment: Evidence for or against adverse impacts and major knowledge gaps. *Environmental Toxicology and Chemistry, 37*, 2776–2796.
172. Crawford, C. B., & Quinn, B. (2017). Plastic production, waste and legislation. In C. B. Crawford & B. Quinn (Eds.), *Microplastic pollutants* (pp. 39–56). Elsevier Science.
173. ter Halle, A., Ladirat, L., Martignac, M., Mingotaud, A. F., Boyron, O., & Perez, E. (2017). To what extent are microplastics from the open ocean weathered? *Environmental Pollution, 227*, 167–174.
174. Mueller, R. J. (2006). Biological degradation of synthetic polyesters-enzymes as potential-catalysts for polyester recycling. *Process Biochemistry, 41*, 2124–2128.
175. Jumaah, O. S. (2017). Screening of plastic degrading bacteria from dumped soil area. *IOSR Journal of Environmental Science, Toxicology and Food Technology, 11*, 93–98.
176. Zhang, X., Li, Y., Ouyang, D., Lei, J., Tan, Q., Xie, L., Li, Z., Liu, T., Xiao, Y., Farooq, T. H., Wu, X., Chen, L., & Yan, W. (2021). Systematical review of interactions between microplastics and microorganisms in the soil environment. *Journal of Hazardous Materials, 418*, 126288. https://doi.org/10.1016/j.jhazmat.2021.126288
177. Artham, T., & Doble, M. (2008). Biodegradation of aliphatic and aromatic polycarbonates. *Macromolecular Bioscience, 8*, 14–24. https://doi.org/10.1002/mabi.200700106
178. Dawson, A., Huston, W., Kawaguchi, S., King, C., Cropp, R., Wild, S., et al. (2018). Uptake and depuration kinetics influence microplastic bioaccumulation and toxicity in Antarctic krill (*Euphausia superba*). *Environmental Science & Technology, 52*(5), 3195–3201. https://doi.org/10.1021/acs.est.7b05759. PMID: 29397707.
179. Gerber, K. (2009). *User's guide and technical documentation KABAM version 1.0 (Kow (based) aquatic bioaccumulation model)*. Report of Environmental Fate and Effects Division Office of Pesticide Programs, US Environmental Protection Agency.
180. Maher, B., Taylor, A., Batley, G., & Simpson, S. (2016). Bioaccumulation. In *Sediment quality assessment: A practical guide* (pp. 123–156). CSIRO Publishing.
181. Wang, J. D., Tan, Z., Peng, J. P., Qiu, Q. X., & Li, M. M. (2016). The behaviors of microplastics in the marine environment. *Marine Environmental Research, 113*, 7–17. https://doi.org/10.1016/j.marenvres.2015.10.014. PMID: 26559150.
182. Kelly, B. C., Ikonomou, M. G., Blair, J. D., Morin, A. E., & Gobas, F. A. (2007). Food web–specific biomagnification of persistent organic pollutants. *Science, 317*(5835), 236–239. https://doi.org/10.1126/science.1138275
183. Eerkes-Medrano, D., Leslie, H. A., & Quinn, B. (2019). Microplastics in drinking water: A review and assessment. *Current Opinion in Environmental Science & Health, 7*, 69–75.
184. Magnusson, K., Eliasson, K., Fråne, A., Haikonen, K., Hultén, J., Mikael, O. M., Stadmark, J., Voisin, A., & Miljöinstitutet, I. S. (2016). *IVL-rapport C 183 Swedish sources and pathways for microplastics to the marine environment*. IVL Svenska Miljöinstitutet.
185. Goldstein, M. C., Titmus, A. J., & Ford, M. (2013). Scales of spatial heterogeneity of plastic marine debris in the Northeast Pacific Ocean. *PLoS One, 8*, 1. https://doi.org/10.1371/journal.pone.0080020
186. UNEP. (2017). *Draft resolution on marine litter and microplastics*. Retrieved December 12, 2018, from https://papersmart.unon.org/resolution/uploads/k1709154.docx
187. Wilcox, C., Van Sebille, E., & Hardesty, B. D. (2015). Threat of plastic pollution to seabirds is global, pervasive, and increasing. *Proceedings of the National Academy of Sciences of the United States of America, 112*, 11899–11190.
188. Padervand, M., Lichtfouse, E., Robert, D., & Wang, C. (2020). Removal of microplastics from the environment. A review. *Environmental Chemistry Letters, 18*, 807–828.
189. Prata, J. C., da Costa, J. P., Lopes, I., Duarte, A. C., & Rocha-Santos, T. (2019). Effects of microplastics on microalgae populations: A critical review. *Science of the Total Environment, 665*, 400–405.

190. Lusher, A. L., Burke, A., O'Connor, I., & Officer, R. (2014). Microplastic pollution in the north East Atlantic Ocean: Validated and opportunistic sampling. *Marine Pollution Bulletin, 88*, 325–333.
191. Shim, W. J., Hong, S. H., & Eo, S. (2018). Book: Microplastic contamination in aquatic environments. In *Marine Microplastics: Abundance, distribution, and composition* (pp. 1–26). Elsevier.
192. Hidalgo-Ruz, V., Gutow, L., Thompson, R. C., & Thiel, M. (2012). Microplastics in the mari- neenvironment: A review of the methods used for identification and quantification. *Environmental Science & Technology, 46*, 3060–3075.
193. Elkhatib, D., & Oyanedel-Craver, V. (2020). A critical review of extraction and identification methods of microplastics in wastewater and drinking water. *Environmental Science & Technology, 54*(12), 7037–7049. https://doi.org/10.1021/acs.est.9b06672. Epub 2020 Jun 2.
194. Pico, Y., Alfarhan, A., & Barcelo, D. (2019). Nano-and microplastic analysis: Focus on their occurrence in freshwater ecosystems and remediation technologies. *TrAC Trends in Analytical Chemistry, 113*, 409–425.
195. Erni-Cassola, G., Gibson, M. I., Thompson, R. C., & Christie-Oleza, J. A. (2017). Lost, but found with Nile red: A novel method for detecting and quantifying Small microplastics (1 mm to 20 μm) in environmental samples. *Environmental Science & Technology, 51*(23), 13641–13648. https://doi.org/10.1021/acs.est.7b04512. Epub 2017 Nov 20.
196. Møller, A. M. J., Delaissé, J. M., Olesen, J. B., Madsen, J. S., Canto, L. M., Bechmann, T., Rogatto, S. R., & Søe, K. (2020). Aging and menopause reprogram osteoclast precursors for aggressive bone resorption. *Bone Research, 8*, 27. https://doi.org/10.1038/s41413-020-0102-7. PMID: 32637185; PMCID: PMC7329827.
197. Munno, K., De Frond, H., O'Donnell, B., & Rochman, C. M. (2020). Increasing the accessibility for characterizing microplastics: Introducing new application-based and spectral libraries of plastic particles (SLoPP and SLoPP-E). *Analytical Chemistry, 92*(3), 2443–2451. https://doi.org/10.1021/acs.analchem.9b03626. Epub 2020 Jan 23.
198. Cabernard, L., Roscher, L., Lorenz, C., Gerdts, G., & Primpke, S. (2018). Comparison of Raman and Fourier transform infrared spectroscopy for the quantification of microplastics in the aquatic environment. *Environmental Science & Technology, 52*(22), 13279–13288. https://doi.org/10.1021/acs.est.8b03438. Epub 2018 Nov 6.
199. Witzig, C. S., Földi, C., Wörle, K., Habermehl, P., Pittroff, M., Müller, Y. K., Lauschke, T., Fiener, P., Dierkes, G., Freier, K. P., & Zumbülte, N. (2020). When good intentions go bad-false positive microplastic detection caused by disposable gloves. *Environmental Science & Technology, 54*(19), 12164–12172. https://doi.org/10.1021/acs.est.0c03742. Epub 2020 Sep 17.
200. Shim, W. J., Hong, S. H., & Eo, S. E. (2017). Identification methods in microplastic analysis: A review. *Analytical Methods, 9*, 1384–1391. https://doi.org/10.1039/C6AY02558G
201. Möller, K., Utz Schäffer, & Frank Verbeeten. (2020). digitalization in management accounting and control: An editorial. *journal of management Control, 31*, 1–8. https://doi.org/10.1007/s00187-020-00300-5
202. Hermabessiere, L., Himber, C., Boricaud, B., Kazour, M., Amara, R., Cassone, A. L., Laurentie, M., Paul-Pont, I., Soudant, P., Dehaut, A., & Duflos, G. (2018). Optimization, performance, and application of a pyrolysis-GC/MS method for the identification of microplastics. *Analytical and Bioanalytical Chemistry, 410*(25), 6663–6676. https://doi.org/10.1007/s00216-018-1279-0. Epub 2018 Jul 27.
203. Mani, T., Frehland, S., Kalberer, A., & Burkhardt-Holm, P. (2019). Using castor oil to separate microplastics from four different environmental matrices. *Analytical Methods, 11*, 1788–1794. https://doi.org/10.1039/C8AY02559B

# Chapter 6
# Nanoplastics and Microplastics: Environmental Hazards and Effect on Organisms

**Kazi Sabnam Banu**

## Introduction

Plastic production increases exponentially day by day due to its huge industrial applications. For this, plastic waste generation also increases astronomically over the decades. These polymeric waste materials have been accepted as not biodegradable, and they are mostly found polymers on different types of ecosystems on our planet [1]. These polymer productions increased to 400 million tons in recent years (2018) from a very few tons (1950s) [2]. This becomes a serious concern among environmentalists, biologists, and common people. Due to various environmental factors such as ablation, photooxidation, various biotc degradation pathways [3], these polymers lost their mechanical integrity and converted to very small particles, viz. MPs(microplastics) and nano plastics(NPs) that create serious threats to the environment [4]. This term ‘microplastic ‘was introduced by Thompson and his group [5]. These small microplasticsoriginated from larger plastics through fragmentation, were identified as major components in the ocean floor debris in the 1970s. Besides this, both microplastics (MPs) and nanoplastics (NPs) have been found to be present in polar regions are encapsulated in ice, in aquatic bodies around the equator, in coastal areas, and below sea level. The literature has shown that there is inconsistency in the categorization of MNPs sizes among various authors and organizations. Many have reported the upper limit of MPs is 5 mm [6–8] and the upper limit for NPs is 100 nm [9, 10]. The International Organization for Standardization has also reported the range of dimensions of NPS ~ 1–100 nm [11]. In this literature, microplastics and nanoplastics are specifically categorized within the 100-nanometer −5 dimensions range, and > 100 nanometers in dimension, respectively. The two categories of microplastics and nanoplastics (MNPs) are

K. S. Banu (✉)
Department of Chemistry, Netaji Nagar Day College, University of Calcutta, Kolkata, India



S. Kulkarni, S. Das (eds.), *Microplastics and Nanoplastics*, Lecture Notes in Nanoscale Science and Technology 49,
https://doi.org/10.1007/978-3-032-17963-0_6

present in the environment, viz., primary and secondary. Primary MNPs originated from different industrial sources and are associated with various consumer goods, such as toothpaste, textile fibres, in the raw materials used for the production of plastic materials [10]. Secondary MNPs originate from the plastic debris, which is degraded by microbial, animal, and various physical processes [12]. Alteration in the physicochemical properties of MNPs increases the toxicity and bio-availability of MNPs [13], and it's a matter of concern amongst the scientists due to their large contamination in aquatic e.g., marine ecosystems [3], freshwater bodies [14, 15], and terrestrial ecosystems [16, 17]which leads potential threat to living organisms. MNPs have heterogeneity due to their organic and inorganic counterparts and form both homo and hetero aggregates [18, 19]. These aggregates create bio-accumulation and bio-amplification phenomena and lead to poisonous effects on living organisms [20] and raising more health concerns. As MNPs are very minute particles, they get easily ingested by various organisms and are easily absorbed and disrupting their cell membranes [21]. It also leads to oxidative stress among the organisms [22]. In their life cycle, at least once, various marine organisms, viz. Echinoderms, polychaetes, copepods, and bivalves consume microplastics and nanoplastics from the environment [23–25]. As MNPs act as transporters of different pesticides such as DDT and hexachlorobenzene, when these MNPs are consumed by living organisms, these toxic pesticides enter food chains and affect human health [26]. Both indoor and outdoor atmospheres transport these MNPs and accumulate in the environment, which also raises public health concerns [27, 28]. Despite growing research on this field, many gray areas persist about the long-term effects of MNPs exposure and their high ecological impacts on the environment. This article gives a deeper understanding of the MNP's impact and its contamination across diverse ecosystems in the environment. This article provides a critical review of recent research on the sources and distribution of MNPs, their negative impact on both living organisms and nonliving components of ecosystems, and the latest strategies aimed at reducing their detrimental impacts. The information presented in this paper is expected to serve as a vital reference for all researchers and policymakers guiding and ineffectively managing the impacts of MNPs in our environment.

## Effect and Occurrence of Micro-and Nanoplastics in Aquatic Biota

MNPs accumulation in water bodies(mainly freshwater and marine) is primarily from the various plastic materials such as PVC, PE. PS, PP (Fig. 6.1). These polymeric materials interact with the different chemical substances and also different types of microorganisms present in those aquatic habitats and get accumulated there as MNPs [29]. These MNPs absorb toxic elements, behave as a magnet for toxic materials, are transported in the environment, and spread detrimental effects amongst different habitats [30, 31]. The enormous amount of MNPs present greatly

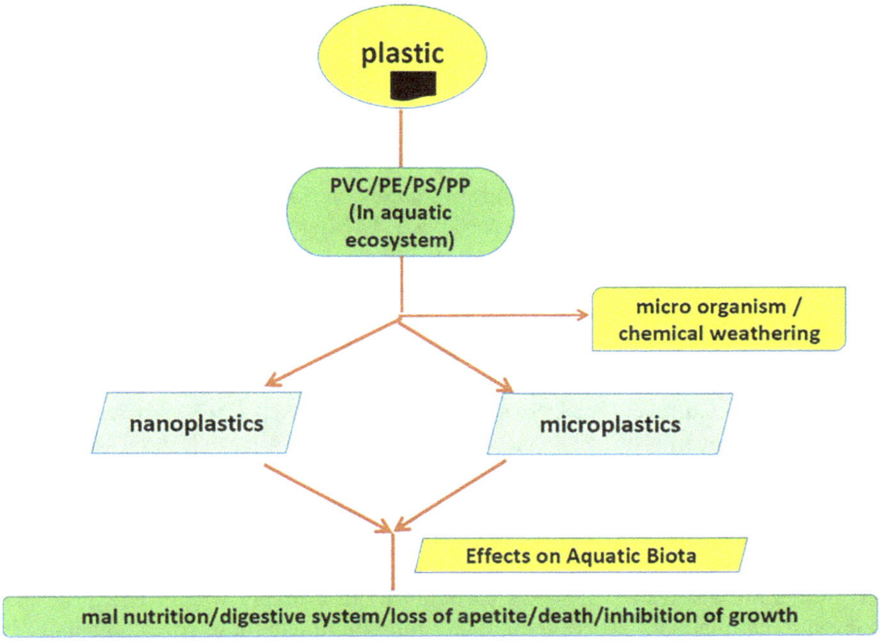


**Fig. 6.1** Availability and effects of MNPs on aquatic biota

influences terrestrial habitat and atmosphere as all are interconnected in the various ecosystems [32, 33]. Different types of marine species could consume these toxic MNPs, enter their food chains, which spreads through food web and leads to poisonous effects on all marine and terrestrial lives on the earth [34].

Aquatic life, e.g., vertebrates such as amphibians, fish, seabirds, etc., and invertebrates like zooplankton, microbes are greatly affected by the poisonous effects of MNPs [35, 36]. The digestive system of aquatic organisms is greatly affected by MNPs, which lead to malnutrition, loss of appetite, and death in some cases [37, 38]. The MNPs also cause inhibition to the development of different aquatic microbes, e.g., fungi, algae, bacteria, etc., which play important roles in the environments [29, 39, 40]. Human lives also face a severe threat when they consume aquatic food products and accumulate MNPs in their bodies. When measuring the impact of MNPs on aquatic life and to ensure the safety of the use of aquatic food products, specific bio-indicator organisms have been chosen as sentinel bio-monitors. In the marine sediment toxicity test, a very strong deposit feeder, the lugworm (Arenicola marina), generally acts as a monitoring agent [41], and for monitoring marine pollution, a sentinel species, Mytilus galloprovincialis (globally acknowledged), is used as a monitoring species [34].

Microplastic pollution exhibits significant variability across aquatic environments such as marine and freshwater, influenced by factors viz. proximity to urban areas, industrial activities, and ocean currents, etc. Here's a summary of key findings of MNPs from various studies:

Marine and Estuarine Environments: In a study by Gray et al. [42], micro-plastic accumulation in the sea surface microlayer averaged 6.6 ± 1.3 particles/L in Charleston Harbor and 30.8 ± 12.1 particles/L in Winyah Bay, South Carolina (USA). Inter tidal sediments showed averages of 413.8 ± 76.7 and 221.0 ± 25.6 prtcls/$m^2$. Micro plastic concentrations in Charleston Harbor averaged 6.6 ± 1.3 particles per liter, ranging from 3 to 11 particles/L/L. Micro plastic concentrations vary significantly across different marine environments. In Winyah Bay, Gray et al. [42] reported values ranging from 6–88 particles per liter, with a mean of 30.8 ± 12.1 particles/L/L. By contrast, notably lower micro-and nano-plastic (MNP) abundances have been documented in various other regions:Barents, Kara, Laptev, and East Siberian Seas: 0–0.045 particles/$m^2$ [43] Lake Winnipeg, Canada: 0.19 particles/$m^2$ [44], Gullmar Fjord, Swedish West Coast: 0.18–0.92 particles/$m^2$ [45], Skagerrak/Kattegat, Baltic Sea, and Gulf of Bothnia: 3.74 particles/$m^2$ [46].

Concerns about the impacts of MNPs on the health of the human body are intensifying day by day, particularly as studies reveal that freshwater sources often contain higher concentrations of these particles compared to seawater. For instance, research in the Czech Republic detected microplastic levels (1473–3605 particles / lit) in natural water and (338–628 particles/lit) in drinking water, indicating that conventional treatment processes may not fully eliminate these contaminants [47]. Similarly, a study in Germany found microplastics in bottled water, with concentrations reaching up to 193 particles/liter [48]. Environmental assessments have also identified significant microplastic pollution in various habitats: Beaches along the southeast coast of India, particularly in Tamil Nadu, exhibited microplastic densities between 2 and 178 particles/square meter, and the North Atlantic subtropical gyre, known for accumulating oceanic debris, showed microplastic concentrations ranging from 13 to 501 particles/square meter [49, 50]. Sediments from the Changjiang (Yangtze) Estuary in China contained microplastic levels between 20 and 340 particles per kilogram [51]; Urban rivers and tidal flats in Shanghai reported microplastic abundances varying from 52 to 1600 particles per kilogram [52]. Coastal sites within the southeastern United States' National Park Service units had microplastic concentrations ranging from 100 to 300 particles per kilogram [53]. These findings underscore the pervasive presence of microplastics across diverse aquatic environments, raising concerns about their potential implications for human health and ecosystem integrity.

## Effect and Occurrence of MNPs in Terrestrial Habitats

Currently, some reports have highlighted direct and indirect harmful impacts of micro-and nanoplastic (MNP) pollution on various terrestrial ecosystems [43, 54–56]. Notably, most plastic waste that ultimately reaches aquatic environments originates from land, where it is initially produced, used, and often carelessly discarded [57]. As a result, terrestrial habitats are now recognized as major reservoirs of micro-and nanoplastics (MNPs), offering multiple pathways of exposure to terrestrial organisms and potentially disrupting geochemical processes and causing environmental toxicity [58] (Fig. 6.2). Consequently, increasing attention has been directed toward day-to-day investigation of terrestrial soils to examine the presence of MNPs and to measure their concentrations across different terrestrial environments. In agricultural soils from southeastern Germany, MNP abundance ranged from 0–1.25 particles per kilogram of dry soil, with an average of 0.34 particles/kg [59]. In contrast, a study conducted in the suburbs of Shanghai reported significantly higher MNP concentrations across four soil types. Floodplain soils contained the highest levels (256.7 ± 62.2 particles/kg), followed by paddy soils (190 ± 31.2 particles/kg), yellow-brown soils (155 ± 95.2 particles/kg), and farmland soils (36.6 ± 41.7 particles/kg) [60]. Similarly, floodplain soils in Switzerland exhibited a mean MNP concentration of 593 particles/kg [61]. A significant number of reports have suggested that China has produced a huge amount of micro-and nano plastics in terrestrial ecosystems, highlighting a global data gap, and it is also needed to expand thorough investigation in different regions of the Earth.

Accumulated micro-and nanoplastics in the soil, having harmful impacts, are difficult to quantify, as these particles often interact with other hazardous substances and organic pollutants, amplifying their effects and posing serious threats to terrestrial biota [62]. Notably, MNPs—capable of persisting in the environment for centuries interact with the organic substances in soil, altering soil's physicochemical properties, contaminating groundwater, and ultimately hindering plant growth and reducing overall ecosystem productivity [63].

Micro-and nanoplastics (MNPs) also exert significant adverse effects on the fauna of soil, particularly nematodes and earthworms, impairing their growth, reproduction, lifespan, and survival through a range of toxic mechanisms. These include bio-accumulation, DNA damage, genotoxicity, gut microbiota imbalances, histopathological alterations, metabolic disruptions, neurotoxicity, oxidative stress,

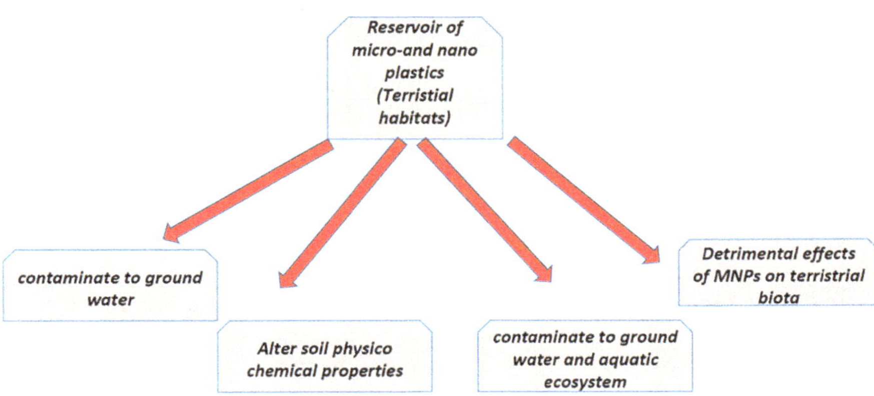


**Fig. 6.2** Effects of MNPs on terrestrial habitat

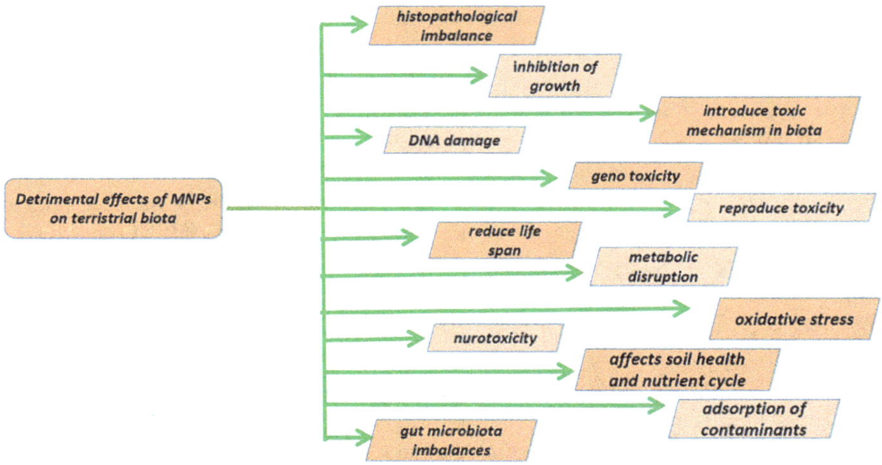


**Fig. 6.3** Various harmful effects of MNPs

and reproductive toxicity [64] (Fig. 6.3). Such impacts can disrupt the essential ecological functions these organisms perform—such as litter decomposition, nutrient cycling, and energy flow—thereby posing serious environmental and public health risks [65]. Additionally, due to their high surface area-to-volume ratio and hydrophobic nature, MNPs may serve as carriers for pathogens and organic pollutants in terrestrial environments. Microorganisms that attach to micro-and nanoplastics (MNPs) also pose environmental risks by facilitating the transfer of these particles from soil to plants, and ultimately to other organisms through the food chain [62].

## Occurrence and Ecological Impacts of Micro-and Nanoplastics in the Atmpsphere

Unlike micro-and nano plastics in terrestrial and aquatic habitats, airborne plastics of these types are continuously inhaled by the organisms, giving rise to various kinds of illnesses among these organisms. Therefore, gaining deeper knowledge about the concentration, sources, and their detrimental effects in these ecosystems is essential [28]. Airborne MNPs have been found to transfer from their sources and concentrate in various lands and aquatic environments, implying multiple detrimental effects to the biosphere [66].

To date, in very few places, airborne MNPs have been detected. Asaluyeh County in Iran; Beijing, Dongguan, and Shanghai in China; California in the United States; the French Pyrenees and Paris in France; Hamburg in Germany; London and Nottingham in the United Kingdom; and Surabaya in Indonesia are examples of such places.

For instance, high-level MNP concentrations (ranging from 132.75–174.97 particles per $m^3$) measured in Surabaya City, Indonesia, create polluted suburban areas [67]. In contrast, other studies reported significantly lower concentrations. In Shanghai, China, the average abundance of suspended MNP in air was 4.18 ptcls/$m^3$ [68], while in Asaluyeh County, Iran, suspended dust contained between 0.3 and 1.1 particles/$m^3$ [69]. Similarly, a study over the western Pacific Ocean found a comparatively minimum amount of presence of MNPs in suspended aerosols in air(~0 to 1.37 particles/$m^3$) [70]. Micro-and nanoplastics (MNPs) present in the atmosphere can be inhaled and eventually accumulate in various land organisms, including humans. Epidemiological studies have linked airborne MNP pollution to serious respiratory and cardiovascular health effects [71].

Several factors influence the atmospheric transport of micro-and nanoplastics, such as direction of wind, micro-and nano plastic size, falling of raindrops, anthropogenic activities, and population density [33], (Figs. 6.4 and 6.5). Studies have shown that the direction of wind and airborne amount of MNP are interconnected [28, 69, 71]. In Germany, wind's direction changes from westerly to southerly winds corresponded with increased MNP contamination [72], and higher concentrations were observed at downwind locations [28]. The physical characteristics of MNPs—particularly their size, shape, and length—also significantly affect their atmospheric transport [73, 74]. A study by Enyoh et al. [74] found that smaller MNPs (less than 25 μm) were more prevalent in samples, while larger particles were less frequently detected.

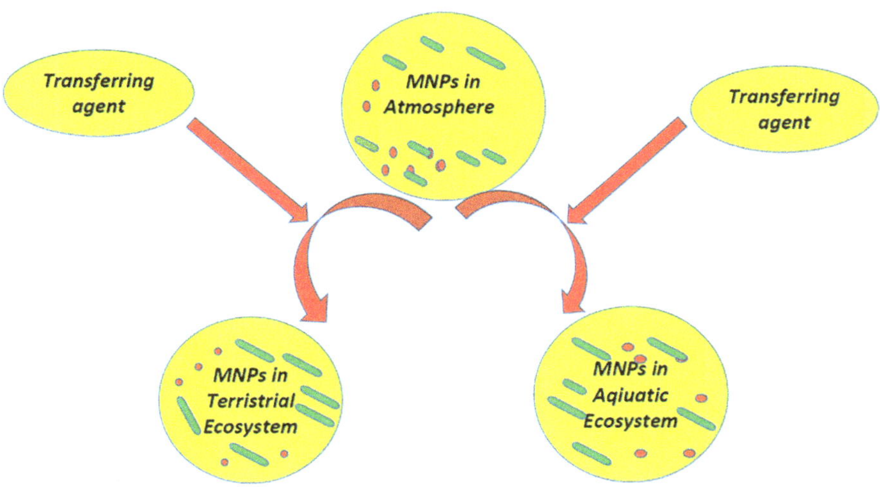


**Fig. 6.4** Transfer of MNPs from the atmosphere to aquatic and terrestrial ecosystems

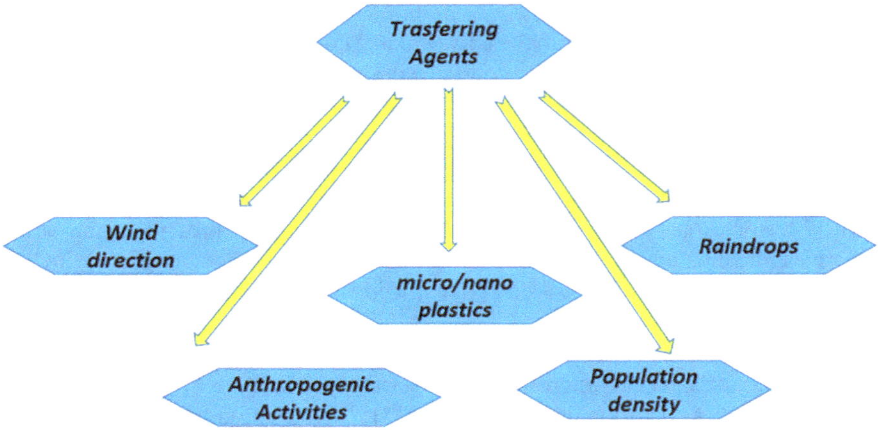


**Fig. 6.5** Transferring agents of MNPs dispersal and distribution

**Fig. 6.6** Various strategies of mitigation

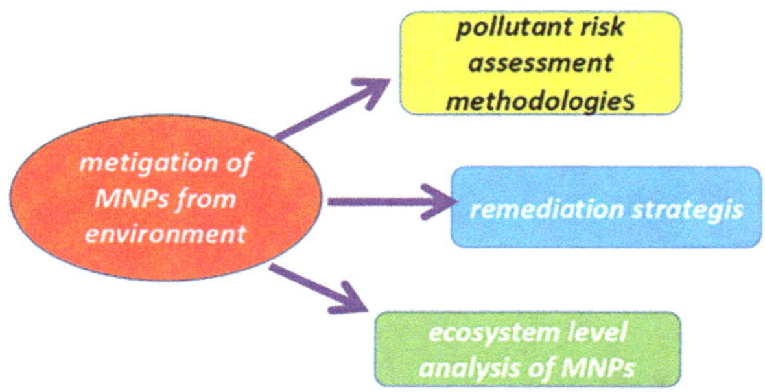


## Mitigating the Impact of Mnps

There is a critical need to expand investigations into the toxicity of micro-and nanoplastics (MNPs) by extending laboratory studies into real-world environmental scenarios. Utilizing enhanced pollutant risk assessment methodologies will improve predictions of MNP exposure and its associated pathophysiological risks. This approach will enable more precise determination of effect concentrations and enhance the accuracy of real-time environmental risk forecasts [75], (Fig. 6.6). Additionally, employing system-level analyses of MNP toxicological effects—integrating both field and laboratory exposure data—will deepen our understanding of remediation strategies and contribute to developing more sustainable ecosystems.

Biotechnology is increasingly being harnessed through innovative strategies to mitigate the environmental accumulation of micro-and nanoplastics (MNPs) (Fig. 6.7). A promising solution lies in bioplastics synthesis—materials obtained from renewable biological substances that are more sustainable and

**Fig. 6.7** Mitigation through the biotechnology field

**Fig. 6.8** Some environmentally friendly bioplastics

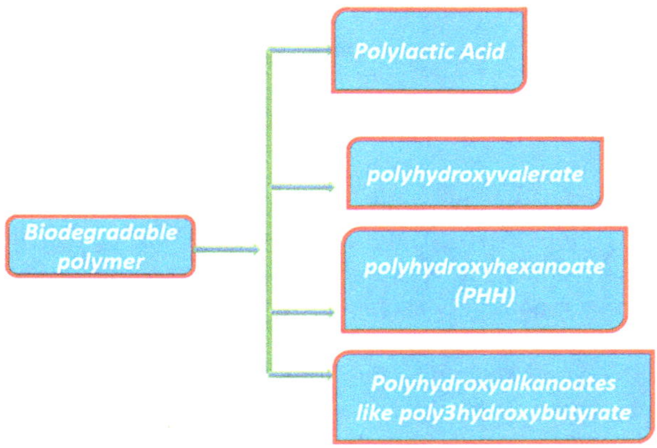


environmentally friendly than conventional plastics. Examples include biodegradable polymers,e.g., Polylactic Acid, Polyhydroxyalkanoates like poly-3-hydroxybutyrate, polyhydroxyvalerate, and polyhydroxyhexanoate (PHH), which are synthesized via microbial fermentation and can be degraded by microorganisms and their enzymes (Fig. 6.8). Additionally, Polyamide 11 (nylon 11) is a biopolymer derived from castor oil; while it is not biodegradable, it offers recyclability and exceptional mechanical and chemical stability for high-performance applications.

Moreover, significant efforts are underway to break down persistent synthetic plastics—currently the primary contributors to global plastic pollution—through mechanisms like biodeterioration, biofragmentation, assimilation, and mineralization. Researchers are actively exploring how various actinomycetes, bacteria, fungi, insects, and specific microbial enzymes can transform these petrochemical-derived polymers and their tiny fragments into less harmful forms [1] (Fig. 6.9).

Beyond these scientific endeavors, broader strategies focus on curbing plastic waste generation and preventing its release into ecosystems. Governments worldwide are implementing comprehensive measures, including phased or outright bans on single-use plastic bags and restrictions on microplastics in cosmetic products.

In 2017, Delhi's National Green Tribunal enforced a ban on all disposable plastics—including bags, cups, and cutlery—with effect from January 1. Earlier, in March 2016, Karnataka implemented a comprehensive prohibition across the state on single-use plastic items—such as carry bags, banners, plates, cups, spoons, cling films, and thermocol goods—regardless of their thickness.

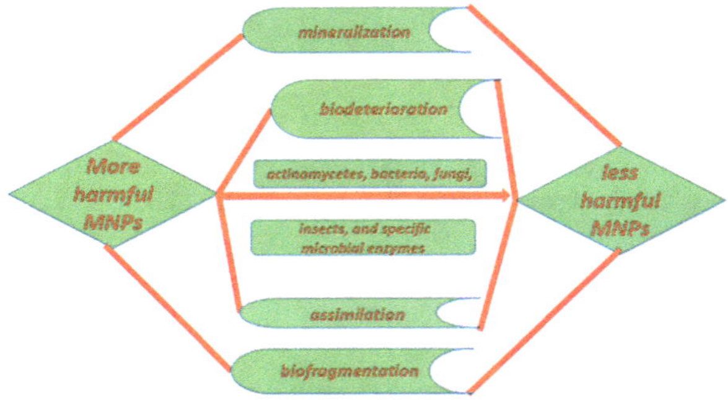


**Fig. 6.9** Conversion of more harmful MNPs to less harmful MNPs

Ecolabeling has emerged as a proactive and powerful strategy for reducing plastic pollution [76]. The primary goals of this ecolabelling are twofold: to lower the environmental footprint of consumer goods and to elevate awareness about the environment among consumers—thereby increasing the likelihood of sustainable purchasing decisions.

Beyond scientific initiatives, both non-government and government organizations have taken the initiative to raise public awareness and educational outreach about the use and impact of MNPs. These are vital for instigating meaningful behavioral and professional shifts that can curb the disposal and excessive use of this polymeric waste. In the industries, eliminating microplastic beads from cosmetics and other products is a key mitigation strategy. Mandating bans on these particles through legislation and regulatory action would amplify both the reach and effectiveness of these efforts [76].

Recycling also offers a sustainable avenue to mitigate the harmful impacts of micro-and nanoplastics (MNPs), particularly since a large volume of discarded plastic eventually fragments into microplastics in landfills and natural environments. By extending the lifetime of plastics through reuse and re-purposing, recycling helps to preserve resources, helps for the reduction of emissions of greenhouse gas, limits pollutant discharge—all of which support a more circular economy (Fig. 6.10).

Implementing taxes on pure plastics, raising consciousness about the environmental advantages of recycled alternatives, and offering government benefits to support recycled-plastic production can significantly bolster the recycled plastics market [77]. By increasing the cost of virgin plastics and lowering financial barriers for recycled options, these strategies help level the economic playing field—making sustainability a competitive choice (Fig. 6.11).

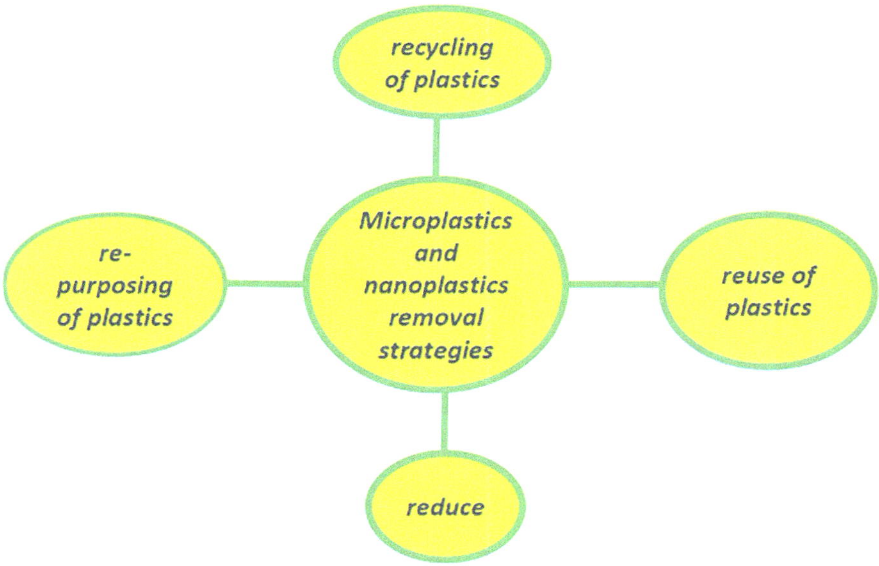


**Fig. 6.10** Removal of MNPs through recycling

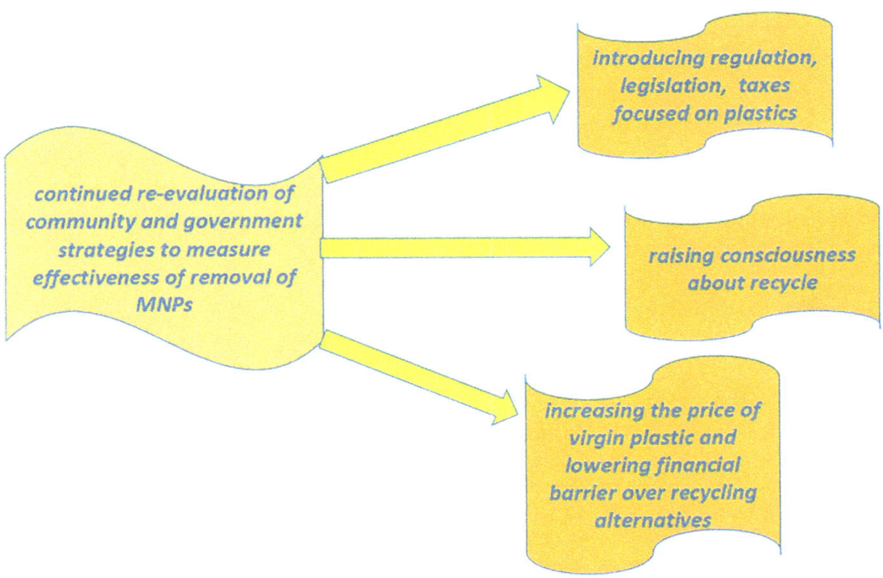


**Fig. 6.11** Various Governmental strategies for the removal of MNPs

## Conclusion

Microplastics and nanoplastics (MNPs), which are tiny fragments, have been found to travel vast areas globally for their unique characteristics, such as their lightweight nature, varied shapes, etc. The main source of micro-and nano plastic in the air and in the aquatic system is terrestrial ecosystems. Their persistence and mobility allow MNPs to accumulate in microorganisms and lower-level organisms within the food web—particularly marine species like phytoplankton—eventually moving up through the trophic levels. Once inside these organisms, MNPs tend to accumulate in various organs and cellular structures, often causing harmful effects primarily through oxidative stress.

Despite growing scientific efforts to mitigate the impact of MNPs on living organisms, progress has been limited due to the rapid rise of MNPs in the environment and the biological complexity of affected food chains. One major challenge remains the extended timescales needed to demonstrate the biodegradation of MNPs. Moreover, limited understanding of their environmental fate and ecotoxicological effects—whether from primary or secondary sources—highlights the need for more efficient scientific approaches to measure the detrimental effects of micro- and nanoplastics on different habitats of various ecosystems.

Researchers are also responsible for accurately measuring the decaying processes of larger polymeric materials into micro-and nanoplastics and for uncovering the mechanisms driving these processes. In addition, it is very necessary for the development as well as enforcement of comprehensive legislation and regulations governing micro-and nanoplastics at large. Standardizing monitoring methods, addressing both technical as well as waste management shortcomings, promoting reproducing initiatives, etc., are the essential steps toward reducing the harmful effects of MNP pollution. This review aims to underscore the scale of the MNP problem and provide valuable insights for managing plastic pollution in both developed and developing nations.

## References

1. Amobonye, A., Bhagwat, P., Singh, S., & Pillai, S. (2020). Plastic biodegradation: Frontline microbes and their enzymes. *Science of the Total Environment, 759*, 143536.
2. Syberg, K., Nielsen, M. B., Clausen, L. P. W., Van Calster, G., Van Wezel, A., Rochman, C., et al. (2021). Regulation of plastic from a circular economy perspective. *Current Opinion in Green and Sustainable Chemistry, 29*, 100462. https://doi.org/10.1016/j.cogsc.2021.100462
3. Wright, S., Ulke, J., Font, A., Chan, K., & Kelly, F. (2020). Atmospheric microplastic deposition in an urban environment and an evaluation of transport. *Environment International, 136*, 105411. https://doi.org/10.1016/j.envint.2019.105411
4. Hartmann, N. B., HüFfer, T., Thompson, R. C., HassellöV, M., Verschoor, A., Daugaard, A. E., et al. (2019). Are we speaking the same language? Recommendations for a definition and categorization framework for plastic debris. *Environmental Science & Technology, 53*, 1039–1047. https://doi.org/10.1021/acs.est.8b05297. CrossRef Full Text | Google Scholar

5. Thompson, R., Olsen, Y., Mitchell, R., Davis, A., Rowland, S., John, A., et al. (2004). Lost at sea: Where is all the plastic. *Science, 304*, 838. https://doi.org/10.1126/science.1094559. PubMed Abstract | CrossRef Full Text | Google Scholar
6. Frère, L., Maignien, L., Chalopin, M., Huvet, A., Rinnert, E., Morrison, H., et al. (2018). Microplastic bacterial communities in the bay of Brest: Influence of polymer type and size. *Environmental Pollution, 242*, 614–625. https://doi.org/10.1016/j.envpol.2018.07.023. CrossRef Full Text | Google Scholar
7. Cai, H., Chen, M., Chen, Q., Du, F., Liu, J., & Shi, H. (2020a). Microplastic quantification affected by structure and pore size of filters. *Chemosphere, 257*, 127198. https://doi.org/10.1016/j.chemosphere.2020.127198
8. Stock, V., Laurisch, C., Franke, J., Dönmez, M. H., Voss, L., Böhmert, L., et al. (2021). Uptake and cellular effects of PE, PP, PET and PVC microplastic particles. *Toxicology In Vitro, 70*, 105021. https://doi.org/10.1016/j.tiv.2020.105021. PubMed Abstract | CrossRef Full Text | Google Scholar
9. Lehner, R., Weder, C., Petri-Fink, A., & Rothen-Rutishauser, B. (2019). Emergence of nanoplastic in the environment and possible impact on human health. *Environmental Science & Technology, 53*, 1748–1765. https://doi.org/10.1021/acs.est.8b05512
10. Gonçalves, J. M., & Bebianno, M. J. (2021). Nanoplastics impact on marine biota: A review. *Environmental Pollution, 273*, 116426. https://doi.org/10.1016/j.envpol.2021.116426. PubMed Abstract | CrossRef Full Text | Google Scholar
11. Boverhof, D. R., Bramante, C. M., Butala, J. H., Clancy, S. F., Lafranconi, M., West, J., et al. (2015). Comparative assessment of nanomaterial definitions and safety evaluation considerations. *Regulatory Toxicology and Pharmacology, 73*, 137–150. https://doi.org/10.1016/j.yrtph.2015.06.001. PubMed Abstract | CrossRef Full Text | Google Scholar
12. Laskar, N., & Kumar, U. (2019). Plastics and microplastics: A threat to environment. *Environmental Technology and Innovation, 14*, 100352. https://doi.org/10.1016/j.eti.2019.100352. PubMed Abstract | CrossRef Full Text | Google Scholar
13. Ma, H., Pu, S., Liu, S., Bai, Y., Mandal, S., & Xing, B. (2020). Microplastics in aquatic environments: Toxicity to trigger ecological consequences. *Environmental Pollution, 261*, 114089. https://doi.org/10.1016/j.envpol.2020.114089. PubMed Abstract | CrossRef Full Text | Google Scholar.
14. Li, C., Busquets, R., & Campos, L. C. (2020a). Assessment of microplastics in freshwater systems: A review. *Science of The Total Environment, 707*, 135578. https://doi.org/10.1016/j.scitotenv.2019.135578. PubMed Abstract | CrossRef Full Text | Google Scholar
15. Dong, Y., Gao, M., Qiu, W., & Song, Z. (2021). Effects of microplastic on arsenic accumulation in Chlamydomonas reinhardtii in a freshwater environment. *Journal of Hazardous Materials, 405*, 124232. https://doi.org/10.1016/j.jhazmat.2020.124232. PubMed Abstract | CrossRef Full Text | Google Scholar
16. Mohajerani, A., & Karabatak, B. (2020). Microplastics and pollutants in biosolids have contaminated agricultural soils: An analytical study and a proposal to cease the use of biosolids in farmlands and utilise them in sustainable bricks. *Waste Management, 107*, 252–265. https://doi.org/10.1016/j.wasman.2020.04.021
17. Tan, M., Liu, L., Zhang, M., Liu, Y., & Li, C. (2021). Effects of solution chemistry and humic acid on the transport of polystyrene microplastics in manganese oxides coated sand. *Journal of Hazardous Materials, 413*, 125410. https://doi.org/10.1016/j.jhazmat.2021.125410
18. Jakubowicz, I., Enebro, J., & Yarahmadi, N. (2020). Challenges in the search for nanoplastics in the environment—A critical review from the polymer science perspective. *Polymer Testing, 93*, 106953. https://doi.org/10.1016/j.polymertesting.2020.106953
19. Sharma, V. K., Ma, X., Guo, B., & Zhang, K. (2021). Environmental factors-mediated behavior of microplastics and nanoplastics in water: A review. *Chemosphere, 271*, 129597. https://doi.org/10.1016/j.chemosphere.2021.129597. PubMed Abstract | CrossRef Full Text | Google Scholar
20. Da Costa, J. P. (2018). Micro-and nanoplastics in the environment: Research and policymaking. *Current Opinion in Environmental Science & Health, 1*, 12–16. https://doi.org/10.1016/j.coesh.2017.11.002. CrossRef Full Text | Google Scholar

21. Ding, Y., Zhang, R., Li, B., Du, Y., Li, J., Tong, X., et al. (2021). Tissue distribution of polystyrene nanoplastics in mice and their entry, transport, and cytotoxicity to GES-1 cells. *Environmental Pollution, 280*, 116974. https://doi.org/10.1016/j.envpol.2021.116974. PubMed Abstract | CrossRef Full Text | Google Scholar
22. Hu, M., & Palić, D. (2020). Micro-and nano-plastics activation of oxidative and inflammatory adverse outcome pathways. *Redox Biology, 37*, 101620. https://doi.org/10.1016/j.redox.2020.101620. PubMed Abstract | CrossRef Full Text | Google Scholar
23. Yu, J., Tian, J. Y., Xu, R., Zhang, Z. Y., Yang, G. P., Wang, X. D., et al. (2020). Effects of microplastics exposure on ingestion, fecundity, development, and dimethylsulfide production in Tigriopus japonicus (Harpacticoida, copepod). *Environmental Pollution, 267*, 115429. https://doi.org/10.1016/j.envpol.2020.115429. PubMed Abstract | CrossRef Full Text | Google Scholar
24. Zhang, F., Man, Y. B., Mo, W. Y., Man, K. Y., & Wong, M. H. (2020b). Direct and indirect effects of microplastics on bivalves, with a focus on edible species: A mini-review. *Critical Reviews In Environmental Science And Technology, 50*, 2109–2143. https://doi.org/10.1080/10643389.2019.1700752
25. Suckling, C. C. (2021). Responses to environmentally relevant microplastics are species-specific with dietary habit as a potential sensitivity indicator. *Science of the Total Environment, 751*, 142341. https://doi.org/10.1016/j.scitotenv.2020.142341. PubMed Abstract | CrossRef Full Text | Google Scholar
26. Jin, M., Wang, X., Ren, T., Wang, J., & Shan, J. (2021). Microplastics contamination in food and beverages: Direct exposure to humans. *Journal Of Food Science, 86*, 2816–2837. https://doi.org/10.1111/1750-3841.15802. PubMed Abstract | CrossRef Full Text | Google Scholar
27. Evangeliou, N., Grythe, H., Klimont, Z., Heyes, C., Eckhardt, S., Lopez-Aparicio, S., et al. (2020). Atmospheric transport is a major pathway of microplastics to remote regions. *Nature Communications, 11*, 1–11. https://doi.org/10.1038/s41467-020-17201-9. CrossRef Full Text | Google Scholar
28. Chen, G., Feng, Q., & Wang, J. (2020a). Mini-review of microplastics in the atmosphere and their risks to humans. *Science of The Total Environment, 703*, 135504. https://doi.org/10.1016/j.scitotenv.2019.135504. PubMed Abstract | CrossRef Full Text | Google Scholar
29. Mammo, F., Amoah, I., Gani, K., Pillay, L., Ratha, S., Bux, F., et al. (2020). Microplastics in the environment: Interactions with microbes and chemical contaminants. *Science of the Total Environment, 743*, 140518. https://doi.org/10.1016/j.scitotenv.2020.140518. CrossRef Full Text | Google Scholar
30. Wang, T., Li, B., Zou, X., Wang, Y., Li, Y., Xu, Y., et al. (2019). Emission of primary microplastics in mainland China: Invisible but not negligible. *Water Research, 162*, 214–224. https://doi.org/10.1016/j.watres.2019.06.042. PubMed Abstract | CrossRef Full Text | Google Scholar
31. Yuan, J., Ma, J., Sun, Y., Zhou, T., Zhao, Y., & Yu, F. (2020). Microbial degradation and other environmental aspects of microplastics/plastics. *Science of The Total Environment, 715*, 136968. https://doi.org/10.1016/j.scitotenv.2020.136968. PubMed Abstract | CrossRef Full Text | Google Scholar
32. Bergmann, M., Mützel, S., Primpke, S., Tekman, M. B., Trachsel, J., & Gerdts, G. (2019). White and wonderful? Microplastics prevail in snow from the Alps to the Arctic. *Science Advances, 5*, eaax1157. https://doi.org/10.1126/sciadv.aax1157. PubMed Abstract | CrossRef Full Text | Google Scholar
33. Zhang, Y., Kang, S., Allen, S., Allen, D., Gao, T., & Sillanpää, M. (2020d). Atmospheric microplastics: A review on current status and perspectives. *Earth-Science Reviews, 203*, 103118. https://doi.org/10.1016/j.earscirev.2020.103118. CrossRef Full Text | Google Scholar
34. Al-Thawadi, S. (2020). Microplastics and nanoplastics in aquatic environments: Challenges and threats to aquatic organisms. *Arabian Journal for Science and Engineering, 45*, 4419–4440. https://doi.org/10.1007/s13369-020-04402-z. CrossRef Full Text | Google Scholar
35. Frydkjær, C. K., Iversen, N., & Roslev, P. (2017). Ingestion and egestion of microplastics by the cladoceran Daphnia magna: Effects of regular and irregular shaped plastic and sorbed phenanthrene. *Bulletin of Environmental Contamination and Toxicology, 99*, 655–661. https://doi.org/10.1007/s00128-017-2186-3. PubMed Abstract | CrossRef Full Text | Google Scholar

36. Sana, S. S., Dogiparthi, L. K., Gangadhar, L., Chakravorty, A., & Abhishek, N. (2020). Effects of microplastics and nanoplastics on marine environment and human health. *Environmental Science and Pollution Research, 27*, 44743–44756. https://doi.org/10.1007/s11356-020-10573-x. CrossRef Full Text | Google Scholar
37. Lee, K.-W., Shim, W. J., Kwon, O. Y., & Kang, J.-H. (2013). Size-dependent effects of micro polystyrene particles in the marine copepod Tigriopus japonicus. *Environmental Science & Technology, 47*, 11278–11283. https://doi.org/10.1021/es401932b. PubMed Abstract | CrossRef Full Text | Google Scholar
38. Van Cauwenberghe, L., Vanreusel, A., Mees, J., & Janssen, C. R. (2013). Microplastic pollution in deep-sea sediments. *Environmental Pollution, 182*, 495–499. https://doi.org/10.1016/j.envpol.2013.08.013. PubMed Abstract | CrossRef Full Text | Google Scholar
39. Nomura, T., Tani, S., Yamamoto, M., Nakagawa, T., Toyoda, S., Fujisawa, E., et al. (2016). Cytotoxicity and colloidal behavior of polystyrene latex nanoparticles toward filamentous fungi in isotonic solutions. *Chemosphere, 149*, 84–90. https://doi.org/10.1016/j.chemosphere.2016.01.091. PubMed Abstract | CrossRef Full Text | Google Scholar.
40. Sun, X., Chen, B., Li, Q., Liu, N., Xia, B., Zhu, L., et al. (2018). Toxicities of polystyrene nano-and microplastics toward marine bacterium Halomonas alkaliphila. *Science of The Total Environment, 642*, 1378–1385. https://doi.org/10.1016/j.scitotenv.2018.06.141. PubMed Abstract | CrossRef Full Text | Google Scholar
41. Besseling, E., Foekema, E. M., Van Den Heuvel-Greve, M. J., & Koelmans, A. A. (2017). The effect of microplastic on the uptake of chemicals by the lugworm Arenicola marina (L.) under environmentally relevant exposure conditions. *Environmental Science & Technology, 51*, 8795–8804. https://doi.org/10.1021/acs.est.7b02286. PubMed Abstract | CrossRef Full Text | Google Scholar
42. Gray, A. D., Wertz, H., Leads, R. R., & Weinstein, J. E. (2018). Microplastic in two South Carolina estuaries: Occurrence, distribution, and composition. *Marine Pollution Bulletin, 128*, 223–233. https://doi.org/10.1016/j.marpolbul.2018.01.030. PubMed Abstract | CrossRef Full Text | Google Scholar
43. Yakushev, E., Gebruk, A., Osadchiev, A., Pakhomova, S., Lusher, A., Berezina, A., et al. (2021). Microplastics distribution in the Eurasian Arctic is affected by Atlantic waters and Siberian rivers. *Communications Earth & Environment, 2*, 1–10. https://doi.org/10.1038/s43247-021-00091-0. CrossRef Full Text | Google Scholar
44. Anderson, P. J., Warrack, S., Langen, V., Challis, J. K., Hanson, M. L., & Rennie, M. D. (2017). Microplastic contamination in Lake Winnipeg. Canada. *Environmental Pollution, 225*, 223–231. https://doi.org/10.1016/j.envpol.2017.02.072. PubMed Abstract | CrossRef Full Text | Google Scholar
45. Karlsson, T. M., Kärrman, A., Rotander, A., & Hassellöv, M. (2020). Comparison between manta trawl and in situ pump filtration methods, and guidance for visual identification of microplastics in surface waters. *Environmental Science & Policy, 27*, 5559–5571. https://doi.org/10.1007/s11356-019-07274-5. PubMed Abstract | CrossRef Full Text | Google Scholar
46. Schönlau, C., Karlsson, T. M., Rotander, A., Nilsson, H., Engwall, M., Van Bavel, B., et al. (2020). Microplastics in sea-surface waters surrounding Sweden sampled by manta trawl and in-situ pump. *Marine Pollution Bulletin, 153*, 111019. https://doi.org/10.1016/j.marpolbul.2020.111019. PubMed Abstract | CrossRef Full Text | Google Scholar
47. Pivokonsky, M., Cermakova, L., Novotna, K., Peer, P., Cajthaml, T., & Janda, V. (2018). Occurrence of microplastics in raw and treated drinking water. *Science of the Total Environment, 643*, 1644–1651. https://doi.org/10.1016/j.scitotenv.2018.08.102. PubMed Abstract | CrossRef Full Text | Google Scholar
48. Schymanski, D., Goldbeck, C., Humpf, H.-U., & Fürst, P. (2018). Analysis of microplastics in water by micro-Raman spectroscopy: Release of plastic particles from different packaging into mineral water. *Water Research, 129*, 154–162. https://doi.org/10.1016/j.watres.2017.11.011. PubMed Abstract | CrossRef Full Text | Google Scholar

49. Ter Halle, A., Jeanneau, L., Martignac, M., Jardé, E., Pedrono, B., Brach, L., et al. (2017). Nanoplastic in the North Atlantic subtropical gyre. *Environmental Science & Technology, 51*, 13689–13697. https://doi.org/10.1021/acs.est.7b03667. PubMed Abstract | CrossRef Full Text | Google Scholar
50. Karthik, R., Robin, R., Purvaja, R., Ganguly, D., Anandavelu, I., Raghuraman, R., et al. (2018). Microplastics along the beaches of southeast coast of India. *Science of The Total Environment, 645*, 1388–1399. https://doi.org/10.1016/j.scitotenv.2018.07.242. PubMed Abstract | CrossRef Full Text | Google Scholar
51. Peng, G., Zhu, B., Yang, D., Su, L., Shi, H., & Li, D. (2017). Microplastics in sediments of the Changjiang estuary, China. *Environmental Pollution, 225*, 283–290. https://doi.org/10.1016/j.envpol.2016.12.064. PubMed Abstract | CrossRef Full Text | Google Scholar
52. Peng, G., Xu, P., Zhu, B., Bai, M., & Li, D. (2018). Microplastics in freshwater river sediments in Shanghai, China: A case study of risk assessment in mega-cities. *Environmental Pollution, 234*, 448–456. https://doi.org/10.1016/j.envpol.2017.11.034. PubMed Abstract | CrossRef Full Text | Google Scholar
53. Yu, X., Ladewig, S., Bao, S., Toline, C. A., Whitmire, S., & Chow, A. T. (2018). Occurrence and distribution of microplastics at selected coastal sites along the southeastern United States. *Science of the Total Environment, 613–614*, 298–305. https://doi.org/10.1016/j.scitotenv.2017.09.100. CrossRef Full Text | Google Scholar
54. Ambrosini, R., Azzoni, R. S., Pittino, F., Diolaiuti, G., Franzetti, A., & Parolini, M. (2019). First evidence of microplastic contamination in the supraglacial debris of an alpine glacier. *Environmental Pollution, 253*, 297–301. https://doi.org/10.1016/j.envpol.2019.07.005. PubMed Abstract | CrossRef Full Text | Google Scholar
55. Zhang, D., Ng, E. L., Hu, W., Wang, H., Galaviz, P., Yang, H., et al. (2020a). Plastic pollution in croplands threatens long-term food security. *Global Change Biology, 26*, 3356–3367. https://doi.org/10.1111/gcb.15043. PubMed Abstract | CrossRef Full Text | Google Scholar
56. Chen, Y., Leng, Y., Liu, X., & Wang, J. (2020b). Microplastic pollution in vegetable farmlands of suburb Wuhan, Central China. *Environmental Pollution, 257*, 113449. https://doi.org/10.1016/j.envpol.2019.113449. PubMed Abstract | CrossRef Full Text | Google Scholar
57. De Souza Machado, A. A., Kloas, W., Zarfl, C., Hempel, S., & Rillig, M. C. (2018). Microplastics as an emerging threat to terrestrial ecosystems. *Global Change Biology, 24*, 1405–1416. https://doi.org/10.1111/gcb.14020. PubMed Abstract | CrossRef Full Text | Google Scholar
58. Allouzi, M. M. A., Tang, D. Y. Y., Chew, K. W., Rinklebe, J., Bolan, N., Allouzi, S. M. A., et al. (2021). Micro (nano) plastic pollution: The ecological influence on soil-plant system and human health. *Science of the Total Environment, 788*, 147815. https://doi.org/10.1016/j.scitotenv.2021.147815. PubMed Abstract | CrossRef Full Text | Google Scholar
59. Piehl, S., Leibner, A., Löder, M. G., Dris, R., Bogner, C., & Laforsch, C. (2018). Identification and quantification of macro-and microplastics on an agricultural farmland. *Scientific Reports, 8*, 1–9. https://doi.org/10.1038/s41598-018-36172-y. CrossRef Full Text | Google Scholar
60. Liu, M., Song, Y., Lu, S., Qiu, R., Hu, J., Li, X., et al. (2019d). A method for extracting soil microplastics through circulation of sodium bromide solutions. *Science of The Total Environment, 691*, 341–347. https://doi.org/10.1016/j.scitotenv.2019.07.144. PubMed Abstract | CrossRef Full Text | Google Scholar
61. Scheurer, M., & Bigalke, M. (2018). Microplastics in Swiss floodplain soils. *Environmental Science & Technology, 52*, 3591–3598. https://doi.org/10.1021/acs.est.7b06003. PubMed Abstract | CrossRef Full Text | Google Scholar
62. Chai, B., Wei, Q., She, Y., Lu, G., Dang, Z., & Yin, H. (2020). Soil microplastic pollution in an e-waste dismantling zone of China. *Waste Management, 118*, 291–301. https://doi.org/10.1016/j.wasman.2020.08.048. PubMed Abstract | CrossRef Full Text | Google Scholar
63. Wahl, A., Le Juge, C., Davranche, M., El Hadri, H., Grassl, B., Reynaud, S., et al. (2021). Nanoplastic occurrence in a soil amended with plastic debris. *Chemosphere, 262*, 127784. https://doi.org/10.1016/j.chemosphere.2020.127784. PubMed Abstract | CrossRef Full Text | Google Scholar

64. Wang, Q., Adams, C. A., Wang, F., Sun, Y., & Zhang, S. (2021a). Interactions between microplastics and soil fauna: A critical review. *Critical Reviews in Environmental Science and Technology, 52*, 1–33. https://doi.org/10.1080/10643389.2021.1915035. CrossRef Full Text | Google Scholar
65. Wang, W., Ge, J., Yu, X., & Li, H. (2020). Environmental fate and impacts of microplastics in soil ecosystems: Progress and perspective. *Science of The Total Environment, 708*, 134841. https://doi.org/10.1016/j.scitotenv.2019.134841. PubMed Abstract | CrossRef Full Text | Google Scholar
66. Mbachu, O., Jenkins, G., Pratt, C., & Kaparaju, P. (2020). A new contaminant superhighway? A review of sources, measurement techniques and fate of atmospheric microplastics. *Water, Air, and Soil Pollution, 231*, 1–27. https://doi.org/10.1007/s11270-020-4459-4. CrossRef Full Text | Google Scholar
67. Asrin, N. R. N., & Dipareza, A. (2019). Microplastics in ambient air (case study: Urip Sumoharjo street and Mayjend Sungkono street of Surabaya City, Indonesia). *IAETSD–Journal for Advanced Research Applied Science, 6*, 54–57.
68. Liu, K., Wang, X., Fang, T., Xu, P., Zhu, L., & Li, D. (2019b). Source and potential risk assessment of suspended atmospheric microplastics in Shanghai. *Science of The Total Environment, 675*, 462–471. https://doi.org/10.1016/j.scitotenv.2019.04.110. PubMed Abstract | CrossRef Full Text | Google Scholar
69. Abbasi, S., Keshavarzi, B., Moore, F., Turner, A., Kelly, F. J., Dominguez, A. O., et al. (2019). Distribution and potential health impacts of microplastics and microrubbers in air and street dusts from Asaluyeh County. Iran. *Environmental Pollution, 244*, 153–164. https://doi.org/10.1016/j.envpol.2018.10.039. PubMed Abstract | CrossRef Full Text | Google Scholar
70. Liu, K., Wu, T., Wang, X., Song, Z., Zong, C., Wei, N., et al. (2019c). Consistent transport of terrestrial microplastics to the ocean through atmosphere. *Environmental Science & Technology, 53*, 10612–10619. https://doi.org/10.1021/acs.est.9b03427. PubMed Abstract | CrossRef Full Text | Google Scholar
71. Browne, M. A., Galloway, T. S., & Thompson, R. C. (2010). Spatial patterns of plastic debris along estuarine shorelines. *Environmental Science & Technology, 44*, 3404–3409. https://doi.org/10.1021/es903784e. PubMed Abstract | CrossRef Full Text | Google Scholar
72. Klein, M., & Fischer, E. K. (2019). Microplastic abundance in atmospheric deposition within the metropolitan area of Hamburg. Germany. *Science of The Total Environment, 685*, 96–103. https://doi.org/10.1016/j.scitotenv.2019.05.405. PubMed Abstract | CrossRef Full Text | Google Scholar
73. Allen, S., Allen, D., Phoenix, V. R., Le Roux, G., Jiménez, P. D., Simonneau, A., et al. (2019). Atmospheric transport and deposition of microplastics in a remote mountain catchment. *Nature Geoscience, 12*, 339–344. https://doi.org/10.1038/s41561-019-0335-5. CrossRef Full Text | Google Scholar
74. Enyoh, C. E., Verla, A. W., Verla, E. N., Ibe, F. C., & Amaobi, C. E. (2019). Airborne microplastics: A review study on method for analysis, occurrence, movement and risks. *Environmental Monitoring and Assessment, 191*, 1–17. https://doi.org/10.1007/s10661-019-7842-0. CrossRef Full Text | Google Scholar
75. Anbumani, S., & Kakkar, P. (2018). Ecotoxicological effects of microplastics on biota: A review. *Environmental Science and Pollution Research, 25*, 14373–14396. https://doi.org/10.1007/s11356-018-1999-x. PubMed Abstract | CrossRef Full Text | Google Scholar
76. Anagnosti, L., Varvaresou, A., Pavlou, P., Protopapa, E., & Carayanni, V. (2021). Worldwide actions against plastic pollution from microbeads and microplastics in cosmetics focusing on European policies. Has the issue been handled effectively? *Marine Pollution Bulletin, 162*, 111883. https://doi.org/10.1016/j.marpolbul.2020.111883. PubMed Abstract | CrossRef Full Text | Google Scholar
77. Calero, M., Godoy, V., Quesada, L., & Martín-Lara, M. Á. (2021). Green strategies for microplastics reduction. *Current Opinion in Green and Sustainable Chemistry, 28*, 100442. https://doi.org/10.1016/j.cogsc.2020.100442. CrossRef Full Text | Google Scholar

# Chapter 7
# Environmental Risks and Impact of Microplastics and Nanoplastics

**Manu Mehrotra, Manish Baboo Agarwal, Vinay Kumar Jadon, and Seema Agarwal**

## Abbreviations

| | |
|---|---|
| ARGs | Antibiotic resistance genes |
| BPA | Bisphenol-A |
| EDCs | Endocrine-disrupting chemicals |
| EPR | Extended producer responsibility |
| MPs | Microplastics |
| NPs | Nanoplastics |
| PAHs | Polycyclic aromatic hydrocarbons |
| PCBs | Polychlorinated biphenyls |
| POPs | Persistent organic pollutants |
| PPCPs | Pharmaceuticals and personal care products |
| ROS | Reactive oxygen species |
| WWTPs | Waste water treatment plants |

M. Mehrotra (✉) · M. B. Agarwal · V. K. Jadon
Sharda School of Basic Sciences and Research, Sharda University Agra, Keetham, Agra, Uttar Pradesh, India

S. Agarwal
Department of Mathematics, SRMIST, Ghaziabad, Uttar Pradesh, India


S. Kulkarni, S. Das (eds.), *Microplastics and Nanoplastics*, Lecture Notes in Nanoscale Science and Technology 49,
https://doi.org/10.1007/978-3-032-17963-0_7

# Introduction

## *Definition of Microplastics (MPs) and Nanoplastics (NPs)*

Microplastics are generally classified as plastic fragments smaller than 5 mm in diameter. They result either from deliberate production of scrubby beads in facial cleansers or grit in factory workshops or from larger items, like sandwich wrappers and fleecy jackets, breaking apart over time.

Nanoplastics sit at a much smaller scale, usually pegged at less than 100 nm. At that dimension, the particles drift more freely in water, slip through membranes that keep out larger molecules and behave erratically under an optical microscope [1].

Both of them show up in ocean currents, river sediment, farmed soil, city air and even bottled drinking water. The durability of plastic, combined with its tendency to colonize toxins, raises alarms among scientists studying wildlife and human health.

## *Sources and Pathways of MPs and NPs in the Environment*

Microplastics and nanoplastics follow a different routes once they leave the point of use. Tire dust sloughs off onto pavement, laundry lint collects in waste-water plants and improperly discarded plastic may degrade under sunlight and freeze-thaw cycles. Mapping these entry points is the first step toward gauging the larger ecological and health fallout [2].

### Primary and Secondary Sources

Microplastics (MPs) and Nanoplastics (NPs) can be divided into two categories. One made deliberately on a microscopic scale and those that emerge after larger plastic goods break apart. Items labeled primary MPs and NPs include cosmetic micro beads, plastic factory pellets and medical-grade microspheres, all engineered to be small from the outset. Because these fragments are already tiny, they often slip through sewage works and show up in rivers and estuaries long before anyone notices. Secondary MPs and NPs arise when everyday abandoned shopping bag or a weather worn fishing net cracks, peels and crumbles under sunlight, rain, grit and bacteria. The leftover shreds still float or drift long after the original objects have vanished from sight.

## Pathways into the Environment

No matter how they are released, the fragments always find a way to travel. They ride storm runoff, blow across a field, or lift on the soles of hikers. Once deposited, MPs and NPs hang around in sediment, cling to soil particles, or remain suspended in air for weeks, even months.

### Aquatic Pathways

Rivers, lakes and coastal seas collectively function as final holding tanks for fragmented plastic debris. Polling points include roadside ditches, sewers and pipes that deliver unprocessed effluent from treatment works. Once in the open ocean, circulating gyres carry microplastics farther away, dispersing the particles from local harbors into the abyssal zone [3].

During laundering, synthetic clothes lose filament strands during the rinse cycle and hardly any of these fibers survive the plant screens that follow. Laundry borne pollution is therefore both incidental and persistent.

### Terrestrial Pathways

Microplastic traces appear in agricultural soil mainly through different routes. One is by the sludge based fertilizers that contain reclaimed bio solids and plastic mulch that eventually breaks into flakes. City street grit enriched with tire dust and packing pellets, flows into drains and is re-deposited on the land. When rainwater percolates through the mass, the resulting leachate then seeps sideways into adjacent earth and in some cases in groundwater.

### Atmospheric Pathways

Air suspended microplastics arise from urban settings, synthetic fibers, wear of tires, fling off polymer grit, fleece blankets and polyester suits and industrial activity. Wind carries the grains farther away lodging them in distant areas and on snow fields in the Arctic or perched cliffs miles above sea level.

Household air is by no means cleaner either. Examinations show that sofas, curtains and even anodized cookware can contaminate the breathing space. People inhale these particles or find them settled on countertops and open food containers.

### Biological Vectors

Biological vectors occasionally carry microplastics and nanoplastics across unexpected distances. An amphibian that swallows contaminated sediment may retain those particles within its tissues and subsequent predators inherit the load through routine feeding. Waste products and decaying carcasses then release the plastics back into soil or water bodies where fresh transport cycles can begin (Fig. 7.1).

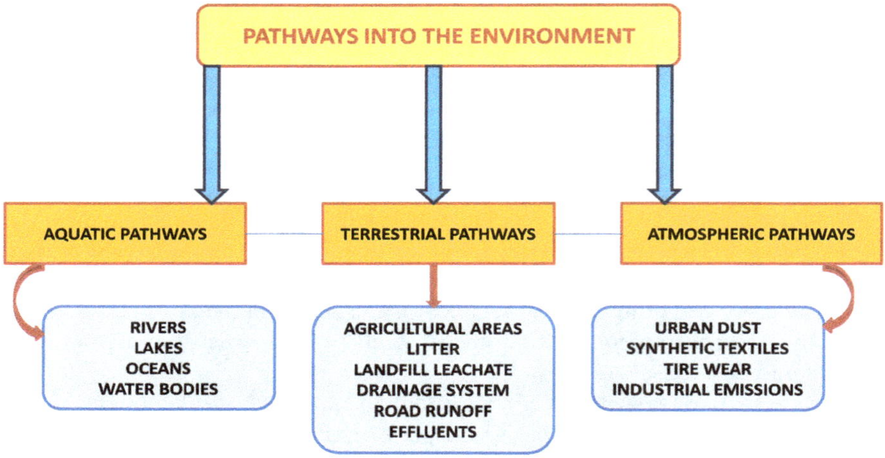


**Fig. 7.1** Various pathways of Microplastic and nanoplastic into the environment

## Environmental Distribution of MPs and NPs

### *Presence in Aquatic Ecosystems (Freshwater & Marine)*

More than any other habitat, aquatic biomes are among the most affected by microplastics and nanoplastics pollution. Peer-reviewed surveys show these plastic flecks in the deepest ocean basins, along raw water intakes and on the surface of arctic sea ice raises alarms about water chemistry, resident species and broader food web stability [4].

#### Freshwater Ecosystems

The freshwater systems such as streams, reservoirs and lakes acts as sources and transport routes for MPs and NPs. Storm water, drains, factory outlets, untreated sewage and carelessly discarded refuse all fall debris into rivers with little filtration. Wastewater treatment plants try to mitigate them, but their level of success is uneven. Centrifuges, filters and clarifiers capture many larger particles, yet the micro and nano fractions slip through because they resemble dissolved salts more than solid grit. By the time the effluent reaches an adjacent stream or lagoon, it contains thousands of invisible plastic shards that seed the water body and continue drifting downstream.

Microplastics in freshwater settings are frequently derived from broken down plastic refuse, released textile microfibers, fragments shed by vehicle tires and cosmetic exfoliants. Once afloat, these particles may linger indefinitely in the water, embed in bottom sediments, or be swallowed by grazing organisms such as zooplankton, invertebrates and several fish species.

#### Marine Ecosystems

Ocean basins mainly act as the final reservoir for much of the world's plastic waste. Debris that washes into salt water via rivers and storm is subsequently broken into finer particles by wave action, ultraviolet light and bacterial colonization. Contemporary sampling results have unveiled that MPs and NPs from surface, slicks all the way to the Mariana Trench and Polar Regions [5].

Microplastics in the ocean are often sorted by how well they float and by the chemistry of the polymer itself. Lighter grades drift on the surface and collect in swirling reservoirs such as the Great Pacific Garbage Patch. Heavier variants sink steadily and carpet the sea bottom, mingling with sediments and disturbing bottom dwelling life. Near the surface, tiny zooplankton may mistake the particles for food, while larger predators like sea birds and marine mammals swallow them higher up the chain. For that reason, MPs serve as a mechanical vector for any toxic additives they picked up along the way.

Nanoplastics raise even greater alarm because their scale allows them to slip past cellular membranes with relative ease. Once inside, they can spark oxidative damage and upset organ function.

#### Transport and Distribution Factors

Distribution of MPs and NPs within oceanic waters is influenced by several factors. Currents governed by wind and the earth's rotation funnel debris too far off areas. Heat and salt content can either prolong fragments buoyancy or accelerate its submersion. Grazing organisms further remix the population, carrying particles both deep into the water and back to surface. Thus, MPs and NPs drift through nearly every inch of freshwater lake and open ocean salt water. Their stubborn presence in these habitats disrupts food webs, diminishes species richness and ultimately threatens human health.

### *Soil Contamination*

Also there is an adverse affect of microplastics and nanoplastics into farm fields and soil. These particles settle into soil and tend to linger for years. Many researchers warn that contaminated soil will shrug off common bioremediation strategies, leaving worms, pollinators and even human foods exposed to persistent threat of plastic debris [6].

## Sources of Microplastics and Nanoplastics in Soil

### Plastic Mulch and Agronomic Waste

Wide strips of plastic mulch are still the preferred to keep strawberry rows damp and weed-free, even though farmers know the film will shred long before harvest. Sunlight and irrigation chisels degrade it that accumulates directly onto cultivated top soil [7].

Local municipalities often use treated sewage sludge as fertilizer which adds MPs and NPs directly onto the agricultural land. Washing machines, chemical plants and laundries filter loads of fleece and polyester that survive routine treatment and add to soil contamination.

### Industrial and Urban Sources

Mountain top, landfills often allow plastics wander on wind gusts or during storm. Old dumps are now filled with plastics that are dispersed into nearby areas. Construction zones and improperly managed road surfaces regularly shed shavings of polymer aggregates, leaving a trail of gritty residue that drifts into nearby lawns and garden plots.

### Atmospheric Deposition

Atmospheric fall out of MPs and NPs also plays a key role. Polyester fleece, synthetic fiber, brake lining composites and stack ash from petrochemical plants release microscopic chips that are thicker over cities than over the open areas. Other sources include worn vehicle tires, non-woven industrial filters, delivery pallets and traffic generated pavement erosion.

## Environmental Behavior and Fate in Soil

Once the particles reach the ground the outcome depends on several factors, including the size of each flake, the chemistry of the base resin, surface ionic charge, current soil water content, pH and temperature [8]. Microplastic flecks can collect at the top horizon or be washed deeper by percolating rain water or pushed lower by earthworms. In addition, they can bond with pollutants such as lead dust or herbicide granules and carry them downward. Over the time plastics may alter microscopic pore sizes, shift cation-exchange capacity, or even inhibit the respiration rates of ground dwelling bacteria.

## Effect on Soil Health and Terrestrial Ecosystems

The introduction of MPs and NPs into the ground layer initiates a chain reaction in the ecosystem. Tests show that plastic debris can reshape the resident microbial community, thinning overall diversity and impairing critical processes such as nitrogen turn over and the breakdown of plant litter. Field observations indicate that MPs compact the material matrix, diminishing pore space and thus cutting both oxygen diffusion and moisture availability that roots depend on. Soil dwelling invertebrates, including earthworms, mistake polymer granules for organic morsels. This result in gut impaction and in some cases, a drop in reproduction rates [9]. The plastic particles themselves can move into edible plant tissues, raising questions of food safety.

As plastics appear in everything from synthetic fertilizers to urban runoff, exposure pathways are widening as plastic production increases. Restoring the integrity of the soil is non-negotiable if we wish to meet global food demands while steering clear of enduring ecological collapse.

## *Atmospheric Distribution*

The study indicates that urban and rural air may contain several hundred micrograms of plastic debris per cubic meter that has elevated anxiety about both ecological integrity and public health. Because the atmosphere functions simultaneously, a site of direct exposure, inhaled particles represent a pathway of immediate concern [10].

### Sources of Atmospheric Microplastics and Nanoplastics

Tiny plastic particles in air are viewed as environmental concerns and have become staple entries in urban air quality reports. Urban centers are sources for airborne plastics, with heavy traffic, cement mixers, demolition crews and processing plants all stirring the air. Car and truck tires shred, loose bits of synthetic rubber, skim the road surface and winds carry them along sidewalks and bicycle lanes. Factories that extrude, mold, or trim plastic goods tend to leak a fine plume of thermoplastic powder no one can see until it coats the edge of a nearby roof top [11].

Synthetic clothing polyester fleeces migrate into the air in the form of micro filaments. Domestic washing machines shed the fiber inside the house and kitchen exhausts push the fibers back outside, in the clouds that look like smoke.

In economies where money is short or regulation is lax, workers set plastic trash alight in back lots. The flames reduces bottles to ash but scatter hazardous particles invisible to the naked eye. Old dumps also blow a curtain of dust when disturbed by mechanical activity or wind contributing to airborne plastic contamination [12].

Every plastic fleck that settles on the grass can be carried back into suspension by a breeze, or by a footstep. Boat wakes and bird wings whip spent pellets into the atmosphere, adding to contamination.

### Transport and Distribution in the Atmosphere

Wind driven aerosols carry airborne MPs and NPs hundreds or even thousands of kilometers from their original release points. Surveys have confirmed the presence of these particles in isolated settings such as the Arctic ice, alpine snow packs and uninhabited tropical areas, providing striking evidence that plastic debris can cover vast terrestrial and oceanic distances with ease [13].

Particle size, specific surface area and bulk density govern how long they will remain airborne. Coarse flecks typically succumb to gravity within hours and accumulate in nearby soil or waterways, while the finer fraction stays aloft for days or weeks and enters human lungs with each breath.

### Deposition and Environmental Impact

Plastics suspended in the atmosphere eventually settle onto land and water, by wet or dry deposition. Wet deposition takes the form of airborne particles being entangled in raindrops or snowflakes before they fall to the ground. Dry deposition is slower and micro fragments drift downward under gravity. Once on the surface, the pieces can leach into soil or streams and become a source of contamination. The plastics may also be inhaled by animals and humans who come into direct contact with the newly contaminated substrate.

### Human Health Implications

Breathing in airborne MPs and NPs creates an immediate link between the environment and human lungs. Case reports suggest symptoms range from mild irritation to chronic inflammation and measurable drops in lung capacity. As nanometer-sized particles are small enough to penetrate alveolar sacs, they may breach vascular membranes and appear in blood or even in distant organs. Moreover, each fragment can shuttle carcinogens or infectious microbes, turning a relatively inert particle into a vehicle for compounded biological risk [14]. Animal also show similar pathways of harm in mammalian species beyond humans. Preventing further releases, ramping up airborne monitoring and transitioning to biodegradable or less volatile materials is required to prevent from this pervasive hazard (Table 7.1).

**Table 7.1** Characteristic of Microplastics (MPs) and Nanoplastics (NPs)

| Features | Microplastics (MPs) | Nanoplastics (NPs) |
|---|---|---|
| Dimensions | Between 1 μm to 5 mm | Between 1 nm to 1 μm |
| Sources | Micro-beads in cosmetics, synthetic fibers, tire wear, fragmentation of larger plastics | Degradation products of MPs |
| Appearance | Visible under light microscope | Visible by electron microscopy, spectroscopy |
| Environmental behavior | Tend to settle in soil, freshwater<br>Adsorbs pollutants<br>Interact with organic matter | Remain suspended in water<br>Due to high surface to volume ratio increases adsorptive capacity<br>Can cross cell membrane |
| Environmental impact | Affect terrestrial and aquatic organisms<br>Bioaccumulate and transfer in food webs | Causes oxidative stress<br>Inflammation<br>Transport toxins |
| Public health | Exposure through inhalation and ingestion | Crosses biological barriers |
| Dispersal | Found in terrestrial, freshwater, marine and atmospheric environments | Found in aquatic environments and in atmosphere |
| Analytical tools | FTIR, Raman spectroscopy, microscopy are mainly used | Detection is difficult<br>Advanced methods required |

## Biological Impacts

### *Ingestion by Aquatic and Terrestrial Species*

Microplastics and nanoplastics enter the bodies of both freshwater and terrestrial animals in consistent numbers. Researchers have recorded the particles in plankton, crustaceans, migratory fish, sea birds, soil arthropods and even beached baleen mammals. In almost every case, ingestion appears to alter behavior, impair growth, or trigger inflammation.

#### Aquatic Species

Littered throughout lakes, rivers and coastal bays, MPs and NPs particles frequently match the size and shape of ordinary zooplankton and pick up enough organic film to mimic food odor. Plankton that ingests microplastics show slowed grazing rates and carries lower energy into subsequent life stages. Mussels and oysters store the particles within their gut folds, triggering localized inflammation and over longer exposure, suppressing both growth and reproduction. Fish, crabs and shrimp consume plastic both directly from the water and indirectly via contaminated prey, leading to mechanical blockages and impaired nutrient uptake. Larger marine vertebrates, including many seabirds, swallow plastics in bulk because the debris resembles large, moving prey. Nanoplastics intensify the problem by slipping through cellular membranes and settling in muscle, nervous, or reproductive tissues where they remain difficult to clear. Once inside a living organism, nanoparticle contaminants can trigger oxidative stress [15]. They also initiate inflammatory pathways and produce direct cytotoxic effects.

#### Terrestrial Species

Microplastics do not remain confined to marine habitats, many terrestrial organisms encounter them as well. Studies have shown that earthworm and several other soil invertebrates consume loose polymer fragments buried in sediment [16]. The polymers disrupt resident microbiomes inside the gut, impair digestive efficiency and further reduce reproductive output. Scavenging birds, along with insectivorous species, frequently ingest plastic filaments when they feed directly or through contaminated prey. Such ingestion can produce intestinal blockages, decrease nutrient assimilation and in some cases lead to death. Cattle and poultry occasionally absorb microplastics incidentally through tainted feed, water, or pasture and that raises alarms about transfer into the human food chain. Studies reveal that small insects, such as crickets and various beetles also ingest plastic debris which alters growth, modifies activity patterns and lowers reproductive success.

#### Trophic Transfer and Bioaccumulation

Trophic transfer describes the vertical movement of MPs and NPs debris through successive alimentary links, a phenomenon first documented several decades ago [17]. When a predator swallows prey already laced with polymer fragments, the plastics attach themselves within the predator's own tissues. There is a gradual bioaccumulation of the particles in the digestive, hepatic and renal compartments of these higher level consumers and hint at possible biomagnifications.

### *Toxicological Effects on Organisms*

Microplastics and nanoplastics have headlined environmental hazard in barely a decade. Acute exposure often inflicts mechanical wounds, while the chronic toxic effect causes kidney disruption in zebrafish and stress hormone surge in seabirds. Per-size particle behavior, polymer composition and their capacity to carry other pollutants make them harmful to the biological systems [18].

### Physical and Chemical Interactions

Plastics in the micro- and nano-size range threaten biological systems through a combination of mechanical and chemical stress. These fragments grind against soft epithelium, creating micro-scale lacerations. Such abrasion is compounded by the occasional compacting of material into a bolus that obstructs passage through the gut. The exposed fishes fail to move food past the blockage, lessen nutrient uptake and forcing a false sense of fullness that starves the animal with time.

Plastic manufacturing uses additives like lubricants, colorants, smoke suppressants and aging inhibitors. When these composite pellets are broken down into micro- and nano-sized fragments, some of the additives diffuse out, subjecting living tissues to hormonally active carcinogenic entities like certain phthalates. The small particles also develop affinity for seawater pollutants, owing to the relatively huge surface area. Heavy metals, organics, agricultural sprays and even trace antibiotics cling to the plastic grain, turning each speck into a localized poison dispenser once swallowed.

### Cellular and Molecular Toxicity

Nanoscale debris poses a different set of challenges because its size can dip below the typical thickness of a mammalian cell membrane. A single particle can slip inside a cytoplasmic compartment, where it begins to prod the cell into an overdrive of reactive oxygen species that ultimately scrambles lipids, genetic loops and protein scaffolds.

Persistent exposure, even at low concentrations, nudges the immune system toward a state of steady, misdirected inflammation, eroding the very tissues that should resist infection. Laboratory experiments underscore this problem in vertebrates, fish and cultured human cell lines [19].

### Developmental and Reproductive Effects

Studies have correlated microplastic and nanoplastic exposure with developmental and reproductive impairments across taxonomic groups. In controlled aquatic settings, crustaceans and bony fishes often experience declines in fertility and overall reproductive yield. Embryos sometimes manifest skeletal deformities and amphibian larvae may display prolonged incubation under polluted regimes. Disruption of endocrine signaling pathways is frequently attributed to carrier additives such as bisphenol-A and various phthalates that leach from the polymer matrix.

#### Species-Specific Responses

Toxicological outcomes are not uniform they fluctuate with both the organism and the precise chemical load. Phytoplankton, copepods and daphnids regularly exhibit stunted growth and diminished clutch sizes when exposure reaches just a few milligrams per liter. In higher vertebrates, lesions can appear in the liver and gill epithelium, while notable shifts in foraging or migratory behavior have been recorded. Organisms already classified as endangered often bear the brunt because their population and available habitats are uncommonly narrow [20]. The particles deliver their toxic effect not only through chemical leachates but also by jamming into tissues and lowering basic mobility.

### *Cellular and Physiological Damage*

MS and NS can cause both cellular and physiological damage to animals that live in water and those that do not. Their small dimensions of the order of a millionth of a millimeter in the case of nanoplastics allow them to pass through membranes and nestle against mitochondria, nuclei, or other sensitive components. Over time, even small amount disrupt normal metabolism and threaten the organism health.

#### Cellular Uptake

Particles smaller than 100 nanometers slip into cells, affecting endocytic vesicles. After crossing the membrane, individual specks can settle in lysosomes, mitochondria, or the nucleus and upset the routine function. They may soften an organelles acid balance, stall the ATP assembly line, or jam chromatin remodeling, long after the initial exposure.

#### Oxidative Stress and Free Radical Formation

Investigators frequently associate the introduction of plastic particles into biological systems with a surge in reactive oxygen species (ROS). Such an increase tilts the redox equilibrium toward oxidative stress, a state that permits free radicals to outnumber customary counter agents [21]. If unchecked, that imbalance can instigate lipid peroxidation, denature structural and enzymatic proteins and fragment genetic material, thereby predisposing cells to mutations and hindering their ability to replicate accurately.

### Inflammatory Response

Plastic fragments often provoke a pro-inflammatory reaction in the immune system. Chronic irritation of that sort can scar tissues, push the immune apparatus out of its normal lane and leave the animal open to other infections. Fish and sessile invertebrates regularly show inflamed guts or gills after short exposures, while mammals develop itchy lungs or upset intestines if they inhale or swallow the same plastics.

### Apoptosis and Necrosis

Persistent contact with micro- or nanoplastics can trigger both the tidy, programmed death known as apoptosis and the messier process we call necrosis. Studies trace those outcomes to wear and tear on mitochondria or to bursts of oxidative stress that short-circuit cellular energy supplies. In fish liver, in gill tissues and along the wall of the intestine, researchers have spotted the apoptotic cell loss [22].

### Hormonal and Metabolic Disruption

Plastic particles seldom drift through water in a pure state. They pick up or carry along them a mixture of chemicals. Compounds such as bisphenol-A, various phthalates and flame retardants can latch onto the plastic and then barge into animal's endocrine system, scrambling reproductive signals and thyroid output. It often includes skewed lipid storage, out-of-whack glucose control and shifts in overall energy burn, symptoms that show up across several species [23].

### Tissue Accumulation and Organ Damage

Plastic debris finds its way into the liver, kidneys, gut and lungs over the course of repeated exposures. Studies on both fish and rodents documents liver fibrosis, disruption of intestinal microvilli and injury to pulmonary tissues. The microscopic plastics induce oxidative stress, trigger inflammation and compromise hormonal balance. Taken together, it leads to damage in multiple organ systems.

## Human Health Concerns

### *Exposure Pathways*

People pick up microplastics and nanoplastics mainly through various pathways like diet, water and airborne dust. These particles refuse to break down and crop up almost everywhere in modern life. Early data show that the contaminants can cross into human tissues, yet definitive studies on long-term harm are still underway [24].

#### Food Consumption

Human diet is a primary channel for microplastic intrusion into the body. Marine products mostly shellfish, crustaceans and certain fish specifically teem with fragments because the open ocean has become a repository for plastics [25]. People who consume these animals inadvertently take in the plastics lodged in their guts. Processed commodities like honey, sugar and canned goods offer another entry point, since packaging materials and factory routinely shed particles. Agricultural produce and livestock are indirectly tainted when sludge fertilizers or plastic mulch tear into micro-fragments and wash into soil or irrigation sources. Once inside the digestive tract, larger particles simply transit through. The smaller ones, however, can breach the gut wall and enter the blood stream or lymph, possibly settling in organs far from the original point of entry.

#### Drinking Water

Microplastic specks in both the bottled water and the tap supply. A survey of bottled water found several hundred to several thousand fragmentary fibers per liter, hinting that they may arise during bottling process and packaging. Municipal systems, even after sedimentation and membrane filtration, carry particles from corroded mains, carry over from source lakes, or from older treatment plants [26].

#### Inhalation of Airborne Microplastics

Airborne MPS and NPs have now been confirmed in living rooms, offices, highway cafeterias and even in the air above remote glaciers [27]. Researchers are beginning to rank inhalation as the principal pathway for exposure. Homes that utilize a high proportion of synthetic fibers, whether shag rugs, velvet upholstery, or gym clothing, typically harbor dense clouds of airborne particles. Simple activity, like vacuuming the hallway carpet or laundering a polyester hoodie, can briefly boost the fiber concentration by an order of magnitude. Outdoors, car tires, factory vents and street repairing scatter their own share of plastic debris into the urban breeze. Once inhaled, particles may lodge in the nasal sinuses or the branching bronchi and can slip across the alveolar membrane and enter the blood circulation.

## Potential Health Effects

Continuous exposure of MPs and NPs is already linked to chronic inflammation, hormonal disturbances and metabolic strain in laboratory mammals, suggesting a similar risk for people who live in heavily contaminated environments. MPs on the kitchen air vent and NPs in the blood stream serve as mobile carriers of chemical hazards during every inhalation [21].

### Inflammation

Microplastic and nanoplastic particles unleash a steady inflammatory reaction the moment they enter the body. People swallow them, breathe them and less frequently, absorb them through the skin. Cytokines flood the blood stream and oxidative stress sensors spike. Lasting inflammation of any organ has long been linked to ailments as diverse as coronary artery disease, malignant growths and deregulated autoimmune responses.

### Hormonal Disruption

A wide range of polymer either leaches or sequesters chemicals that behave like endocrine disrupting compounds (EDCs) [28]. Bisphenol-A, various phthalates, nonylphenol and a handful of chlorinated biphenyls slip into body fluids and imitate, mute, or hijack natural hormones such as estrogen, testosterone and thyroid secretions. Hormonal receptors that normally transmit cellular orders mistakenly register these mimics, opening the door to altered patterns of growth, delayed puberty and diminished fertility. EDC exposure from plastic result in reduced sperm motility, erratic menstrual cycles, early onset of puberty and a heightened likelihood of hormone-sensitive tumors, such as those lining the breast or prostate are very frequent.

### Metabolic Disorders

Microplastics exposure also results in metabolic disorder. Chemicals residing within the debris have been dubbed obesogens because they appear to nudge stem cells toward the adipocyte lineage and then coax those cells to swell. Glucose handling does not escape unscathed either. Both micro-and nanoplastic contaminants have shown the ability to jam insulin signaling [28].

The liver plays a major role in detoxification, has of late revealed signs of oxidative stress, early scarring and wobbling lipid profiles when littered with these particles.

## Environmental Interactions

### *Adsorption of Heavy Metals and Persistent Organic Pollutants (POPs)*

Microplastics and nanoplastics cause environmental pollution by carrying industrial toxins through estuaries and across food webs [18]. The plastics long-lived, hydrophobic surfaces suck up heavy metals and myriad of POPs from their vicinity. Once lodged in sediment or ingested by zooplankton, the specks release their payload, amplifying both chemical exposure and ecological risk.

#### Mechanisms of Adsorption

Microplastics and nanoplastics usually have non-polar, hydrophobic polymer surfaces. These smooth surfaces latch onto a mix of organic and inorganic pollutants present. Environmental wear and tear, sunlight, oxidation and creeping bio-films, roughens the particles and sprinkles new functional groups across the coating, boosting adsorption strength. Factors like pH, salinity and temperature also matter, as does the base resin itself (polyethylene, polystyrene, and polypropylene) as each operate with a distinct adsorption profile.

#### Heavy Metals

Lead, cadmium, mercury, chromium and copper comprise of some of the heavy metals and even trace amounts can prove toxic to living systems. Industrial outfalls, mine drainage, fertilizer run-off and untreated urban water keep them in constant circulation [29]. Aging or oxidizing a plastic fragment tends to amplify its affinity for these cations, locking them onto the polymer matrix. Unfortunately, the bond is not permanent, the shift in temperature or ionic strength can release the metals, turning the particles into mobile hot spots. When an aquatic organism swallows a contaminated fragment, the metal often transfers directly to internal tissues, where it may trigger neurotoxicity, organ damage, reproductive failure, or even cancer down the line

### Persistent Organic Pollutants (POPs)

POPs are synthetic chemicals designed to resist degradation. Once released into the environment, these compounds become entrenched in soil, sediment and biota [30]. Examples include polychlorinated biphenyls, various polycyclic aromatic hydrocarbons, dioxins and classic organochlorine pesticides such as DDT and aldrin. Because POPs are lipophilic, they adhere readily to hydrophobic surfaces, including many types of plastic. Surveys of river and marine environment show POP concentrations that exceed those of the ambient water column by orders of magnitude. When ingested, contaminated debris off loads its chemical payload into fish, birds, or mammals and thereby elevates the overall toxic burden at both individual and population scales.

The co-occurrence of plastics and adsorbed pollutants frequently produces effects that are greater than the sum of their parts. Fragment of polyethylene or polypropylene act like Trojan horses, delivering concentrated toxins deep into the soft tissue [31]. Once inside a gut or gill, the contaminants become far more available, increasing cellular uptake and cause systemic harm.

## *Role of Plastics as Contaminant Carriers*

These fragments are now viewed more as delivery vehicles whisking toxins across marine and terrestrial settings. Due to their physiological properties they snag everything from lead ions to household antibiotics, POPs, pathogens and other residues and once loaded they release them in unexpected hotspots.

### Physicochemical Properties Enabling Contaminant Transport

Polymeric debris exhibits a unique combination of hydrophobicity and low density that allows it to carry toxins across unexpected distances. The surfaces of polyethylene, polypropylene and polystyrene sheets repel aqueous media, thereby welcoming many non polar solvents, industrial intermediates and biogenic contaminants. Ultraviolet radiation, photochemical degradation and sheer mechanical wear subsequently fracture these coatings into micro-craters and new oxygen based moieties lock guest molecules. Particles left to drift either on the sea surface or aloft in the wind can migrate hundreds of kilometers before settling down [32].

### Categories of Contaminants Carried by Plastics

Plastic flakes drifting through water can take on variety of pollutants with them. Examples include polychlorinated biphenyls (PCBs), polycyclic aromatic hydrocarbons (PAHs) and the dioxins. Lead, cadmium, arsenic and mercury often cling to older fragments because these metals linger in rivers and ponds long after industrial discharges have ceased. The same surfaces that dull with time also become prime adsorption sites for heavy ions. Contamination also comes from Pharmaceuticals and Personal Care Products (PPCPs) such as trace antibiotics, synthetic hormones and cosmetic preservatives. Sewage treatment streams and factory outfalls leave these compounds behind as the water evaporates or is chemically cleaned.

#### Implications on Ecosystems and Food Chains

Chemically loaded plastics pose a two-fold risk to any organism that swallows them, first as gritty indigestible matter, second as a slow-release reservoir of toxic compounds [33]. Predators prey on items laced with pollutants, transfer those exactly the way lipids migrate up a food web, thus promoting bioaccumulation and biomagnification. Fragments that settle into riverbeds or agricultural soils can upset resident microbial communities, knock back species richness and blunt the natural processes that recycle nutrients.

#### Human Health Risks

When tiny flecks of plastic lodge in seafood, drinking water, or produce, they can carry along a various endocrine disruptors, probable carcinogens and heavy metals. Nanoplastics easily slip through biological membranes, scattering these pollutants throughout tissues and organs causing systemic toxicity [34].

## Agricultural and Land Use Implications

### *Impact on Soil Microbiology and Fertility*

MPs and NPs are frequently found in farmland and pasture soils [35]. Contamination mostly tracks back to farming inputs like plastic mulch, greenhouse covers, compost blends, bio-solid fertilizer and even some irrigation water. Because these materials can choke microbial communities, disrupt nutrient cycles and diminish long-term soil fertility, their steady accumulation poses a fundamental challenge to sustainable agriculture.

**Alteration of Soil Microbial Communities**: Soil microbial communities govern nutrient turnover, the breakdown of litter and the physical texture of tilled soil. Microparticles, whether plastic threads or nanoscale flecks, can tilt that balance toward fewer thriving organisms. A sudden decline in species richness translates directly into the loss of many routine ecosystem chores [36].

**Effects on Soil Enzymatic Activity**: Enzymes such as urease and phosphatase act as indicators of biological activity and are vital processes within the soil biology. Exposure of MPs results in decrease in soil enzymatic activity in nitrogen and phosphorus cycle. The decrease in enzyme output reduces the overall biological productivity of the soil, decreases the rate of breakdown of organic matter, affect the nutrient distribution and overall fertility of the soil with time.

**Physical Impact on Soil Properties**: MPs and NPs pieces causes' mechanical stress resulting in biochemical shifts in the soil biota. Polymer fragments block pores and force clumping, which chokes off water entry and limits the gas diffusion and growth of roots. Thinner strands tend to suspend in the matrix, making it difficult for roots to grow as oxygen becomes scarce. Disrupted soil structure increases soil erosion resulting in loss of soil nutrients further.

**Impact on Plant Growth and Crop Yield**: Altered soil biology and disrupted texture show up first in seedlings and in crops that push growth cells into the ground, such as carrots or beets. Some polymer blends, leach plasticizers or trap pre-existing contaminants, letting those compounds enter into vascular tissue and end up accumulating in the edible parts, increasing food safety concerns.

**Soil Fertility Decline**: Microplastics settle into soil layers as a slow-moving pollutant that cannot be detected easily. Studies indicate that the particles resist breakdown, leaving behind a soil matrix that performs poorly, biologically and physically over the decades [9]. Farmers soon discover that the natural nutrient release from humus has dropped, prompting them to supplement with synthetic fertilizers. Even then yields and land value often decreases.

## *Plastic Mulch and Sludge-Fertilizer Contamination*

Plastic Mulch sheets and recycled-sewage nutrient are now days used to increase the productivity of the crops. But both of them shred micro-scale fragments that become part of the top soil, resulting in their accumulation. They become the major source for plastics found in land ecosystem, mainly in commercial agriculture where machinery is used daily.

**Mulch Films: A Growing Problem:** Thin polyethylene mulch sheets are mainly used in vegetable and berry production because they decrease water evaporation and weed germination, maintain the temperature of the soil and improve the crop productivity. However under sunlight, the plastic loses strength, shreds and breaks into micro and nanometer sized particles [7].

These particles mix into the soil matrix and seep into adjoining pores, where they clog air channels and hinder the root growth. This accumulated debris lowers both water holding capacity and the fertility of the soil. Efforts are being made to produce biodegradable films that can easily break down in the soil.

**Sludge-Based Fertilizers**: Municipal wastewater treatment plants recycle sewage sludge, as a low cost fertilizer rich in phosphorus and nitrogen concentrations, for the soil to reduce the expenses of the farmers. It improves soil moisture retention and biological activity. Nevertheless, during its treatment process some microplastics and nanoplastics particles are not removed and are left behind. Contamination enters the sludge through household laundry, cosmetic microbeads, food-packaging shreds and airborne or wash-water drift from plastics processing factories [37].

Once incorporated into the soils the polymer fragments carry over additives altering the physical habitat for bacteria and fungi essential to nitrogen mineralization. Microorganisms ingest these particles and transfer the absorbed pollutants into root zones where agricultural crops can take them up and can travel up the human food chain. These fragments disrupt the natural microbiome and affect the crop safety as well as food-chain integrity. Switch to biodegradable covers, strict regulations and frequent monitoring on sludge spreads can reduce the sludge based contamination.

## Airborne Microplastics

Tiny airborne plastic particles present in the atmosphere have raised concern about their consequences on public health and nature.

### *Sources and Spread*

Airborne microplastic particles arise from diverse sources which include shredded fleece, worn tires, factory material, dust from scaffolded concrete, road side litter, etc. Household activities like washing machine spins, a vacuum, a dryer and household plastic products also contribute notably to it. Farmers who use plastic film or spread sewage sludge raise clouds of lofted fragments when tilling or harvesting. During gusty weather these particles are carried too far off places, like high-altitude peaks, oceans and polar-regions where they settle down during rain or snow [38]. These airborne MPs when inhaled or ingested pose various health issues like oxidative stress, respiratory problem and other chronic diseases. These MPs persist over a longer period of time in air widely affecting the ecological balance.

### *Respiratory Risks to Humans and Animals*

Microplastics suspended in air poses respiratory risks to humans and animals due to their very small size [39]. Once inhaled, the particles can settle in the alveolar sacs and quietly start-off the damage. Their routine exposure may provoke chronic inflammation, elevate oxidative stress and ultimately led to chronic lung infection. As microplastics adsorb hazardous additives, like phthalates and bisphenol-A, they release these chemicals once lodged in human or animal physiology.

Pets, farm animals and even scavenging wildlife frequently ingest or inhale plastic debris carried by breezes, dust clouds, or animal bedding. These embedded plastic particles provoke pulmonary inflammation, affect immune systems and modify genes in them.

Microplastics also harm human health as they can settle almost anywhere. As kids and older adults have more fragile respiratory tracts, the danger is likely heavier on these groups. The increase in microplastics release makes it urgent for health officials to figure out methods to lessen the intensity of the risk. Strict regulations and smarter ways to clean the air inside homes and workplaces, could shield everyone from further exposure.

## Policy and Mitigation Strategies

### *Waste Management Improvements*

Improper waste handling of discarded plastic items is the main cause of pollution caused by MPs and NPs. Treating the problem at source by using compostable or reusable substitutes with outright bans on carry bags and other items will help to mitigate the problem [40]. This can be achieved by imposing strict regulations and general awareness programs.

Improved sorting of trash at both the kitchen counter and at municipal level will practically keep most plastics out of open air and water. Use of conveyor belts, optical sorters, chemical and mechanical recycling technologies will increase plastic recovery and reduce landfill. Waste pickers should be provided with proper training, hand tools, gloves, masks and containers for segregating the waste properly to improve the environmental outcome. Manufacturer should manage the end-of-life disposal of plastic waste.

Thus waste management policy typically focuses on its reduction at source, proper recycling, efficient disposal and public engagement. Collectively, these elements create a barrier against airborne microplastics by interrupting the pollution chain at multiple points.

### *Reduction in Plastic Use*

With the passage of time these plastics fragment into even smaller particles. These particles are then easily carried by the wind into the air. Decreasing the plastic consumption will decrease the airborne microplastics transport [41]. By the use of biodegradable material, compost ready packaging, goods woven from hemp, jute, or paper can help to decrease the underlying pollution problem almost at once. Regulatory measures should be taken by imposing fines, taxes and bans on single-use items. Developed countries have already imposed ban on the use of single use plastic, resulting in decrease in plastic waste generation.

Extended Producer Responsibility (EPR) laws will effectively decrease waste equation, making manufacturers answerable for a products entire lifecycle, from design to landfill. Public agencies can quicken the pace of transformation by favoring bidders whose proposals ditch plastic altogether when awarding procurement contracts.

## *Public Awareness and Education*

Public awareness programs should be carried out to make consumers aware about the harmful effect of MPs and NPs on the environment. Along with regulatory measures, consumer eco-conscious living can be a driving force for change. Outreach campaigns that highlight the environmental and health costs of single use plastics should be encouraged. Schools, businesses, and neighborhood groups can launch zero-waste programs in the classroom, office and community center [42].

Reductions in plastics can only be achieved if government authorities, corporate producers and everyday consumers coordinate their actions. By controlling the plastic supply chain at its origin, it can limit the flow of airborne microfibers while safeguarding the ecosystems and public health.

Increased awareness also converts scientific warnings into personal responsibility, thus amplifying the impact of legislative and industry efforts. When residents know the pathways through which microscopic polymers invade the atmosphere, they are more likely to support stringent rules and modify their own purchasing habits. Short videos, graphics, community workshops and targeted social media messages can convert passive viewers into active participants in the quest for cleaner air.

Collaborative efforts among governmental bodies, non-governmental organizations and universities can help to organize clean-ups, month-long zero-waste competitions, or dedicated awareness weeks that encourage citizens to cut plastic use and manage waste more thoughtfully. Clear labeling schemes and third-party certifications for plastic-free or low-footprint items offer shoppers practical shorthand for making quicker, more informed purchases. Instructional messages that spotlight simple habits relying on reusable containers, source separation of recyclables and the outright avoidance of open plastic burning are necessary for meeting the environmental benchmarks. When such tailored education campaigns combine with public outreach, they leverage awareness into durable behavioral shifts thereby safeguarding the community health.

## Summary

Microplastics and nanoplastics contaminants are present in every ecological stage from surface water and river sediment to Arctic ice. Their small size allows them to transport across vast distances and in entirely remote areas. Two distinct pathways are responsible for their environmental dispersion. First is, by primary microplastics, that are blended into goods such as personal-care scrubs, polish abrasives, or polyester fleece and second is by secondary particles that break free from larger plastic debris through photodegradation, mechanical wear, or chemical weathering.

They are the main source of modern water pollution. Aquatic animals occasionally ingest them, mistaking them for food. The result is minute cuts inside the gut, a deceptive sense of fullness, and, in some cases, it blocks the digestion altogether. The smaller nanoplastic fragments slip through cell membranes causing inflammation, oxidative stress and genetic mutation.

Terrestrial ecosystems are by no means spared from plastic pollution. Microplastic fragments have made their way into farmland via sewage sludge, contaminated irrigation water and the slow decay of mulch film. Once lodged into the ground, these particles can stiffen soil aggregates, decrease nitrogen availability and cause imbalance of microbial communities. It may retard root development and threaten the stability of the global food supply.

These minute plastic debris also adsorb harmful chemicals present in the soil. These contaminants when ingested by animals enter the food chain, leading to bioaccumulation and impacting the entire predator community.

Airborne plastic fragments that are carried by the breeze, can drift thousands of kilometers and settle both into cities and isolated areas. The cumulative effect is environment contamination affecting people and wildlife who inhale them regularly.

Organic debris decomposes within few months or years while MPs and NPs refuse to break down completely. They simply break into even smaller pieces which persist in soil, sediment, or water for many decades becoming a threat to ecosystem health.

Thus MPs and NPs have an overall long lasting environmental impact. Global efforts should be made on reducing plastic waste and promoting alternatives to mitigate their environmental footprints. Further delay in it will lead to permanent losses in biodiversity.

## References

1. Garcia-Muñoz, P., Robert, D., Ruppert, A. M., & Keller, N. (2022). *Microplastics (MPs) and nanoplastics (NPs): Introduction* (pp. 1–32). *Elsevier eBooks*. https://doi.org/10.1016/b978-0-323-99908-3.00013-0
2. Yang, T., & Wang, J. (2023). Exposure sources and pathways of micro- and nanoplastics in the environment, with emphasis on potential effects in humans: A systematic review. *Integrated*

*Environmental Assessment and Management, 19*(6), 1422–1432. https://doi.org/10.1002/ieam.4742
3. Bhardwaj, L. K., Rath, P., Yadav, P., & Gupta, U. (2024). Microplastic contamination, an emerging threat to the freshwater environment: A systematic review. *Environmental Systems Research, 13*(1), 8. https://doi.org/10.1186/s40068-024-00338-7
4. Habumugisha, T., Zhang, Z., Uwizewe, C., Yan, C., Ndayishimiye, J. C., Rehman, A., & Zhang, X. (2024). Toxicological review of micro- and nano-plastics in aquatic environments: Risks to ecosystems, food web dynamics and human health. *Ecotoxicology and Environmental Safety, 278*, 116426. https://doi.org/10.1016/j.ecoenv.2024.116426
5. Allen, D., Allen, S., Abbasi, S., Baker, A., Bergmann, M., Brahney, J., Butler, T., Duce, R. A., Eckhardt, S., Evangeliou, N., Jickells, T., Kanakidou, M., Kershaw, P., Laj, P., Levermore, J., Li, D., Liss, P., Liu, K., Mahowald, N., et al. (2022). Microplastics and nanoplastics in the marine-atmosphere environment. *Nature Reviews Earth & Environment, 3*(6), 393–405. https://doi.org/10.1038/s43017-022-00292-x
6. Campanale, C., Galafassi, S., Savino, I., Massarelli, C., Ancona, V., Volta, P., & Uricchio, V. F. (2021). Microplastics pollution in the terrestrial environments: Poorly known diffuse sources and implications for plants. *The Science of the Total Environment, 805*, 150431.
7. Ramanayaka, S., Zhang, H., & Semple, K. T. (2024). Environmental fate of microplastics and common polymer additives in non-biodegradable plastic mulch applied agricultural soils. *Environmental Pollution, 363*, 125249. https://doi.org/10.1016/j.envpol.2024.125249
8. Wang, Y., Xiang, L., Amelung, W., Elsner, M., Gan, J., Kueppers, S., Christian, L., Jiang, X., Adu-Gyamfi, J., Heng, L., Ok, Y. S., Ivleva, N. P., Luo, Y., Barceló, D., Schäffer, A., & Wang, F. (2023). Micro- and nanoplastics in soil ecosystems: Analytical methods, fate, and effects. *TrAC - Trends in Analytical Chemistry, 169*, 117309. https://doi.org/10.1016/j.trac.2023.117309
9. Chang, N., Chen, L., Wang, N., Cui, Q., Qiu, T., Zhao, S., He, H., Zeng, Y., Dai, W., Duan, C., & Fang, L. (2024). Unveiling the impacts of microplastic pollution on soil health: A comprehensive review. *The Science of the Total Environment, 951*, 175643. https://doi.org/10.1016/j.scitotenv.2024.175643
10. Ramsperger, A. F., Bergamaschi, E., Panizzolo, M., Fenoglio, I., Barbero, F., Peters, R., Undas, A., Purker, S., Giese, B., Lalyer, C. R., Tamargo, A., Moreno-Arribas, M. V., Grossart, H., Kühnel, D., Dietrich, J., Paulsen, F., Afanou, A. K., Zienolddiny-Narui, S., Hammer, S. E., . . . Laforsch, C. (2022). Nano- and microplastics: A comprehensive review on their exposure routes, translocation, and fate in humans. NanoImpact, 29, 100441. https://doi.org/10.1016/j.impact.2022.100441.
11. Torres-Agullo, A., Karanasiou, A., Moreno, T., & Lacorte, S. (2022). Airborne microplastic particle concentrations and characterization in indoor urban microenvironments. *Environmental Pollution, 308*, 119707. https://doi.org/10.1016/j.envpol.2022.119707
12. Pathak, G., Nichter, M., Hardon, A., Moyer, E., Latkar, A., Simbaya, J., Pakasi, D., Taqueban, E., & Love, J. (2023). Plastic pollution and the open burning of plastic wastes. *Global Environmental Change, 80*, 102648. https://doi.org/10.1016/j.gloenvcha.2023.102648
13. Rednikin, A. R., Frank, Y. A., Rozhin, A. O., Vorobiev, D. S., & Fakhrullin, R. F. (2024). Airborne microplastics: Challenges, prospects, and experimental approaches. *Atmosphere, 15*(11), 1380. https://doi.org/10.3390/atmos15111380
14. Inamuddin, Altalhi, T., & Fernandes, V. C. (2024). *Toxic effects of micro- and nanoplastics: Environment, food and human health*. Wiley eBooks. https://doi.org/10.1002/9781394238163
15. Arowojolu, I. M., De Oliveira Alves, I., Benson, N. U., & Sodré, F. F. (2023). Microplastics in aquatic environments: A growing, unresolved concern. *Chemistry of the Total Environment (COTE), 3*(1), 8–22. https://doi.org/10.52493/j.cote.2023.1.76
16. Li, J., Yu, Y., Craig, N. J., He, W., & Su, L. (2023). Interactions between microplastics and insects in terrestrial ecosystems—A systematic review and meta-analysis. *Journal of Hazardous Materials, 462*, 132783. https://doi.org/10.1016/j.jhazmat.2023.132783

17. Saikumar, S., Mani, R., Ganesan, M., Dhinakarasamy, I., Palanisami, T., & Gopal, D. (2023). Trophic transfer and their impact of microplastics on estuarine food chain model. *Journal of Hazardous Materials, 464*, 132927. https://doi.org/10.1016/j.jhazmat.2023.132927
18. Tsochatzis, E. D., Gika, H., Theodoridis, G., Maragou, N., Thomaidis, N., & Corredig, M. (2024). Microplastics and nanoplastics: Exposure and toxicological effects require important analysis considerations. *Heliyon, 10*(11), e32261. https://doi.org/10.1016/j.heliyon.2024.e32261
19. Winiarska, E., Jutel, M., & Zemelka-Wiacek, M. (2024). The potential impact of nano-and microplastics on human health: Understanding human health risks. *Environmental Research, 251*, 118535. https://doi.org/10.1016/j.envres.2024.118535
20. Li, X., Chen, Y., Zhang, S., Dong, Y., Pang, Q., Lynch, I., Xie, C., Guo, Z., & Zhang, P. (2023). From marine to freshwater environment: A review of the ecotoxicological effects of microplastics. *Ecotoxicology and Environmental Safety, 251*, 114564. https://doi.org/10.1016/j.ecoenv.2023.114564
21. Kadac-Czapska, K., Ośko, J., Knez, E., & Grembecka, M. (2024). Microplastics and oxidative stress—Current problems and prospects. *Antioxidants, 13*(5), 579. https://doi.org/10.3390/antiox13050579
22. Wu, Q., Cao, J., Liu, X., Zhu, X., Huang, C., Wang, X., & Song, Y. (2024). Micro(nano)-plastics exposure induced programmed cell death and corresponding influence factors. *The Science of the Total Environment, 921*, 171230. https://doi.org/10.1016/j.scitotenv.2024.171230
23. Jayavel, S., Govindaraju, B., Michael, J. R., & Viswanathan, B. (2024). Impacts of micro and nanoplastics on human health. *Bulletin of the National Research Centre/Bulletin of the National Research Center, 48*(1), 110. https://doi.org/10.1186/s42269-024-01268-1
24. Jahedi, F., & Fard, N. J. H. (2025). Micro- and nanoplastic toxicity in humans: Exposure pathways, cellular effects, and mitigation strategies. *Toxicology Reports, 14*, 102043. https://doi.org/10.1016/j.toxrep.2025.102043
25. Barboza, L. G. A., Vethaak, A. D., Lavorante, B. R., Lundebye, A., & Guilhermino, L. (2018). Marine microplastic debris: An emerging issue for food security, food safety and human health. *Marine Pollution Bulletin, 133*, 336–348. https://doi.org/10.1016/j.marpolbul.2018.05.047
26. Muhib, M. I., Uddin, M. K., Rahman, M. M., & Malafaia, G. (2022). Occurrence of microplastics in tap and bottled water, and food packaging: A narrative review on current knowledge. *The Science of the Total Environment, 865*, 161274. https://doi.org/10.1016/j.scitotenv.2022.161274
27. Sheraz, M., Kim, J., & Kim, J. (2023). Nano/microplastics in indoor air: A critical review of synthesis routes for toxicity testing and preventative measure strategies. *Process Safety and Environmental Protection, 180*, 274–304. https://doi.org/10.1016/j.psep.2023.10.002
28. Ahn, C., & Jeung, E. (2023). Endocrine-disrupting chemicals and disease endpoints. *International Journal of Molecular Sciences, 24*(6), 5342. https://doi.org/10.3390/ijms24065342
29. Liu, S., Shi, J., Wang, J., Dai, Y., Li, H., Li, J., Liu, X., Chen, X., Wang, Z., & Zhang, P. (2021). Interactions between microplastics and heavy metals in aquatic environments: A review. *Frontiers in Microbiology, 12*, 1. https://doi.org/10.3389/fmicb.2021.652520
30. Kumar, J. A., Krithiga, T., Sathish, S., Renita, A. A., Prabu, D., Lokesh, S., Geetha, R., Namasivayam, S. K. R., & Sillanpaa, M. (2022). Persistent organic pollutants in water resources: Fate, occurrence, characterization and risk analysis. *The Science of the Total Environment, 831*, 154808. https://doi.org/10.1016/j.scitotenv.2022.154808
31. Hu, L., Zhao, Y., & Xu, H. (2022). Trojan horse in the intestine: A review on the biotoxicity of microplastics combined environmental contaminants. *Journal of Hazardous Materials, 439*, 129652. https://doi.org/10.1016/j.jhazmat.2022.129652
32. Cássio, F., Batista, D., & Pradhan, A. (2022). Plastic interactions with pollutants and consequences to aquatic ecosystems: What we know and what we do not know. *Biomolecules, 12*(6), 798. https://doi.org/10.3390/biom12060798

33. Sun, N., Shi, H., Li, X., Gao, C., & Liu, R. (2022). Combined toxicity of micro/nanoplastics loaded with environmental pollutants to organisms and cells: Role, effects, and mechanism. *Environment International, 171*, 107711. https://doi.org/10.1016/j.envint.2022.107711
34. Li, Y., Chen, L., Zhou, N., Chen, Y., Ling, Z., & Xiang, P. (2024). Microplastics in the human body: A comprehensive review of exposure, distribution, migration mechanisms, and toxicity. *The Science of the Total Environment, 946*, 174215. https://doi.org/10.1016/j.scitotenv.2024.174215
35. Astner, A. F., Gillmore, A. B., Yu, Y., Flury, M., DeBruyn, J. M., Schaeffer, S. M., & Hayes, D. G. (2023). Formation, behavior, properties and impact of micro- and nanoplastics on agricultural soil ecosystems (a review). *NanoImpact, 31*, 100474. https://doi.org/10.1016/j.impact.2023.100474
36. Li, T., Cui, L., Xu, Z., Liu, H., Cui, X., & Fantke, P. (2023). Micro- and nanoplastics in soil: Linking sources to damage on soil ecosystem services in life cycle assessment. *Science of the Total Environment, 904*, 166925. https://doi.org/10.1016/j.scitotenv.2023.166925
37. Hassan, F., Prasetya, K. D., Hanun, J. N., Bui, H. M., Rajendran, S., Kataria, N., Khoo, K. S., Wang, Y., You, S., & Jiang, J. (2023). Microplastic contamination in sewage sludge: Abundance, characteristics, and impacts on the environment and human health. *Environmental Technology & Innovation, 31*, 103176. https://doi.org/10.1016/j.eti.2023.103176
38. Noorimotlagh, Z., Hopke, P. K., & Mirzaee, S. A. (2024). A systematic review of airborne microplastics emissions as emerging contaminants in outdoor and indoor air environments. *Emerging Contaminants, 10*(4), 100372. https://doi.org/10.1016/j.emcon.2024.100372
39. Saha, S. C., & Saha, G. (2024). Effect of microplastics deposition on human lung airways: A review with computational benefits and challenges. *Heliyon, 10*(2), e24355. https://doi.org/10.1016/j.heliyon.2024.e24355
40. Kibria, M. G., Masuk, N. I., Safayet, R., Nguyen, H. Q., & Mourshed, M. (2023). Plastic waste: Challenges and opportunities to mitigate pollution and effective management. *International Journal of Environmental Research, 17*(1), 20. https://doi.org/10.1007/s41742-023-00507-z
41. Kumar, M., Chaudhary, V., Kumar, R., Chaudhary, V., & Srivastav, A. L. (2025). Microplastics, their effects on ecosystems, and general strategies for mitigation of microplastics: A review of recent developments, challenges, and future prospects. *Environmental Pollution and Management, 2*, 87. https://doi.org/10.1016/j.epm.2025.03.001
42. Mistry, M., & Joshi, S. G. (2025). Unveiling microplastic ignorance: A study on knowledge and awareness among Pune's urban population—A mixed method approach. *Nature Environment and Pollution Technology, 24*(1), B4142. https://doi.org/10.46488/nept.2025.v24i01.b4142

# Chapter 8
# Soil Contamination by Micro- and Nano Plastics: Emerging Environmental Threats

**Suganya Ilango, Gopi Devarajan, and K. Malathi**

## Introduction

Even though the amount of plastic released into soil annually is four to twenty-three times more than that discharged into the sea, research on plastic pollution in the oceans was conducted before that on soil contamination. Since it was found that plastics may accumulate and remain in the environment for hundreds of years under low-light, low-oxygen circumstances, there has been an increase in concern and interest in plastic pollution [1].

Microplastics can influence the social and ecological environment, making it essential to investigate their physical and chemical properties, sources of release, effects on ecosystems, and their accumulation in food chains—particularly within the human food web—as well as their potential health impacts. These microplastics are small plastic fragments measuring less than 5 mm in diameter. They are made of thermoplastics and thermosets and vary in color depending on where they are emitted. Due to inappropriate plastics disposal, fragmentation, and/or degradation, MNPs are now more common in the environment. These comprise secondary plastics, which are produced when bigger plastic objects decompose, and primary

---

S. Ilango (✉)
Department of Biochemistry, Faculty of Science and Humanities, SRM Institute of Science and Technology, Chengalpattu, Tamil Nadu, India
e-mail: suganyai@srmist.edu.in

G. Devarajan
Department of Economics, Faculty of Science and Humanities, SRM Institute of Science and Technology, Chengalpattu, Tamil Nadu, India
e-mail: gopid1@srmist.edu.in

K. Malathi
Department of Economics, University of Madras, Chennai, Tamil Nadu, India



S. Kulkarni, S. Das (eds.), *Microplastics and Nanoplastics*, Lecture Notes in Nanoscale Science and Technology 49,
https://doi.org/10.1007/978-3-032-17963-0_8

plastics, which are intentionally produced at the micro- and nanoscale level. Because of their propensity to build up in water bodies and harm aquatic life, the advent of these microscopic plastic particles has raised worries. As a result, its pollution and the harm it causes to ecosystems have grown to be a major worldwide concern. Even though the effects of MNPs are well established, their presence and potential risks in the environment have received a lot of attention lately. These tiny plastic particles, known as MPs (less than 5 mm) and nano plastics (less than 100 nm), are already prevalent contaminants in aquatic habitats and have the potential to jeopardize aquatic life and the overall health of the ecosystem [2].

Human health and well-being are fundamentally influenced by soils. However, the intensification of agricultural production systems brought about by population growth and rising food demand has stressed soils and increased human exposure to soil pollutants. In scientific studies, the links between soil health and human health are beginning to become more apparent. Heavy metals (HMs), one of the well-known human-induced pollutants, have long been found in agricultural soils. Others, such antibiotic-resistant bacteria (ARB) or micro(nano)plastics (MNPs), provide new and developing risks. As a result, HMs were the primary focus of early soil contamination research, but MNP and ARB research has exploded in recent years [3].

Plastic mulching, irrigation, flooding, atmospheric deposition, and the use of compost and sewage sludge as fertilizer are just a few of the many ways that microplastics can enter the soil. Top-down irrigation carries microplastics and nano plastics (MPs/NPs) into the soil and soil voids, where they may eventually end up in groundwater. During the environmental migration process, MPs and NPs will unavoidably come into contact with hazardous and dangerous metal contaminants. A number of variables are taken into account to affect microplastic transport, which can reveal the environmental risk of MPs in soil settings. These variables include the characteristics of the packing material, pore water forms, and the microplastics themselves. Absorption and migration are two crucial aspects of MPs/NPs' environmental behavior. As a carrier, microplastic or nano plastic particles adsorb pollutants and alter their movement. In the soil environment, MPs are transferred through non-living transport as well as vertical and horizontal migration [4].

The soil-plant system, which sustains the global biosphere and supplies food for people, is one of the most important environmental compartments in the terrestrial environment. MPs were expected to be more common in soils with more plastics sources on the ground, especially agricultural soils. Furthermore, MNPs might enter the human food chain after being absorbed by plants from soils. Despite disagreement, the fate of MNPs in the soil-plant system is a serious issue that is directly tied to human health and food contamination [5].

Microplastics (MPs) can be ingested and accumulated by numerous organisms, including those commonly consumed by humans (Zhang et al., 2020). Evidence for the transfer of MPs through the food chain has been obtained through data collection, controlled experiments, and simulated models demonstrating their movement across trophic levels. One pioneering study traced the movement of 24 nm polystyrene nanoparticles from moss to zooplankton and subsequently to a particular fish species. The results revealed that exposure to these particles significantly affected the fish—zooplankton consumption doubled, and notable metabolic changes were observed, including weight loss, altered triglyceride-to-cholesterol ratios, and shifts in cholesterol distribution within the liver and muscles. These findings confirm that MPs can be transferred from prey to predator within the food web.

Some MPs have long-term stable carbon components that increase the amount of organic carbon in the soil. The carbon cycle is impacted by increased carbon dioxide emissions in the soil, their impacts on plant development, soil microbial activity, and the decomposition of waste items in the soil. The addition of polyethylene MPs significantly reduces the soil's levels of potassium, total phosphorus, and total nitrogen. Furthermore, there is a correlation between the amount of soil organic matter and the quantity of MP particles of polyethylene, polypropylene, and polystyrene [5, 6].

The majority of the more than 6500 scientific publications on micro- and nanoplastics often discuss the presence of MPs in certain aquatic environments, their absorption in the marine food chain, and sediment inventories; NP is covered in significantly fewer of these studies. Determining the scope and gravity of the issue has been made easier by these investigations. The necessity of consistent procedures for sample collection, preparation, identification, and reporting units has only been brought up in a few numbers of publications [6].

Every area of the environment contains microplastics. Therefore, a crucial component of determining the degree of food contamination is comprehending the pathways via which microplastics contaminate our foods and drinks. First, human activity in terrestrial environments is mostly responsible for plastic contamination. Except for a few naturally occurring polymers such as latex (natural rubber), most widely used plastics are entirely human-made. The primary sources of plastic waste on land include open landfills, untreated sewage, wind-dispersed debris, and human-generated refuse like bottles, plastic bags, containers, road marking paints, footwear, and vehicle tires from cars, buses, and trucks [7].

The movement of MPs in the soil food chain is not well understood. There are two theories regarding the presence of MP in the feces and gravel of hens fed a diet devoid of MP. According to the first theory, MPs might have been created during digestion from larger polymers, and the second theory suggests that MPs might have originated from hens eating earthworms. MPs may be transferred across the soil food chain if the second theory is correct. Furthermore, a toxicological test that examined the presence of MPs in mice revealed that the MPs build up in organs such as the gut, liver, and kidney, suggesting that the mice internalize and store the MPs in these body tissues. These tissues are consumed by other organisms further along the food chain [7].

This chapter offers a thorough examination of microplastic pollution, emphasizing its pervasive effects on soil ecosystems, the food chain, and the marine environment. By providing a workable, reasonably priced, and environmentally friendly method of preventing microplastics from entering the food chain, it also demonstrates how the findings might be used in the real world.

## Microplastic Pollution: Pervasive Threats to Soil Ecosystem

Globally, a sizable portion of residual plastics are landfilled or dumped into the land and sea, even if some are recycled or burned for electricity. Microplastic pollution, or garbage smaller than 5 mm, is considered a growing global contaminant of concern (Table 8.1). However, throughout the last ten years, the majority of studies on the effects of microplastic pollution have concentrated on the marine environment, with very few studies examining the terrestrial environment. Microplastic pollutants are both significantly sucked up by and transported to the aquatic environment via soil. The majority of microplastics in the soil environment come from a number of sources, after which they are broken up and dispersed both horizontally and vertically across the nearby ecosystems. Additionally, there are detrimental effects on soil biota, which could have an impact on the food chain and lead to health issues for people (Table 8.2) [8].

Microplastics negatively influence the soil's biophysical properties by modifying factors such as pH, bulk density, water-holding capacity, porosity, soil aggregation, and hydraulic conductivity. Furthermore, the tendency of MPs to aggregate can affect the vertical movement of contaminants within the soil layers, while changes in soil porosity caused by MPs may influence the dispersion of volatile pollutants. As a result, the ecological processes linked to soil fertility and health gradually deteriorate. All of these changes eventually affect soil sustainability and ecology. Another potential source of microplastics in the food chain includes soil pollution, in addition to contamination of the air and water. There are multiple ways that soil

**Table 8.1** Soil contamination by micro- and nanoplastics

| Aspect | Details | References |
|---|---|---|
| Primary sources | Airborne deposition, compost/fertilizer contamination, plastic mulching films, sewage sludge application, and wastewater irrigation. | [9] |
| Secondary sources | Larger plastic debris fragmentation, tire abrasion particles, agricultural plastic deterioration, littering, and leachates from landfills. | [10] |
| Physical impacts | Changes the bulk density, aeration, water retention, and porosity of the soil; hinders root systems. | [11] |
| Chemical impacts | Adsorption and transportation of pesticides and heavy metals, as well as the leaching of chemicals (such as phthalates, BPA, and flame retardants). | [12] |
| Biological impacts | Impacts the health of earthworms, interferes with decomposers, lowers soil microbial activity, and modifies nutrient cycles. | [13] |
| Food chain entry | Ingestion by soil organisms, uptake by plants (roots, leaves), and biomagnification via both terrestrial and aquatic channels. | [14] |
| Human health risks | Indirect exposure through buildup in the food chain, soil-to-water transfer, and contaminated crops. | [15] |
| Mitigation approaches | Policy controls, microbial/enzymatic degradation, biochar additives, restricted sludge use, and biodegradable mulching. | [16] |

**Table 8.2** Summary of microplastic pollution and its threats to soil ecosystems

| Aspect of microplastic pollution | Key findings | References |
|---|---|---|
| Sources of soil microplastics | Soil contamination include air deposition, wastewater/ sludge application, and agricultural practices (such as mulching). | [17] |
| Impacts on soil physicochemical properties | Changes the pH, water retention, aggregation, bulk density, porosity, and nutrient availability of soil. | [18] |
| Effects on soil biochemical properties & microbiota | Impairs soil function by influencing dissolved organic carbon, pH, nutrient availability, enzyme activity (such as urease, catalase, and β-glucosidase), and soil respiration. | [19] |
| Influence on soil fauna (e.g., earthworms) | Microplastic fibers and nanoplastics are ingested and retained by earthworms; ingestion can change egestion, cause accumulation, and vary in dose and size. | [20] |
| Accumulation and combined effects (earthworm toxicity) | MPs can worsen the absorption of heavy metals (like cadmium), which can impact DNA integrity, reproduction, and gut flora. | [21] |
| Microplastics in plants and food chain | MPs are absorbed by plants and impede photosynthesis and growth, potentially entering the food chain and jeopardizing food security. | [22] |
| Mitigation strategies | To lessen MP input and enhance soil health, alternatives include the use of biodegradable plastics, biochar amendment, and better waste management. | [23] |

microplastic contamination happens. Landfills, soil treatment, fertilizing soil with sewage sludge, using wastewater for irrigation, using compost and organic fertilizers, mulching foil residue, tire wear, and atmospheric gradient are a few examples. The presence of microplastics in soil degrades its quality and signals that it will continue to fragment [24].

The chemical and physical properties of soil, such as pH, bulk density, hydraulic conductivity, electrical conductivity, porosity, aggregate stability, microbial community structure, and enzyme activity, can all be changed by MPs. These changes will also impact the physiological and biochemical indexes of plants as well as the structure of the plant community [25].

One important possible source of MNP growth in soil is the presence of plastic trash in the fields. Depending on the AP materials and application, there are different risks of soil contamination by plastic and the emergence of NMP. Single-use conventional AP that is intended to be mixed into the soil or used in close proximity to it poses more risks. An strategy based on the source-pathway-receptor-consequence (SPRC) model was published in (FAO, 2021) for the environmental risk assessment of soil contamination by AP. The Red-Amber-Green (RAG) rating system was used to analyze the qualitative soil pollution risk assessment of a few chosen AP products (low risk: green (1), medium risk: amber (2), and high risk: red (3)). The evaluation criteria for each product were based on the SPRC model and

**Table 8.3** Risk assessment of soil contamination by Agricultural Plastics (AP) using SPRC–RAG model

| SPRC category | Evaluation criteria | RAG rating scale | Explanation |
|---|---|---|---|
| Source | Annual durability (single-use vs. several seasons) | Red (3) = Single-use plastics; Amber (2) = Few seasons; Green (1) = Durable, reusable | Single-use plastics are more likely to become contaminated. |
| Pathway | Potential leakage to environment (3D concept: Air, soil, water movement) | Red (3) = High leakage risk; Amber (2) = Moderate; Green (1) = Low | Leakage makes it easier for MNPs to fragment. |
| Receptor | Ecosystem exposure (soil, water, biodiversity) | Red (3) = Direct soil contact/ incorporation; Amber (2) = Indirect exposure; Green (1) = Minimal exposure | The likelihood of contamination increases with direct inclusion. |
| Consequence | Potential harm (MNP generation, ecosystem impacts) | Red (3) = High MNP release and toxicity; Amber (2) = Moderate; Green (1) = Low | Reflects the toxicity and degradability of plastic. |
| Overall AP risk score | Normalized sum of RAG scores for all categories | Medium risk (Amber, 6–9), low risk (green, ≤5), and high risk (red, ≥10) | Gives the AP product's qualitative risk profile. |

included possible harm (potential for MNP generation, impacts), ecosystem (soil, water), possibility for environmental leakage (route, 3D concept), and annual durability (single use, multiple seasons). The normalized sum of the RAG rates assigned to the AP being evaluated for each SPRC category was used to calculate the overall AP score (Table 8.3) [26, 27].

## Microplastic Pollution: Pervasive Threats to the Food Chain

In every facet of human existence, plastic has shown to be an immensely practical and essential material. Numerous developments in business, technology, and medicine would not have been feasible without it. However, due to its accessibility and simplicity of use, it has been misused globally. Fundamentally, plastics may be made rather easily with a variety of chemical agents, but they are regrettably not recyclable or reusable and are instead burned, dumped in landfills, or thrown away as trash. Plastic begins to decompose into small fragments once it is in the environment. Because many species confuse plastic for food, it can enter a freshwater, marine, or terrestrial food system. These microplastics are chemically inert, but they have been shown to act as microscopic "bio-sponges" for dangerous compounds found in the environment, changing plastic particles from harmless to potentially dangerous. It was thought that microparticles would have no biological impact and would just move through both human and animal gastrointestinal tracts. Nonetheless, research has indicated that they can occasionally be absorbed and dispersed throughout the lymphatic and circulatory systems, as well as retained in the adipose tissues of many creatures. Their absorption resulted in endocrine disruption, liver dysfunction, and possible carcinogenic consequences [28].

One of the most significant pathways for MPs to be absorbed in the food chain seems to be through fish and shellfish that live in the sea and oceans. Furthermore, food safety is compromised when MPs are found in table salt and drinking water. Unfortunately, increased exposure to MP can also lead to health problems. These include, for instance, cancer, neurotoxicity, changes in microbiota, respiratory disorders, and reproductive issues. Consequently, it is important to acknowledge this risk and adopt the appropriate safety measures. In this regard, even modest actions are required. It is crucial that people become more conscious of this problem and that nations implement different policies to stop this menace. The quality of the soil and the application of fertilizers are the first steps in food production. Microparticles like MPs that may be present in soil are among the environmental factors that plants react to. Water quality and, consequently, aquatic species' living conditions are influenced by the exchange of pollutants and microparticles between sediment and water. MPs in drinking water increase the likelihood that MPs will be found in food items used on a daily basis [29].

Since microplastics harm the environment, it has been demonstrated that they are present in the food chain (Table 8.4) at the marine environment, microplastics have been discovered at lower trophic levels of copepods, slaps, ichthyoplankton, zooplankton, and chaetognath. Microplastic contamination also affects invertebrates (polychaetas, crustaceans, echinoderms, and bivalves) and vertebrates (fish, seabirds, and mammals) at higher trophic levels. Either directly or by trophic transmission, they are exposed to plastic particles [30].

The health consequences of consuming microplastics include cancer, harm to the respiratory system and organs, and issues with reproduction. The capacity of

**Table 8.4** Summary of microplastic pollution: pervasive threats to the food chain

| Aspect | Key findings | References |
|---|---|---|
| Sources of microplastics in food chain | MPs enter through crop absorption, packaged food, salt, shellfish, and drinking water. | [31] |
| Bioaccumulation in aquatic organisms | MPs are consumed by fish, shellfish, and plankton, resulting in trophic transfer. | [32] |
| Impacts on nutritional safety | MPs reduce feeding efficiency, nutrient absorption, and food quality. | [33] |
| Human exposure through seafood | MPs enter the human diet through fish, mussels, and oysters. | [34] |
| Transfer through terrestrial food products | MPs are absorbed by crops and animal feed, which allows them to enter the terrestrial food chain. | [35] |
| Potential human health risks | Toxic substances that cause oxidative stress, inflammation, and disturbance of the microbiota may be carried by MPs. | [36] |
| Mitigation and policy strategies | MP dangers are decreased by wastewater treatment, biodegradable plastics, and restrictions. | [37] |

microplastics to enter the metabolism, adsorption into the circulatory system for translocation, and difficulty, if not impossibility, of removal are the main causes of their toxicity [38].

Cancer, immunotoxicity, intestinal, pulmonary, cardiovascular, and inflammatory illnesses, as well as loss of male and female sterility, unfavorable pregnancy outcomes, and maternal exposure to offspring, are among the possible health hazards associated with microplastic in the food chain. Without conducting extensive, multidisciplinary research, it is hard to determine the specific impacts of microplastics and nano plastics on human health [39].

## Microplastic Pollution: Pervasive Threats to the Marine Environment

MPs may enter a variety of pathways, including water, sediments, and primary producers, as evidenced by the numerous studies that have shown their presence in a variety of marine species with a range of feeding techniques. Additionally, MPs are easily absorbed by animals and have the potential to discharge poisonous plastic additives into the water, they pose an ecotoxicological risk to marine biota [40].

In coastal regions, macrophytes and microalgae have significant ecological responsibilities. As primary producers and contributors to sediment formation, macroalgae offer a variety of species food and habitat in addition to shielding benthic organisms (epifauna and infauna) from predators. However, as primary producers and suppliers of necessary nutrients for the development and survival of larval and adult animals, microalgae are an important part of marine ecosystems. According to reports, MPs have the ability to stick to the surfaces of algae and put pressure on these habitats and the biota that is linked with them. For marine ecosystems, the existence and impact of microplastics on algae are therefore vital [41].

The buildup of microplastics in marine species' cells and tissues causes long-term biological damage. The ingestion of microplastics may cause chromosomal alterations that increase the risk of cancer, obesity, and infertility in humans. Considering the growing threat microplastics pose to human health and marine life, it is essential to regulate the extensive use of plastic additives and implement laws and policies aimed at reducing the sources of plastic pollution [42].

Species that are extremely specialized, have a limited range, have little potential for adaptation, and are therefore vulnerable to pollution and changes in their local habitat are referred to as sensitive species. The species that are least likely to be able to move or adapt to external environmental forces are those that are biome-specific and range-restricted. Rare and regionally limited species are frequently essential to maintaining ecosystem function and processes in different populations. Their unique functional characteristics, which frequently enhance those of other species in the community and preserve its resilience, are the cause of this. Even if there are examples of species that are sensitive to one or more features in every environment, the overall sensitivity of a community increases as the fraction of sensitive species within it increases. For every group of species, vulnerability can be considered to be a combination of exposure, sensitivity to stress, dispersal capacity, ecological resilience and resistance to stress, and adaptive aptitude and opportunity (including effective population size and distribution) [43].

Ship waste disposal and the direct release of lost, misplaced, or abandoned fishing gear into the ocean are two examples of ocean-based sources of microplastics. 600,000 plastic containers are used every day for shipping waste worldwide. Plastic material discharge volumes are higher in underdeveloped nations where proper garbage disposal is sometimes inadequate. Natural disasters including hurricanes, tsunamis, severe floods, and rain can also produce plastic waste at sea [44].

Microplastics build up in marine ecosystems and occur in widely differing concentrations. Persistent organic pollutants (POPs) that attach to these microplastics can be taken up and metabolized by fish and other aquatic organisms. Among these are PBDEs, which have been detected in the tissues of fish, marine amphipods, and *Allorchestes compressa*, and are known to cause physical harm to marine life. Additionally, microplastics with sharp edges can damage intestinal and gill tissues [45].

Coastal tourism, recreational activities, agricultural practices, untreated sewage discharge, and the disposal of marine trash from residential and municipal use near shorelines are all potential land-based sources of microplastics in the environment. As the world's population continues to rise, solid items such as plastic, glass, metals, paper, rubber, textiles, processed lumber, cigarettes, caps and lids, beverage bottles, straws and stirrers, and more are routinely discarded into the ocean due to poor or malfunctioning trash disposal. Because the majority of plastic materials, including synthetic polymers, are buoyant, their particles are more likely to float and are therefore commonly dragged or washed ashore [46].

By absorbing pollutants and aiding in their bioaccumulation, microplastics have been shown to serve as vectors, especially in marine habitats, organisms, and subsequently food webs. Microplastic has been discovered in the stomachs of fish, bigger mammals from various trophic levels, benthic invertebrates, and the deep ocean biota. Because ingested microplastic spreads throughout the food chain, worries about the dangers to aquatic biota are growing. In marine ecosystems and species, microplastic can build up and provide risks (Table 8.5) [47].

## Sustainable Methods to Prevent Microplastics from Entering the Food Chain

To lessen the detrimental effects of microplastics on the terrestrial ecology, alternative plastic mulch solutions must be used in place of biodegradable plastics. These biodegradable plastics may be made from plant-based polymers or starches. Bioremediation, which can speed up the metabolism route of soil bacteria that break

**Table 8.5** Summary of microplastic pollution: pervasive threats to the marine environment

| Aspect | Key findings | References |
|---|---|---|
| Sources of microplastics in marine systems | Inputs from textiles, tires, packaging, wastewater, fishing gear, and the decomposition of macro plastics. | [48] |
| Distribution and transport | From surface waters to deep-sea trenches, MPs are carried by winds, rivers, and ocean currents. | [49] |
| Ingestion by marine organisms | Plankton, fish, seabirds, turtles, and marine animals eat it, which reduces feeding and obstructs digestion. | [50] |
| Bioaccumulation and trophic transfer | MPs and additives build up in higher predators as they go up the food chain. | [51] |
| Physical impacts | Marine animals suffer harm, malnutrition, stunted growth, and mortality as a result of entanglement and ingestion. | [52] |
| Chemical interactions | MPs emit chemicals like BPA and phthalates and absorb heavy metals and persistent organic pollutants (POPs). | [53] |
| Microbial colonization (Plastisphere) | MPs change marine microbiomes by acting as substrates for microbial biofilms, which include invasive species and diseases. | [54] |
| Impacts on coral reefs and benthic ecosystems | MPs disrupt sediment ecosystems, lower coral health, and hinder symbiotic photosynthesis. | [55] |
| Nano plastics | MPs are less able to infiltrate cells, traverse biological barriers, and cause oxidative stress than nano plastics (<1000 nm). | [56] |
| Climate change links | MPs may have an impact on the exchange of $CO_2$ between the ocean and atmosphere, impact carbon cycle, and decrease plankton photosynthesis. | [57] |
| Global and human implications | With negative socioeconomic and health effects, MPs endanger fisheries, aquaculture, biodiversity, and seafood safety. | [58] |
| Mitigation strategies | Global policies (e.g., MARPOL), biodegradable materials, plastic bans, enhanced wastewater treatment, and prevention at the source. | [59] |

down plastic, is another promising tactic. Additionally, the release of MPs into soil and water environments can be reduced by upgrading wastewater treatment facilities with reliable technologies like coagulation-flocculation or membrane filtration. As a result, waste management systems that incorporate effective waste release sorting and recycling must be strengthened. This is an economical way to prevent the direct dumping of plastic garbage at the source, hence lowering environmental contamination. In summary, effective wastewater treatment and its management with certain biological approaches, such as bioremediation, bioaugmentation, and biodegradation by bacteria that devour plastic, may be necessary to reduce MP pollution in soil [60].

The type of matrix determines the MP monitoring techniques. Monitoring items in the food chain, which may include feed, is essentially different from evaluating bioaccumulation in the tissues of, say, animals in in vivo research. This part assesses food and feed product monitoring, which may involve destroying the sample matrix to release MPs. The process basically consists of a number of steps, starting with the sampling of a specific quantity and kind of material and ending with the final result in terms of counts or concentrations, which are specified for the types of plastic included. Physical particle isolation and sample treatment are the first steps, which are followed by a detection method and a final identification process [60].

The four Rs are the foundation of waste management: reduce, reuse, recycle, and recover. The best strategy to prevent the production of microplastics is to reduce plastic waste. Reducing the manufacture of plastic materials, substituting them with materials like cardboard, glass, or other recyclable or biodegradable materials, and recycling used plastics by turning them into new products are some ways to do this [61].

The fragmentation of bigger plastic waste, textile synthetic fibers, and microbeads used in personal care products are some of the sources of microplastics. The development of comprehensive and uniform preventive measures is challenging since each source necessitates different prevention and control tactics. Designing efficient management measures is further complicated by the need to comprehend how microplastics travel and accumulate in various environmental compartments, such as soil, water, and air. The lack of defined methods for measuring and sampling microplastics presents another difficulty. Standardized procedures must be established for microplastics sampling, characterization, and analysis in order to guarantee reliable and consistent data for use in decision-making. Adopting environmentally friendly practices and preventative measures is often hampered by socioeconomic reasons. Making the shift to more sustainable options and methods frequently calls for large financial outlays, adjustments in consumer behavior, and collaboration between communities, governments, and businesses. To achieve broad acceptance of microplastic management methods, it is essential to overcome resistance to change, guarantee the availability and affordability of alternatives, and develop incentives for sustainable activities. It is feasible to create sustainable and efficient methods to reduce the accumulation and discharge of microplastics, safeguarding ecosystems and enhancing human welfare, by tackling these issues (Table 8.6) [62].

In an attempt to become more sustainable, governments and businesses have embraced the Sustainable Development Goals (SDGs) since the United Nations (UN) announced them in 2015. The 17 Sustainable Development Goals (SDGs) have 169 targets, and they can be measured using 247 distinct indicators. Despite the fact that (micro)plastic contamination is widespread throughout the world, Goal 14 only has one indicator (14.1.1b) that is specifically connected to lessening the effects of (micro)plastics. The pervasiveness of (micro)plastic pollution may exacerbate the challenges governments and organizations have in implementing the 247 SDG indicators if they are not sufficiently monitored [63].

A sustainable, economical, and successful approach to preventing microplastics (MPs) from entering the food chain is to combine technological, policy-driven, and natural solutions that are practical for daily use. Using biodegradable mulch instead of conventional plastic mulch, binding MPs in soil with biochar additives, and filtering wastewater with artificial wetlands are all economical and environmentally

**Table 8.6** Sustainable methods to prevent microplastics from entering the food chain

| Method | Approach | Sustainability & economy | Effectiveness | References |
|---|---|---|---|---|
| Biodegradable alternatives | Replace single-use plastics with biodegradable or compostable materials (PLA, PHA, starch-based polymers). | Sustainable and cuts down on long-term waste; while it may cost more up front, it will save money over time. | Stops the creation of fresh MP at its source. | [64] |
| Biochar soil amendment | Biochar binds MPs in soils, improving fertility and reducing plant uptake. | Inexpensive, environmentally friendly (produced from agri-waste), and improves soil health. | Decreases the amount of MPs that crops can absorb. | [65] |
| Constructed wetlands & phytoremediation | Wetland plants and microbes trap/filter MPs from wastewater before reaching rivers/seas. | Low maintenance, nature-based, and cost-effective solution. | Effective in lowering wastewater discharge's MPs load. | [66] |
| Advanced wastewater treatment | Membrane bioreactors, sand filtration, and electrocoagulation to capture MPs. | Initially more expensive, but keeps MPs out of food and aquatic systems. | Eliminates 95–99% of MPs from wastewater. | [67] |
| Improved agricultural practices | Reduce use of plastic mulch, adopt biodegradable mulching films, and proper sludge management. | Economical on a farm size; lessens soil pollution. | Stops MPs from going straight into farms. | [68] |
| Microbial biodegradation | Use of plastic-degrading microbes/enzymes (e.g., *Ideonella sakaiensis* PETase). | Environmentally benign, yet still in its infancy; has industrial use potential. | Eliminates MPs before they build up in food chains. | [69] |
| Policy & circular economy | Bans on microbeads, promoting recycling, extended producer responsibility (EPR). | Affordable and sustainable on a societal level; lessens the leakage of plastic. | Sustained progress in reducing the production of MPs. | [31] |

friendly methods. These methods reduce the number of MPs that enter aquatic and agricultural crops while also enhancing soil fertility, water quality, and ecosystem resilience. Furthermore, when supported by strong regulatory frameworks and community involvement, microbial biodegradation and improved waste management practices ensure long-term success. The effectiveness of these techniques has been demonstrated by experimental studies and field applications, which also show that they are applicable in real-world scenarios and offer scalable solutions to safeguard food security and human health.

## Future Study

Future studies on microplastics and nano plastics in agroecosystems should concentrate on a few key topics. It is crucial to first create uniform methods for the detection, measurement, and characterization of MPs and NPs in soil, water, and food matrices in order to ensure the comparability of results. Second, because the majority of existing research is short-term and laboratory-based, long-term field experiments are necessary to better understand the long-term effects of MPs on soil fertility, plant development, microbial populations, and nutrient cycling. Third, the cellular and molecular mechanisms of MP/NP toxicity, particularly their effects on toxicant uptake and interactions with microbial enzymes, root systems, and digestive tracts, should be the main focus of future research. Fourth, sustainable options such as phytoremediation, biochar treatments, biodegradable polymers, and microbial biodegradation technologies need to be investigated in real agricultural environments to verify their cost and feasibility. Additionally, there is opportunity to research regenerative agriculture practices and their potential to reduce plastic pollution and increase food security. Last but not least, multidisciplinary research combining ecotoxicology, agronomy, biotechnology, and policy frameworks is essential to creating complete strategies that safeguard ecosystem and human health while lowering the entry of microplastics into the food chain.

## Conclusion

Microplastic and nano plastic contamination pose a major threat to soil health, agroecosystems, marine environments, and ultimately the food chain. Because of their tiny size, ubiquity, chemical complexity, and persistence, they can disrupt nutrient cycle, plant development, soil microbial activity, and the physiology of both terrestrial and aquatic organisms. Food security is compromised and food yield is reduced as a result. The figures show that these pollutants increase the hazards to the environment and human health by acting as transporters of dangerous substances and physical stressors.

Sustainable techniques including regenerative agriculture, the use of biodegradable and bio-based plastics, the application of biochar, clean biotechnologies, and improved water and waste management, however, offer workable, affordable, and scalable solutions to lessen these risks. Reducing the quantity of microplastics that enter the food chain, restoring ecological resilience, and protecting human health are all achievable by incorporating such strategies into food production systems.

This analysis highlights how vital it is to deploy sustainable solutions and validates their applicability in real-world situations.

## References


1. Pathan, S. I., Arfaioli, P., Bardelli, T., Ceccherini, M. T., Nannipieri, P., & Pietramellara, G. (2020). Soil pollution from micro- and nanoplastic debris: A hidden and unknown biohazard. *Sustainability, 12*, 7255. https://doi.org/10.3390/su12187255
2. Ali, M., Xu, C., & Li, K. (2025). Micro- and nanoplastics as emerging threats to both terrestrial and aquatic animals: A comprehensive review. *Vet. Sci., 12*, 688. https://doi.org/10.3390/vetsci12080688
3. Perković, S., Paul, C., Vasić, F., & Helming, K. (2022). Human health and soil health risks from heavy metals, micro(nano)plastics, and antibiotic resistant bacteria in agricultural soils. *Agronomy, 12*, 2945. https://doi.org/10.3390/agronomy12122945
4. Reddy, K. V., Ranjit, P., Alvarado, J. I. H., Garcia, J. H. F., & Maddela, N. R. (2023). Chapter 1. In N. R. Maddela, K. V. Reddy, & P. Ranjit (Eds.), *Micro and nanoplastics in soil*. Springer. https://doi.org/10.1007/978-3-031-21195-9_1
5. He, D., Zhang, Y., & Gao, W. (2021). *Current Opinion in Food Science, 41*, 116. https://doi.org/10.1016/j.cofs.2021.04.001
6. Uddin, S., Fowler, S. W., Habibi, N., & Behbehani, M. (2022). Micro-nano plastic in the aquatic environment: Methodological problems and challenges. *Animals, 12*, 297. https://doi.org/10.3390/ani12030297
7. Waring, R. H., Harris, R. M., & Mitchell, S. C. (2018). Plastic Contamination of the food chain: A threat to human health? *Maturitas, 115*, 64. https://doi.org/10.1016/j.maturitas.2018.06.010
8. Kim, Y. N., Yoon, J. H., & Kim, K. H. J. (2021). Microplastic contamination in soil environment—A review. *Soil Science Annual, 71*, 300. https://doi.org/10.37501/soilsa/131646
9. Mai, L., Bao, L. J., Wong, C. S., & Zeng, E. Y. (2024). *Microplastic contamination in aquatic environments* (pp. 229–247). Elsevier. https://doi.org/10.1016/B978-0-443-15332-7.00012-0
10. Salehi, Z., Hashemi, S. H., & Flury, M. (2023). Micro- and mesoplastics in farmlands with different irrigation water sources. *Water, Air, and Soil Pollution, 234*, 267. https://doi.org/10.1007/s11270-023-06289-6
11. Thiemann, T. (2024). *microplastics in environment* (pp. 113–129). CRC Press. https://doi.org/10.1201/9781003468141-6
12. Preininger, C., Hackl, E., & Stagl, V. (2024). NETmicroplastic in agricultural soil and its impact on soil properties. *European Journal of Soil Science, 75*, e13496. https://doi.org/10.1111/ejss.13496
13. Besseling, E., Redondo-Hasselerharm, P., Foekema, E. M., & Koelmans, A. A. (2019). Quantifying ecological risks of aquatic micro- and nanoplastic. *Critical Reviews in Environmental Science and Technology, 49*, 32. https://doi.org/10.1080/10643389.2018.1531688
14. Banerjee, A., & Shelver, W. L. (2021). Micro- and nanoplastic induced cellular toxicity in mammals: A review. *Science of The Total Environment, 755*, 142518. https://doi.org/10.1016/j.scitotenv.2020.142518
15. Ng, E. L., Lwanga, E. H., Eldridge, S. M., Johnston, P., Hu, H. W., Geissen, V., & Chen, D. (2018). An overview of microplastic and nanoplastic pollution in agroecosystems. *Science of The Total Environment, 627*, 1377. https://doi.org/10.1016/j.scitotenv.2018.01.341
16. Blair, R. M., Waldron, S., Phoenix, V., & Gauchotte-Lindsay, C. (2017). Micro- and nanoplastic pollution of freshwater and wastewater treatment systems. *Springer Science Reviews, 5*, 19. https://doi.org/10.1007/s40362-017-0044-7

17. Yu, H., Zhang, Y., Tan, W., & Zhang, Z. (2022). Microplastics as an emerging environmental pollutant in agricultural soils: Effects on ecosystems and human health. *Frontiers in Environmental Science, 10*, 855292.
18. Das, E. J., & Alam, A. K. M. R. (2024). Effects of microplastics polluted soil on the growth of *Solanum lycopersicum* L. *Environmental Systems Research, 13*, 36. https://doi.org/10.1186/s40068-024-00367-2
19. Cheng, Y., Wang, F., Huang, W., & Liu, Y. (2024). Response of soil biochemical properties and ecosystem function to microplastics pollution. *Scientific Reports, 14*, 28328. https://doi.org/10.1038/s41598-024-80124-8
20. Lahive, E., Cross, R., Saarloos, A. I., Horton, A. A., Svendsen, C., Hufenus, R., & Mitrano, D. M. (2022). Earthworms ingest microplastic fibres and nanoplastics with effects on egestion rate and long-term retention. *Science of the Total Environment, 807*, 151022. https://doi.org/10.1016/j.scitotenv.2021.151022
21. Cui, W., Gao, P., Zhang, M., Wang, L., Sun, H., & Liu, C. (2022). Adverse effects of microplastics on earthworms: A critical review. *Science of The Total Environment, 850*, 158041. https://doi.org/10.1016/j.scitotenv.2022.158041
22. Sun, H., Shi, Y., Zhao, P., Long, G., Li, C., Wang, J., et al. (2023). Effects of polyethylene and biodegradable microplastics on photosynthesis, antioxidant defense systems, and arsenic accumulation in maize (*Zea mays* L.) seedlings grown in arsenic-contaminated soils. *Science of the Total Environment, 868*, 161557. https://doi.org/10.1016/j.scitotenv.2023.161557
23. Adekiya, A. O., Olayanju, T. M. A., Ejue, S. W., Alori, E. T., & Adegbite, K. A. (2020). *Soil health* (pp. 99–113). Springer. https://doi.org/10.1007/978-3-030-44364-1_6
24. Rai, M., Pant, G., Pant, K., Aloo, B. N., Kumar, G., Singh, H. B., & Tripathi, V. (2023). Microplastic pollution in terrestrial ecosystems and its interaction with other soil pollutants: A potential threat to soil ecosystem sustainability. *Resources, 12*, 67. https://doi.org/10.3390/resources12060067
25. Walker, T. R. (2021). (Micro)plastics and the UN Sustainable Development Goals. *Current Opinion in Green and Sustainable Chemistry, 30*, 100497. https://doi.org/10.1016/j.cogsc.2021.100497
26. Briassoulis, D. (2023). Agricultural plastics as a potential threat to food security, health, and environment through soil pollution by microplastics: Problem definition. *Science of The Total Environment, 892*, 164533. https://doi.org/10.1016/j.scitotenv.2023.164533
27. He, D., Luo, Y., Lu, S., Liu, M., Song, Y., & Lei, L. (2018). Microplastics in soils: Analytical methods, pollution characteristics and ecological risks. *TrAC—Trends in Analytical Chemistry, 109*, 163. https://doi.org/10.1016/j.trac.2018.10.006
28. Lehel, J., & Murphy, S. (2021). In P. de Voogt (Ed.), *Reviews of environmental contamination and toxicology* (Vol. 259, pp. 1–29). Springer. https://doi.org/10.1007/398_2021_77
29. Çıtar Dazıroğlu, M. E., & Bilici, S. (2024). The hidden threat to food safety and human health: microplastics. *Environment, Development and Sustainability, 26*, 21913. https://doi.org/10.1007/s10668-023-03565-7
30. Cverenkárová, K., Valachovičová, M., Mackuľak, T., Žemlička, L., & Bírošová, L. (2021). Microplastics in the food chain. *Life, 11*, 1349. https://doi.org/10.3390/life11121349
31. Cox, K. D., Covernton, G. A., Davies, H. L., Dower, J. F., Juanes, F., & Dudas, S. E. (2019). Human consumption of microplastics. *Environmental Science & Technology, 53*, 7068. https://doi.org/10.1021/acs.est.9b01517
32. Bailly, E., Levi, Y., & Karolak, S. (2013). Calibration and field evaluation of polar organic chemical integrative sampler (POCIS) for monitoring pharmaceuticals in hospital wastewater. *Environmental Pollution, 174*, 100. https://doi.org/10.1016/j.envpol.2012.10.025
33. Barboza, L. G. A., Vethaak, A. D., Lavorante, B. R., Lundebye, A. K., & Guilhermino, L. (2018). Marine microplastic debris: An emerging issue for food security, food safety and human health. *Marine Pollution Bulletin, 133*, 336. https://doi.org/10.1016/j.marpolbul.2018.05.047
34. Van Cauwenberghe, L., & Janssen, C. R. (2014). Microplastics in bivalves cultured for human consumption. *Environmental Pollution, 193*, 65. https://doi.org/10.1016/j.envpol.2014.06.010

35. Conti, G. O., Ferrante, M., Banni, M., Favara, C., Nicolosi, I., Cristaldi, A., et al. (2020). Micro- and nano-plastics in edible fruit and vegetables. The first diet risks assessment for the general population. *Environmental Research, 187*, 109677. https://doi.org/10.1016/j.envres.2020.109677
36. Filiciotto, L., & Rothenberg, G. (2021). Biodegradable plastics: Standards, policies, and impacts. *ChemSusChem, 14*, 56. https://doi.org/10.1002/cssc.202002044
37. Mazhandu, Z. S., Muzenda, E., Mamvura, T. A., Belaid, M., & Nhubu, T. (2020). Integrated and consolidated review of plastic waste management and bio-based biodegradable plastics: Challenges and opportunities. *Sustainability, 12*, 8360. https://doi.org/10.3390/su12208360
38. Eze, C. G., Nwankwo, C. E., Dey, S., et al. (2024). Food chain microplastics contamination and impact on human health: A review. *Environmental Chemistry Letters, 22*, 1889. https://doi.org/10.1007/s10311-024-01734-2
39. Chalannavar, R. K., Malabadi, R. B., Divakar, M. S., Swathi, S., Karamchand, K. S., Kamble, A. A., & Munhoz, A. N. R. (2025). Microplastic in food chain-major health issues-An update. *World Journal of Advanced Research and Reviews, 26*, 4067. https://doi.org/10.30574/wjarr.2025.26.1.1531
40. Salomone, V. N., Passucci, V., & Areco, M. M. (2023). Microplastic pollution in marine environments: Exploring sources, sinks, and consequences with a focus on algal interactions. *Regional studies in marine science, 68*, 103270. https://doi.org/10.1016/j.rsma.2023.103270
41. Sharma, S., & Chatterjee, S. (2017). Microplastic pollution, a threat to marine ecosystem and human health: A short review. *Environmental Science and Pollution Research, 24*, 21530. https://doi.org/10.1007/s11356-017-9910-8
42. Tanaka, K., Takada, H., Yamashita, R., Mizukawa, K., Fukuwaka, M. A., & Watanuki, Y. (2013). Accumulation of plastic-derived chemicals in tissues of seabirds ingesting marine plastics. *Marine Pollution Bulletin, 69*, 219. https://doi.org/10.1016/j.marpolbul.2012.12.010
43. Horton, A. A., & Barnes, D. K. (2020). Microplastic pollution in a rapidly changing world: Implications for remote and vulnerable marine ecosystems. *Science of the Total Environment, 738*, 140349. https://doi.org/10.1016/j.scitotenv.2020.140349
44. NOAA Marine Debris Program. (2021). *Detecting Japan Tsunami Marine Debris at sea: A synthesis of efforts and lessons learned.* Available online: https://marinedebris.noaa.gov/sites/default/files/JTMD_Detection_Report.pdf.
45. Onyena, A. P., Aniche, D. C., Ogbolu, B. O., Rakib, M. R. J., Uddin, J., & Walker, T. R. (2022). Governance strategies for mitigating microplastic pollution in the marine environment: A review. *Microplastics, 1*, 15. https://doi.org/10.3390/microplastics1010003
46. Rillig, M. C., Ingraffia, R., & De Souza Machado, A. A. (2017). Microplastic Incorporation into Soil in Agroecosystems. *Frontiers in Plant Science, 8*, 1805. https://doi.org/10.3389/fpls.2017.01805
47. Amelia, T. S. M., Khalik, W. M. A. W. M., Ong, M. C., et al. (2021). Marine microplastics as vectors of major ocean pollutants and its hazards to the marine ecosystem and humans. *Progress in Earth and Planetary Science, 8*, 12. https://doi.org/10.1186/s40645-020-00405-4
48. Browne, M. A., Crump, P., Niven, S. J., Teuten, E., Tonkin, A., Galloway, T., & Thompson, R. (2011). Accumulation of microplastic on shorelines worldwide: Sources and sinks. *Environmental Science & Technology, 45*, 9175. https://doi.org/10.1021/es201811s
49. Kane, I. A., & Clare, M. A. (2019). Dispersion, accumulation, and the ultimate fate of microplastics in deep-marine environments: A review and future directions. *Frontiers in Earth Science, 7*, 80.
50. Cole, M., Lindeque, P., Fileman, E., Halsband, C., Goodhead, R., Moger, J., & Galloway, T. S. (2013). Microplastic ingestion by zooplankton. *Environmental Science & Technology, 47*, 6646. https://doi.org/10.1021/es400663f
51. Carbery, M., O'Connor, W., & Palanisami, T. (2018). Trophic transfer of microplastics and mixed contaminants in the marine food web and implications for human health. *Environment International, 115*, 400. https://doi.org/10.1016/j.envint.2018.03.007

52. Laist, D. W. (1997). In J. M. Coe & D. B. Rogers (Eds.), *Marine debris: Sources, impacts, and solutions* (p. 99). Springer. https://doi.org/10.1007/978-1-4613-8486-1_10
53. Kinigopoulou, V., Pashalidis, I., Kalderis, D., & Anastopoulos, I. (2022). *Journal of Molecular Liquids, 350*, 118580. https://doi.org/10.1016/j.molliq.2022.118580
54. Debroy, A., George, N., & Mukherjee, G. (2022). *Journal of Chemical Technology and Biotechnology, 97*, 3271. https://doi.org/10.1002/jctb.6978
55. Liang, J., et al. (2025). Polystyrene microplastics exhibit toxic effects on the widespread coral symbiotic Cladocopium goreaui. *Environmental Research, 268*, 120750. https://doi.org/10.1016/j.envres.2025.120750
56. Liang, B., et al. (2021). Underestimated health risks: polystyrene micro- and nanoplastics jointly induce intestinal barrier dysfunction by ROS-mediated epithelial cell apoptosis. *Particle and Fibre Toxicology, 18*, 20. https://doi.org/10.1186/s12989-021-00414-1
57. Kvale, K. (2022). Implications of plastic pollution on global marine carbon cycling and climate. *Emerging Topics in Life Sciences, 6*, 359. https://doi.org/10.1042/ETLS20220013
58. Fred-Ahmadu, O. H., et al. (2024). Investigation of meso- and microplastics in commercially sold dried pink shrimp in Ekiti State, South West Nigeria. *Environmental Sciences Europe, 36*, 181. https://doi.org/10.1186/s12302-024-01024-2
59. Chang, Y. C., & Saqib, M. (2025). International legal systems in tackling the marine plastic pollution: A critical analysis of UNCLOS and MARPOL. *Water (Basel), 17*, 1547. https://doi.org/10.3390/w17101547
60. Bhattacharjee, U., Baruah, K. N., & Shah, M. P. (2025). Exploring sustainable strategies for mitigating microplastic contamination in soil, water, and the food chain: A comprehensive analysis. *Environmental Chemistry and Ecotoxicology*.
61. Calero, M., Godoy, V., Quesada, L., & Martín-Lara, M. Á. (2021). Green strategies for microplastics reduction. *Current Opinion in Green and Sustainable Chemistry, 28*, 100442. https://doi.org/10.1016/j.cogsc.2020.100442
62. Alam, M. J., & Rahman, M. M. (2025). *Microplastics* (p. 441). Elsevier.
63. Iacovidou, E., Hahladakis, J., Deans, I., Velis, C., & Purnell, P. (2018). Technical properties of biomass and solid recovered fuel (SRF) co-fired with coal: Impact on multidimensional resource recovery value. *Waste Management, 73*, 535. https://doi.org/10.1016/j.wasman.2017.07.001
64. Elbasiouny, H., et al. (2023). *Agronomy, 13*, 149. https://doi.org/10.3390/agronomy13010149
65. Sharma, N., Singh, A., & Batra, N. (2018). *Ecological wisdom inspired restoration engineering* (p. 223). https://doi.org/10.1007/978-981-13-0149-0_13
66. Bayo, J., López-Castellanos, J., & Olmos, S. (2020). Membrane bioreactor and rapid sand filtration for the removal of microplastics in an urban wastewater treatment plant. *Marine Pollution Bulletin, 156*, 111211. https://doi.org/10.1016/j.marpolbul.2020.111211
67. Kasirajan, S., & Ngouajio, M. (2012). Polyethylene and biodegradable mulches for agricultural applications: A review. *Agronomy for Sustainable Development, 32*, 501. https://doi.org/10.1007/s13593-011-0068-3
68. Dhali, S. L., Parida, D., Kumar, B., & Bala, K. (2024). Recent trends in microbial and enzymatic plastic degradation: A solution for plastic pollution predicaments. *Biotechnology for Sustainable Materials, 1*, 11. https://doi.org/10.1186/s44316-024-00011-0
69. Jokiaho, J., Rezaie, S., & Vanhuyse, F. (2025). *Marine plastic pollution and extended producer responsibility in the G20*. SEI report. Stockholm Environment Institute. https://doi.org/10.51414/sei2025.036

# Chapter 9
# Microplastics in Soil: Threats to Plant-Based Food

**S. Surendarnath, G. Ramesh, and G. Flora**

## Introduction

MP particles have been detected in nearly all terrestrial and marine environments, including Antarctica and mountains. However, according to a number of recent research, agricultural farms are among the places in terrestrial systems with the highest levels of MP contamination.

As the world becomes more dependent on plant-based foods because of their essential nutritional value and recommendations from public health organizations like the WHO, the contamination of these foods by MPs has become a growing concern. By getting into the structures of plants and animals, microplastics can both bioaccumulate and go up the food chain. Micro and nanoparticle accumulation in fruits, vegetables, and crops poses a serious risk to public health and food safety. In recent years, the deposition of microplastics and plastic additives has made agricultural soil vulnerable. The main way that MPs enter soil ecosystems is through crop-targeting agricultural techniques, which cause MPs to accumulate on agricultural soil through air deposition. A little amount of sewage sludge from residential and commercial wastewater collected in wastewater treatment plants, human activity, composting, the movement of agricultural plastic mulch, and other processes can all cause MPs to enter soil.

S. Surendarnath
Department of Mechanical Engineering, DVR & Dr. HS MIC College of Technology (Autonomous), Vijayawada, India

G. Ramesh
Department of Metallurgical and Materials Engineering, Rajiv Gandhi University of Knowledge Technologies, R K Valley Campus, Kadapa, India

G. Flora (✉)
Department of Botany, St. Mary's College (Autonomous), Thoothukudi, India


S. Kulkarni, S. Das (eds.), *Microplastics and Nanoplastics*, Lecture Notes in Nanoscale Science and Technology 49,
https://doi.org/10.1007/978-3-032-17963-0_9

Microplastics can enter plant cells, seeds, roots, stems, leaves, and fruits, depending on their size and kind. Very small MPs have been said to be absorbed by plants under controlled settings. Due to the fact that microscopic MPs can penetrate plant cell walls and membrane barriers or block cell pores. Since these microplastics have collected in these plant components and are consumed by humans, they are promptly absorbed [1]. It is very challenging to comprehend the intricate relationships between the MPs on soil physical, chemical, and biological systems and crop yield, health, and rhizosphere health; however, a lot of research has recently been conducted to do so, and data is being generated to help clarify this to ensure food security. Microparticles smaller than 5 mm can enter soil ecosystems in a variety of ways, such as through contaminated biosolid fertilizers, plastic mulching, water treatment plant slurry used as compost, irrigated water contaminated with microplastics, and macroplastics, which are already degraded in agricultural soil. When microplastics get into the soil, they change its physical, chemical characteristics, moisture retention, and aeration. Microscopic organisms and animals that live there also consume the microplastics as their food, which hinders their ability to thrive. The plants' ability to grow depends on these soil characteristics. The latest research also suggests that microplastics in the soil were taken up by the plants through their roots and entered the plant body, potentially lowering agricultural production by interfering with respiration, photosynthesis, and other metabolic processes [2]. Given the alarming accumulation of microplastics in agricultural soils, it is imperative to investigate the intricate effects of these particles on plant-soil interactions. By affecting crucial factors like bulk density, water retention, and soil aggregation, these interactions directly affect root development and nutrient uptake [3] [4]. Furthermore, plasticizers such as pthalates and non-pthalates, fame retardants in microplastics, and other plastics permeate the soil matrix and negatively impact the soil biome's growth and development, which is essential for the cycling of nutrients and the breakdown of organic matter that promotes plant growth [5]. MPs act as carriers of harmful elements, absorbing organic contaminants and heavy metals that could worsen soil toxicity and damage plant health. Their presence alters soil aggregation, clogs pores, and decreases water infiltration—all of which are critical for root development and soil fertility [6]. Despite growing awareness, little is understood about the ways in which microplastics impact the morphological, physiological, and behavioral levels of species that live in soil, as well as microbial biodiversity, the soil biogeochemical cycle, enzymatic processes, root structure, and oxidative damage. Both soil fertility and agricultural productivity are indirectly impacted by these. Keeping this in mind, there are currently studies and limited commercial availability of biodegradable plastic as an alternative to plastic packaging materials. However, there is a lack of thorough research on the byproducts they generate during degradation. Therefore, research is concentrating on creating standardized detection and evaluation techniques, addressing current information gaps, and creating informed agricultural policies in order to assure sustainable food production in the face of growing microplastic pollution. The impact of soil microplastics on plant-based food systems is examined in this chapter, along with the types, traits, origins, uptake, mobility, and impacts on microbial activity, soil structure, and water dynamics. Potential health risks, detection techniques, mitigation strategies, and research gaps are also covered.

## Origins and Routes of Microplastics into Soil

Due to the overuse of single-use plastics and other plastic products, microplastics are becoming a more significant environmental problem. There are several ways that Mps can enter agricultural soil systems, but they are mostly brought in by agricultural practices like composting, sewage irrigation, sewage sludge application, and the breakdown of plastic mulch films. Microplastics are created during the physicochemical degradation process of these plastic products, and occasionally they leak out of personal care products as well [7]. Tire abrasion and airborne deposition are two other sources [8]. In North America and Europe, sewage sludge alone adds tens of thousands of tons of MPs to farmlands annually [9]. MP concentrations are noticeably greater in soils watered with wastewater than in soils irrigated with clean water [10]. MPs, which are made up of synthetic polymers that are insoluble in water, can break up into even smaller particles once they are in the soil. They can then survive and interact with the soil environment [11]. These several input pathways demonstrate how MPs are increasingly contaminating terrestrial ecosystems.

In nations like China, where plastic mulching is widely employed to increase crop productivity and water use efficiency (WUE), it is a significant source of microplastics in agricultural soils. Especially in arid areas like Northwest China, plastic films—such black mulching films are used to improve moisture retention, alter soil conditions, and encourage crop development. Despite the fact that these plastic films greatly increase crop yields and WUE, plastic residues build up in the soil as a result of their slow recovery and gradual breakdown. Soil pollution is a result of residual plastic sheets dissolving into microplastic particles [12]. Despite the fact that plastic mulching films are frequently used in agriculture to boost crop yields, they are a major source of soil microplastics due to their low recovery rates. China alone only collects more than 60% of these films once they are used [13]. The UN Food and Agriculture Organization estimates that around 30 million hectares of cropland are covered by two million tons of plastic mulch spread year worldwide. Regretfully, only around 30% of spent plastic mulch is recovered, which leads to significant amounts of leftover plastic building up in soils at rates of 50–100 kg per hectare per year [14]. These films break up and build up in soil over time; research indicates that mulched soils have noticeably greater levels of microplastics than unmulched soils. Depending on the area and farming method, longer-term application, such as more than 24 years leads to larger particle concentrations per gram of soil [15]. Most plastic mulches are made of durable polymers like polyethylene (PE) and polyvinyl chloride (PVC) that don't decompose naturally. White pollution, which is caused by wastes decomposing into microplastics, can alter the composition of the soil, interfere with the movement of nutrients and moisture, and interfere with crop root formation, lowering maize and wheat yields by 9–23% [16]. Plastic debris from mulching can endanger aquatic ecosystems and animal health by getting into water bodies. Experimental research indicates that microplastics like polyethylene can alter the chemistry, pH, and water retention of soil [17]. The health of plants and soil may be further impacted by these changes, which may make toxic metals like cadmium and chromium more mobile and bioavailable [18].

Particularly when surface waters and treated wastewater are utilized for irrigation, microplastics play a major role in soil contamination. Even though wastewater treatment facilities remove 88% of microplastics, rainwater runoff and large rivers still contain high concentrations of microplastics [19]. Microplastics in soils are primarily caused by untreated or recycled wastewater, especially from irrigation and desalination, according to research from dry farming areas in the Canary Islands, Spain. Regenerated wastewater (RWW) has higher levels of these microfibers than desalinated water, and they are mostly polyester and cellulosic [20]. The main source of microplastic introduction into terrestrial ecosystems is agricultural irrigation, as evidenced by the higher levels of microplastic pollution in the upper layers of soil irrigated with RWW.

Additional soil contamination occurs in fields that receive irrigation from MP-contaminated water sources, such as lakes, rivers, or groundwater. Flooding makes the issue worse by bringing vast amounts of plastic debris into agricultural areas from landfills and water bodies, raising the possibility of microplastic contamination in soils and crops [21]. Microplastics (MPs) build up in sewage sludge as a result of sewage treatment plants' efficient removal of MPs from wastewater. Because this sludge is commonly utilized as fertilizer in agricultural fields, it indirectly introduces micro- and nanoplastics into the soil [22]. Likewise, it has been demonstrated that the application of compost on agricultural soil aids in the assimilation of microplastics. The topsoil (0–10 cm) in soils from different Beijing agricultural facilities had the greatest MP concentrations, which progressively decreased with depth. This suggests that surface-level inputs like fertilizer and irrigation are the main sources. The majority of MPs were white, fiber-shaped, and primarily made of polypropylene and polyethylene, which was indicative of their use in synthetic fertilizers, irrigation tubing, and plastic films [23]. MPs in these soils are generally small and fragile, according to the conditional fragmentation hypothesis, which suggests that bigger plastic fragments can generate secondary microplastics. The main vectors bringing MPs into facility soils were found to be irrigation with treated or recycled wastewater and organic fertilizers, which are frequently made from sewage sludge or compost that contains plastic residues.

The sources and pathways of microplastic contamination in agricultural soil are shown in Fig. 9.1. MP particles have been found in commercial organic fertilizers in a number of investigations, underscoring their unintentional contribution to the introduction of plastic pollutants into the soil ecosystem. For instance, a study conducted in Bangladesh measured the amount of MP contamination in organic fertilizers and found that the average contribution was from 971.31 to 1387.37 particles kg/m$^3$, with contributions ranging from 275.16 to 3144.27 items/m$^3$. These MPs can bind with heavy metals and other harmful substances after being applied to the soil, increasing their ecological danger. According to data from Beijing, China's facility

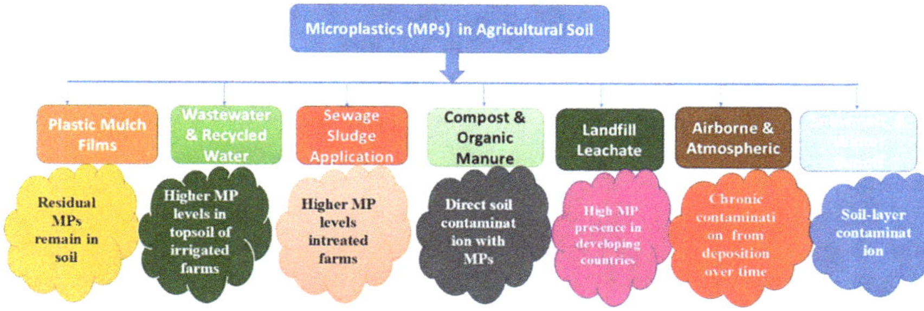


**Fig. 9.1** Sources of microplastic accumulation in agricultural soil

agriculture, the main sources of MP were organic fertilizer and irrigation water, with surface soil contamination levels as high as 2120 items/kg [24]. Furthermore, it has been demonstrated that earthworms like *Lumbricus terrestris* consume and move MPs farther into the soil profile, promoting bioaccumulation and possible groundwater contamination. MPs' environmental durability and possible penetration into the terrestrial food chain are made worse by their fragmentation and biogenic movement.

An another important source of microplasticcont amination in agricultural soils is landfill leachate, which is becoming more well acknowledged. Leachate, the liquid runoff generated by the breakdown of solid waste in landfills, contains both primary microplastics, which are produced on a tiny size, and secondary microplastics, which are produced when larger plastic garbage fragments are present [25]. The extensive use of plastic products and their propensity to fragment into tiny bits have led to the prevalence of microplastics in soil, especially fibers and fragments of polyethylene (PE), polypropylene (PP), and polystyrene (PS) [26]. Microplastics (MPs) can contaminate soil and perhaps make their way into the food chain when landfill leachate, which contains MPs and other pollutants, is misused in agriculture, especially in poorer countries with poor waste management. Limited water supplies and inadequate waste disposal infrastructure make this worse [27]. Landfills and wastewater treatment facilities have the potential to contaminate adjacent agricultural land through surface and groundwater, endangering soil and perhaps affecting the food chain. MPs' mobility in soil is influenced by their size, shape, and color; their small size enables them to pass across soil layers, potentially upsetting microbial activity and soil biota [28]. The risk is increased by the absence of standardized leachate monitoring and MP treatment, underscoring the urgent need for efficient waste management regulations to prevent microplastics from getting into agricultural areas.

Particularly in peri-urban and urban-fringe settings, deposition from snowmelt and airborne particles can have a substantial impact on microplastic contamination of agricultural soils. Recent studies that reveal alarming levels of microplastics in snow cover, such as an examination of sledding hills in Krakow, Poland, demonstrate that recreational winter activities can be a major source of microplastic pollution. The microplastics discovered in the snow samples from different sledding slopes differed in sort (e.g., polyester, hydrocarbon resin, polypropylene, and polyurethane), size (0.6–5.0 mm), shape (fragments, threads, flakes), and color. Microplastic debris is created and accumulated as a result of the mechanical abrasion from plastic sleds, shoes, and outerwear during high-intensity use [29]. Snowmelt then carries this debris into nearby soil ecosystems. Furthermore, airborne microplastics can land on snow surfaces and be taken to soils when the snow melts. These microplastics can be delivered by urban breezes or come from neighboring highways and industrial activity. Particularly susceptible to such deposition are agricultural fields next to urban green spaces or recreational areas, particularly during winter runoff episodes. Chronic contamination of soil systems can result from this, impacting nutrient dynamics, microbial activity, soil physical structure, and overall fertility. Microplastic (MP) deposition in agricultural soils is increasingly ascribed to a variety of environmental vectors, most notably precipitation, snowmelt, and gas (airborne dust). MPs are transported over these routes from industrial and urban areas into rural and peri-urban agricultural areas. A major contribution is made via airborne transportation through atmospheric fallout, in which microplastics from synthetic textiles, automobile wear, the breakdown of plastic trash, and industrial pollutants are suspended in the atmosphere and eventually land on agricultural soils through dust and dry or wet deposition [30]. Snow acts as a makeshift sink for microplastics during the winter. MPs of different sizes, shapes, and polymers are among the flying particles that snow collects as it builds up. These trapped particles are released into the top soil layers when the snow melts.

## Types and Properties of Microplastics in Soil

Soil ecosystems contain a variety of microplastics (MPs) with varying forms and characteristics that are mostly determined by their origin, polymer composition, exposure to the environment, and degradation processes. Depending on their size, MPs can be further divided into three categories: nanoplastics (<100 nm), small MPs (0.3–1 mm), and giant MPs (1–5 mm) [31]. These shapes affect MPs' ecological interactions, motility, and persistence in the soil environment. Polyethylene (PE), polypropylene (PP), polyvinyl chloride (PVC), polystyrene (PS), and polyethylene terephthalate (PET) are the most prevalent microplastic (MP) polymers found in soil. These polymers are extensively utilized in many different products and packaging, and MP contamination in soil is greatly exacerbated by their degradation [32]. As a common polymer, polyethylene (PE) is utilized extensively in agriculture as plastic mulch. Because it is inexpensive, simple to prepare, and has advantages for crop production like weed control, moisture retention, and elevated soil temperature, it is widely used [33]. The density, crystallinity, hydrophobicity, and thermal stability of microplastics (MPs) in soil are among their physical and chemical characteristics that have a major impact on how they degrade and interact with other soil constituents. These characteristics influence the degradation of MPs, their ability to adsorb contaminants such as pesticides, and their interactions with microbes and soil particles. Weathering-induced changes in MPs' physical properties may result in the formation of pits and fractures in the microplastic structure as well as an increase in surface area and improved interaction with the soil environments [34]. Variability in MP density and shape can affect potential uptake by soil organisms as well as vertical transfer within soil layers. All microplastic shapes and polymer types had a negative impact on soil aggregation, according to the study, with shape and polymer composition showing up as the main factors influencing variation. Soil aggregation was significantly reduced by roughly 29% with fibers, 25% with films, 20% with foams, and 27% with fragments when compared to controls. The greatest impact was caused by PET films, which reduced soil aggregation by around 35%. LDPE foams and PP and PC fragments came in second and third, respectively, at about 32% and 31%. In all concentrations and polymer types, microplastic fibers consistently caused soil aggregates to become unstable. On the other hand, LDPE foams shown a decreasing negative impact with increasing concentration, whereas films demonstrated a concentration-dependent negative impact. The effects of fragments differed: at lower concentrations, PET and PP fragments decreased aggregation; however, this effect decreased at 0.4%. The highest concentration of PC pieces considerably decreased soil stability, exhibiting a non-linear trend [35]. Furthermore, when it came to impacting plant and soil properties, shape was more important than polymer type. The impact of microplastics on soil microbial activity varied according to their form, concentration, and kind of polymer. Microbial activity generally decreased when microplastics were added, particularly PP fragments (−20%) and LDPE films (−17%). This was associated with reduced soil agglomeration and impeded water and oxygen flow. Because of their pollutant concentration and hazardous monomers, foams such as PU and PS further decreased activity. In fact, adding high concentrations of polyethylene terephthalate (PES) fibers and low concentrations (0.1% and 0.4%) of polypropylene (PP) films to soil increased microbial activity. This is probably because the plastics had a positive interaction with soil microorganisms, which caused them to release more DOC into the environment [36] [37] [38]. Microplastics can have a major effect on ecosystems and soil health because of their physical characteristics. Smaller particles interact with soil constituents like organic matter and microorganisms more easily because of their increased surface area. These interactions, as well as the physical and chemical properties of different kinds of microplastics (including bigger fragments), can change the porosity and structure of soil and may have detrimental effects on plant development and the general functioning of ecosystems [39].

## Microplastics' Effect on Soil Properties

Changes in bulk density, aggregate stability, and structure were particularly apparent when various microplastics, such as PEHD, PES, PET, PP, and PS, were added. The majority of microplastics raised density in the rhizosphere but decreased bulk density in bulk soil, suggesting zone-specific reactions. Water-stable aggregates dramatically declined with PA, PES, and PS, although the rhizosphere exhibited comparatively greater aggregation. Despite increasing evaporation rates, microplastics typically improved water availability by improving water holding capacity, and these structural alterations had an impact on water dynamics, where evapotranspiration increased significantly up to 50% with PES. The presence of plants, however, mitigated these effects and frequently decreased the amount of water available in soils that had been damaged by microplastics. Furthermore, microbial activity exhibited plastic-specific reactions: PA, PEHD, and PES increased microbial metabolism; however, plant–plastic interactions, namely with PA, PEHD, and PET, offset this advantage. These results suggest that microplastics have the ability to change the physical and biological characteristics of soil, hence reshaping its environment [40]. Notably, the kind of microplastic, its concentration, the presence of vegetation, and interactions within the rhizosphere all have a significant impact on these alterations. Understanding these complex interactions is necessary to evaluate the long-term impacts of microplastic contamination on soil ecosystems and plant health.

MPs differ in content and characteristics from soil particles that contain organic and mineral components. MP buildup in soil will unavoidably have a direct and indirect impact on soil characteristics. Most MPs have hydrophobic surfaces that can change the soil's capacity to retain, move, and release water [40]. Microplastics (MPs) have a high adsorption capacity, which allows them to drastically change soil characteristics and nutrient cycling. They may alter contaminants' and nutrients' adsorption and desorption patterns, which could impact the soil's capacity to absorb them. This can then affect the toxicity of contaminants and the fertility of the soil [41]. Soil health can be determined primarily by three factors: physical, chemical, and microbiological. The ability of the soil to sustain plant development, control the cycles of water and nutrients, and preserve ecosystem stability is influenced by these characteristics taken together [42]. By harming soil fauna such as nematodes and earthworms, microplastics (MPs) in soil can disturb natural processes including bioturbation and the breakdown of organic materials, hence impairing soil characteristics. Water retention, nutrient availability, and soil structure can all be altered by this interference [43]. As pollutants deteriorate and experience surface oxidation, MPs can change the activity of soil enzymes, diminish the symbioses of beneficial microorganisms, and interact with them, increasing their toxicity and possibly spreading antibiotic resistance.

Recent experimental data indicates that microplastics (MPs) can significantly alter the biophysical properties and microbiological activity of soil. After five weeks in soil, stereomicroscopic analysis showed that MPs made of polyacrylic, polyester, and polyethylene mostly maintained their structure. Polyethylene fragments remained loosely attached to the matrix, while linear fibers from polyester and polyacrylic helped to form larger soil aggregates. Interestingly, polyamide beads could not be distinguished from soil particles, highlighting the shortcomings of visual-based MP detection techniques. Polyester fibers produced a concentration-dependent reduction ($F = 6.32$, $p < 0.05$), perhaps because of their low density and shape, while also uniquely boosting water holding capacity ($F = 7.07$, $p < 0.05$). All MP types evaluated had an impact on soil bulk density ($p < 0.001$). Hydraulic conductivity did not significantly change in spite of these modifications. Water-stable aggregates were also significantly reduced by MPs, especially polyester and polyacrylic fibers, which compromised the structural stability of the soil in a concentration-dependent manner (polyester: $F = 12.04$, $p < 0.01$). Nonmonotonic responses, where lower MP concentrations elicited more pronounced effects than higher ones, accompanies these structural shifts, which are primarily attributed to the entanglement capacity of linear MPs. These responses are most likely the result of changes in the dominant interaction regimes between MPs and soil constituents. Furthermore, total MP concentration correlated with microbial activity ($F = 6.14$, $p < 0.05$) but specific MP types did not, and linear MPs were linked to decreased microbial function. Microbial activity highly correlated with aggregate formation in certain situations, such as exposure to polyethylene ($F = 22.42$, $p < 0.001$), indicating that MPs alter microbial–soil interactions via both structural and metabolic mechanisms [44]. These findings highlight the urgent need for further mechanistic study carried out in more ecologically relevant contexts, as they show the complex, type- and shape-specific effects of MPs on soil health. Standardized methodology and comprehensive study across a range of soil types and environmental circumstances are essential to better understand and mitigate the ecological risks of microplastic pollution.

## Mechanisms of Microplastic Absorption and Translocation in Plants and Their Effects on Growth and Yield

The primary way that micro- and nanoplastics (MNPs) can enter plant systems is through root uptake, though foliar absorption has also been documented. In the rhizosphere, both apoplastic and symplastic pathways allow nanoplastics (less than 100 nm) to pierce the root epidermis. If not blocked by the Casparian strip, the apoplastic route enables MNPs to get through cell walls and intercellular gaps without going through the plasma membrane, frequently arriving at the vascular tissues. In contrast, the symplastic pathway involves intercellular migration through plasmodesmata and MNP uptake by endocytosis [45]. The presence of fluorescently labeled polystyrene nanoparticles (PS-NPs) in model plants is confirmed by TEM and CLSM imaging studies, which show that these particles move throughout the plant systemically through xylem-mediated upward transport that is powered by transpiration. PS-NPs were discovered in the chloroplasts and leaves of lettuce and rice. Apart from root uptake, MNPs have also been reported to enter the leaves, especially when there is foliar spray or air deposition. Evidence from microscopy demonstrates that MNPs can enter leaf tissues more easily by lodging in stomatal apertures or penetrating via tiny fissures in the cuticle. MNPs cause a series of physiological and biochemical abnormalities after they are absorbed. Reactive oxygen species (ROS) generation and the plant's antioxidant defense system are out of balance in oxidative stress, which is frequently brought on by environmental variables like MPs. greater levels of ROS, such as hydrogen peroxide ($H_2O_2$), superoxide anions ($O2^-$), and hydroxyl radicals (OH•), as well as enhanced lipid peroxidation, as seen by greater levels of malondialdehyde (MDA), are caused by this imbalance [46]. Root elongation, leaf growth, and biomass accumulation can all be morphologically inhibited by exposure to 100 nm MP. in seedlings of maize [47]. Additionally, MNPs interfere with photosynthesis by lowering the amount of chlorophyll and reducing the effectiveness of photosystem II (Fv/Fm) [48]. Photosynthetic inhibition is further supported by PAM fluorometry experiments, which show a decrease in chlorophyll fluorescence. By adsorbing necessary ions or changing transporter function, MNPs can also disrupt nutritional uptake, leading to imbalances in calcium, phosphorus, potassium, nitrogen, and trace metals like iron and zinc. Inductively coupled plasma mass spectrometry (ICP-MS) and synchrotron-based X-ray fluorescence (XRF) studies support these findings in crops like barley and wheat [46]. MNPs have been demonstrated to cause genotoxic effects, decrease cell division, and hinder mitotic activity at the cellular level. After being exposed to MNPs, DNA strand breakage and chromosomal abnormalities are detected by comet assays and micronucleus tests in onion root tips. Analysis of gene expression shows that genes related to auxin and cytokinin pathways, in particular, are down-regulated in hormone signaling. When taken as a whole, these results show that MNPs not only penetrate and move through plant tissues but also compromise physiological, biochemical, and genetic integrity [49]. This raises significant questions regarding the possible effects of MNPs on food quality, crop productivity, and ecosystem health. Plastic buildup and translocation in plants under various exposure situations have been shown in a number of studies. After 28 days of soil exposure, it was discovered that polystyrene (PS) particles with amino and carboxyl functionalizations accumulated in the leaves of *Zea mays* and were subsequently carried to the roots [50]. PS nanoparticles (NPs) at concentrations of 0.1 and 1 mg $L^{-1}$ were absorbed by the stomata of the leaves and moved downward through the plant in *Lactuca sativa* after 35 days of soil exposure [51]. After 65 days in a hydroponic environment, *Cucumis sativus* exposed to PS particles ranging in size from 100 to 700 nm showed uptake by the roots, followed by translocation to leaves, flowers, and fruits via the stem [52]. *Allium cepa* showed PS uptake in different cellular compartments following a 72-h hydroponic exposure to PS particles (0.01–1 g $L^{-1}$) [53]. Positively charged PS NPs ($SO_3H$, $NH_2$) were more effectively absorbed and internalized by *Arabidopsis thaliana* roots over the course of 35 days of soil exposure than negatively charged particles [54]. In *Triticum aestivum* and *L. sativa*, PS NPs (0.2 and 2 μm) entered roots through crack-entry channels and were transported to shoots through transpiration after exposure periods of 20 and 10 days, respectively, in sandy soil and hydroponics [55]. Finally, microplastics and nanoplastics ($10^3$–$10^7$ particles $mL^{-1}$) accumulated in *L. sativum* seed pores over the course of a 72-h exposure on filter paper, postponing germination [56]. Field tests conducted in high-contamination areas, like the urban areas of Tianjin, China, showed that plants like lettuce gathered significantly more nanoplastics when grown outdoors as opposed to in greenhouse conditions [57]. These findings raise significant concerns regarding ecosystem health and food safety since they demonstrate how plants might serve as entry points for MPs and NPs into the food chain. Mechanisms of microplastic absorption and translocation in plants and their effects on growth and yield are shown in Fig. 9.2.

## Hazards to Human Health and Food Safety Issues

Concerns regarding food safety and human health are raised by recent research that found microplastics in frequently consumed fruits and vegetables. According to an Italian study, the levels of MP contamination in vegetables like lettuce, broccoli, and carrots ranged from 72,175 to 130,500 particles per kilogram, while the levels in fruits like apples and pears ranged from 52,600 to 307,750 particles per kilogram. The microplastics that were found had particle sizes ranging from 1.51 to 2.52 μm. Analysis using scanning electron microscopy (SEM) and energy-dispersive X-ray spectroscopy (EDX) confirmed the presence and composition of MPs in the edible tissues [58]. These results point to the possibility of internalization and transfer of microplastics within plant tissues, most likely due to vascular transport and root uptake, in addition to surface contamination. The existence of MNPs in frequently consumed produce, such as apples, pears, carrots, lettuce, tomatoes, onions, and cucumbers, has been verified by studies. People who eat such tainted food run the immediate danger of developing health problems. Growing evidence shows that MNPs are present in human tissues, blood, placenta, breast milk, and even carotid plaques, suggesting links to cardiovascular diseases, endocrine disruption, metabolic disorders, neurotoxicity, reproductive dysfunction, and developmental abnormalities, even though it is still difficult to quantify the precise intake [59]. There are also serious health concerns about infants when MNPs are found in breast milk. These toxicological conclusions are supported by experimental research employing human cell cultures and animal models. Even though they are invisible, MNPs are becoming a serious hazard to public health. Adding to the increasing amount of data, it is now evident that both microplastics (MPs) and nanoplastics (MNPs) enter the human body through a variety of exposure routes, such as ingestion, inhalation, and dermal absorption. This presents significant questions regarding the long-term and cumulative health impacts of these contaminants. Airborne microplastics, which are largely produced by synthetic textiles, deteriorating building materials, plastic abrasion, landfills, and waste incineration, can expose people to up to 272 MP particles every day [60]. These particles may settle in the lungs after inhalation, where they are taken up by macrophages and go into the lymphatic and circulatory systems, where they cause systemic distribution and immunological reactions. However, ingestion continues to be the main exposure route, with an estimated 39,000 to 52,000 MPs being consumed each year through food and drinking water [61]. Polystyrene nanospheres as small as 0.2 μm have been demonstrated to be internalized by vegetables like lettuce, exposing consumers to the particles directly through their stems and leaves [62]. Furthermore, because of their minuscule size,

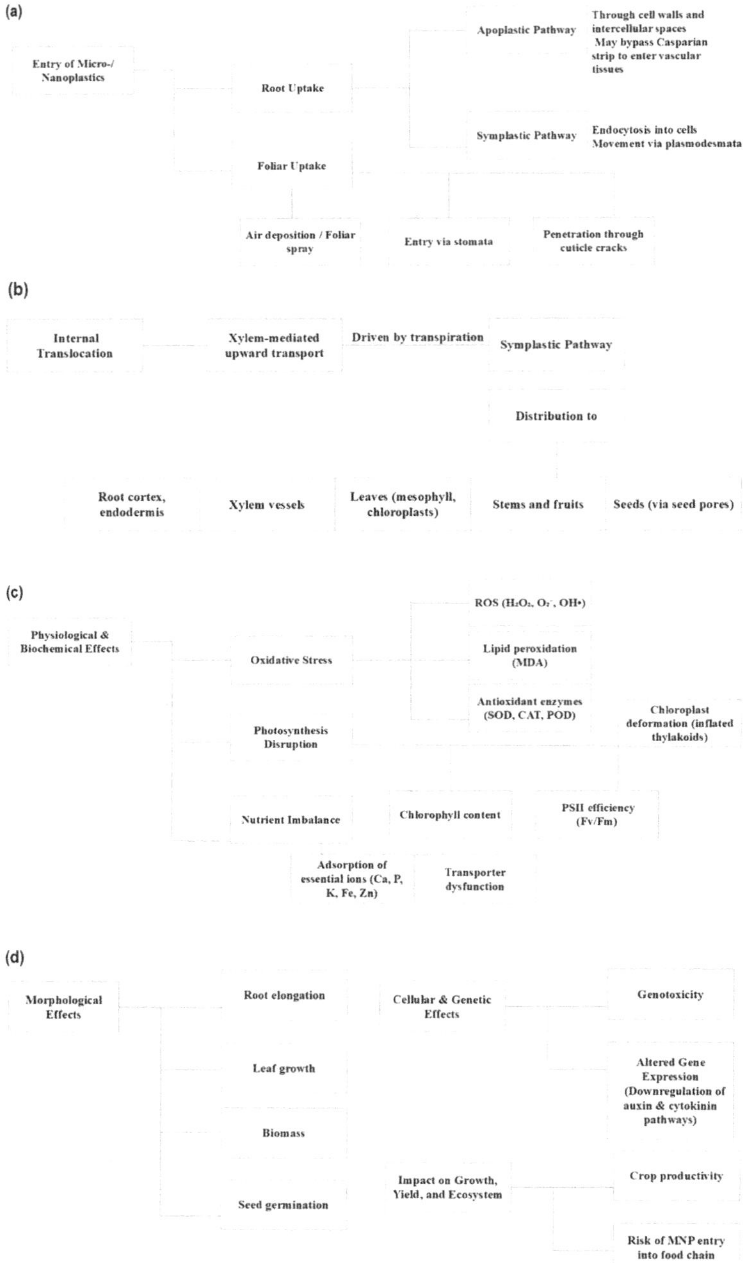


**Fig. 9.2** Mechanisms of microplastic absorption and translocation in plants and their effects on growth and yield

microplastics found in personal care items including toothpaste, dental products, hand sanitizers, and face cleansers can penetrate the skin and cause both local and systemic toxicity [63]. MPs can accumulate in the human body's key organs, including the kidneys, liver, intestines, and even brain tissue. Animal experiments have shown that this buildup causes oxidative stress, inflammation, changes in gene expression, and cell damage. There is also mounting evidence that MPs are linked to immunological dysfunction, neurotoxicity, endocrine disruption, and perhaps cancer [64]. The combinatorial toxicity posed by MPs serving as carriers for dangerous chemicals like heavy metals and phthalates is especially concerning since these substances can intensify endocrine-disrupting effects and alter immunological responses, which may result in chronic disease or autoimmunity [65]. Additives including phthalates, bisphenol A (BPA), and flame retardants are commonly present in plastic formulations. They are not covalently bound and easily leak from broken-down particles. Following their release, these substances function as endocrine-disrupting compounds (EDCs), imitating or interfering with normal hormones to change physiological processes in a variety of organ systems [66]. Furthermore, they can penetrate cell membranes, congregate in organelles such as mitochondria and nuclei, and induce oxidative stress, DNA damage, and apoptosis [67]. Fish exposed to nanoplastics that penetrate the brain have already shown altered behavior, according to animal studies, including those referenced by the Leibniz Institute of Freshwater Ecology and Inland Fisheries. Given that identical routes and mechanisms may function in people to cause neurological illnesses or developmental abnormalities [68], these early findings highlight the critical need for long-term human investigations.

## Techniques for Detection and Analysis of Microplastics in Plants and Soil

Accurate detection and analysis are necessary to comprehend the ecological effects, transportation, and environmental fate of microplastics (MPs) in soil and plants. Traditional detection techniques include pyrolysis-gas chromatography-mass spectrometry (Py-GC/MS), scanning electron microscopy (SEM), Raman spectroscopy, Fourier-transform infrared (FTIR) spectroscopy, and thermogravimetric analysis (TGA). These techniques enable the classification and measurement of polymer types according to their appearance, thermal characteristics, and molecular fingerprints [69]. To separate MPs from organic and inorganic matter in soil samples, pretreatment procedures such density separation, oxidative digestion (e.g., using $H2O_2$ or Fenton's reagent), and filtration are usually necessary [70]. In the tissues of plants, enzymatic digestion procedures or chemical treatments that degrade cell walls while retaining plastic particles are used to extract MPs. Staining (such as Nile Red) and fluorescence microscopy are then used for viewing. By utilizing the material-specific spectral signatures of MPs, emerging methods such as terahertz

(THz) time-domain spectroscopy provide quick, non-destructive examination with little sample preparation. This technique has demonstrated potential for distinguishing between polymers like PVC and LDPE in soils from different geographical areas [71]. Furthermore, μFTIR and μRaman spectroscopy offer high-resolution imaging and are especially useful for identifying and describing MPs at the microscale in rhizosphere samples or plant tissues. Emerging techniques for detecting microplastics in soils include quantitative NMR spectroscopy, LC-based methods (LC/MS, GPC, SEC), and spectroscopic instruments including NIR, SEM-EDS, and TOF-SIMS in addition to FTIR, Raman, and thermoanalysis. For polymers such as PET, PE, and PBS, NMR provides great specificity; nonetheless, it necessitates costly deuterated solvents and polymer-specific solubility. Molecular weight data is provided using LC methods; however, matrix interference may occur and large polymer concentrations are required. Although NIR and other surface methods are helpful for quick screening, their sensitivity (g $kg^{-1}$ range) is restricted, and they are influenced by the composition of the soil [72]. Even if these methods show promise, they still need to be improved for regular, precise, and scalable environmental monitoring.

Conventional techniques for identifying MPs in soil and plant systems, like spectroscopy, chemical digestion, and optical sorting, are frequently laborious, prone to human error, and inadequate for complicated matrix processing. Promising alternatives for improving the identification, classification, and quantification of MPs with increased speed, accuracy, and automation are provided by recent developments in artificial intelligence (AI), namely machine learning (ML) and deep learning (DL). Through the integration of AI algorithms with sophisticated imaging and spectroscopic techniques including μFTIR, μRaman, hyperspectral imaging, and 3D modeling, researchers can accurately map MP distribution in plant-soil systems, process massive datasets, and identify polymer kinds. By facilitating high-throughput, real-time analysis that supports pollution reduction and sustainable land management plans, this AI-powered approach has the potential to completely transform environmental monitoring.

The study emphasizes the monitoring of soil microplastic contamination using Least Squares Support Vector Machine (LS-SVM) modeling and terahertz time-domain spectroscopy. This non-invasive technique provides a scalable, accurate, and scalable alternative for rapid field evaluations while overcoming region-specific limitations [73]. For the detection of microplastics (MPs) in plant and soil systems, machine learning (ML) and deep learning (DL) are becoming viable substitutes for conventional techniques. These AI-powered methods for analyzing and identifying MPs in complicated environmental samples are quicker, more precise, and possibly more effective. These approaches combine AI algorithms with imaging techniques to process large datasets, identify polymer types, and map MP distribution accurately, potentially revolutionizing environmental monitoring. Further clustering and dimensionality reduction techniques have demonstrated promising capabilities, building on the application of unsupervised machine learning (ML) in microplastic (MP) and nanoplastic (NP) studies. An variation of DBSCAN, for instance, is Hierarchical Density-Based Spatial Clustering of Applications with Noise

(HDBSCAN), which increases clustering performance and noise identification, particularly in heterogeneous spectrum datasets. It is useful for environmental MP spectra with unequal distribution and background noise because it manages data variability well without requiring a predetermined number of clusters [74] [75]. Another widely utilized technology is Principal Component Analysis (PCA), which is a dimensionality reduction technique rather than a clustering method per se. PCA projects high-dimensional data onto a lower-dimensional space while preserving maximum variance, simplifying complex spectra for further categorization. PCA increases classification clarity and computational efficiency when used in conjunction with clustering methods such as HCA or K-means [75–77]. A more recent method that retains both global and local data structures better than PCA is Uniform Manifold Approximation and Projection (UMAP). It improves the interpretability of spectral fingerprints, especially when using FTIR or Raman datasets, by providing improved cluster separation and display of spectral data from microplastic mixes [78–80]. Given the vast variety of plastic kinds, degradation degrees, and environmental matrices, these unsupervised learning technologies enable researchers to investigate the intrinsic structure of complicated microplastic data without the need for predefined labels. The expanding trend toward automation and scalability in MP/NP detection workflows is reflected in the increased usage of such AI-based technologies, which greatly aid in standardizing analytical techniques and enhancing environmental monitoring initiatives.

## Mitigation and Remediation Strategies

A combination of preventive, reduction, and treatment techniques are used in mitigation and remediation strategies for microplastics (MPs) in soil and plants in order to minimize their release into the environment and eliminate or neutralize any contamination that already exists. Through sustainable plastic use and waste management techniques, such as encouraging a circular economy that prioritizes reducing, reusing, and recycling plastics, prevention aims to lower the input of MPs [81]. To enhance packaging design, limit the use of single-use plastics, encourage biodegradable alternatives will lessen plastic waste [82]. In order to influence consumer behavior and encourage the responsible use of plastic, education and awareness campaigns are also crucial [83]. Physical removal methods like soil washing and filtration, as well as biological methods like phytoremediation, in which plants are used to absorb or immobilize MPs and related contaminants, are examples of remediation strategies for soils that are already contaminated with MPs [84]. Additionally, microplastics may be broken down or their bioavailability in soil systems reduced by new technologies like microbial degradation and the use of biochar [85]. Combining these repair and mitigation strategies can make MPs in plant systems and agricultural soils less hazardous to the environment and human health, enhancing soil quality and leading to more secure food production. Following up on the previous discussion of mitigation and remediation strategies for microplastics

(MPs), it is important to note that because of the diversity of MP contaminants and their varying physico-chemical characteristics, abatement and treatment strategies must be customized for the particular environmental medium, such as soil, water, or sediment. The best preventive approach is still source reduction, which aims to restrict MP infiltration through better waste management, sustainable consumption, and regulatory changes. However, a variety of therapeutic technologies are needed after MPs have entered natural systems. Alongside more well-established methods like photocatalysis, electrocoagulation, and magnetic extraction, newer technologies like acoustic wave separation and sophisticated oxidation processes also show promise in aquatic systems. In order to capture, break down, or isolate MPs, these methods use chemical, physical, and biological principles; nevertheless, their effectiveness is frequently medium-dependent and limited by elements including particle size, water chemistry, and interaction with co-contaminants [81]. Because of soil heterogeneity and the numerous ways that MPs interact with organic matter and soil biota, treatment is more complicated in soils. Although there is yet little wide-scale applicability, biodegradation employing microbial consortia, electrochemical oxidation, and pyrolysis is being investigated at the lab or pilot scale. Despite its effectiveness, burning contaminated biosolids or sediments raises questions regarding costs, energy consumption, and emissions [86]. Physical methods including dredging, landfilling, and density separation are common in sediment cleanup; however, their use is frequently limited by cost, environmental impact, and logistical considerations [87]. Although they require rigorous environmental risk assessment, supplemental approaches like capping and biodegradation can be used as containment or degradation solutions. The two main methods for remediating soil microplastics (MP) are bioremediation by soil animals and rhizoremediation by plants and microorganisms. While hydrophytes and plant leaves adsorb MPs by electrostatic, hydrogen bonding, and hydrophobic interactions, plant roots have the potential to immobilize MPs, lowering their bioavailability and toxicity. MP immobilization and plant resistance are improved by mycorrhizal fungi like AMF and ectomycorrhizal species [88]. Additionally, mucilage and root exudates released by plants capture MPs and promote the growth of microorganisms that break down MPs. The carbon cycle can be impacted by rhizospheric bacteria and fungi that can mineralize polymers like PS, PET, and PLA into $CO_2$ and $CH_4$ or enzymatically break them down into biomass. Potential options for phytoremediation include some plants, such as *Mirabilis jalapa* [89], which exhibit increased antioxidant responses to MP stress. However, rhizoremediation's effectiveness may be constrained by elements like soil type, nutrient availability, microbial competition, and environmental circumstances. Improvements using chemical agents, charcoal, and genetic engineering are being investigated. At the same time, MP decomposition is aided by soil fauna such wax moths, snails (Achatina fulica) [90], mealworms (Tenebrio molitor) [91], superworms (Zophobas atratus) [92], and earthworms [93]. These organisms break down and metabolize MPs into simpler chemicals or $CO_2$ with the aid of the gut microbiome. Monooxygenases, peroxidases, and dehydrogenases are examples of enzymes that help break down polymers and turn them into biomass or energy [94]. Earthworms also promote soil health and act as markers of MP toxicity. When combined, these biological techniques present viable, environmentally benign methods for MP remediation in soils; however, additional refinement is required for widespread implementation. The long-term destiny of MPs during therapy, their interactions with co-contaminants, and the possible generation of harmful breakdown by-products remain largely unknown despite these advancements. Another issue is whether or not these technologies can be deployed on a big scale economically, especially in underdeveloped nations. Understanding the dynamics of degradation in intricate matrices, creating efficient and selective removal methods, and combining several treatment stages for complementary results must be the top priorities of future study. A multi-barrier strategy including source control, technological innovation, and ecosystem-specific treatment modalities is required to successfully reduce the environmental impacts of MPs.

## Future Directions and Research Gaps

Food safety is becoming a concern due to the contamination of terrestrial ecosystems by microplastics (MPs). Nonetheless, there are still unanswered questions about MP accumulation in plants and soils. To comprehend MP absorption, soil conditions, and interactions with contaminants, interdisciplinary research is required. There are also no standardized methods for detecting and quantifying MP. In addition to investigating the molecular and physiological reactions of plants to MP exposure, future research must give top priority to the development of sensitive and standardized detection methods. Finding crop varieties that are less likely to absorb MP may also help direct more environmentally friendly farming methods. Long-term field research is also necessary to evaluate the cumulative exposure and any health hazards linked to long-term ingestion of food contaminated with MP.

In order to further this knowledge, systemic issues that restrict thorough MP risk assessment in agricultural systems must be addressed. Accurate evaluation and comparison among agroecosystems are currently hampered by the lack of global and region-specific data inventories on MP contamination in farmlands. Data discrepancy is exacerbated by the lack of standardized techniques for MP sample and analysis. Determining baseline contamination levels and ecological consequences therefore requires standardizing methodology and carrying out comprehensive assessments under various agricultural systems and management practices. Additionally, the kind, size, shape, and chemical composition of MPs vary widely, all of which affect the interactions between soil, plants, and microbes. Most current studies focus on individual MPs rather than the complex impacts of MP mixtures, especially when mixed with other environmental pollutants such heavy metals and pesticides. To enable thorough environmental risk evaluations, research must increasingly concentrate on the antagonistic or synergistic effects of these co-contaminants. Although significant attention has been paid to how MPs change the architecture of microbial communities, less is known about how they affect microbial processes that are crucial for biogeochemical cycling, like greenhouse gas management, phosphorus solubilization, and nitrogen fixation. Long-term food security, agricultural production, and soil health all depend on an understanding of these functional effects. Future studies in this area must also investigate creative, environmentally friendly remedial techniques that complement sustainable farming objectives. Using green nanoparticles made from natural biopolymers like chitosan, cellulose, and starch is one possible approach [95]. These substances provide safe, eco-friendly ways to boost nutrient delivery, strengthen soil health, and perhaps lessen MP contamination. The ability of these green nanotechnologies to interact with, break down, or immobilize MPs is still unknown, despite their demonstrated promise in plant nutrition and pest control. Examining their function in plastic remediation may lead to novel approaches to lowering MP buildup in soil-plant systems.

MP detection, tracking, and exclusion procedures at the field level must be given top priority by the agriculture sector as global food safety rules continue to tighten, particularly in industries like horticulture. Despite mounting evidence of microplastics in fruits and vegetables and the health hazards they pose, screening for them is rarely part of current food safety measures. In order to preserve market access and safeguard public health, microplastic monitoring must be incorporated into agricultural production and certification processes. In order to create sustainable, plastic-resilient food systems and guarantee the long-term safety and integrity of agricultural goods, it will be essential to address these future directions, from foundational research to field-level implementation.

## Conclusion

Soil contamination by microplastics poses an increasingly pernicious danger to food safety, agricultural production, and terrestrial ecosystems. As discussed in this chapter, microplastics disturb important microbial populations that support nutrient cycling and plant health in addition to changing the physicochemical characteristics of soil. Serious worries regarding long-term health impacts, such as endocrine and reproductive toxicity, are raised by the possibility that microplastics could be absorbed by crops and make their way into the human food chain. These substances have the potential to cause developmental, reproductive, and metabolic problems by upsetting the body's endocrine system and interfering with hormone production and function. Critical information gaps about the behavior, destiny, and ecological consequences of soil-borne microplastics still exist despite growing awareness of these dangers. These gaps must be filled by interdisciplinary research in order to develop effective detection, mitigation, and remediation strategies. Sustainable farming practices, policy implementation, and scientific research must all be used in tandem to effectively address microplastic pollution in agricultural soils. The safety of food production and the preservation of ecosystem health depend on this all-encompassing approach.

## References

1. Liang, Y., Cao, X., Mo, A., Jiang, J., Zhang, Y., Gao, W., & He, D. (2023). Micro(nano)plastics in plant-derived food: Source, contamination pathways and human exposure risks. *TrAC Trends in Analytical Chemistry, 165*, 117138. https://doi.org/10.1016/j.trac.2023.117138
2. Murphy, D. J. (2025, March 20). Microplastics: Are they poisoning crops and jeopardising food production? *Down To Earth*. https://www.downtoearth.org.in/waste/microplastics-are-they-poisoning-crops-and-jeopardising-food-production#:~:text=Microplastic%20exposure%2C%20they% 20found%2C%20reduces,between%204%25%20and%2014%25.
3. Hasan, M. M., & Tarannum, M. N. (2025). Adverse impacts of microplastics on soil physico-chemical properties and crop health in agricultural systems. *Journal of Hazardous Materials Advances, 17*, 100528. https://doi.org/10.1016/j.hazadv.2024.100528
4. Yang, L., Zhang, Y., Kang, S., Wang, Z., & Wu, C. (2021). Microplastics in soil: A review on methods, occurrence, sources, and potential risk. *Science of The Total Environment*, 146546. https://doi.org/10.1016/j.scitotenv.2021.146546
5. Aralappanavar, V. K., Mukhopadhyay, R., Yu, Y., Liu, J., Bhatnagar, A., Praveena, S. M., Li, Y., Paller, M., Adyel, T. M., Rinklebe, J., Bolan, N. S., & Sarkar, B. (2024). Effects of microplastics on soil microorganisms and microbial functions in nutrients and carbon cycling—A review. *Science of the Total Environment, 924*, 171435. https://doi.org/10.1016/j.scitotenv.2024.171435
6. Verla, A. W., Enyoh, C. E., Verla, E. N., & Ehikhamenor, F. A. (2019). Microplastic–toxic chemical interaction: A review study on quantified levels, mechanism and implication. *SN Applied Sciences, 1*, Article 1400. https://doi.org/10.1007/s42452-019-1352-0
7. Petroody, S. S. A., Hashemi, S. H., & van Gestel, C. A. M. (2021). Transport and accumulation of microplastics through wastewater treatment sludge processes. *Chemosphere, 278*, 130471. https://doi.org/10.1016/j.chemosphere.2021.130471
8. Selonen, S., Dolar, A., Jemec Kokalj, A., Sackey, L. N. A., Skalar, T., Fernandes, V. C., Rede, D., Delerue-Matos, C., Hurley, R., Nizzetto, L., & van Gestel, C. A. M. (2021). Exploring the impacts of microplastics and associated chemicals in the terrestrial environment—Exposure of soil invertebrates to tire particles. *Environmental Research, 201*, 111495. https://doi.org/10.1016/j.envres.2021.111495
9. Gionfra, S. (2018). Plastic pollution in soil. *Institute for European Environmental Policy*, 1–6.
10. van den Berg, P., Huerta-Lwanga, E., Corradini, F., & Geissen, V. (2020). Sewage sludge application as a vehicle for microplastics in eastern Spanish agricultural soils. *Environmental Pollution, 261*, 114198. https://doi.org/10.1016/j.envpol.2020.114198
11. Järlskog, I., Strömvall, A.-M., Magnusson, K., Galfi, H., Björklund, K., Polukarova, M., Garção, R., Markiewicz, A., Aronsson, M., Gustafsson, M., Norin, M., Blom, L., & Andersson-Sköld, Y. (2021). Traffic-related microplastic particles, metals, and organic pollutants in an urban area under reconstruction. *Science of the Total Environment, 774*, 145503. https://doi.org/10.1016/j.scitotenv.2021.145503
12. Gao, H., Yan, C., Liu, Q., & Ding, W. (2018). Effects of plastic mulching and plastic residue on agricultural production: A meta-analysis. The Science of the Total Environment, 651(Pt 1). https://doi.org/10.1016/j.scitotenv.2018.09.105.
13. Zhaorong, M., Yousheng, L., Qianqian, Z., & Guangguo, Y. (2020). The usage and environmental pollution of agricultural plastic film. *Asian Journal of Ecotoxicology, 15*, 21–32.
14. Li, Y., Yang, R., Guo, L., Gao, W., Su, P., Xu, Z., Xiao, H., Ma, Z., Liu, X., Gao, P., et al. (2022). The composition, biotic network, and assembly of plastisphere protistan taxonomic and functional communities in plastic-mulching croplands. *Journal of Hazardous Materials, 430*, 128390. https://doi.org/10.1016/j.jhazmat.2022.128390
15. Huang, Y., Liu, Q., Jia, W., Yan, C., & Wang, J. (2020). Agricultural plastic mulching as a source of microplastics in the terrestrial environment. *Environmental Pollution, 260*, 114096. https://doi.org/10.1016/j.envpol.2020.114096

16. Yan, H., Ma, J., Zhang, J., Wang, G., Zhang, C., Akhlaq, M., Huang, S., & Yu, J. (2022). Effects of film mulching on the physiological and morphological parameters and yield of cucumber under insufficient drip irrigation. *Irrigation and Drainage, 71*(3), 897–911. https://doi.org/10.1002/ird.2665
17. Gan, C.-D., Liao, Y.-L., Liu, H.-B., Yang, J.-Y., & Nikitin, A. (2025). Microplastic-induced changes in Cd and Cr behavior in the agricultural soil-wheat system: Insights into metal bioavailability and phytotoxicity. *Journal of Hazardous Materials, 482*, 136592. https://doi.org/10.1016/j.jhazmat.2024.136592
18. Wang, B., Wang, P., Zhao, S., Shi, H., Zhu, Y., Teng, Y., Jiang, G., & Liu, S. (2023). Combined effects of microplastics and cadmium on the soil-plant system: Phytotoxicity, Cd accumulation and microbial activity. *Environmental Pollution, 333*, 121960. https://doi.org/10.1016/j.envpol.2023.121960
19. Pérez-Reverón, R., González-Sálamo, J., Hernández-Sánchez, C., González-Pleiter, M., Hernández-Borges, J., & Díaz-Peña, F. J. (2022). Recycled wastewater as a potential source of microplastics in irrigated soils from an arid-insular territory (Fuerteventura, Spain). *Science of the Total Environment, 817*, 152830. https://doi.org/10.1016/j.scitotenv.2021.152830
20. Aydin, S., Mucevher, O., Ulvi, A., et al. (2025). Effects of long-term wastewater irrigation on microplastics pollution in agricultural soil. *Environmental Science and Pollution Research, 32*, 12340–12359. https://doi.org/10.1007/s11356-025-36452-x
21. Zhou, B., Wang, J., Zhang, H., Shi, H., Fei, Y., Huang, S., et al. (2020). Microplastics in agricultural soils on the coastal plain of Hangzhou Bay, East China: Multiple sources other than plastic mulching film. *Journal of Hazardous Materials, 388*, 121814. https://doi.org/10.1016/j.jhazmat.2019.121814
22. Corradini, F., Meza, P., Eguiluz, R., Casado, F., Huerta-Lwanga, E., & Geissen, V. (2019). Evidence of microplastic accumulation in agricultural soils from sewage sludge disposal. *Science of the Total Environment, 671*, 411–420. https://doi.org/10.1016/j.scitotenv.2019.03.368
23. Guo, S., Zhang, J., Liu, J., Guo, N., Zhang, L., Wang, S., Wang, X., Zhao, M., Zhang, B., & Chen, Y. (2023). Organic fertilizer and irrigation water are the primary sources of microplastics in the facility soil, Beijing. *Science of The Total Environment, 895*, 165005. https://doi.org/10.1016/j.scitotenv.2023.165005
24. Rana, M. M., Haque, M. R., Tasnim, S. S., & Rahman, M. M. (2023). The potential contribution of microplastic pollution by organic fertilizers in agricultural soils of Bangladesh: Quantification, characterization, and risk appraisals. *Frontiers in Environmental Science, 11*, 1205603. https://doi.org/10.3389/fenvs.2023.1205603
25. Kabir, M. S., Wang, H., Luster-Teasley, S., Zhang, L., & Zhao, R. (2023). Microplastics in landfill leachate: Sources, detection, occurrence, and removal. *Environmental Science and Ecotechnology, 16*, 100256. https://doi.org/10.1016/j.ese.2023.100256
26. Ashrafy, A., Liza, A. A., Islam, M. N., Billah, M. M., Arafat, S. T., Rahman, M. M., & Rahman, S. M. (2023). Microplastics pollution: A brief review of its source and abundance in different aquatic ecosystems. *Journal of Hazardous Materials Advances, 9*, 100215. https://doi.org/10.1016/j.hazadv.2022.100215
27. Lakhiar, I. A., Yan, H., Zhang, J., Wang, G., Deng, S., Bao, R., Zhang, C., Syed, T. N., Wang, B., Zhou, R., & Wang, X. (2024). Plastic pollution in agriculture as a threat to food security, the ecosystem, and the environment: An overview. *Agronomy, 14*(3), 548. https://doi.org/10.3390/agronomy14030548
28. Zaman, F., Rahman, M. A., Haque, M. M., Akbor, M. A., & Tareq, S. M. (2024). Pervasiveness and classification of microplastics in landfill leachate: Impacts, risks, and treatment efficiency. *Journal of Hazardous Materials Advances, 16*, 100502. https://doi.org/10.1016/j.hazadv.2024.100502
29. Lasota, J., Piaszczyk, W., Tabor, S., & Błońska, E. (2023). Microplastic in the snow on Sledding Hills in green areas of Krakow. *Sustainability, 15*(17), 12995. https://doi.org/10.3390/su151712995

30. Fakour, H., Lo, S.-L., Yoashi, N. T., Massao, A. M., Lema, N. N., Mkhontfo, F. B., Jomalema, P. C., Jumanne, N. S., Mbuya, B. H., Mtweve, J. T., & Imani, M. (2021). Quantification and analysis of microplastics in farmland soils: Characterization, sources, and pathways. *Agriculture, 11*(4), 330. https://doi.org/10.3390/agriculture11040330
31. Sajjad, M., Huang, Q., Khan, S., Khan, M. A., Liu, Y., Wang, J., Lian, F., Wang, Q., & Guo, G. (2022). Microplastics in the soil environment: A critical review. *Environmental Technology & Innovation, 27*, 102408. https://doi.org/10.1016/j.eti.2022.102408
32. Kumar, M., Chaudhary, V., Kumar, R., Chaudhary, V., & Srivastav, A. L. (2025). Microplastics, their effects on ecosystems, and general strategies for mitigation of microplastics: A review of recent developments, challenges, and future prospects. *Environmental Pollution and Management, 2*, 87–105. https://doi.org/10.1016/j.epm.2025.03.001
33. Somanathan, H., Sathasivam, R., Sivaram, S., Kumaresan, S. M., Muthuraman, M. S., & Park, S. U. (2022). An update on polyethylene and biodegradable plastic mulch films and their impact on the environment. *Chemosphere, 307*(Part 3), 135839. https://doi.org/10.1016/j.chemosphere.2022.135839
34. Astner, A. F., Gillmore, A. B., Yu, Y., Flury, M., DeBruyn, J. M., Schaeffer, S. M., & Hayes, D. G. (2023). Formation, behavior, properties and impact of micro- and nanoplastics on agricultural soil ecosystems (A review). *NanoImpact, 31*, 100474. https://doi.org/10.1016/j.impact.2023.100474
35. Lozano, Y. M., Lehnert, T., Linck, L. T., Lehmann, A., & Rillig, M. C. (2021). Microplastic shape, polymer type, and concentration affect soil properties and plant biomass. *Frontiers in Plant Science, 12*, 616645. https://doi.org/10.3389/fpls.2021.616645
36. Six, J., Bossuyt, H., Degryze, S., & Denef, K. (2004). A history of research on the link between (micro)aggregates, soil biota, and soil organic matter dynamics. *Soil and Tillage Research, 79*(1), 7–31. https://doi.org/10.1016/j.still.2004.03.008
37. Fei, Y., Huang, S., Zhang, H., Tong, Y., Wen, D., Xia, X., et al. (2020). Response of soil enzyme activities and bacterial communities to the accumulation of microplastics in an acid cropped soil. *Science of the Total Environment, 707*, 135634. https://doi.org/10.1016/j.scitotenv.2019.135634
38. Rillig, M. C., Lehmann, A., Ryo, M., & Bergmann, J. (2019). Shaping up: Toward considering the shape and form of pollutants. *Environmental Science & Technology, 53*(14), 7925–7926. https://doi.org/10.1021/acs.est.9b03520
39. Yoon, J.-H., Kim, B.-H., & Kim, K.-H. (2024). Distribution of microplastics in soil by types of land use in metropolitan area of Seoul. *Applied Biological Chemistry, 67*(15). https://doi.org/10.1186/s13765-024-00877-0
40. Machado, A. A. d. S., Lau, C. W., Kloas, W., Bergmann, J., Bachelier, J. B., Faltin, E., Becker, R., Görlich, A. S., & Rillig, M. C. (2019). Microplastics can change soil properties and affect plant performance. *Environmental Science & Technology, 53*(10), 6044–6052. https://doi.org/10.1021/acs.est.9b01339
41. Jin, T., Tang, J., Lyu, H., Wang, L., Gillmore, A. B., & Schaeffer, S. M. (2022). Activities of microplastics (MPs) in agricultural soil: A review of MPs pollution from the perspective of agricultural ecosystems. *Journal of Agricultural and Food Chemistry, 70*(14), 4182–4201. https://doi.org/10.1021/acs.jafc.1c07849
42. Yu, Y., Battu, A. K., Varga, T., Denny, A. C., Zahid, T. M., Chowdhury, I., & Flury, M. (2023). Minimal impacts of microplastics on soil physical properties under environmentally relevant concentrations. *Environmental Science & Technology, 57*(13), 5296–5304. https://doi.org/10.1021/acs.est.2c09822
43. Bodor, A., Feigl, G., Kolossa, B., Mészáros, E., Laczi, K., Kovács, E., Perei, K., & Rákhely, G. (2024). Soils in distress: The impacts and ecological risks of (micro)plastic pollution in the terrestrial environment. *Ecotoxicology and Environmental Safety, 269*, 115807. https://doi.org/10.1016/j.ecoenv.2023.115807

44. Machado, A. A. d. S., Lau, C. W., Till, J., Kloas, W., Lehmann, A., Becker, R., & Rillig, M. C. (2018). Impacts of microplastics on the soil biophysical environment. *Environmental Science & Technology, 52*(17), 9656–9665. https://doi.org/10.1021/acs.est.8b02212
45. Jamil, A., Ahmad, A., Moeen-ud-din, M., Zhang, Y., Zhao, Y., Chen, X., Cui, X., Tong, Y., & Liu, X. (2025). Unveiling the mechanism of micro-and-nano plastic phytotoxicity on terrestrial plants: A comprehensive review of omics approaches. *Environment International, 195*, 109257. https://doi.org/10.1016/j.envint.2025.109257
46. Yu, Z., Xu, X., Guo, L., Jin, R., & Lu, Y. (2024). Uptake and transport of micro/nanoplastics in terrestrial plants: Detection, mechanisms, and influencing factors. *Science of The Total Environment, 907*, 168155. https://doi.org/10.1016/j.scitotenv.2023.168155
47. Wang, F., Zhang, X., Zhang, S., Zhang, S., & Sun, Y. (2020). Interactions of microplastics and cadmium on plant growth and arbuscular mycorrhizal fungal communities in an agricultural soil. *Chemosphere, 254*, 126791. https://doi.org/10.1016/j.chemosphere.2020.126791
48. Kalhor, M. S., Aliniaeifard, S., Seif, M., Asayesh, E. J., Bernard, F., Hassani, B., et al. (2018). Title: Enhanced salt tolerance and photosynthetic performance: Implication of γ-amino butyric acid application in salt-exposed lettuce (*Lactuca sativa* L.) plants. *Plant Physiology and Biochemistry, 130*, 157–172. https://doi.org/10.1016/j.plaphy.2018.07.003
49. Mahmud, F., Sarker, D. B., Jocelyn, J. A., & Sang, Q. A. (2024). Molecular and cellular effects of microplastics and nanoplastics: Focus on inflammation and senescence. *Cells, 13*(21), 1788. https://doi.org/10.3390/cells13211788
50. Sun, H., Lei, C., Xu, J., & Li, R. (2021). Foliar uptake and leaf-to-root translocation of nanoplastics with different coating charge in maize plants. *Journal of Hazardous Materials, 416*, 125854. https://doi.org/10.1016/j.jhazmat.2021.125854
51. Lian, J., Liu, W., Meng, L., Wu, J., Chao, L., Zeb, A., & Sun, Y. (2021). Foliar-applied polystyrene nanoplastics (PSNPs) reduce the growth and nutritional quality of lettuce (*Lactuca sativa* L.). *Environmental Pollution, 280*, 116978. https://doi.org/10.1016/j.envpol.2021.116978
52. Li, Z., Li, Q., Li, R., Zhou, J., & Wang, G. (2021). The distribution and impact of polystyrene nanoplastics on cucumber plants. *Environmental Science And Pollution Research, 28*, 16042–16053. https://doi.org/10.1007/s11356-020-11702-2
53. Giorgetti, L., Spanò, C., Muccifora, S., Bottega, S., Barbieri, F., Bellani, L., & Castiglione, M. R. (2020). Exploring the interaction between polystyrene nanoplastics and *Allium cepa* during germination: Internalization in root cells, induction of toxicity and oxidative stress. *Plant Physiology and Biochemistry, 149*, 170–177. https://doi.org/10.1016/j.plaphy.2020.02.014
54. Sun, X.-D., Yuan, X.-Z., Jia, Y., Feng, L.-J., Zhu, F.-P., Dong, S.-S., Liu, J., Kong, X., Tian, H., Duan, J.-L., et al. (2020). Differentially charged nanoplastics demonstrate distinct accumulation in *Arabidopsis thaliana*. *Nature Nanotechnology, 15*, 755–760. https://doi.org/10.1038/s41565-020-0707-4
55. Li, L., Luo, Y., Li, R., Zhou, Q., Peijnenburg, W. J. G. M., Yin, N., Yang, J., Tu, C., & Zhang, Y. (2020). Effective uptake of submicrometre plastics by crop plants via a crack-entry mode. *Nature Sustainability, 3*, 929–937. https://doi.org/10.1038/s41893-020-0567-9
56. Bosker, T., Bouwman, L. J., Brun, N. R., Behrens, P., & Vijver, M. G. (2019). Microplastics accumulate on pores in seed capsule and delay germination and root growth of the terrestrial vascular plant *Lepidium sativum*. *Chemosphere, 226*, 774–781. https://doi.org/10.1016/j.chemosphere.2019.03.163
57. Down To Earth. (2025). *Microplastics in air enter plant leaves, pass to animals & humans: Study*. https://chatgpt.com/c/683ec9f2-b238-8013-8427-bc42bc2931a6.
58. Oliveri Conti, G., Ferrante, M., Banni, M., Favara, C., Nicolosi, I., Cristaldi, A., Fiore, M., & Zuccarello, P. (2020). Micro- and nano-plastics in edible fruit and vegetables: The first diet risk assessment for the general population. *Environmental Research, 187*, 109677. https://doi.org/10.1016/j.envres.2020.109677
59. Shourie, A. (2024, December 4). *How microplastics are destroying soil and human health*. 360info. https://360info.org/how-microplastics-are-destroying-soil-and-human-health/

60. Duis, K., & Coors, A. (2016). Microplastics in the aquatic and terrestrial environment: Sources (with a specific focus on personal care products), fate and effects. *Environmental Sciences Europe, 28*, 2. https://doi.org/10.1186/s12302-015-0069-y
61. Barboza, L. G. A., Vethaak, A. D., Lavorante, B. R. B. O., Lundebye, A. K., & Guilhermino, L. (2018). Marine microplastic debris: An emerging issue for food security, food safety and human health. *Marine Pollution Bulletin, 133*, 336–348. https://doi.org/10.1016/j.marpolbul.2018.05.047
62. Okeke, E. S., Chukwudozie, K. I., Addey, C. I., Okoro, J. O., Ezeorba, T. P. C., Atakpa, E. O., Okoye, C. O., & Nwuche, C. O. (2023). Micro and nanoplastics ravaging our agroecosystem: A review of occurrence, fate, ecological impacts, detection, remediation, and prospects. *Heliyon, 9*(2), e13296. https://doi.org/10.1016/j.heliyon.2023.e13296
63. Bikiaris, N., Nikolaidis, N. F., & Barmpalexis, P. (2024). Microplastics (MPs) in cosmetics: A review on their presence in Personal-Care, Cosmetic, and Cleaning Products (PCCPs) and sustainable alternatives from biobased and biodegradable polymers. *Cosmetics, 11*(5), 145. https://doi.org/10.3390/cosmetics11050145
64. Bora, S. S., Gogoi, R., Sharma, M. R., Anshu, Borah, M. P., Deka, P., Bora, J., Naorem, R. S., Das, J., & Teli, A. B. (2024). Microplastics and human health: Unveiling the gut microbiome disruption and chronic disease risks. *Frontiers in Cellular and Infection Microbiology, 14*, 1492759. https://doi.org/10.3389/fcimb.2024.1492759
65. Kowalczyk, A., Wrzecińska, M., Czerniawska-Piątkowska, E., Araújo, J. P., & Cwynar, P. (2022). Molecular consequences of the exposure to toxic substances for the endocrine system of females. *Biomedicine & Pharmacotherapy, 155*, 113730. https://doi.org/10.1016/j.biopha.2022.113730
66. Dalamaga, M., Kounatidis, D., Tsilingiris, D., Vallianou, N. G., Karampela, I., Psallida, S., & Papavassiliou, A. G. (2024). The role of endocrine disruptors bisphenols and phthalates in obesity: Current evidence, perspectives and controversies. *International Journal of Molecular Sciences, 25*(1), 675. https://doi.org/10.3390/ijms25010675
67. Yang, Z., DeLoid, G. M., Zarbl, H., Baw, J., & Demokritou, P. (2023). Micro- and nanoplastics (MNPs) and their potential toxicological outcomes: State of science, knowledge gaps and research needs. *NanoImpact, 32*, 100481. https://doi.org/10.1016/j.impact.2023.100481
68. United Nations Development Programme. (2021). *December 22*. How tiny plastic particles are polluting our soil. https://www.undp.org/news/plastic-planet-how-tiny-plastic-particles-are-polluting-our-soil
69. Mushtak, F., Prakash, J., & Katoch, S. S. (2024). Microplastics in complex soil matrix: Recovery, identification and removal using micro nano techniques. *Micro and Nano Engineering, 22*, 100237. https://doi.org/10.1016/j.mne.2023.100237
70. Rani, M., Ducoli, S., Depero, L. E., Prica, M., Tubić, A., Ademovic, Z., Morrison, L., & Federici, S. (2023). A complete guide to extraction methods of microplastics from complex environmental matrices. *Molecules, 28*(15), 5710. https://doi.org/10.3390/molecules28155710
71. Guan, A.-H., & Chao, Y.-Y. (2019). Quantitative analysis of alum based on terahertz time-domain spectroscopy technology and support vector machine. *Optik, 193*, 163017. https://doi.org/10.1016/j.ijleo.2019.163017
72. Zhang, Z., Geng, Y., Zhou, W., Shao, X., Lin, H., & Zhou, Y. (2024). Development of a multi-spectroscopy method coupling μ-FTIR and μ-Raman analysis for one-stop detection of microplastics in environmental and biological samples. *Science of the Total Environment, 917*, 170396. https://doi.org/10.1016/j.scitotenv.2024.170396
73. Li, Y., Yao, J., Nie, P., Feng, X., & Liu, J. (2021). An effective method for the rapid detection of microplastics in soil. *Chemosphere, 276*, 128696. https://doi.org/10.1016/j.chemosphere.2020.128696
74. Ankerst, M., Breunig, M. M., Kriegel, H.-P., & Sander, J. (1999). OPTICS: Ordering points to identify the clustering structure. *Acm Sigmod Record, 28*(2), 49–60. https://doi.org/10.1145/304181.304187

75. Bro, R., & Smilde, A. K. (2014). Principal component analysis. *Analytical Methods, 6*(9), 2812–2831. https://doi.org/10.1039/C3AY41907J
76. Granato, D., Santos, J. S., Escher, G. B., Ferreira, B. L., & Maggio, R. M. (2018). Use of principal component analysis (PCA) and hierarchical cluster analysis (HCA) for a multivariate association between bioactive compounds and functional properties in foods: A critical perspective. *Trends in Food Science & Technology, 72*, 83–90. https://doi.org/10.1016/j.tifs.2017.12.006
77. da Silva, V. H., Lopes, J. F., Oliveira, F. R., dos Santos, J. G., & de Oliveira, H. P. M. (2020). Classification and quantification of microplastics (<100 μm) using a focal plane array–Fourier transform infrared imaging system and machine learning. *Analytical Chemistry, 92*(18), 13724–13733. https://doi.org/10.1021/acs.analchem.0c02638
78. Gruvaeus, G., & Wainer, H. (1972). Two additions to hierarchical cluster analysis. *British Journal of Mathematical and Statistical Psychology, 25*(2), 200–206. https://doi.org/10.1111/j.2044-8317.1972.tb00491.x. arxiv.org
79. Primpke, S., Lorenz, C., Rascher-Friesenhausen, R., & Gerdts, G. (2017). An automated approach for microplastics analysis using focal plane array (FPA) FTIR microscopy and image analysis. *Analytical Methods, 9*(9), 1499–1511. https://doi.org/10.1039/C6AY02476A. pubs.rsc.org
80. Hufnagl, B., Stibi, M., Martirosyan, H., Wilczek, U., Möller, J. N., Löder, M. G. J., Laforsch, C., & Lohninger, H. (2022). Computer-assisted analysis of microplastics in environmental samples based on μFTIR imaging in combination with machine learning. *Environmental Science & Technology Letters, 9*(1), 90–95. https://doi.org/10.1021/acs.estlett.1c00851
81. Matavos-Aramyan, S. (2024). Addressing the microplastic crisis: A multifaceted approach to removal and regulation. *Environmental Advances, 17*, 100579. https://doi.org/10.1016/j.envadv.2024.100579
82. Moshood, T. D., Nawanir, G., Mahmud, F., Mohamad, F., Ahmad, M. H., & AbdulGhani, A. (2022). Sustainability of biodegradable plastics: New problem or solution to solve the global plastic pollution? *Cleaner and Responsible Consumption, 4*, 100273. https://doi.org/10.1016/j.crgsc.2022.100273
83. de Sousa, F. D. B. (2023). Consumer awareness of plastic: An overview of different research areas. *Circular Economy and Sustainability, 3*, 2083–2107. https://doi.org/10.1007/s43615-023-00263-4
84. Kafle, A., Timilsina, A., Gautam, A., Adhikari, K., Bhattarai, A., & Aryal, N. (2022). Phytoremediation: Mechanisms, plant selection and enhancement by natural and synthetic agents. *Environmental Advances, 8*, 100203. https://doi.org/10.1016/j.envadv.2022.100203
85. Wu, C., Ma, Y., Shan, Y., Song, X., Wang, D., Ren, X., Hu, H., Cui, J., & Ma, Y. (2024). Exploring the potential of biochar for the remediation of microbial communities and element cycling in microplastic-contaminated soil. *Chemosphere, 362*, 142698. https://doi.org/10.1016/j.chemosphere.2024.142698
86. Pondala, S., & Botsa, S. M. (2025). Physical, thermal, chemical and biological approaches for plastics degradation–A review. *Cleaner Chemical Engineering, 11*, 100162. https://doi.org/10.1016/j.clce.2025.100162
87. Maisto, M., Ranauda, M. A., & Zuzolo, D. (2022). Influence of sediment texture on HDPE microplastics recovery by density separation. *Journal of Environmental Management, 317*(1–2), 115363. https://doi.org/10.1016/j.jenvman.2022.115363
88. Thapliyal, C., Priya, A., Singh, S. B., & Bahuguna, V. (2024). Potential strategies for bioremediation of microplastic contaminated soil. *Environmental Chemistry and Ecotoxicology, 6*. https://doi.org/10.1016/j.enceco.2024.05.001
89. Wang, M., Xiao, Y., Li, Y., & Liu, J. (2023). Optimistic effects of galaxolide and polystyrene microplastic stress on the physio-biochemical characteristics and metabolic profiles of an ornamental plant. *Plant Physiology and Biochemistry, 196*. https://doi.org/10.1016/j.plaphy.2023.01.062

90. Song, Y., Qiu, R., Hu, J., Li, X., Zhang, X., Chen, Y., Wu, W. M., & He, D. (2020). Biodegradation and disintegration of expanded polystyrene by land snails *Achatina fulica*. *Science of the Total Environment, 746*. https://doi.org/10.1016/j.scitotenv.2020.141289
91. Brandon, A. M., Gao, S. H., Tian, R., Ning, D., Yang, S. S., Zhou, J., Wu, W. M., & Criddle, C. S. (2018). Biodegradation of polyethylene and plastic mixtures in mealworms (larvae of *Tenebrio molitor*) and effects on the gut microbiome. *Environmental Science & Technology, 52*. https://doi.org/10.1021/acs.est.8b02301
92. Wang, Y., Luo, L., Li, X., Wang, J., Wang, H., Chen, C., Guo, H., Han, T., Zhou, A., & Zhao, X. (2022). Different plastics ingestion preferences and efficiencies of superworm (*Zophobas atratus* fab.) and yellow mealworm (*Tenebrio molitor* Linn.) associated with distinct gut microbiome changes. *Science of the Total Environment, 837*. https://doi.org/10.1016/j.scitotenv.2022.155719
93. Khaldoon, S., Lalung, J., Maheer, U., Kamaruddin, M. A., Yhaya, M. F., Alsolami, E. S., Alorfi, H. S., Hussein, M. A., & Rafatullah, M. (2022). A review on the role of earthworms in plastics degradation: Issues and challenges. *Polymers, 14*(21), 4770. https://doi.org/10.3390/polym14214770
94. Mamtimin, T., Han, H., Khan, A., et al. (2023). Gut microbiome of mealworms (Tenebrio molitor larvae) show similar responses to polystyrene and corn straw diets. *Microbiome, 11*, 98. https://doi.org/10.1186/s40168-023-01550-w
95. Bhattacharjee, J., & Roy, S. (2023). Green remediation of microplastics using bionanomaterials. In *Remediation of plastic and microplastic waste* (pp. 240–260). CRC Press. https://doi.org/10.1201/9781003449133-14

# Chapter 10
# Microplastics in Human Consumption: Pathways, Risks, and Public Health Implication

**Sonali Bhosale and Aparna Gunjal**

## Introduction

Microplastics are the plastics with the particle size less than 5 mm, and have emerged as an omnipresent environmental contaminant due to the global pervasiveness of plastic pollution [1]. There is growing concern about microplastics entering the human food chain and the possible health risks they may pose. This chapter examines how microplastics occur, the ways they enter into our food chain and the potential health impacts on humans [2].

### *Sources and Characteristics of Microplastics*

Microplastics (MPs) are generally defined as plastic particles smaller than 5 mm in diameter, although their size can range down to a few nanometers [3]. Based on their origin and method of introduction, these plastic particles are broadly categorized into primary and secondary microplastics.

Primary microplastics are intentionally produced in small sizes for specific uses. They are commonly found in personal care products like exfoliating scrubs, toothpaste, cosmetics, and sunscreens [4]. Industries also contribute by using tiny plastic pellets, called as "nurdles," as raw materials in plastic production. Additionally, processes like air-blasting used for industrial cleaning and surface preparation release notable amounts of microplastic particles into the environment [5].

S. Bhosale · A. Gunjal (✉)
Department of Environmental Science, Dr. D. Y. Patil, Arts, Commerce & Science College, Pimpri, Pune, Maharashtra, India
e-mail: aparna.gunjal@dypvp.edu.in



S. Kulkarni, S. Das (eds.), *Microplastics and Nanoplastics*, Lecture Notes in Nanoscale Science and Technology 49,
https://doi.org/10.1007/978-3-032-17963-0_10

Secondary microplastics are created when bigger plastic items break apart into smaller pieces over time. This happens through exposure to sunlight, constant rubbing or movement like ocean waves and the activity of microorganisms [6]. Common sources include everyday plastic waste such as bags, bottles, fishing gear, packaging, and synthetic clothing materials [7].

Microplastics are found in various forms. They can look like tiny fibers, broken fragments, thin films, round beads, or foam [8]. Their color plays a role in how they affect wildlife brightly colored pieces are more likely to be eaten by fish and other marine animals because they look like food [9]. The most common types of plastic found in microplastics include materials like polyethylene (PE), polypropylene (PP), polystyrene (PS), PVC, and PET, which are often used in everyday products like bags, bottles, and packaging [10]. Microplastics have hydrophobic surfaces that allow them to attract and carry harmful chemicals, such as Persistent Organic pollutants, heavy metals, and additives like phthalates and BPA [11]. When animals or humans swallow these microplastics, the attached toxins can be released inside the digestive system, making them more easily absorbed by the body and potentially more harmful [12].

Humans are a major source of microplastic pollution. For example, washing synthetic clothes releases countless micro fibers into the water, many of which isn't caught by treatment plants and end up in rivers and oceans [13]. Another common but less known source is the tiny particles that come off car tires and roads (Tyre and Road Wear Particles) as vehicles drive these are especially common in cities and urban areas [14]. Latest studies have identified atmospheric fallout as a substantial pathway of microplastic exposure. Microplastic present in air can be inhaled or deposited on food during processing and packaging [15]. The presence of airborne MPs in urban environments is increasingly recognized as a route of human intake that demands further study [16].

## *Environmental Distribution and Pathways to Humans*

Microplastics (MPs) are found almost everywhere e.g. in the ocean, rivers, soil, and even the air because they have low density, durable and are resistant to degradation [17]. As a result, people can be exposed to them in many ways, mainly by eating or drinking contaminated food and water, breathing in airborne particles, or through skin contact. Oceans are recognized as major sinks for microplastic (MP) pollution, while rivers serve as transport pathways, carrying plastic waste from land-based sources to marine ecosystems [18]. Microplastics have been found in seawater, marine sediments, and throughout the marine food web from microscopic plankton to top predators [19]. A key concern is the trophic transfer of MPs through aquatic organisms commonly consumed by humans, including fish, bivalves, and crustaceans, which may pose potential risks to human health [20].

In freshwater environments such as lakes, rivers, and drinking water reservoirs, microplastics (MPs) primarily originate from sources like municipal wastewater, agricultural runoff, and urban stormwater [21]. Although wastewater treatment plants (WWTPs) are generally effective at capturing larger microplastic particles, they still discharge considerable amounts into freshwater systems because current filtration technologies are not fully capable of removing smaller MPs [22].

Agricultural soils are now considered important reservoirs of microplastics (MPs), mainly due to the use of sewage sludge, compost, and plastic mulching films in farming practices [23]. Once microplastics enter the soil, they tend to persist over time and can impact soil health by disrupting microbial communities, potentially reducing soil fertility, and even being taken up by food crops [24]. Plastic waste also builds up in landfills and urban areas, where it gradually breaks down into secondary microplastics (MPs). These smaller particles can then enter stormwater drainage systems or become suspended in the air. In addition, particles generated from tire and road surface wear are a major source of MPs in urban runoff, which often carries them into nearby rivers, lakes, and other water bodies, contributing to environmental pollution [25].

Airborne microplastics come from various sources, including fibers released from synthetic textiles, urban dust, and the wear and tear of vehicle tires and construction materials [26]. These tiny particles can be carried over long distances by wind and have been found in both indoor and outdoor air even in remote areas like the Arctic [27]. Inhalation of airborne microplastics is now being closely studied as a potential pathway for human exposure, particularly in densely populated and industrialized areas. Indoor spaces may present a greater risk of microplastic (MP) exposure because of limited airflow and the buildup of fibers from items like clothing, carpets, and furniture. Research has found that microplastic concentrations in indoor air are often much higher than those measured outdoors [28].

Microplastics have been identified in many common food items and beverages, including seafood, honey, salt, sugar, milk, fruits, and vegetables [29]. Bottled water and soft drinks often show higher levels of microplastic contamination than tap water, most likely due to the leaching of particles from plastic packaging or during the manufacturing process [30]. Microplastic contamination of food may happen at various stages such as processing, packaging, and distribution. Packaging plastics can deteriorate or release small particles, and airborne microplastics can settle onto food items during handling and transport [31]. Eating contaminated food and drinking water is the most popular way people take in microplastics, with estimates indicating that individuals may consume thousands of these particles each year [32]. Breathing in microplastics is a less obvious but important route of exposure, particularly in busy urban or industrial locations [33]. Absorption through the skin is generally thought to be minimal because of the skin's protective barrier, although exposure through cuts or during the use of cosmetic products is still being studied [34].

## *Occurrence of Microplastics in Food and Beverages*

The presence of microplastics (MPs) in food and beverages represents a common and direct source of human exposure. Numerous studies worldwide have detected MPs not only in seafood but also in agricultural products, table salt, and drinking water [35]. This widespread contamination results from environmental pollution as well as from processes involved in food production, packaging, and distribution.

Marine animals, particularly filter feeders and species that live on the seafloor, are key organisms studied for microplastic ingestion. Bivalves like mussels, oysters, and clams are especially prone to accumulating microplastics because they filter large volumes of water and often live in areas with high sediment content [36]. A global review of bivalves reported an average of 0.36 to 0.47 microplastic particles per gram of wet tissue, with differences across regions depending on the extent of coastal pollution [20]. Fish are exposed to microplastics both by directly ingesting them from the surrounding water and by eating prey that has already consumed these particles. Research has identified microplastic fibers and fragments in the stomachs and intestines of commonly consumed fish species such as anchovies, sardines, and mackerel [37]. Although these internal organs are typically removed before the fish reaches consumers, exposure can still happen through the consumption of whole fish or through contamination during processing and preparation.

Table salt, whether it comes from the sea or is mined from rock, is consistently found to contain microplastics. Research has shown that salt from different brands around the world can have anywhere from 120 to over 400 microplastic particles per kilogram [38]. This contamination may happen during the collection of salt from polluted seawater or from airborne microplastics settling on the salt during processing and packaging. Granulated foods like sugar have also been found to contain microplastics, likely introduced through contact with plastic packaging or exposure to airborne particles during manufacturing [39]. Even honey, though considered a natural product, is not free from contamination; microplastics may enter it through airborne particles that settle during bee foraging or during processing and packaging stages [40].

Drinking water is one of the most frequent and direct sources of microplastic exposure in daily life. There is a study which has shown that bottled water has more amounts of microplastics than tap water. A report has shown that bottled water contains plastics such as polypropylene and nylon [30]. This contamination is believed to result from the plastic packaging particularly bottle caps and containers as well as the bottling process itself. Varying levels of microplastic contamination have been detected in commercially bottled beverages such as soft drinks, milk, and beer. This contamination is influenced by both the packaging materials used and the conditions within the processing and bottling environments, which contribute to the presence of microplastic particles [41].

Recent investigations have revealed the presence of microplastics in fresh produce, including apples, carrots, broccoli, and lettuce [29]. These contaminants can enter the plants during irrigation with water contaminated by microplastics or through atmospheric deposition onto the surfaces of the plants. Additionally, micro- and nanoplastics have been shown, in controlled greenhouse studies, to be absorbed by plant roots and subsequently translocated into the edible tissues [42]. Microplastics have also been detected in grains and cereals, particularly when these products are stored or transported in plastic containers or bags [43]. Prolonged storage and mechanical processing of grains can cause abrasion, resulting in the release of plastic particles from the containers into the food.

Processed and packaged foods are susceptible to contamination by microplastics through direct contact with synthetic materials like plastic films, trays, and wrappers. Mechanical processes involved in food preparation, such as cutting and blending, can dislodge plastic particles, which often go unnoticed without the use of advanced filtration and spectroscopic techniques [44]. Additionally, exposure to heat, particularly in microwaveable plastic containers, has been demonstrated to accelerate the release and breakdown of plastic fragments into the food [45]. The occurrence of microplastics in food varies geographically, with elevated concentrations commonly observed near urbanized or industrial coastal regions [46]. Seasonal factors also impact microplastic levels by influencing plastic degradation rates, marine activity, and the amount of pollutant runoff. Additionally, variations in packaging standards and water treatment systems across different countries further contribute to this variability [47].

## *Ingestion, Absorption, and Bioaccumulation in Humans*

Human ingestion of microplastics (MPs) has become a global health issue, given their widespread contamination in a variety of foods, drinks, and inhalable air. Once consumed, MPs can interact with the body by being absorbed, accumulating in different tissues, and potentially migrating to other organs [48]. Although the full range of health effects is still being explored, emerging research suggests that these particles are not biologically inert and may provoke adverse biological reactions. Quantitative studies indicate that humans ingest thousands of microplastic particles annually. Cox et al. (2019) showed that yearly humans consume 39,000 to 52,000 microplastic particles, which depend on age, food, and locality, with excess amount due to inhalation [32]. Furthermore, individuals who mainly consume bottled water may ingest over 90,000 microplastic particles per year from drinking water alone [30]. Dietary patterns play a major role in microplastic exposure, with populations that consume large quantities of seafood, salt, and packaged foods facing a higher risk. Additionally, infants and children may be more susceptible due to their greater intake relative to body weight and the frequent use of plastic-based feeding products [49].

Following ingestion, microplastics (MPs) move through the gastrointestinal (GI) tract. Most particles exceeding 150 μm are typically eliminated without being absorbed [39]. However, smaller particles, especially those below 10 μm and nano-sized plastics, have been shown to cross the intestinal barrier by mechanisms including endocytosis, paracellular diffusion, and transcytosis through Peyer's patches in the ileum [50]. Experimental models have revealed that microplastic particles can build up in Gut-Associated Lymphoid Tissues (GALT), causing localized inflammatory responses [51]. Furthermore, in vitro studies with human intestinal epithelial cells have demonstrated that microplastic exposure results in elevated oxidative stress, disruption of cell membranes, and modifications in cytokine production [52]. After intestinal absorption, microplastics (MPs) may enter the blood or lymphatic systems. From there, they can be transported to secondary organs such as the liver, kidneys, lungs, and potentially the brain [53]. Laboratory experiments in mice using polystyrene particles have shown their accumulation in the liver, spleen, and placenta, raising concerns about possible fetal exposure and associated developmental toxicity [54].

There are reports which have shown microplastics (MPs) can enter even through the mother's placenta. Ragusa et al. (2021) identified microplastics in the placental tissues of healthy women, demonstrating the ability of these particles to traverse the maternal-fetal barrier [55]. This finding underscores the necessity for additional studies to assess the potential developmental and reproductive toxicity associated with MP exposure. Unlike heavy metals and persistent organic pollutants, microplastics are less likely to bioaccumulate in the body due to their partial elimination. However, ongoing exposure combined with inadequate clearance mechanisms can lead to the accumulation of particles, especially those that are smaller or have surface modifications [56]. The microplastics are found in human tissues in a large amount. A 2022 study detected microplastics in lung biopsy samples, demonstrating their prolonged retention in the tissue [57]. Moreover, the identification of MPs in blood samples supports their systemic circulation and ongoing exposure [58].

There are reports which have shown that microplastics (MPs) can affect and damage the gut microbiota which causes dysbiosis. Studies in rodent models exposed to MPs have shown a decline in microbial diversity alongside a rise in pro-inflammatory bacterial species [59]. Such microbial imbalances are linked to metabolic dysfunction, immune system disorders, and neurobehavioral impairments. Although human data remain limited, the critical role of gut microbiota in maintaining health suggests that long-term ingestion of microplastics (MPs) could potentially contribute to a range of pathophysiological effects [60].

## *Toxicological Effects and Health Risks*

Microplastics (MPs), while chemically inert in their original polymeric state, can induce various toxic effects because of their physical characteristics, the presence of chemical additives, and their ability to adsorb environmental contaminants. These properties enable MPs to interact with human biological systems, triggering cellular, systemic, and immune responses, which may pose substantial health risks with prolonged exposure. Microplastics (MPs) induce toxicity through both physical and chemical mechanisms. Physically, ingested MPs may cause irritation, abrasion, or obstruction within the gastrointestinal tract, as demonstrated in animal studies involving high exposure levels [61]. While such physical impacts are less well documented in humans, there is ongoing concern for vulnerable populations, including infants and individuals with preexisting gastrointestinal conditions.

Microplastics (MPs) frequently contain chemical additives such as phthalates, bisphenol A (BPA), polybrominated diphenyl ethers (PBDEs), and heavy metals, many of which function as endocrine disruptors [62]. These chemicals are therefore dangerous. Furthermore, MPs serve as carriers for environmental contaminants, including persistent organic pollutants (POPs), antibiotics, and heavy metals, by adsorbing these substances from their surroundings [63]. Upon ingestion, these adsorbed pollutants can detach from microplastics under the acidic and enzymatic environment of the gastrointestinal tract, which may increase their bioavailability and contribute to systemic toxicity [64].

Exposure to microplastics (MPs) has been linked to oxidative stress, mitochondrial impairment, and inflammatory responses. In vitro experiments using human epithelial and immune cells have demonstrated that contact with MPs, especially fragments of polystyrene and polyethylene terephthalate (PET), leads to elevated production of reactive oxygen species (ROS), increased lipid peroxidation, and enhanced cytokine secretion [65]. Studies in animals support these results, showing that oral exposure to microplastics causes widespread inflammation, changes in antioxidant enzyme activity, and DNA damage in organs such as the liver, kidneys, and intestines [66].

The body's immune reaction to microplastics is still being studied. Research indicates that microplastics may trigger the innate immune system, shown by elevated amounts of IL-6, TNF-α, and other inflammation-related cytokines [67]. Long-term exposure might cause the immune system to become weakened or unbalanced, making the body more prone to infections and autoimmune diseases. Microplastics might act as adjuvants, boosting the immune system's reaction to other antigens present at the same time. This effect raises concerns, especially regarding food allergies and hypersensitivity responses [68].

Endocrine disruption related to microplastics mainly results from additives such as BPA, phthalates, and nonylphenol that leach out. These chemicals interfere with hormone receptors and metabolic processes. Research has connected microplastic exposure to changes in estrogen, thyroid, and androgen functions [69]. Research on animals has shown that exposure to microplastics can harm reproductive health, causing lower sperm quality, irregular menstrual cycles, and hormone disruptions [70]. Microplastics and nanoplastics can potentially pass through the blood-brain barrier and build up in brain tissues. Studies in rodents reveal that exposure to polystyrene microplastics leads to altered neurotransmitter levels, memory problems, and behaviors resembling anxiety [71]. These effects are thought to result from oxidative stress, inflammation, and possible death of nerve cells. While human data is still limited, these results highlight possible dangers, especially for fetuses and children whose nervous systems are still developing [68]. No direct evidence currently proves that microplastic exposure causes cancer in humans, but several findings indicate a possible cancer risk. Microplastics may trigger chronic inflammation, oxidative DNA damage, and genomic instability, processes commonly linked to the formation of cancer [72]. Moreover, chemicals added to plastics, like vinyl chloride (a known cancer-causing agent), and harmful compounds such as polycyclic aromatic hydrocarbons (PAHs) that can bind to microplastics, have been linked to cancer through other forms of exposure [73]. This suggests that long-term contact with microplastics, particularly those loaded with toxic substances, might raise the risk of developing cancer over time.

Groups such as children, pregnant women, the elderly, and those with pre-existing health conditions are at higher risk from microplastic exposure. For instance, microplastics can pass through the placenta, potentially affecting fetal development [55]. Infants are also at risk of ingesting microplastics from formula prepared in plastic bottles, which may release millions of microplastic particles when exposed to heat [49].

## *Detection Methods and Analytical Challenges*

Accurately identifying, measuring, and analyzing microplastics (MPs) in biological samples and food is essential for assessing health risks and creating regulations. However, the wide variety of MPs differing in shape, size, color, polymer composition, and chemical additives makes this task very challenging. Detecting microplastics in complex materials such as human tissues, bodily fluids, and food items demands advanced and multi-stage analytical techniques. Accurate detection relies on careful sampling and precise sample preparation. Contamination from airborne plastic particles and laboratory materials poses a significant risk. To minimize false positives, it is essential to use non-plastic tools, conduct work in laminar flow hoods, and incorporate procedural blanks during testing [74]. Sample preparation for microplastic (MP) analysis typically involves digesting organic matter to isolate the particles. Common digestion approaches include:

- Enzymatic digestion: Utilizes enzymes such as proteinase-K or lipase to gently degrade biological tissues while preserving the integrity of MPs [75].
- Chemical digestion: Employs reagents like hydrogen peroxide ($H_2O_2$), potassium hydroxide (KOH), or nitric acid ($HNO_3$) to break down organic material efficiently; however, these chemicals can sometimes damage sensitive types of polymers [76].
- Density separation: Relies on differences in density between MPs and other debris by using high-density salt solutions such as zinc chloride ($ZnCl_2$) or sodium iodide (NaI), allowing MPs to float and be separated from heavier particles [77].

Achieving a balance between effective removal of organic matter and maintaining the structural integrity of microplastics remains a significant challenge in sample preparation protocols.

Microscopic techniques play a crucial role in the initial evaluation of microplastic samples:

- Optical microscopy: Suitable for observing particles larger than approximately 50 μm, though it cannot determine the chemical composition of the particles [8].
- Fluorescence microscopy: Enhances visualization by staining hydrophobic polymers with dyes like Nile Red, which fluoresce under ultraviolet light [78].
- Scanning Electron Microscopy (SEM): Provides detailed, high-resolution images of particle surfaces, but does not offer direct chemical identification [79].

Since microscopy alone cannot confirm polymer identity, it is often combined with spectroscopic analyses for comprehensive characterization.

Spectroscopic methods are considered the definitive techniques for verifying the composition of microplastics:

- Fourier Transform Infrared Spectroscopy (FTIR): Commonly used to identify polymer types in particles larger than 10 μm. Micro-FTIR enables spatial mapping of MPs within complex matrices such as tissues or food samples [80].
- Raman Spectroscopy: Effective for detecting smaller MPs below 10 μm and capable of analyzing colored or pigmented plastics. Raman imaging provides detailed spatial distribution information within samples [81].

Both FTIR and Raman spectroscopy depend on reference spectral libraries to accurately classify polymers. Disadvantages include interference from sample fluorescence and low signals due to sample matrices. Both Fourier Transform Infrared Spectroscopy (FTIR) and Raman Spectroscopy depend on extensive reference libraries of polymer spectra to ensure accurate identification of microplastics. However, these techniques face challenges such as fluorescence interference and signal attenuation, especially when analyzing complex sample matrices.

Thermal analytical methods are widely employed for the quantitative measurement of total microplastic content:

- Pyrolysis-Gas Chromatography-Mass Spectrometry (Py-GC-MS): This technique thermally decomposes microplastics into characteristic chemical fragments, which are then separated and identified by gas chromatography-mass spectrometry [82].
- Thermogravimetric Analysis (TGA): Measures the thermal decomposition profiles of polymers, providing an estimate of polymer mass within a sample [83].

While these approaches offer highly sensitive and quantitative data, they are inherently destructive and do not provide morphological information about the particles. Emerging advanced analytical methods include:

- Laser Direct Infrared (LDIR) Imaging: Combines fast scanning capabilities with high spectral resolution, allowing detailed microplastic analysis directly within tissue samples [84].

- Time-of-Flight Secondary Ion Mass Spectrometry (ToF-SIMS): Enables chemical mapping at the nanoscale, which is particularly useful for investigating the behavior of nanoplastics [85].
- Nano-FTIR and Tip-Enhanced Raman Spectroscopy (TERS): These cutting-edge techniques push the boundaries of detection to below 100 nanometers, providing unparalleled chemical resolution at the nanoscale [86].

Although promising, these methods are still in developmental stages and not yet widely adopted for routine micro- and nanoplastic analysis.

Several key challenges continue to limit the effectiveness and comparability of microplastic analyses:

- Lack of Standardization: Variability in protocols across studies hampers direct comparison of results, highlighting the urgent need for internationally accepted standard methods [87].
- Size Detection Limits: Detecting nanoplastics smaller than 1 μm remains difficult due to current technological constraints [88].
- Complex Sample Matrices: Biological tissues (such as blood, placenta, and organs) and processed food samples contain complex backgrounds that interfere with MP isolation and detection [89].
- Polymer Degradation: Harsh digestion processes and prolonged analysis times may alter or degrade microplastics, potentially leading to underestimations in quantification [90].

Given these limitations, no single analytical technique can comprehensively characterize the full range of micro- and nanoplastic contamination. Consequently, combining microscopy, spectroscopy, and thermal analysis represents the most reliable approach at present.

Looking ahead, future research and technological development should prioritize:

- The creation of miniaturized, portable sensors capable of rapid, on-site microplastic screening.
- The development of non-destructive, label-free imaging methods to enable real-time, in vivo tracking of microplastics.
- The integration of artificial intelligence (AI) algorithms for automated recognition and classification of particles, improving throughput and accuracy [91].

Collaboration across scientific disciplines will be essential to harmonize analytical protocols and establish validated reference materials for microplastic research.

## *Risk Assessment and Toxicological Implications*

Determining the toxicological impact of microplastics (MPs) on human health is a critical and ongoing challenge in environmental toxicology. Although evidence increasingly shows that MPs are present in foods and biological tissues consumed by humans, assessing the true health risk is complicated by the interplay of particle characteristics, exposure pathways, biological defenses, and individual variability.

#### Exposure Assessment

Risk assessment starts by evaluating how much humans are exposed to microplastics, which depends on factors like where people live, their daily habits, and what they eat or drink. Research indicates that people might consume tens of thousands of microplastic particles each year through food and drinking water, with even more inhaled from the air [32]. For example, bottled water can contain as many as 10,000 microplastic particles per liter [30]. Additionally, airborne fibers, especially those released from synthetic clothing and urban dust, are a significant source of respiratory exposure [15].

#### Bioavailability and Uptake

Understanding how many microplastics (MPs) can pass through biological membranes and reach the bloodstream is essential for evaluating their risk. Smaller particles, particularly those under 10 μm and in the nanoscale size range, are more likely to be absorbed because they can cross epithelial barriers through processes like endocytosis or movement between cells (paracellular diffusion) [92]. Research using mouse models and human cell cultures in the lab shows that MPs can build up in organs such as the liver, spleen, and kidneys, and they are even capable of crossing the placental barrier [55]. The size, shape, and surface properties of particles affect how they are absorbed by the body. Irregularly shaped fragments can cause more physical harm, while chemicals attached to their surfaces, such as pollutants or additives, can increase their toxic effects [65]. Hydrophobic plastics, which repel water, often attract and hold environmental contaminants like polychlorinated biphenyls (PCBs), polycyclic aromatic hydrocarbons (PAHs), and heavy metals; these substances may be released when the particles are ingested [93].

#### Cellular and Molecular Effects

Microplastics (MPs) can trigger various harmful cellular effects:

- Oxidative stress: MPs promote the production of reactive oxygen species (ROS), which can damage cell components by causing lipid peroxidation, protein modification, and DNA damage [94].
- Inflammatory response: Exposure to MPs activates immune signaling, leading to the release of pro-inflammatory molecules like interleukin-6 (IL-6) and tumor necrosis factor-alpha (TNF-α), potentially resulting in ongoing inflammation [34].

- Cell toxicity and apoptosis: At high concentrations, MPs can lead to cell death by disrupting mitochondrial function and activating enzymes such as caspases that initiate programmed cell death [95].
- Hormonal disruption: Chemicals added to plastics, such as bisphenol A (BPA), phthalates, and flame retardants can leach out and interfere with the body's hormonal systems [96].

These effects have been seen in various human cell types, including intestinal, liver, and placental cells, suggesting that MPs may pose a risk to multiple organ systems.

### Organ-Level and Systemic Effects

Animal research has shown that microplastics (MPs) can affect specific organs in distinct ways:

- Digestive system: When ingested, MPs can disrupt the balance of gut microbiota and interfere with the absorption of nutrients [54].
- Liver and kidneys: MPs tend to build up in these organs, leading to tissue damage and changes in enzyme function [51].
- Nervous system: Recent findings indicate that nanoplastics may be able to cross the blood-brain barrier, potentially impacting brain function [97].

While long-term, low-level exposure to MPs is not yet fully understood, ongoing accumulation over a lifetime may increase the risk of metabolic, immune, and reproductive health problems.

### Vulnerable Populations

Certain groups, such as infants, pregnant individuals, and those with existing health conditions are especially vulnerable to the effects of microplastics (MPs). The presence of MPs in human placental tissue has raised concerns about potential exposure to the developing fetus [55]. Additionally, infants who are bottle-fed using plastic containers may consume substantially higher amounts of MPs, as heating during formula preparation can release particles from the plastic [98].

### Data Gaps and Uncertainties

Several key limitations currently impede accurate risk assessment of microplastics (MPs):

- Inconsistent measurement units: Exposure levels are reported in various forms such as particle count, mass, or surface area which makes it difficult to compare results across studies [58].

- Simplified experimental conditions: Many laboratory studies use clean, unaltered MPs rather than aged particles that have interacted with environmental pollutants, reducing real-world relevance [99].
- Limited data on long-term exposure: Most toxicological research focuses on short-term or moderate exposure, with a lack of studies examining the effects of chronic, low-dose exposure over time [100].
- Challenges in applying animal data to humans: Much of the current evidence comes from rodent studies or in vitro experiments, and human-specific data remain scarce [34].

Because of these uncertainties, regulatory bodies have not yet defined health-based safety thresholds for microplastic exposure.

### *Toward Comprehensive Risk Frameworks*

A comprehensive risk assessment of microplastics (MPs) should incorporate multiple key components:

- Exposure through different pathways: This includes ingestion, inhalation, and skin contact.
- Toxicokinetics: Understanding how MPs are absorbed, distributed, metabolized, and eliminated by the body.
- Mechanistic insights: Using advanced tools like omics technologies and systems biology to uncover how MPs affect biological systems at the molecular level.
- Combined risks: Evaluating the cumulative impact of microplastics along with chemical additives, environmental pollutants, and total particle load.

To effectively model and predict health outcomes from MP exposure, collaboration among toxicologists, epidemiologists, and environmental scientists is critical [101].

## Concluding Remarks

The growing presence of microplastics (MPs) in human consumption patterns presents an urgent environmental and public health challenge that demands concerted action. While much has been learned about the prevalence of MPs in food, water, and air, their toxicological effects, bioavailability, and long-term health implications remain areas of active research. The evidence presented throughout this chapter underscores the need for a multi-disciplinary approach to mitigate the risks associated with microplastics, from improved detection methods to more stringent regulatory frameworks.

**Acknowledgement** None Declare

**Conflict of Interest** None Declare

## References


1. Hompson, R. C., Olsen, Y., Mitchell, R. P., Davis, A., Rowland, S. J., John, A. W. G., & Russell, A. E. (2004). Lost at sea: Where is all the plastic? *Science, 304*(5672), 838.
2. Smith, M., Love, D. C., Rochman, C. M., & Neff, R. A. (2018). Microplastics in seafood and the implications for human health. *Current Environmental Health Reports, 5*(3), 375–386.
3. Frias, J. P. G. L., & Nash, R. (2019). Microplastics: Finding a consensus on the definition. *Marine Pollution Bulletin, 138*, 145–147.
4. Napper, I. E., & Thompson, R. C. (2016). Release of synthetic microplastic plastic fibres from domestic washing machines: Effects of fabric type and washing conditions. *Marine Pollution Bulletin, 112*(1–2), 39–45.
5. Fendall, L. S., & Sewell, M. A. (2009). Contributing to marine pollution by washing your face: Microplastics in facial cleansers. *Marine Pollution Bulletin, 58*(8), 1225–1228.
6. Andrady, A. L. (2011). Microplastics in the marine environment. *Marine Pollution Bulletin, 62*(8), 1596–1605.
7. Cole, M., Lindeque, P., Halsband, C., & Galloway, T. S. (2011). Microplastics as contaminants in the marine environment: A review. *Marine Pollution Bulletin, 62*(12), 2588–2597.
8. Hidalgo-Ruz, V., Gutow, L., Thompson, R. C., & Thiel, M. (2012). Microplastics in the marine environment: A review of the methods used for identification and quantification. *Environmental Science & Technology, 46*(6), 3060–3075.
9. Boerger, C. M., Lattin, G. L., Moore, S. L., & Moore, C. J. (2010). Plastic ingestion by planktivorous fishes in the North Pacific Central Gyre. *Marine Pollution Bulletin, 60*(12), 2275–2278.
10. Browne, M. A., Crump, P., Niven, S. J., Teuten, E., Tonkin, A., Galloway, T., & Thompson, R. (2011). Accumulation of microplastic on shorelines worldwide: Sources and sinks. *Environmental Science & Technology, 45*(21), 9175–9179.
11. Teuten, E. L., Saquing, J. M., Knappe, D. R., Barlaz, M. A., Jonsson, S., Björn, A., & Takada, H. (2009). Transport and release of chemicals from plastics to the environment and to wildlife. *Philosophical Transactions of the Royal Society B: Biological Sciences, 364*(1526), 2027–2045.
12. Wright, S. L., Thompson, R. C., & Galloway, T. S. (2013). The physical impacts of microplastics on marine organisms: A review. *Environmental Pollution, 178*, 483–492.
13. Henry, B., Laitala, K., & Klepp, I. G. (2019). Microfibres from apparel and home textiles: Prospects for including microplastics in environmental sustainability assessment. *Science of the Total Environment, 652*, 483–494.
14. Kole, P. J., Löhr, A. J., Van Belleghem, F. G., & Ragas, A. M. (2017). Wear and tear of tyres: A stealthy source of microplastics in the environment. *International Journal of Environmental Research and Public Health, 14*(10), 1265.
15. Dris, R., Gasperi, J., Mirande, C., Mandin, C., Guerrouache, M., Langlois, V., & Tassin, B. (2017). A first overview of textile fibers, including microplastics, in indoor and outdoor environments. *Environmental Pollution, 221*, 453–458.
16. Allen, S., Allen, D., Phoenix, V. R., Le Roux, G., Jiménez, P. D., Simonneau, A., et al. (2019). Atmospheric transport and deposition of microplastics in a remote mountain catchment. *Nature Geoscience, 12*(5), 339–344.
17. Rochman, C. M. (2015). The complex mixture, fate and toxicity of chemicals associated with plastic debris in the marine environment. In M. Bergmann, L. Gutow, & M. Klages (Eds.), *Marine anthropogenic litter* (pp. 117–140). Springer.
18. Lebreton, L. C. M., van der Zwet, J., Damsteeg, J. W., Slat, B., Andrady, A., & Reisser, J. (2017). River plastic emissions to the world's oceans. *Nature Communications, 8*, 15611.
19. Setälä, O., Fleming-Lehtinen, V., & Lehtiniemi, M. (2014). Ingestion and transfer of microplastics in the planktonic food web. *Environmental Pollution, 185*, 77–83.
20. Van Cauwenberghe, L., & Janssen, C. R. (2014). Microplastics in bivalves cultured for human consumption. *Environmental Pollution, 193*, 65–70.

21. Mintenig, S. M., Int-Veen, I., Löder, M. G. J., Primpke, S., & Gerdts, G. (2017). Identification of microplastic in effluents of waste water treatment plants using focal plane array-based micro-Fourier-transform infrared imaging. *Water Research, 108*, 365–372.
22. Murphy, F., Ewins, C., Carbonnier, F., & Quinn, B. (2016). Wastewater treatment works (WwTW) as a source of microplastics in the aquatic environment. *Environmental Science & Technology, 50*(11), 5800–5808.
23. Nizzetto, L., Futter, M., & Langaas, S. (2016). Are agricultural soils dumps for microplastics of urban origin? *Environmental Science & Technology, 50*(20), 10777–10779.
24. Machado, A. A. D. S., Lau, C. W., Till, J., Kloas, W., Lehmann, A., Becker, R., & Rillig, M. C. (2018). Impacts of microplastics on the soil biophysical environment. *Environmental Science & Technology, 52*(17), 9656–9665.
25. Wagner, S., Hüffer, T., Klöckner, P., Wehrhahn, M., Hofmann, T., & Reemtsma, T. (2018). Tire wear particles in the aquatic environment—A review on generation, analysis, occurrence, fate and effects. *Water Research, 139*, 83–100.
26. Liu, K., Wang, X., Fang, T., Xu, P., Zhu, L., & Li, D. (2019). Source and potential risk assessment of suspended atmospheric microplastics in Shanghai. *Science of the Total Environment, 675*, 462–471.
27. Bergmann, M., Mützel, S., Primpke, S., Tekman, M. B., Trachsel, J., & Gerdts, G. (2019). White and wonderful? Microplastics prevail in snow from the Alps to the Arctic. *Science Advances, 5*(8), eaax1157.
28. Prata, J. C. (2018). Airborne microplastics: Consequences to human health? *Environmental Pollution, 234*, 115–126.
29. Oliveri Conti, G., Ferrante, M., Banni, M., Favara, C., Nicolosi, I., Cristaldi, A., & Fiore, M. (2020). Micro- and nano-plastics in edible fruit and vegetables. *Scientific Reports, 10*, 17552.
30. Mason, S. A., Welch, V. G., & Neratko, J. (2018). Synthetic polymer contamination in bottled water. *Frontiers in Chemistry, 6*, 407.
31. Catarino, A. I., Macchia, V., Sanderson, W. G., Thompson, R. C., & Henry, T. B. (2018). Low levels of microplastics (MP) in wild mussels indicate that MP ingestion by humans is minimal compared to exposure via household fibers fallout during a meal. *Environmental Pollution, 237*, 675–684.
32. Cox, K. D., Covernton, G. A., Davies, H. L., Dower, J. F., Juanes, F., & Dudas, S. E. (2019). Human consumption of microplastics. *Environmental Science & Technology, 53*(12), 7068–7074.
33. Vianello, A., Jensen, R. L., Liu, L., & Vollertsen, J. (2019). Simulating human exposure to indoor airborne microplastics using a breathing thermal manikin. *Scientific Reports, 9*(1), 8670.
34. Yong, C. Q. Y., Valiyaveettil, S., & Tang, B. L. (2020). Toxicity of microplastics and nanoplastics in mammalian systems. *International Journal of Environmental Research and Public Health, 17*(5), 1509.
35. Karami, A., Golieskardi, A., Keong Choo, C., Larat, V., Galloway, T. S., & Salamatinia, B. (2017). The presence of microplastics in commercial salts from different countries. *Scientific Reports, 7*, 46173.
36. Van Cauwenberghe, L., Vanreusel, A., Mees, J., & Janssen, C. R. (2013). Microplastic pollution in deep-sea sediments. *Environmental Pollution, 182*, 495–499.
37. Barboza, L. G. A., Vieira, L. R., Branco, V., Figueiredo, N., Carvalho, F., Carvalho, C., & Guilhermino, L. (2020). Microplastics cause neurotoxicity, oxidative damage and energy-related changes and interact with the bioaccumulation of mercury in the European seabass, *Dicentrarchus labrax* (Linnaeus, 1758). *Aquatic Toxicology, 219*, 105311.
38. Karami, A., Golieskardi, A., Keong Choo, C., Larat, V., & Salamatinia, B. (2017). Microplastic and mesoplastic contamination in canned sardines and sprats. *Science of The Total Environment, 612*, 1380–1386.

39. EFSA Panel on Contaminants in the Food Chain (CONTAM). (2016). Presence of microplastics and nanoplastics in food, with particular focus on seafood. *EFSA Journal, 14*(6), e04501.
40. Liebezeit, G., & Liebezeit, E. (2013). Non-pollen particulates in honey and sugar. *Food Additives & Contaminants: Part A, 30*(12), 2136–2140.
41. Schymanski, D., Goldbeck, C., Humpf, H. U., & Fürst, P. (2018). Analysis of microplastics in water by micro-Raman spectroscopy: Release of plastic particles from different packaging into mineral water. *Water Research, 129*, 154–162.
42. Li, L., Luo, Y., Li, R., Zhou, Q., Peijnenburg, W. J. G. M., Yin, N., & Zhang, Y. (2020). Effective uptake of submicrometre plastics by crop plants via a crack-entry mode. *Nature Sustainability, 3*(11), 929–937.
43. Gündoğdu, S., & Çevik, C. (2017). Microplastic pollution in table salts from Turkey. *Food Additives & Contaminants. Part A, Chemistry, Analysis, Control, Exposure & Risk Assessment, 34*(10), 1856–1864.
44. Toussaint, B., Raffael, B., Angers-Loustau, A., Gilliland, D., Kestens, V., Petrillo, M., & Van den Eede, G. (2019). Review of micro- and nanoplastic contamination in the food chain. *Food Additives & Contaminants: Part A, 36*(5), 639–673.
45. Zimmermann, L., Dierkes, G., Ternes, T. A., Völker, C., & Wagner, M. (2019). Benchmarking the in vitro toxicity and chemical composition of plastic consumer products. *Environmental Science & Technology, 53*(19), 11467–11477.
46. Zhang, Q., Xu, E. G., Li, J., Chen, Q., Ma, L., Zeng, E. Y., & Shi, H. (2020). A review of microplastics in table salt, drinking water, and air: Direct human exposure. *Environmental Science & Technology, 54*(7), 3740–3751.
47. Barboza, L. G. A., & Gimenez, B. C. G. (2015). Microplastics in the marine environment: Current trends and future perspectives. *Marine Pollution Bulletin, 97*(1–2), 5–12.
48. Wright, S. L., Ulke, J., Font, A., Chan, K. L., & Kelly, F. J. (2020). Atmospheric microplastic deposition in an urban environment and an evaluation of transport. *Environment International, 136*, 105411.
49. Koelmans, A. A., Nor, N. H. M., Hermsen, E., Kooi, M., Mintenig, S. M., & De France, J. (2022). Microplastics in freshwaters and drinking water: Critical review and assessment of data quality. *Water Research, 223*, 118974.
50. Walczak, A. P., Hendriksen, P. J. M., Woutersen, R. A., van der Zande, M., Undas, A. K., Helsdingen, R., & Bouwmeester, H. (2015). Bioavailability and biodistribution of differently charged polystyrene nanoparticles upon oral exposure in rats. *Journal of Nanoparticle Research, 17*(5), 231.
51. Deng, Y., Zhang, Y., Lemos, B., & Ren, H. (2017). Tissue accumulation of microplastics in mice and biomarker responses suggest widespread health risks of exposure. *Scientific Reports, 7*, 46687.
52. Stock, V., Böhmert, L., Lisicki, E., Block, R., Cara-Carmona, J., Pack, L. K., & Sieg, H. (2019). Uptake and effects of orally ingested polystyrene microplastic particles in vitro and in vivo. *Archives of Toxicology, 93*, 1817–1833.
53. Galloway, T. S., & Lewis, C. N. (2016). Marine microplastics spell big problems for future generations. *Proceedings of the National Academy of Sciences of the United States of America, 113*(9), 2331–2333.
54. Jin, Y., Lu, L., Tu, W., Luo, T., & Fu, Z. (2019). Impacts of polystyrene microplastic on the gut barrier, microbiota and metabolism of mice. *Science of the Total Environment, 649*, 308–317.
55. Ragusa, A., Svelato, A., Santacroce, C., Catalano, P., Notarstefano, V., Carnevali, O., & Giorgini, E. (2021). Plasticenta: First evidence of microplastics in human placenta. *Environment International, 146*, 106274.
56. Sharma, S., & Chatterjee, S. (2017). Microplastic pollution, a threat to marine ecosystem and human health: A short review. *Environmental Science and Pollution Research, 24*(27), 21530–21547.

57. Jenner, L. C., Rotchell, J. M., Bennett, R. T., Cowen, M., Tentzeris, V., & Sadofsky, L. R. (2022). Detection of microplastics in human lung tissue using μFTIR spectroscopy. *Science of the Total Environment, 831*, 154907.
58. Leslie, H. A., van Velzen, M. J. M., Brandsma, S. H., Vethaak, A. D., Garcia-Vallejo, J. J., & Lamoree, M. H. (2022). Discovery and quantification of plastic particle pollution in human blood. *Environment International, 163*, 107199.
59. Lu, L., Wan, Z., Luo, T., Fu, Z., & Jin, Y. (2018). Polystyrene microplastics induce gut microbiota dysbiosis and hepatic lipid metabolism disorder in mice. Science of The Total Environment, 631–632, 449–458.
60. Jiang, C., Yin, L., Li, Z., Wen, X., Luo, X., Hu, S., & Zhou, Z. (2020). Microplastics in human health: Current status and perspectives. *Environmental Science and Pollution Research, 27*, 24829–24845.
61. Hirt, N., & Body-Malapel, M. (2020). Immunotoxicity and intestinal effects of nano- and microplastics: A review of the literature. *Particle and Fibre Toxicology, 17*(1), 57.
62. Hermabessiere, L., Dehaut, A., Paul-Pont, I., Lacroix, C., Jezequel, R., Soudant, P., & Duflos, G. (2017). Occurrence and effects of plastic additives on marine environments and organisms: A review. *Chemosphere, 182*, 781–793.
63. Hartmann, N. B., Rist, S., Bodin, J., Jensen, L. H. S., Schmidt, S. N., Mayer, P., & Baun, A. (2017). Microplastics as vectors for environmental contaminants: Exploring sorption, desorption, and transfer to biota. *Integrated Environmental Assessment and Management, 13*(3), 488–493.
64. Koelmans, A. A., Bakir, A., Burton, G. A., & Janssen, C. R. (2016). Microplastic as a vector for chemicals in the aquatic environment: Critical review and model-supported reinterpretation of empirical studies. *Environmental Science & Technology, 50*(7), 3315–3326.
65. Schirinzi, G. F., Pérez-Pomeda, I., Sanchís, J., Rossini, C., Farré, M., & Barceló, D. (2017). Cytotoxic effects of commonly used nanomaterials and microplastics on cerebral and epithelial human cells. *Environmental Research, 159*, 579–587.
66. Hou, B., Wang, F., Liu, T., Wang, Z., & Yang, M. (2021). Polystyrene microplastics induce hepatotoxicity and disrupt lipid metabolism in mice. *Science of the Total Environment, 747*, 141370.
67. Lim, X. (2021). Microplastics are everywhere—But are they harmful? *Nature, 593*, 22–25.
68. Prüst, M., Meijer, J., & Westerink, R. H. S. (2020). The plastic brain: Neurotoxicity of micro- and nanoplastics. *Particle and Fibre Toxicology, 17*, 24.
69. Wang, Y. L., Lee, Y. H., Hsieh, T. H., Huang, Y. S., & Huang, C. C. (2021). Microplastic exposure causes reproductive toxicity through endocrine disruption and oxidative stress in zebrafish. *Environmental Pollution, 280*, 116920.
70. Hou, L., Xu, M., Yu, Y., Lv, D., Zhang, Y., & Wu, X. (2021). The impact of microplastic exposure on female reproduction in mice. *Environmental Science and Pollution Research, 28*, 67484–67493.
71. Chen, Q., Wang, F., Sun, J., Zhang, H., & Li, Y. (2020). Neurobehavioral effects of polystyrene microplastics in mice: A pilot study. *Environmental Science and Pollution Research, 27*, 33642–33648.
72. Li, B., Ding, Y., Cheng, X., Sheng, D., Xu, Z., Rong, Q., & Sun, F. (2021). Microplastics interfere with the immune system and induce inflammatory responses in mice: Potential environmental health risks. *Environmental Science & Technology, 55*(9), 6289–6300.
73. Groh, K. J., Backhaus, T., Carney-Almroth, B., Geueke, B., Inostroza, P. A., Lennquist, A., & Muncke, J. (2019). Overview of known plastic packaging-associated chemicals and their hazards. *Science of The Total Environment, 651*, 3253–3268.
74. Cabernard, L., Roscher, L., Lorenz, C., Gerdts, G., & Primpke, S. (2021). A critical review of analytical methods for the quantification of microplastics in environmental matrices. *Environmental Science & Technology, 55*(4), 2279–2290.

75. Löder, M. G. J., Kuczera, M., Mintenig, S., Lorenz, C., & Gerdts, G. (2015). Focal plane array detector-based micro-Fourier-transform infrared imaging for the analysis of microplastics in environmental samples. *Environmental Chemistry, 12*(5), 563–581.
76. Mintenig, S. M., Int-Veen, I., Löder, M. G. J., Primpke, S., & Gerdts, G. (2017). Identification of microplastic in effluents of wastewater treatment plants using focal plane array-based micro-Fourier-transform infrared imaging. *Water Research, 108*, 365–372.
77. Claessens, M., Van Cauwenberghe, L., Vandegehuchte, M. B., & Janssen, C. R. (2013). New techniques for the detection of microplastics in sediments and field collected organisms. *Marine Pollution Bulletin, 70*(1–2), 227–233.
78. Maes, T., Jessop, R., Wellner, N., Haupt, K., & Mayes, A. G. (2017). A rapid-screening approach to detect and quantify microplastics based on fluorescent tagging with Nile Red. *Scientific Reports, 7*, 44501.
79. Yu, F., Yang, C., Zhu, Z., Bai, X., & Ma, J. (2019). Adsorption behavior of organic pollutants on microplastics. Ecotoxicology and Environmental Safety, 173, 417–423.
80. Primpke, S., Lorenz, C., Rascher-Friesenhausen, R., & Gerdts, G. (2017). An automated approach for microplastic analysis using focal plane array (FPA) FTIR microscopy and image analysis. *Analytical Methods, 9*(9), 1499–1511.
81. Käppler, A., Fischer, M., Scholz-Böttcher, B. M., Oberbeckmann, S., Labrenz, M., Fischer, D., & Eichhorn, K. J. (2018). Comparison of μ-ATR-FTIR spectroscopy and Raman spectroscopy for the identification of microplastics in the aquatic environment. *Environmental Science & Technology, 52*(22), 13279–13288.
82. Fries, E., Zarfl, C., Wagner, M., & Schmitt-Jansen, M. (2013). Identification of polymer types and additives in marine microplastic particles using pyrolysis-GC/MS and scanning electron microscopy. *Environmental Science: Processes & Impacts, 15*(10), 1949–1956.
83. Anger, P. M., Prechtl, L., Elsner, M., Niessner, R., & Ivleva, N. P. (2018). TGA-FTIR analysis for microplastic identification and quantification in complex environmental samples. *Analytical Methods, 10*(37), 4360–4367.
84. Dong, M., Song, Y., Zhu, L., & Hu, J. (2020). Analysis and characterization of microplastics and nanoplastics: Emerging methods and their limitations. *Chemical Engineering Journal, 401*, 126810.
85. Li, J., Liu, H., & Chen, J. P. (2018). Microplastics in freshwater systems: A review on occurrence, environmental effects, and methods for microplastics detection. *Water Research, 137*, 362–374.
86. Petersen, E. J., Flores-Cervantes, D. X., Bucheli, T. D., Elliott, L. C. C., Fagan, J. A., Gogos, A., & Hanna, S. (2022). Detection and characterization of nanoplastics in the environment. *Environmental Science & Technology, 56*(3), 1124–1136.
87. Cowger, W., Booth, A. M., Hamilton, B. M., Thaysen, C., Primpke, S., Munno, K., & Rochman, C. M. (2020). Reporting guidelines to increase the reproducibility and comparability of research on microplastics. *Applied Spectroscopy, 74*(9), 1066–1077.
88. Gigault, J., Halle, A. T., Baudrimont, M., Pascal, P. Y., Gauffre, F., Phi, T. L., & El Hadri, H. (2018). Current opinion: What is a nanoplastic? *Environmental Pollution, 235*, 1030–1034.
89. Catarino, A. I., Macchia, V., Sanderson, W. G., Thompson, R. C., & Henry, T. B. (2018). Low levels of microplastics (MP) in wild mussels indicate that MP ingestion by humans is minimal compared to exposure via household fibres fallout during a meal. *Environmental Pollution, 237*, 675–684.
90. Huang, Y., Liu, Q., Jia, W., Yan, C., & Wang, J. (2020). Agricultural plastic mulch film residues in farmland soils and their influencing factors: A review. *Science of The Total Environment, 757*, 143935.
91. Luo, H., Liu, C., He, D., Xu, J., Sun, J., Li, Y., & Pan, X. (2020). Potential applications of artificial intelligence in microplastic identification and characterization: Current progress and future perspectives. *Environmental Pollution, 263*, 114540.

92. Stock, V., Franke, J., Dönmez, M. H., Schlösser, T., & Voss, L. (2021). Uptake and effects of orally ingested polystyrene microplastic particles in in vitro models and rats. *Archives of Toxicology, 95*, 395–411.
93. Koelmans, A. A., Besseling, E., & Foekema, E. M. (2014). Leaching of plastic additives to marine organisms. *Environmental Pollution, 187*, 49–54.
94. Hwang, J., Choi, D., Han, S., Choi, J., & Hong, J. (2019). An assessment of the toxicity of polypropylene microplastics in human derived cells. *Science of the Total Environment, 684*, 657–669.
95. Wu, B., Wu, X., Liu, S., Wang, Z., & Chen, L. (2019). Size-dependent effects of polystyrene microplastics on cytotoxicity and efflux pump inhibition in human Caco-2 cells. *Chemosphere, 221*, 333–341.
96. Talsness, C. E., Andrade, A. J., Kuriyama, S. N., Taylor, J. A., & vom Saal, F. S. (2009). Components of plastic: Experimental studies in animals and relevance for human health. Philosophical Transactions of the Royal Society B: Biological Sciences, 364(1526), 2079–2096.
97. Chen, Q., Gundlach, M., Yang, S., Jiang, J., Velki, M., Yin, D., & Hollert, H. (2017). Quantitative investigation of the mechanisms of microplastics and nanoplastics toward zebrafish larvae locomotor activity. Science of The Total Environment, 584–585, 1022–1031.
98. Li, D., Shi, Y., Yang, L., Xiao, L., Kehoe, D. K., Gun'ko, Y. K., & Zhou, Y. (2020). Microplastic release from the degradation of polypropylene feeding bottles during infant formula preparation. *Nature Food, 1*, 746–754.
99. Koelmans, A. A., Redondo-Hasselerharm, P. E., Mohamed Nor, N. H., & De Ruijter, V. N. (2021). Is in vitro toxicity of pristine microplastics a realistic proxy for effects of weathered microplastics? *Environmental Science & Technology Letters, 8*(7), 530–537.
100. Campanale, C., Massarelli, C., Savino, I., Locaputo, V., & Uricchio, V. F. (2020). A detailed review study on potential effects of microplastics and additives of concern on human health. *International Journal of Environmental Research and Public Health, 17*(4), 1212.
101. Vethaak, A. D., & Legler, J. (2021). Microplastics and human health. *Science, 371*(6530), 672–674.

# Chapter 11
# Micro Plastics in Human Consumption: Routes of Exposure and Its Effects on Human Health

**Priyanka Jayachandran, Suganya Ilango, Aishwarya Vetrivel, K. V. Shalini, and Jeba Sweetly Dharmadhas**

## Introduction

Plastic materials are produced in ever-increasing quantities, are widely used, and are present in the environment. Tiny pieces of plastic are found in food, drinking water, and in the environment everywhere, and they are causing greater concern globally. Depending on their size, these particulate pollutants are known as micro plastics (MPs) or nano plastics (NPs). They have a noticeable, long-term effect on ecosystem and human health [1]. MPs are considered to be plastic particles with a size range of 1 μm to 5 mm. Plastic materials are used in a wide range of products, including consumer goods, automotive and electrical equipment, packaging and food-contact materials. These goods include coatings, labels, and glues, as well as flexible, durable plastics, plastic bags, containers, cups, trays and films [2].

Since plastics are very resistant to breaking down and can stay in the environment for a longer period of time, micro plastics can easily enter the human food chain. Both animal and plant-based foods, additives in food, refreshments, and food packaging materials made of plastic expose people to micro plastics. Micro plastics

P. Jayachandran (✉) · K. V. Shalini · J. S. Dharmadhas
Department of Biochemistry and Biotechnology, Avinashilingam Institute for Home Science and Higher Education for Women, Coimbatore, Tamil Nadu, India
e-mail: priyanka_bc@avinuty.ac.in; shalini_bc@avinuty.ac.in; jeba_bc@avinuty.ac.in

S. Ilango
Department of Biochemistry, Faculty of Science and Humanities, SRM Institute of Science and Technology, Chennai, Tamil Nadu, India
e-mail: suganyai@srmist.edu.in

A. Vetrivel
Department of Biotechnology, Dr. G.R. Damodaran College of Science, Coimbatore, Tamil Nadu, India
e-mail: aishwarya.v@grd.edu.in


S. Kulkarni, S. Das (eds.), *Microplastics and Nanoplastics*, Lecture Notes in Nanoscale Science and Technology 49,
https://doi.org/10.1007/978-3-032-17963-0_11

have the potential to accumulate in the cells and tissues of live organisms which could have long-term biological effects and endanger human health leading to various problems including chromosome changes, cancer, respiratory issues, immune system issues, gastrointestinal diseases and infertility [3].

This chapter focusses on various routes of exposure of micro plastics and its hazardous effects on human health.

## Sources and Route of Exposure of Micro Plastics

Micro plastics invade the human body either through various routes that includes food or inhalation or dermal contact etc. The possible routes of exposure and sources of micro plastics was given in Fig. 11.1

## Micro Plastics in Food

Recently published studies showing MPs in human placenta, breast milk, stool samples, and blood reinforce the idea that MPs might readily contaminate food products intended for human consumption due to their increased levels in terrestrial and aquatic ecosystems [4]. Some researchers also found micro plastics contamination on commonly consumed protein items, including plant-based proteins, shellfish, and terrestrial meats and also in typical food products such beverages, plant-based foods, marine products, and common additives like sugar and salt. MPs may enter our bodies through the consumption of these food items [5]. MPs have mostly been found in the gastrointestinal system of fish items. The quantity and size of the fragments discovered vary here as well; in many instances, the analyses are only

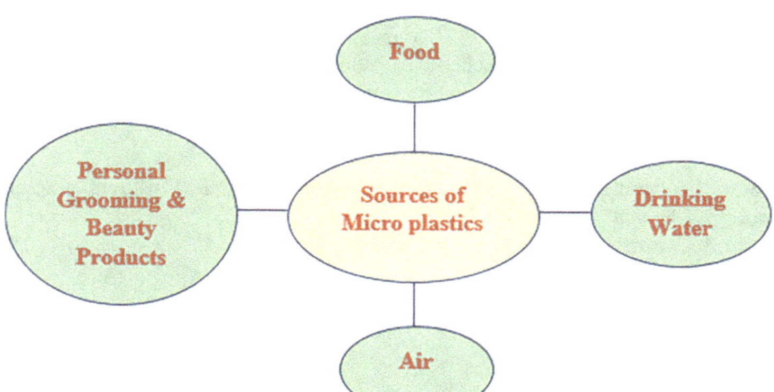


**Fig. 11.1** Sources and routes of exposure of micro plastics

qualitative. Since the majority of these food products are destroyed before consumption, MPs in the digestive track should not be dangerous to the general public [6].

## Micro Plastics in Drinking Water

In recent years, scientists from all around the world have demonstrated that MPs can be found in bottled water that comes in a variety of packaging and tap water that comes from a number of sources including surface, ground, or desalinated water. These packaging options include glass, beverage cartons, reusable plastic, and single-use plastic. Although there are many different sources of MPs in drinking water, waste water treatment plant effluents are thought to be a significant contributor to the contamination of freshwater drinking water sources. As drinking water exposure to MPs increased, worries about the possible health hazards to humans grew [7].

## Micro Plastics in Air

MPs have been found in outdoor air aerosols and deposition throughout the last 10 years in a variety of locations across the globe, including the marine boundary layer, urban and highly industrialized areas, remote alpine areas, and indoor habitats [8]. Considering how common MPs are in the atmosphere, there are a lot of questions regarding whether and how much of MPs we are breathing in from indoor and outdoor air, with interior air probably contributing the most to human MP exposure [9]. The main sources of MPs in the air are textile fibre release, material abrasion (from tires and buildings, for instance), and MP resuspension on surfaces. The size, shape, and chemical makeup of airborne MPs vary, which affects how they interact with the respiratory system and what kinds of adverse consequences they may cause [10].

## Micro Plastics in Personal Grooming and Beauty Products

A significant portion of primary micro plastics originate directly from tiny plastic particles that are added to personal grooming and beauty products (PGBPs) in daily life, often known as microbeads in the cosmetics sector [11]. Because micro plastics are stable, long-lasting, inexpensive, and simple to make, they are frequently utilized in PGBPs (such as shower gel, washing paste, toothpaste, cosmetics for eyes, and lipstick) as alternatives to several natural ingredients [12]. Because of their microscopic size, these plastic microbeads in PGBPs cannot be filtered or recycled; instead, they wash straight into sewers. Because they are difficult to decompose and invariably provide an ecological problem, they eventually gather in surface water bodies [13].

## Life Style Changes and Micro Plastics Exposure

Because high-fat diets and the intake of fatty foods, seafood, processed foods, and bottled water enhance exposure to micro plastics (MPs) and the chemical pollutants they are linked to, lifestyle and dietary choices have a substantial impact on MP exposure. Habits like wearing synthetic garments contribute to MP emission into the environment [14].

Using filtered tap water instead of plastic bottles, glass or stainless steel food containers instead of plastic ones, avoiding microwaving food in plastic, wearing natural fibre clothes instead of synthetic ones, using reusable shopping bags, and choosing cleaning and personal care products free of microbeads are all ways to lessen your exposure to micro plastics. Filtering your home's water and air, cutting less on plastic dust, and eating whole, fresh foods rather than packaged or processed ones can all significantly reduce your intake [15].

## Global Pattern of Micro Plastics Consumption

Xiang Zhao and Fengqi studied the micro plastics uptake pattern in 109 countries and the results are as follows: 57% of plastic particles found in food originate from marine sources, according to the first MP absorption map, which was created in 109 nations on 5 continents. These MPs are ingested through seafood in rapidly industrializing countries like Vietnam and Indonesia, where a sizable amount of industrial value was added to the GDP in 2018.

From 1990 to 2018, a study conducted in temporal and regional areas for the investigation of daily MP human uptake in 5-year increments. The findings showed that the micro plastic accumulation rates *in vivo* increased by more than 59 times in recently industrialized nations such as Malaysia and Indonesia. This is a result of an increase in the use of plastic and the discharge of waste into lakes, marshes, watersheds, and floodplains, and in the areas, which creates plastic debris both locally and in adjacent marine habitats [16].

Human MP exposure can be reduced by 55% by removing the aquatic plastic debris through water management for surface water quality control in lakes, wetlands, and freshwater watersheds, as well as ocean clean ups or efficient management of solid waste. Along important coastlines in 109 nations on 5 continents, these actions can help preserve public health and water quality [17].

## Effect of Micro Plastics on Human Health

Micro plastics exposure can leads to lot of health complications. These hazards include tissue damage, oxidative stress, and inflammation. In particular, research have connected exposure to micro plastic to negative impacts on male fertility and sperm quality, which may threaten a healthy pregnancy. According to certain research, micro plastics may accumulate in the body over time and have detrimental long-term impacts on health. For example, inflammation brought on by micro plastics can result in a variety of various health issues, including cancer, autoimmune disorders, and heart disease [18, 19].

Furthermore, oxidative stress—a condition that can harm DNA and cells—can be produced by micro plastics. Numerous health issues, such as neurological disorders and reproductive issues, may result from this. Chemical additives like phthalate esters and polychlorinated biphenyls that are added to micro plastics during production usually make these health impacts worse. As a result, micro plastics are acknowledged as a new environmental and public health issue that could have an impact on both the human health and environment [20, 21]. The various adverse effects exerted by micro plastics were illustrated in Fig. 11.2.

## Impact of Micro Plastics on the Immune System

Cells have the ability to absorb MPs, which can disrupt intracellular signalling cascades, change immunological homeostasis, and ultimately harm tissues and organs. The primary toxicological mechanism following MP exposure is the generation of ROS (Reactive oxygen species) which can also trigger the production of danger-associated molecular patterns (DAMPs) and link to cytokine production, disruption

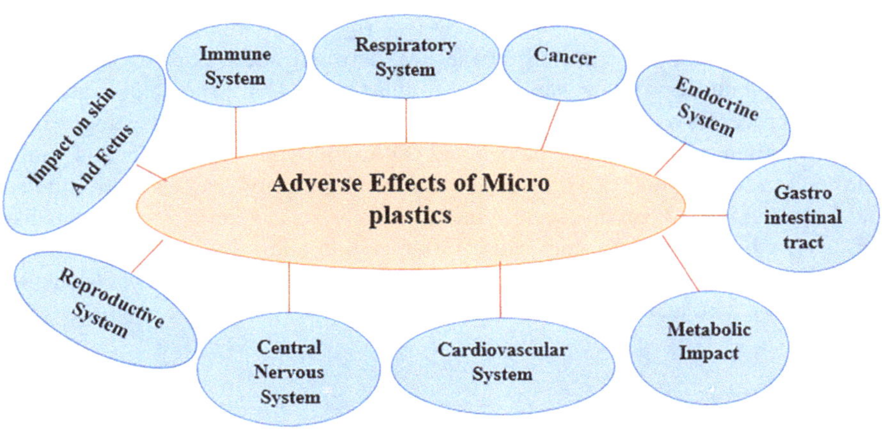


**Fig. 11.2** Adverse effects of micro plastics on human health

of toll-like receptors and the immune cell's inflammatory responses. MPs combine with chemicals, external contaminants, and infections to produce higher toxicity and potent fatal effects. They bind efficiently with internal proteins or cell membranes to form a protein-corona [22, 23].

## Impact of Micro Plastics on the Respiratory System

By causing inflammation, oxidative stress, and lung function impairment, micro plastics can cause respiratory problems when inhaled and may eventually result in chronic respiratory conditions including asthma and fibrosis. They have the capacity to enter the lungs, make their way to the alveoli, where exchange of gases takes place, and interfere with lung surfactants, which impairs the lung's capacity to expel particles. Although studies on occupational exposure have linked it to lung abnormalities, further research is needed to completely comprehend the long-term adverse effects of exposure at lower environmental levels [24, 25].

## Micro Plastics and Cancer

The primary way that micro plastics may cause cancer is by causing stress, which can harm DNA and result in mutations. Furthermore, heavy metals, chemicals that interfere with the endocrine system and may accelerate the development of cancer, and carcinogens like polycyclic aromatic hydrocarbons (PAHs) are frequently found in micro plastics. According to research, people frequently ingest micro plastics through their food, water, and air. Once within the body, micro plastics can cause damage in a number of ways [26, 27].

According to research, chronic inflammation brought on by micro plastic exposure may foster the growth of tumours by fostering an environment that is favourable to them. Although research on the link between micro plastics and human cancer is still under progress, studies and cellular models have shown indications of possible processes via which these particles may contribute to the development of cancer. Certain micro plastics can damage DNA, alter gene expression, and stimulate cell development, according to several laboratory and animal studies. Furthermore, it has been discovered that micro plastics can build up in organs like the liver, which over time may cause inflammation and raise the risk of cancer. According to new research, micro plastics may be a subtle but important contributor to the development of cancer by causing the body to experience ongoing stress [28, 29].

## Impact of Micro Plastics on Endocrine System

The thyroid, adrenal, pituitary, ovaries, testes, hypothalamus, and other mammalian endocrine organs are all negatively impacted by MPs and NPs. The hazardous chemicals that are frequently added to plastics, including polybrominated diphenyl ether, organic contaminants, bisphenols, phthalates, organotin, perfluorinated compounds, polycyclic aromatic hydrocarbons, polychlorinated biphenyl ether, dioxins, and heavy metals, are absorbed and transported by MPs and NPs [30, 31].

Because they are not covalently bound to plastics, the endocrine disrupting chemicals can readily seep into water, milk, and other liquids, which can then have an adverse effect on mammals' endocrine systems. Because smaller particles have a bigger surface area and a superior absorption capacity, they release more hazardous and endocrine disrupting compounds, which is why MPs and NPs cause toxicity that varies with size. Because they share similarities in structure with certain hormone receptors, a variety of endocrine disruptive substances transported or contained by micro plastics interfere with hormone receptors which are in normal state changing the endocrine glands' hormonal activity [32, 33].

## Gastro Intestinal and Metabolic Impacts of Micro Plastics

According to a study on the effects of food packaging container on people, micro- and nano plastics enter the GI tract via the mouth. They also alter the oral microbiota. Additionally, they directly interact with the oral epithelium and mucus layer by inhibiting alpha amylase activity in sputum. Experiments using artificial mucus and saliva have demonstrated that the result is a decrease in mucus viscosity and an increase in saliva viscosity. Both of them are linked to the spreading of pathogens: viscous saliva is linked to dental cavities and periodontal disease, whereas the mucus which is less viscous allows the microbes to reach the epithelial cell surface. Reduced bacterial co-aggregation, pathogen proliferation, oral microbiome dysbiosis and poor oral clearance, are all consequences of increase in viscosity [34, 35].

Swallowing moves microplastics from the oral cavity to the oesophagus, where they may have adverse consequences by causing buildup of ROS species and inflammation and which can kill oesophageal cell lines [36]. Based on the particle size, type of plastic, aging, surface properties, electric charge and many other factors, MPs have a wide range of effects when they reach the stomach. According to studies conducted on human and mice gastric cell lines micro- and nano plastics can reduce antioxidant capacity, interfere with the function of the gastrointestinal barrier, and impede the generation of mucus and gastric juice [37].

Additionally, they can bind and inhibit the activity of Pepsin, as well as bind dietary proteins that can evade digestion or prolong their retention period in the stomach. These effects were demonstrated in vitro by simulating human digestion. MPs can create surface coronas around themselves by directly interacting with various amino acids depending on the kind of plastic, binding proteins, lipids, or both to further expand their potential spectrum of action and capabilities [38].

The presence of endogenous proteins and various dietary matrices affect the composition of the corona. Coronas increase particle size, but they also alter the rates and characteristics of uptake, enabling MPs to be absorbed by cells. In the stomach, micro- and nanoplastics can then have biological effects. MPs can be internalized by gastric epithelial cells, resulting in increased oxidative stress, DNA damage, mitochondrial malfunction, and ultimately, depending on particle size and concentration, apoptosis or necrosis [39].

## Impact of Micro Plastics on Cardiovascular System

According to recent data, micro- and nanoplastic (MNP) particles, which are mostly produced by the mechanical or chemical breakdown of plastics, may be a risk factor for cardiovascular diseases. According to experimental findings in preclinical models, MPs can promote a variety of cardiovascular and metabolic changes that can result in illness and early mortality, as well as platelet aggregation, cell senescence, oxidative stress, and inflammatory responses in immune cells and endothelium. Human atherosclerotic plaques and other cardiovascular tissues, such as the epicardial adipose tissues, pericardial adipose tissues, myocardia, left atrial appendages and pericardia have been found to include MPs made from a variety of polymers, including polyethylene and polyvinyl chloride. MPs appear to concentrate preferentially in regions of vascular lesions and are detectable in thrombi. A higher incidence of cardiovascular events is linked to their presence in carotid plaques [40, 41].

## Impact of Micro Plastics in the Central Nervous System

MPs can negatively impact neuronal health through a number of methods, primarily based on exposure period and dose. MPs generate oxidative stress and stimulate the immune system by accumulating in tissue and overcoming biological barriers. Inhibitory protein gene expression and inflammatory mediator levels rise as a result, whereas regulatory protein expression and function decline [42]. These consequences are linked to increased lipid peroxidation, which damages DNA, misfolded protein accumulation, apoptotic pathway activation, and neuro inflammation. Likewise, even while the expression of proteins implicated in oxidative stress rises in several brain regions and neural cells, MPs may decrease the activity of antioxidant enzymes and the expression of genes related to neurotransmitters. The expression of proteins essential for brain development, which control the structure and plasticity of the central nervous system, may be severely disrupted as a result. Given their role and possible impact on neurodegenerative illness and neurodevelopmental disorders, MPs have the ability to reduce key cognition-related genes, which is the primary cause of cognitive decline [43].

## Impact of Micro Plastics in the Reproductive System

The process by which sperm are produced from spermatogonial stem cells is called as spermatogenesis, and it is incredibly intricate and ongoing. Many cells including peritubular myoid cells, Leydig cells, and Sertoli cells, as well as paracrine substances and hormonal cues, and appropriate epigenetic and genetic control, are required for the maintenance of differentiation and self-renewal in SSCs. Furthermore, these substances are essential for the process of spermatogenesis such as spermatid maturation and spermatocyte meiosis regulation. Male infertility may result from the problem in any of the aforementioned steps. An increasing quantity of research has shown that micro plastics can build up in the testes and are harmful to reproduction process [44].

## Impact of Micro Plastics in the Foetus

The discovery of compounds linked to plastics that are known to affect hormones in human amniotic fluid and the human placenta suggests that plastic particles may also cross the blood-placenta barrier. MPs, which range in size from 5 to 10 μm, have been found in the chorioamniotic membranes, on the maternal side, and on the fetal side of the human placenta. Pericytes and endothelial cells in the chorionic villi were found to contain MPs using a transmission electron microscope in a recent study, suggesting that MPs can enter the fetal blood circulation [45, 46].

## Impact of Micro Plastics on Skin

There is ample evidence that MPs may cause cutaneous changes, trigger inflammatory reactions, and disrupt the physiological homeostasis of the skin. Their diminutive size increases their capacity to penetrate the epidermal barrier, emphasising the concerns of extended exposure. Previous studies have shown that micro plastics can induce skin irritation, inflammation, and disruption of normal skin function, all of which have a detrimental effect on skin health [47]. When MPs come into touch with the skin, cellular receptors recognize patterns linked to damage and pathogens, which triggers an immune response. T cells, Langerhans cells, Keratinocytes, macrophages, melanocytes and dendritic cells, all have these receptors, which might cause a reaction that jeopardizes the integrity of the skin's barrier [48].

## Methods for Detecting Micro Plastics in Human Specimens

Given the concern over MPs' impacts on human and animal health, it is imperative that proven procedures for MP detection and characterization be used. There are three types of methods that have been used to analyse MPs in human samples: (i) Microscopy, including optical, stereo, scanning electron, and fluorescence microscopes following Red Nile staining (ii) FTIR and Raman spectroscopy (iii) Separation (such as high performance liquid chromatography with tandem mass spectrometry) (HPLC–MS/MS) [49, 50].

## Approaches to Reduce MP Exposure in Humans

People should use glass containers and metal bottles instead of plastic-packaged and single-use plastics, filter tap water instead of plastic-bottled water, choose natural fabrics, and use caution when handling hot food and plastic-bottled beverages in order to reduce their exposure to micro plastics (MPs). Other tactics include storing food in glass containers rather than plastic ones, avoiding plastic tea bags, and utilizing air filters to improve the quality of the air indoors [51].

## Conclusion

Humans are exposed to micro plastics through a variety of pathways, and the resulting health impacts are intricate and diverse. Micro plastics are present everywhere. Inhaling indoor air and drinking water from plastic bottles may be the primary exposure pathway. The primary source of exposure to micro plastics is probably food consumption. The harmful consequences of micro plastics on human health, which can differ based on their nature, size, shape, and concentration. Metabolic disorders, organ malfunction, DNA damage, immunological response, neurotoxicity, Oxidative stress, and toxicity to reproduction and development are some of the adverse effects. There are numerous mitigating techniques available to counteract the harmful impacts that these micro plastics cause.

## References

1. Ziani, K., Ioniță-Mîndrican, C. B., Mititelu, M., Neacșu, S. M., Negrei, C., Moroșan, E., Drăgănescu, D., & Preda, O. T. (2023). Microplastics: A real global threat for environment and food safety: A state of the art review. *Nutrients, 15*(3), 617. https://doi.org/10.3390/nu15030617

2. Razaviarani, V., Saudagar, A., Gallage, S., et al. (2024). Comprehensive investigation on microplastics from source to sink. *Clean Technologies and Environmental Policy, 26*, 1755–1782. https://doi.org/10.1007/s10098-024-02738-w
3. Lalrinfela, P., Vanlalsangi, R., Lalrinzuali, K., & Babu, P. J. (2024). Microplastics: Their effects on the environment, human health, and plant ecosystems. *Environmental Pollution and Management, 1*, 248–259. https://doi.org/10.1016/j.epm.2024.11.004
4. Chia, R. W., Atem, N. V., Lee, J. Y., & Cha, J. (2025). Microplastic and human health with focus on pediatric well-being: A comprehensive review and call for future studies. *Clinical and Experimental Pediatrics, 68*(1), 1–15. https://doi.org/10.3345/cep.2023.01739
5. Kumar, V., Sharma, N., Umesh, M., Gupta, P., Sharma, P., Basheer, T., Huligowda, L. K. D., Thomas, J., Bhagat, S. K., & Pasrija, R. (2024). Microplastics in food: Occurrence, toxicity, green analytical detection methods and future challenges. *Green Analytical Chemistry, 11*, 100152. https://doi.org/10.1016/j.greeac.2024.100152
6. Aunurohim, A., Raufanda, M. S., Saptarini, D., Muzaki, F. K., Prabowo, R. E., Pamungkas, M. S. A., Putra, S. C. P., Ekawati, E., Maulidhaturrafida, N., Dewi, S. C., & Danilyan, E. (2024). Microplastic-contamination in the flesh and gastrointestinal tract of nile tilapia (Oreochromis niloticus) reared in floating net cages at lake Ranu Grati, Pasuruan, East Java, Indonesia. *Environmental Advances, 17*, 100587. https://doi.org/10.1016/j.envadv.2024.100587
7. Al-Mansoori, M., Harrad, S., & Abou-Elwafa Abdallah, M. (2025). MPs in drinking water and beverages: Concentrations, characteristics and implications for human exposure. *Emerging Contaminants, 11*, 100552. ISSN 2405-6650. https://doi.org/10.1016/j.emcon.2025.100552
8. Kernchen, S., Löder, M. G. J., Fischer, F., Fischer, D., Moses, S. R., Georgi, C., Nölscher, A. C., Held, A., & Laforsch, C. (2022). Airborne microplastic concentrations and deposition across the Weser River catchment. *Science of the Total Environment, 818*, 151812. https://doi.org/10.1016/j.scitotenv.2021.151812
9. Ahmad, M., Chen, J., Khan, M. T., Yu, Q., Phairuang, W., Furuuchi, M., Ali, S. W., Nawab, A., & Panyametheekul, S. (2023). Sources, analysis, and health implications of atmospheric microplastics. *Emerging Contaminants, 9*(3), 100233. https://doi.org/10.1016/j.emcon.2023.100233
10. Saha, S. C., & Saha, G. (2024). Effect of microplastics deposition on human lung airways: A review with computational benefits and challenges. *Heliyon, 10*(2), e24355. https://doi.org/10.1016/j.heliyon.2024.e24355
11. Giustra, M., Sinesi, G., Spena, F., De Santes, B., Morelli, L., Barbieri, L., Garbujo, S., Galli, P., Prosperi, D., & Colombo, M. (2024). Microplastics in cosmetics: Open questions and sustainable opportunities. *ChemSusChem, 17*(22), e202401065. https://doi.org/10.1002/cssc.202401065
12. Bikiaris, N., Nikolaidis, N. F., & Barmpalexis, P. (2024). Microplastics (MPs) in cosmetics: A review on their presence in personal-care, cosmetic, and cleaning products (PCCPs) and sustainable alternatives from biobased and biodegradable polymers. *Cosmetics, 11*(5), 145. https://doi.org/10.3390/cosmetics11050145
13. Lamichhane, G., Acharya, A., Marahatha, R., Modi, B., Paudel, R., Adhikari, A., Raut, B. K., Aryal, S., & Parajuli, N. (2023). Microplastics in environment: Global concern, challenges, and controlling measures. *International Journal of Environmental Science and Technology: IJEST, 20*(4), 4673–4694. https://doi.org/10.1007/s13762-022-04261-1
14. Mir, M. A., Khan, M. A. A., Banik, B. K., Hasnain, S. M., Alzayer, L., Andrews, K., & Abba, S. I. (2025). Microplastics in food products: Prevalence, artificial intelligence based detection, and potential health impacts on humans. *Emerging Contaminants, 11*(2), 100477. ISSN 2405-6650. https://doi.org/10.1016/j.emcon.2025.100477
15. Zhang, M., Zhang, Y., Liu, T., An, C., & Sun, Y. (2025). Microplastic exposure in daily life and the risk of pregnancy-induced hypertension: A study on the association between environmental pollutants and maternal-fetal health outcomes. *Journal of Hazardous Materials, 494*, 138654. https://doi.org/10.1016/j.jhazmat.2025.138654
16. Zhao, X., & You, F. (2024). Microplastic human dietary uptake from 1990 to 2018 grew across 109 major developing and industrialized countries but can be halved by plastic debris

removal. *Environmental Science & Technology, 58*(20), 8709–8723. https://doi.org/10.1021/acs.est.4c00010
17. Kumar, R., Verma, A., Shome, A., Sinha, R., Sinha, S., Jha, P. K., Kumar, R., Kumar, P., Shubham, Das, S., Sharma, P., & Vara Prasad, P. V. (2021). Impacts of plastic pollution on ecosystem services, sustainable development goals, and need to focus on circular economy and policy interventions. *Sustainability, 13*(17), 9963. https://doi.org/10.3390/su13179963
18. Zurub, R. E., Cariaco, Y., Wade, M. G., & Bainbridge, S. A. (2024). Microplastics exposure: Implications for human fertility, pregnancy and child health. *Frontiers in Endocrinology, 14*, 1330396. https://doi.org/10.3389/fendo.2023.1330396
19. Song, X., Du, L., Sima, L., Zou, D., & Qiu, X. (2023). Effects of micro(nano)plastics on the reproductive system: A review. *Chemosphere, 336*, 139138. https://doi.org/10.1016/j.chemosphere.2023.139138
20. Kadac-Czapska, K., Ośko, J., Knez, E., & Grembecka, M. (2024). Microplastics and oxidative stress-current problems and prospects. *Antioxidants (Basel), 13*(5), 579. https://doi.org/10.3390/antiox13050579
21. Campanale, C., Massarelli, C., Savino, I., Locaputo, V., & Uricchio, V. F. (2020). A detailed review study on potential effects of microplastics and additives of concern on human health. *International Journal of Environmental Research and Public Health, 17*(4), 1212. https://doi.org/10.3390/ijerph17041212
22. Bridgeman, L., Cimbalo, A., López-Rodríguez, D., Pamies, D., & Frangiamone, M. (2025). Exploring toxicological pathways of microplastics and nanoplastics: Insights from animal and cellular models. *Journal of Hazardous Materials, 490*, 137795. https://doi.org/10.1016/j.jhazmat.2025.137795
23. Das, A. (2023). The emerging role of microplastics in systemic toxicity: Involvement of reactive oxygen species (ROS). *Science of the Total Environment, 895*, 165076. https://doi.org/10.1016/j.scitotenv.2023.165076
24. Kang, H., Huang, D., Zhang, W., Wang, J. Y., Liu, Z., Wang, Z., Jiang, G., & Gao, A. (2024). Inhaled polystyrene microplastics impaired lung function through pulmonary flora/TLR4-mediated iron homeostasis imbalance. *Science of the Total Environment, 946*, 174300. https://doi.org/10.1016/j.scitotenv.2024.174300
25. Gou, Z., Wu, H., Li, S., et al. (2024). Airborne micro- and nanoplastics: Emerging causes of respiratory diseases. *Particle and Fibre Toxicology, 21*, 50. https://doi.org/10.1186/s12989-024-00613-6
26. Deng, X., Gui, Y., & Zhao, L. (2025). The micro(nano)plastics perspective: Exploring cancer development and therapy. *Molecular Cancer, 24*, 30. https://doi.org/10.1186/s12943-025-02230-z
27. Lagunas-Rangel, F. A., Linnea-Niemi, J. V., Kudłak, B., Williams, M. J., Jönsson, J., & Schiöth, H. B. (2022). Role of the synergistic interactions of environmental pollutants in the development of cancer. *GeoHealth, 6*(4), e2021GH000552. https://doi.org/10.1029/2021GH000552
28. Chen, Y., Zhang, Z., Ji, K., Zhang, Q., Qian, L., & Yang, C. (2025). Role of microplastics in the tumor microenvironment (review). *Oncology Letters, 29*(4), 193. https://doi.org/10.3892/ol.2025.14939
29. Jayavel, S., Govindaraju, B., Michael, J. R., et al. (2024). Impacts of micro and nanoplastics on human health. *Bulletin of the National Research Centre, 48*, 110. https://doi.org/10.1186/s42269-024-01268-1
30. Li, Y., Yang, G., Wang, J., Lu, L., Li, X., Zheng, Y., Zhang, Z., & Ru, S. (2022). Microplastics increase the accumulation of phenanthrene in the ovaries of marine medaka (Oryzias melastigma) and its transgenerational toxicity. *Journal of Hazardous Materials, 424*(Part D), 127754. https://doi.org/10.1016/j.jhazmat.2021.127754
31. Zhao, X., Liu, Z., Ren, X., & Duan, X. (2021). Parental transfer of nanopolystyrene-enhanced tris(1,3-dichloro-2-propyl) phosphate induces transgenerational thyroid disruption in zebrafish. *Aquatic Toxicology, 236*, 105871. https://doi.org/10.1016/j.aquatox.2021.105871

32. Ullah, S., Ahmad, S., Guo, X., Ullah, S., Ullah, S., Nabi, G., & Wanghe, K. (2023). A review of the endocrine disrupting effects of micro and nano plastic and their associated chemicals in mammals. *Frontiers in Endocrinology, 13*, 1084236. https://doi.org/10.3389/fendo.2022.1084236
33. Wang, Z., Liang, G., Jiang, S., Wang, F., Li, H., Li, B., Zhu, H., Lu, A., & Gong, W. (2024). Understanding the environmental impact and risks of organic additives in plastics: A call for sustained research and sustainable solutions. *Emerging Contaminants, 10*(4), 100388. https://doi.org/10.1016/j.emcon.2024.100388
34. Covello, C., Di Vincenzo, F., Cammarota, G., & Pizzoferrato, M. (2024). Micro(nano)plastics and their potential impact on human gut health: A narrative review. *Current Issues in Molecular Biology, 46*(3), 2658–2677. https://doi.org/10.3390/cimb46030168
35. Bruno, A., Dovizio, M., Milillo, C., Aruffo, E., Pesce, M., Gatta, M., Chiacchiaretta, P., Di Carlo, P., & Ballerini, P. (2024). Orally ingested micro- and nano-plastics: A hidden driver of inflammatory bowel disease and colorectal cancer. *Cancers, 16*(17), 3079. https://doi.org/10.3390/cancers16173079
36. Dzierżyński, E., Gawlik, P. J., Puźniak, D., Flieger, W., Jóźwik, K., Teresiński, G., Forma, A., Wdowiak, P., Baj, J., & Flieger, J. (2024). Microplastics in the human body: Exposure, detection, and risk of carcinogenesis: A state-of-the-art review. *Cancers, 16*(21), 3703. https://doi.org/10.3390/cancers16213703
37. Feng, Y., Tu, C., Li, R., Wu, D., Yang, J., Xia, Y., Peijnenburg, W. J. G. M., & Luo, Y. (2023). A systematic review of the impacts of exposure to micro- and nano-plastics on human tissue accumulation and health. *Eco-Environment & Health, 2*(4), 195–207. https://doi.org/10.1016/j.eehl.2023.08.002
38. Tan, H., Yue, T., Yan, X., Zhao, J., & Xing, B. (2020). Microplastics reduce lipid digestion in simulated human gastrointestinal system. *Environmental Science & Technology, 54*(19), 12285–12294. https://doi.org/10.1021/acs.est.0c02608
39. Yang, Z., DeLoid, G. M., Zarbl, H., Baw, J., & Demokritou, P. (2023). Micro- and nanoplastics (MNPs) and their potential toxicological outcomes: State of science, knowledge gaps and research needs. *NanoImpact, 32*, 100481. https://doi.org/10.1016/j.impact.2023.100481
40. Prattichizzo, F., Ceriello, A., Pellegrini, V., La Grotta, R., Graciotti, L., Olivieri, F., Paolisso, P., D'Agostino, B., Iovino, P., Balestrieri, M. L., Rajagopalan, S., Landrigan, P. J., Marfella, R., & Paolisso, G. (2024). Micro-nanoplastics and cardiovascular diseases: Evidence and perspectives. *European Heart Journal, 45*(38), 4099–4110. https://doi.org/10.1093/eurheartj/ehae552
41. Thapliyal, C., Negi, S., Nagarkoti, S., et al. (2025). Mechanistic insight into potential toxic effects of microplastics and nanoplastics on human health. *Discover Applied Sciences, 7*, 645. https://doi.org/10.1007/s42452-025-07214-8
42. Sadique, S. A., Konarova, M., Niu, X., Szilagyi, I., Nirmal, N., & Li, L. (2025). Impact of microplastics and nanoplastics on human health: Emerging evidence and future directions. *Emerging Contaminants, 11*(3), 100545. ISSN 2405-6650. https://doi.org/10.1016/j.emcon.2025.100545
43. Budhwar, M., Mehra, S., Sharma, M., Ahsan, A. U., & Chopra, M. (2025). Unveiling micro-nanoplastics (MNPs) induced developmental toxicity, transgenerational transport and associated signaling pathways. *Journal of Hazardous Materials Advances, 17*, 100581. ISSN 2772-4166. https://doi.org/10.1016/j.hazadv.2024.100581
44. Yang, J., Kamstra, J., Legler, J., & Aardema, H. (2023). The impact of microplastics on female reproduction and early life. *Animal Reproduction, 20*(2), e20230037. https://doi.org/10.1590/1984-3143-AR2023-0037
45. Sharma, R. K., Kumari, U., & Kumar, S. (2024). Impact of microplastics on pregnancy and fetal development: A systematic review. *Cureus, 16*(5), e60712. https://doi.org/10.7759/cureus.60712
46. Nair, S. R., Nihad, M., Sudheer Shenoy, P., Gupta, S., & Bose, B. (2025). Unveiling the effects of micro and nano plastics in embryonic development. *Toxicology Reports, 14*, 101954. ISSN 2214-7500. https://doi.org/10.1016/j.toxrep.2025.101954

47. Aristizabal, M., Jiménez-Orrego, K. V., Caicedo-León, M. D., Páez-Cárdenas, L. S., Castellanos-García, I., Villalba-Moreno, D. L., Ramírez-Zuluaga, L. V., Hsu, J. T. S., Jaller, J., & Gold, M. (2024). Microplastics in dermatology: Potential effects on skin homeostasis. *Journal of Cosmetic Dermatology, 23*(3), 766–772. https://doi.org/10.1111/jocd.16167
48. Ramsperger, A. F. R. M., Bergamaschi, E., Panizzolo, M., Fenoglio, I., Barbero, F., Peters, R., Undas, A., Purker, S., Giese, B., Lalyer, C. R., Tamargo, A., Victoria Moreno-Arribas, M., Grossart, H.-P., Kühnel, D., Dietrich, J., Paulsen, F., Afanou, A. K., Zienolddiny-Narui, S., Hammer, S. E., ... Laforsch, C. (2023). Nano- and microplastics: A comprehensive review on their exposure routes, translocation, and fate in humans. *NanoImpact, 29*, 100441. ISSN 2452-0748. https://doi.org/10.1016/j.impact.2022.100441
49. Dey, S., Sen, K., Saha, N. C., & Saha, S. (2025). Analytical approaches for quantifying and characterizing microplastics: Environmental impacts and bioaccumulation in aquatic systems. *Green Analytical Chemistry, 12*, 100191. ISSN 2772-5774. https://doi.org/10.1016/j.greeac.2024.100191
50. Adhikari, S., Kelkar, V., Kumar, R., & Halden, R. U. (2022). Methods and challenges in the detection of microplastics and nanoplastics: A mini-review. *Polymer International, 71*, 543–551. https://doi.org/10.1002/pi.6348
51. Nawab, A., Ahmad, M., Khan, M. T., Nafees, M., Khan, I., & Ihsanullah, I. (2024). Human exposure to microplastics: A review on exposure routes and public health impacts. *Journal of Hazardous Materials Advances, 16*, 100487. ISSN 2772-4166. https://doi.org/10.1016/j.hazadv.2024.100487

# Chapter 12
# A Holistic Assessment of Microplastic Pollution in Freshwater Bodies: Sources, Loads, and Biological Consequences

**Mamilla Vijaya Kumar**

## Introduction

The latter half of the twentieth century unveiled remarkable inventions that changed the world, one being plastics. Their major turning point came in 1907, when Leo Baekeland created the first fully synthetic plastic, which could undergo mass production, 'Bakelite.' Today, most modern industries, including healthcare, technology, and telecommunications, rely heavily on them [1]. People favour plastics as they are much more durable as compared to other materials, easily mouldable and come at a low cost [2]. In the case of transportation, altering structural resilience and weight provides an added advantage. Because of their distinct multifunctionality, plastics play a significant role in our society [3]. Of specific importance are synthetic polymers PET and PBT, which serve various industries due to their high durability and thermal stability [4, 5].

Types of plastics differ by polymeric structure and complexity, with some categories includes photodegradable bioplastics and compostable ones which disintegrate more quickly due to light exposure, microbial activity, or during composting [6]. Plastics are classified into five categories based on size: nanoplastics, microplastics, mesoplastics, macroplastics and megaplastics [7]. Microplastics pose critical ecological and health risks as they are smaller than 5 mm [8].

Microplastics are classified into two categories: primary microplastics and secondary microplastics. Primary microplastics are intentionally made to be less than 5 mm in size and are found in products such as synthetic textiles, pharmaceuticals, and personal care items like exfoliating scrubs [9, 10]. They can easily enter aquatic ecosystems through wastewater, rivers, atmospheric deposition, and runoff [11]. In

M. V. Kumar (✉)
Department of Civil Engineering, KSRM College of Engineering,
Kadapa, Andhra Pradesh, India





S. Kulkarni, S. Das (eds.), *Microplastics and Nanoplastics*, Lecture Notes in
Nanoscale Science and Technology 49,
https://doi.org/10.1007/978-3-032-17963-0_12

contrast, secondary microplastics, which come from the breakdown of larger plastics [12, 13], like weathering and mechanical abrasion [14]. They originate from sources such as abandoned fishing gear and plastic waste [15]. Most microplastic pollution in aquatic environments comes from these secondary sources [16, 17], and ongoing plastic waste influx further exacerbates the issue [9].

Plastic pollution has emerged as a significant environmental crisis in the twenty-first century. Global plastic production has dramatically increased, soaring from approximately 2 million tons in 1950 to around 400 million tons in 2022. Each year, about 400 million metric tons of plastic waste are generated, yet only 9% of this waste is recycled. Alarmingly, only about 9% of this waste undergoes recycling, while nearly 80% is either mismanaged or relegated to landfills, ultimately making its way into terrestrial and aquatic ecosystems [18–20]. The United States is the largest contributor to plastic waste, generating an average of 0.35 kilograms per person daily. If current trends persist, global plastic waste could reach 500 million metric tons by 2030 and 800 million metric tons by 2050 [21].

Industrialization, urban expansion, and intensive agricultural practices have transformed ecosystems by introducing various contaminants. Among these, microplastics (MPs) have emerged as a widespread pollutant [22]. Due to their small size and durability, MPs pose a serious threat to terrestrial and aquatic ecosystems, bioaccumulating in organisms like plankton, fish, and marine mammals [23]. Composed of synthetic polymers, MPs can adsorb harmful organic contaminants that may be released upon ingestion, raising ecological health concerns [24–26]. Typically, smaller than 5 mm and resistant to biodegradation, their presence disrupts marine food webs and harms marine organisms, particularly in benthic zones [27, 28].

Freshwater systems can accumulate significant amounts of microplastic particles and fibres, but research on microplastic pollution in these environments is less extensive than in marine ecosystems. Freshwater bodies can act as sources (e.g., wastewater treatment plant effluents), transport pathways (e.g., rivers), or sinks (e.g., lakes) for microplastics, depending on hydrological, geographical, and human factors [29]. The distribution and dynamics of microplastics in freshwater can differ greatly from marine environments due to varying plastic inputs and environmental conditions [30]. Addressing microplastics is challenging due to their size, complicating detection and management. Their ability to be ingested increases the risk of toxin accumulation and biomagnification, threatening ecological integrity [21, 31–33]. Micro and nanoplastics are persistent pollutants that create risks for the environment and public health [20].

## Sources and Pathways of Microplastics

Microplastic pollution represents a multifaceted environmental challenge influenced by several factors, including size, shape, density, and environmental conditions such as wind and water currents [34]. Due to their durability, the extensive presence of microplastics in aquatic ecosystems raises considerable ecological and toxicological concerns [35].

Microplastics arise from the breakdown of larger plastic debris and infiltrate freshwater ecosystems through various pathways, including drinking water treatment processes, discharges from wastewater treatment plants, atmospheric deposition, industrial effluents, and surface runoff [36]. Major sources of plastic pollution include urban expansion, municipal sewage discharge, and intensive agricultural practices [37–39]. Additionally, human activities such as using thermoplastic road markings, packaging waste, discarded plastic bottles, and synthetic textile fibers generate microplastics [40, 41]. The diverse colours of microplastic particles serve as visual indicators of their synthetic origins [42].

## *Microplastic Emissions from Drinking Water and Wastewater Systems*

Drinking water treatment systems offer some protection against microplastic contamination; however, plastic components can degrade and release microplastics into treated water [43]. Notably, plastic bottles and caps significantly contribute to increased microplastic levels in drinking water supplies [44]. Sewage treatment plants are also significant sources of microplastic pollution in freshwater ecosystems [43, 45]. Wastewater effluent resulting from everyday human activities serves as a primary pathway for microplastics [40]. During heavy rainfall, discharges may overflow, bypassing filtration systems and introducing microplastics into rivers and coastal waters. While wastewater treatment plants play a crucial role in managing microplastic pollution, they still permit some microplastics to escape through treated effluent and sludge that is often used as fertilizer [46, 47]. Although advanced technologies can achieve more than 99% removal efficiency for plastic particles [48, 49], substantial amounts still enter aquatic environments and can contaminate land and water through the application of sewage sludge. Additionally, sewage sludge and industrial processes involving plastic granules or resin pellets serve as direct sources of microplastic contamination [50, 51].

## *Atmospheric and Runoff-Driven Pathways of Microplastic Mobilization*

Microplastics can be transported over considerable distances through the atmosphere, allowing them to settle in remote locations, such as mountain ranges and polar regions, due to both wet and dry deposition processes [52, 53]. These particles primarily originate from industrial activities and urban dust, highlighting the atmosphere as a critical pathway for microplastics entering aquatic systems [54]. Urban areas exacerbate this issue, as rainwater runoff transports plastic debris from streets into freshwater bodies. Various human activities, including waste incineration, building emissions, traffic abrasion, and agricultural runoff, also contribute to the release of microplastics into both air and water. Additionally, street dust, often contaminated with heavy metals, serves as a reservoir for microplastics, further endangering ecosystems [55]. Urban dust contains a variety of microplastics from diverse human activities [56], and extreme weather events enhance the movement of microplastics into freshwater ecosystems, thereby increasing pollution variability [57].

## *Land-Based Sources*

Land-based sources are responsible for approximately 80–90% of microplastic pollution in aquatic ecosystems [58]. These sources include everyday materials such as plastic packaging, personal care products, construction materials, and textiles. Additionally, the incineration of plastic waste contributes to microplastics through residues found in bottom ash. Macroplastic debris from various activities degrades into secondary microplastics [59, 60]. Furthermore, microplastics are increasingly being found in cosmetics and pharmaceuticals [61], with common products like facial cleansers and makeup identified as contributors [62].

Synthetic textiles release microfibers during laundering [63], and tyre wear particles from road traffic represent another significant source [64]. Single-use items, such as plastic bottles, straws, and cutlery, also contribute significantly to the issue due to their limited recyclability [65]. The COVID-19 pandemic has further worsened the situation with an increase in disposable face masks made from synthetic materials, adding to microplastic pollution in both terrestrial and aquatic ecosystems [66].

Tire abrasion constitutes a significant source of microplastic pollution, producing both fragmentary and fibrous particles primarily within the 2.5–350 μm size range. Emissions from infrastructure and tyre wear considerably elevate microplastic levels in freshwater systems [67]. Research indicates that tire-derived rubber particles account for 5–28% of particulate matter smaller than 10 μm in roadside dust, highlighting their substantial contribution to airborne microplastics [68]. In addition to atmospheric transport, river flows are crucial for the distribution of microplastics, leading to contamination of downstream aquatic systems, including lakes, reservoirs, and marine environments [69–71].

## *River and Ocean-Based Sources*

Buoyant macroplastic debris plays a significant role in the formation of microplastics in marine environments through both physical and chemical degradation, which fractures larger items into smaller particles that can endure in aquatic ecosystems

[55]. This degradation occurs due to abrasion, weathering, and biological activity, primarily involving common products such as plastic bags and bottles [72]. Microplastics are now present in all major ocean basins as well as numerous freshwater bodies [35]. They enter these environments via direct discharge and the breakdown of larger plastics, influenced by factors like UV radiation and mechanical wear [73]. This degradation not only increases the surface area of the pollutants but also enhances their capacity to absorb toxic substances, thereby worsening their ecological impact [74].

Synthetic textiles are significant contributors to microplastic contamination in river systems [75]. In developing countries, factors such as population growth, rapid urbanization, and shifting lifestyles have resulted in increased municipal solid waste generation [76]. The laundering of synthetic fabrics releases microplastic fibers into marine environments, with the amount released varying depending on the textile type [77, 78].

While land-based sources are the primary contributors to microplastic pollution, ocean-based activities account for approximately 10–20% of microplastics entering marine environments [79, 80]. These activities include coastal tourism, commercial fishing, and shipping. Additionally, abandoned fishing gear, such as monofilament lines and nylon nets, can degrade into microplastics, negatively impacting marine ecosystems [81]. Offshore industries, particularly those involved in petrochemical extraction, also contribute to plastic contamination in marine environments [82, 83].

## Types of Microplastics or Classification of Microplastics

### *Types of Microplastics*

Microplastics are innovatively categorized into two distinct types based on their origin and formation processes: primary and secondary microplastics (MPs). Primary and secondary MPs have symmetrical differences where primary takes a more symmetrical approach while secondary loses the symmetry, transforming into a more asymmetrical form [84].

To assist in distinguishing various plastic materials, the American Society of Plastics Industry SPI created the Resin Identification Code (RIC) system. This method classifies around 50 different types of plastics into 7 major groups: Polyethene terephthalate (PET), high-density polyethylene (HDPE), low-density polyethylene (LDPE), polyvinyl chloride (PVC), polypropylene (PP), polystyrene (PS), and a catch-all category that includes other polymers such as polyvinidene fluoride (PVDF) and polylactic acid (PLA) [85]. In India, the National Institutes of Health (NIH) documents five distinct forms of microplastics as fragments, fibers, foams, pellets, and films [86]. These microplastics contain several polymers, which include PE, PS, PP, PU, PVC, and PET [87–89]. This classification system is important for assessing the problem of microplastics. Figure 12.1 below shows the classification of microplastics.

### Primary Microplastics

Primary microplastics are present in various industrial products, significantly contributing to environmental pollution [90]. These microplastics are intentionally manufactured and come in the form of resin pellets used in plastic production and microbeads found in personal care items, such as shower gels and exfoliants [58, 91]. Resin pellets can occasionally spill directly from factories into freshwater rivers [82], while microbeads may enter waterways through both industrial and domestic wastewater [15]. Additionally, wastewater treatment plants can introduce primary microplastics into aquatic ecosystems [92].

### Secondary Microplastics

Secondary microplastics are formed when larger plastic items, such as bags, bottles, and synthetic clothing, etc. degrade through mechanical and photo-oxidative processes [93]. This degradation occurs unintentionally due to physical abrasion, chemical weathering, and biological activities, with common sources including tyre wear and deteriorating plastic waste [94]. These microplastics arise from the disintegration of macroplastics and mesoplastics, primarily driven by environmental

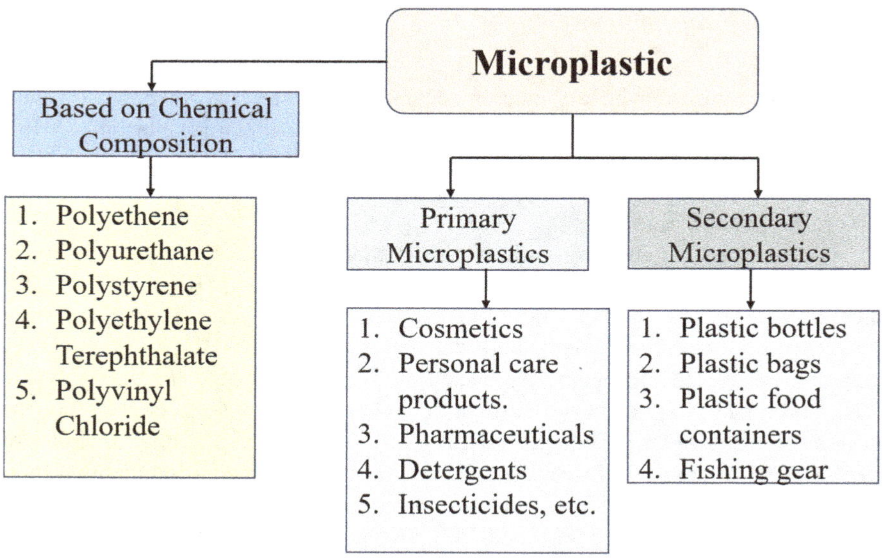


**Fig. 12.1** Various forms of microplastics

factors like photodegradation and biodegradation [95]. Typically, secondary microplastics display rough surfaces and irregular shapes [96], reflecting the effects of weathering processes such as photo-oxidation and breakdown from wave action [97].

Smaller plastic fragments can enter rivers through non-point source pollution from urban and agricultural activities [98]. These microplastics typically come from persistent plastic waste in aquatic environments. Unlike mineralization, which completely breaks down plastics into inorganic molecules, fragmentation is the physical disintegration of polymers at the molecular level [99, 100]. While mineralization is rare for most plastics, fragmentation is common and influenced by the polymer's properties, such as chemical composition and crystallinity. Photodegradation, mainly due to ultraviolet (UV) radiation, is the primary mechanism for plastic degradation in sunlit environments like shorelines and surface waters [101, 102].

## *Most Common Type of Microplastic*

The major types of microplastics and their characteristics are explained in this chapter.

### Plastic Pellets (Raw Materials)

Plastic pellets are small, granular pieces of plastic, typically 2–5 mm in diameter. Plastic pellets, categorized as primary microplastics (MPs), are frequently found in areas with high population density and substantial industrial activity [103]. Used to manufacture various products, they are mainly made from petroleum and coal to produce materials like ethylene and styrene. Plastics are categorized as thermoplastics and thermoset plastics, with most thermoplastics coming from virgin pellets, also known as nurdles, produced during polymer manufacturing or recycling. These pellets are prevalent in everyday life and are found in household appliances, clothing, construction, and medical equipment. However, they are persistent in the environment, degrading slowly and posing risks to wildlife and, ultimately, human health due to their small size, which makes them easy to ingest [104].

### Microbeads

Microbeads, a prevalent type of primary microplastic, are commonly found in personal care products such as facial cleansers and are significant contributors to microplastic pollution in both aquatic and terrestrial environments [105]. Various Analyses indicate that the spherical microplastics detected in nature likely stem from these cosmetic formulations [16]. The most commonly used polymers in exfoliating agents include polyethene (PE), polypropylene (PP), and polystyrene (PS), which are favoured for their stability, affordability, and ease of manufacturing.

These microbeads typically measure between 74 and 420 μm, making them suitable for dermal applications [58]. Moreover, wastewater treatment plants often retain considerable quantities of microbeads, highlighting the inadequacy of conventional treatment processes to eliminate them. This allows these microplastics to enter natural ecosystems [106, 107].

#### Fibers

The increasing prevalence of microplastic (MP) fibers in aquatic environments highlights the considerable impact of the textile industry and associated human activities on pollution. Insufficient treatment of both domestic and industrial wastewater has been linked to elevated concentrations of fiber-based microplastics in freshwater ecosystems [108]. A key pathway for the introduction of synthetic fibers into the environment is through laundry practices, during which microfibers are shed and released into wastewater [109, 110]. Accordingly, laundry effluents play a significant role in microfiber pollution in urban rivers [111], and often flow into surface water bodies without adequate filtration [108]. In densely populated informal settlements with inadequate sanitation, activities such as bathing and washing clothes directly in rivers exacerbate fiber pollution [110]. Moreover, sewage sludge, a byproduct of wastewater treatment, serves as an additional reservoir for fibrous microplastics [112]. In summary, wastewater treatment plants in urban environments are critical conduits for fiber microplastics entering natural water systems, due to their limited effectiveness in fully removing these particles during standard treatment processes [113].

#### Films

Film-shaped microplastics (MPs) are a subtype of secondary MPs that result from the degradation of larger plastic items, such as bags, wrappers, and agricultural films [114].

It can also be noted that these materials are embedded by thin flexible filaments and are mostly made out of MPs, made of PE [115, 116]. As MP film waste enters the environment with detectable periods of expansion [117]. The components of microplastic (MPs) film are thin and flexible due to their composition being polyethylene, which is commonly used for packaging materials [115, 116]. The existence of these MPs shows an alarming rate of environmental pollution. Both urban and agricultural practices contribute to this level of pollution in a significant manner. However, plastic mulching for farming is considered one of the greatest contributors of raw substances [118]. Weathering and fragmentation of this material result in the emission of byproducts into the surrounding soils and water ecosystems. Moreover, the deterioration of agricultural goods, ex. seed coatings aggravate the emission of film MPs [119]. Film coloured MP is chiefly associated with urban sewage [120], through washes away particles from the paint and coating industries, where they pose a threat due to careless disposal.

#### Foam

Foam-type microplastics (MPs) are distinct for their granular appearance, low density, and soft, compressible structure [91, 116]. These particles primarily originate from expanded or extruded polystyrene foams, commonly known as Styrofoam, which is widely utilized in thermal insulation, packaging, and crafting due to its excellent shock-absorbing properties [105]. The prevalence of foam MPs in freshwater ecosystems is primarily linked to the use of polystyrene materials in food distribution, particularly in insulated boxes and disposable containers for fresh produce and takeout meals. Improper disposal and littering contribute to the fragmentation and environmental spread of these materials, significantly exacerbating microplastic pollution in urban waterways [121]. In rural regions, where the variety of plastic inputs may be limited, environmental samples frequently demonstrate that granular MPs are predominantly foam-based. These particles often exhibit irregular shapes due to the processes of environmental degradation [122].

#### Fragments

The fragmentation and degradation of larger plastic debris, driven by various processes, lead to the emergence of secondary microplastics known as fragments [123–125]. Distinct from primary microplastics, these particles are characterized by their irregular shapes and heterogeneous compositions and are closely associated with human activities, particularly in densely populated and recreational areas [98, 126]. Studies conducted on urban and peri-urban areas unveil increased density of plastic fragments within sediments concentrated around recreational tourism, shopping malls, as well as other commercial centres [127].

## Distribution and Environmental Behaviour of Microplastics

Microplastics' distribution and environmental behaviour pertain to their movement, transformation, and persistence within ecosystems. Understanding these dynamics is essential for evaluating exposure risks and formulating effective remediation strategies to mitigate their environmental impact.

## *Dynamics of Microplastic Distribution in Freshwater Ecosystems*

The distribution of microplastics in freshwater ecosystems occurs due to a combination of physical, chemical, and biological factors. Some prominent sources are urban runoff, wastewater discharge, agriculture, and even the atmosphere. After microplastics have been introduced into an environment, they are transported based on flow velocity, turbulence, and sedimentation forces. Size, shape, density, and surface characteristics impact their behaviours. Plastics of low density are expected to float with the current, while more dense particles may sink and end up in sediments. Some additional environmental factors, such as aggregation associated with organic matter, will also influence the settling of microplastics. The fate of microplastics is also influenced by the water column above them as a result of the absorption of pollutants that can change their settling rates. In turn, organisms contribute to this process by ingestion and excretion, which further enables the movement of microplastics through freshwater systems. Microplastic distribution patterns differ across ecosystems as rivers serve as transport channels while lakes and reservoirs act as long-term sinks. Understanding these dynamics is vital for assessing ecological risks related to microplastics and devising effective strategies tailored toward their pollution in freshwater environments.

Freshwater environments have emerged as a critical player in the global plastic cycle. This is especially true for microplastics, because larger fragments of plastics get processed into micro- and nano-plastics [128]. The pollutants mentioned before are harmful to the environment and possess distinct physical properties that impact how they conduct themselves in the world around them. Microplastics are not evenly dispersed throughout freshwater ecosystems but rather follow a certain pattern, such as polymer type, size of the particle, density, and surface chemistry [58]. It has been observed that water "Plastics" with densities below 1 g/cm$^3$ will float or be suspended, while those above 1 g/cm$^3$ will collect on sea beds, which acts as a long-term greenhouse for microplastic pollution [129–131].

When micro and nano particles from technologies enter water bodies, there is some tendency to aggregation, supported by both Homoaggregation—similar particles aggregate to one another; and Heteroaggregation—incorporating different particles, especially natural colloids, to aggregate to more complex particles. Observations indicate that smaller "microplastic" parts are more reactive (and thus more prone to aggregation) than larger "microplastic" parts due to their greater area-to-volume proportion ratio. Several processes interchange, merging effects disorders to small organized substance properties, changing particulars to particle dimension, shape envelope features, and weight. These manifestations of aggregation processes change the physical and chemical properties of the more complex particle aggregate matter. They alter the size and shape—its total surface features—and weights of aggregates. These changes impact the density of the aggregates or complex aggregate particles, displaying differences in flotation, stability, and motions in the water conditions. This has deleterious consequences for particle transport pathways and physicochemical fate in freshwater ecosystems [132].

When micro- and nanoplastics enter the environment or are eaten by aquatic organisms, they quickly interact with nearby biomolecules. This interaction leads to the formation of a surface layer called the eco-corona or bio-corona [133, 134]. The eco-corona has a strong influence on the physical and chemical characteristics of plastic particles including size, surface charge, and reactivity [135]. It provides a biochemical surface that changes how plastic micro- and nanoparticle behaves in the water column and may influence the availability and toxicity and subsequent transfer through the food web via a protective barrier against direct contact with cellular constituents [136].

## *Multifactorial Drivers Governing Microplastic Distribution in Freshwater Ecosystems*

Microplastics have a substantial impact on freshwater systems due to their physical properties that define their distribution, transport, and accumulation. The state and movement of microplastics will depend upon factors such as particle size, shape, density, surface charge, and polymer type. Microplastic distribution in rivers and lakes is also influenced by humans and biological factors (e.g., hydrological conditions, weather patterns, and topography), particularly in urban areas. Due to the complexity of the microplastic problem, it is essential to implement multisectoral strategies that incorporate the assessment and monitoring of microplastic pollution.

Microplastics existence and relocation of microplastics are influenced by seasonal changes and climate conditions such as heavy rain and flooding, which can push land-based microplastics into river ecosystems [137]. River features, such as flow velocity and turbulence, also influence the number of microplastics present [138, 139]. The physical and chemical properties of microplastics, including polymer type and density, affect their buoyancy and how they disperse. Additionally, changes in microplastic distribution relate to the specific morphological and hydraulic processes in each ecosystem, leading to patterns that vary by site [140]. Human activities, particularly dam building and river management, complicate this by altering natural flow and sediment dynamics, which in turn affect how microplastics are retained and transported in waterways [141].

Wastewater treatment plants are major sources of microplastics in freshwater environments. While modern treatment technologies can remove up to 99% of microplastics, the large volumes of effluent processed each day still led to significant amounts being released into aquatic systems. This ongoing discharge adds to the buildup of microplastics in rivers and lakes [49]. Additionally, a large portion of microplastics gets trapped in sewage sludge, which is often used as fertilizer or sent to landfills. This sludge can then release microplastics through surface runoff, allowing them to flow back into freshwater networks and potentially reach marine ecosystems. Therefore, it is essential to develop effective management strategies that address both liquid effluents and solid waste within the wastewater treatment system.

### Seasonal Variations

Seasonality, especially the timing of sampling, greatly impacts the distribution of microplastics (MP) in freshwater systems. Research suggests that MP levels are often elevated during the wet season, particularly during storms and periods of heavy rain [127]. These high MP levels are generally the result of surface runoff displacing plastic waste from farms and urban areas into river systems [110]. MPs rapidly migrate downstream and towards estuaries, aided by enhanced stream flows, as well as discharge from municipal wastewater treatment facilities [127]. Storms also propel plastic debris from land sources into rivers and coastal environments and reduce existing MP concentrations upstream [115]. Therefore, hydrometeorological conditions and spatial hydrodynamics must be considered when studying MP occurrence over time and space [142].

### Water Hydrodynamics

It is recognized that hydrological dynamics and processes (e.g., channel velocity, river geomorphology) influence the distribution of microplastics (MPs) within a river; these dynamic processes can also interact with anthropogenic determinants, such as land use or waste management practices [143]. The processes of sedimentation, which are a function of the flow characteristic of a river and changing shorelines, impact the settling of MPs in benthic and littoral sediments [144]. Surface currents also play a role in transporting and accumulating MPs in areas that are stagnant or have no flow [144]. Smaller freshwater systems can experience higher contamination as these smaller waterbodies possess less dilution capacity based on their hydrologic characteristic [145]. Moreover, MPs will be retained sequestered in sediments at a much slower temporal scale compared to the fluvial surface waters, resulting in more generalized contamination over long periods of time, and episodic due to the hydrologic event of large runoff [146]. It is noted that each microplastic reservoir, including dam reservoirs, would likely have a sediment concentration with more microplastics than the surface water of the dam reservoir. This happens mainly because reduced flow speeds promote sediment settling and encourage long-term buildup within the sediment [141].

### Particle Properties

The distribution of microplastics (MPs) in freshwater ecosystems is heavily influenced by their type and density [147]. Most MPs have a density lower than that of water, which enhances their buoyancy and makes them more prevalent in surface waters [110]. These low-density particles are readily transported by hydrodynamic forces such as wind and currents. Interestingly, the diversity of polymers found in freshwater sediments frequently surpasses that in surface waters [148], reflecting the intricate processes of deposition and transport. In calm environments, high-density MPs (greater than 1 g/cm$^3$) tend to settle to the bottom, while low-density MPs (less than 1 g/cm$^3$) usually remain suspended in the water column [149]. Additionally, as particle size decreases, their density often increases, which is an exception to the usual trend [150]. Consequently, larger and denser MPs are more likely to settle into benthic sediments, whereas smaller MPs are more widely distributed throughout the water column and sediment layers [110]. Additionally, smaller particles are transported by surface runoff more easily, which leads to a larger portion of particles less than 0.5 mm being found in freshwater systems [151]. These examples highlight the influence of particle characteristics on the distribution and fate of microplastics in freshwater systems.

Microplastics (MPs) are an emerging pollutant in aquatic systems and especially the water column, wherein size is a key factor in their transport and ingestion by aquatic organisms. Smaller microplastics are the most transportable and therefore present the greatest ecological risk. In lake ecosystems, there are typically many environmental impacts (human impacts, aquaculture, and fisheries) that represent a contributing source to microplastics. Microplastics enter the lake ecosystem through scenarios such as surface runoff and atmospheric deposition [152]. Ultimately, this kind of complexity places a demand for effective, continuous monitoring and a plan to enhance our protection for both freshwater resources and aquatic biodiversity.

### Other Factors

In aquatic systems, many factors other than polymer density can affect the vertical distribution of MPs [153]. There are important processes that may influence reported instances of MPs in river systems: photodegradation, biotic uptake, hydrodynamic dilution, gravitational sedimentation, and effects of surface-skimming processes [149]. In addition, aggregation and biofouling can change the effective density of MPs, even to the point where buoyant particles settle into sediments [150]. The MP size also will affect distribution in lakes, and environmental conditions also play a role: for example, smaller-sized MPs will easily disperse due to hydrodynamic forces and wind, which implies that smaller MPs will likely be distributed throughout the entire water column compared to larger MPs, which will settle and accumulate within sediments [154]. The size of particles containing the MPs suspended within ice and snow may be smaller as well, due to having no vertical dispersal mechanism. Aquatic organisms tend to ingest MPs, which are typically within the size class of 0.5–1 mm, which may be partially driven by their feeding behaviour. Overall, particle size is an important consideration when assessing the ecological implications and potential exposure to MPs in lake systems [155].

## Toxicological Effects of Pollution

The existence of microplastics in aquatic ecosystems and terrestrial ecosystems presents grave risk to the environment, upon aquatic and terrestrial ecosystems, as well as impacting human health. In order to address these challenges, we must evaluate their ecological toxicity and the effects over time in the (marine or terrestrial) food web.

### *Implications of Microplastic Contamination in the Marine Environment*

Micropollutants have a profound impact on environmental systems, particularly within terrestrial and aquatic ecosystems. These contaminants arise from industrial, agricultural, and domestic sources, disrupting cellular functions and disturbing ecological balance [156, 157]. Microplastics, despite their diminutive size, pose a substantial threat to marine ecosystems. Species such as fish, sea turtles, and seabirds often ingest these particles, leading to physical blockages and malnutrition [158]. Their small size enables them to carry toxic substances and provide habitats for harmful microorganisms, thereby heightening ecological risks [159]. Aquatic organisms frequently consume microplastics, especially fibrous particles, which present serious dangers to marine life, particularly fish and mussels. The toxicity of these microplastics is influenced by various factors, including their physicochemical properties, the duration of exposure, and their capacity to transport harmful pollutants into biological tissues [160].

Microplastics (MPs) significantly affect both aquatic and terrestrial organisms, entering ecosystems through ingestion, inhalation, or skin contact. They can cause growth suppression, oxidative stress, inflammation, organ damage, and reproductive issues at the cellular level [161]. This highlights the need for multidisciplinary research on their impact on ecosystem integrity. In aquatic environments, MPs raise concerns about marine biodiversity [162]. Organisms like zooplankton, bivalves, fish, and seabirds ingest MPs, resulting in internal injuries and increased harmful contaminants in the food web, which may pose risks to human health. Similarly, freshwater ecosystems are contaminated by urban runoff and industrial sources, leading to disruptions in community structure and threatening biodiversity and ecological resilience [163].

Micropollutants significantly alter water chemistry in aquatic ecosystems, disrupting the reproductive, behavioural, and metabolic functions of organisms, which threatens biodiversity and food webs [156, 157]. Microplastics adversely impact a range of species, from plankton to marine mammals [158, 164], causing mechanical damage, gastrointestinal blockages, and inflammation while also carrying harmful contaminants that compromise ecosystem health [165]. They disrupt hormonal signalling and impair gametogenesis in marine organisms, leading to reduced fertility [166]. Ingesting microplastics can damage reproductive tissues and hinder nutrient absorption [167, 168], and they also accumulate heavy metals like cadmium and lead, increasing ecological risks [169, 170].

Microplastics (MPs) threaten aquatic food webs through biomagnification. Their small size enables lower trophic level organisms, such as small fish, to ingest them, leading to accumulation in their digestive systems [171]. As larger fish consume these smaller organisms, MPs move up the food chain, affecting apex predators, including humans [172], and disrupting ecological balance [173]. Humans are exposed to MPs through personal care products and contaminated seafood, raising public health concerns [95]. Furthermore, MPs can cause health issues in marine organisms and transfer harmful chemicals throughout the food web, jeopardizing ecosystem stability [174]. They also absorb pollutants from seawater, becoming more toxic when ingested, and damage sensitive habitats like coral reefs, compromising vital ecological functions [175].

## *Implications of Microplastic Contamination in the Freshwater Ecosystems*

Microplastic contaminants pose a significant threat to freshwater fauna by causing biochemical disruptions, including genotoxicity, increased reactive oxygen species (ROS), liver damage, and inhibited growth, which indicate physiological stress [176]. Their small size allows for ingestion and exposure to harmful pollutants, and inhalation is a critical route of exposure for aquatic and semi-aquatic animals. Microplastics also significantly impact freshwater fish, affecting cellular functions and ecosystem health [177]. Exposure can result in apoptosis, metabolic dysregulation, oxidative stress, and histopathological changes, reducing population viability and species diversity. Fish absorb microplastics through both active ingestion and passive intake, influenced by species behaviours and plastic properties [178].

The use of plastics significantly threatens freshwater ecosystems, with microplastics as major pollutants [179]. Their small size allows easy ingestion by aquatic organisms, leading to biomagnification [128]. Microplastics can also absorb harmful chemicals and pathogens, causing various biological effects. Despite awareness, the long-term health implications of microplastics in freshwater systems remain under-researched. They contaminate vital water sources, including bottled and tap water, and are found in the gastrointestinal tracts of freshwater invertebrates [180].

Ingestion can lead to reduced food assimilation, delayed development, oxidative stress, behavioural changes, and impaired growth [181].

Specific types of microplastics, such as polyvinyl chloride and polyethene, adversely affect fish immune systems and accumulate in their tissues, raising concerns for ecological health and human consumption [182]. Additionally, microplastics are linked to physiological issues in aquatic organisms, including stunted growth and shorter lifespans [183]. Gastrointestinal blockages caused by these particles can hinder food intake, leading to energy deficits that disrupt vital functions and result in reproductive complications.

## *Ecotoxicological Effects of Microplastics on Human Health*

Micropollutants pose a significant threat to human health due to their widespread presence and exposure through ingestion, inhalation, and skin absorption. They can damage biological systems and can influence endocrine function, reproductive health, and increase risk for cancer and developmental disorders, which shows a need for risk assessments [184]. Microplastics are particularly important as they are a greater exposure and toxicity source, with ingestion as the primary route of exposure [185]. Microplastics can be moved by air, water, and soil, which also emphasizes the significance of working on this area of pollution for public health [186]. Microplastics can absorb environmental contaminants, and their small size permits ingestion of plastics by organisms, creating a greater chemical exposure load and increasing the risk of bioaccumulation and biomagnification of toxins [187, 188].

Micro- and nanoplastics (MNP) pollution is a growing global concern. It affects ecosystems and human health through ingestion, inhalation, and skin absorption [189]. Microplastics (0.1–10 μm) can breach barriers like the gastrointestinal lining and blood-brain barrier, while nanoplastics (smaller than 0.1 μm) can cross cell membranes, posing risks to developmental and neurological health [190]. Smaller microplastics are often more toxic due to greater bioavailability, and those from medical devices have been found in human tissues. Inhalation of microplastics can lead to respiratory issues, and dermal exposure from personal care products raises concerns about toxic buildup [160]. Ingestion may increase systemic exposure as microplastics evade digestive clearance [189]. Variability in toxicity based on polymer type, size, and concentration poses significant health risks, with recent studies detecting microplastics in human biological samples, highlighting their prevalence and the need for further research on long-term effects [168, 191].

Research has shown that microplastics pose various toxicological risks to human health. In vitro studies primarily focus on cellular interactions [114, 192–194], while in vivo studies investigate marine organisms [195] and mammalian models [196]. Key effects of microplastic exposure include cytotoxicity, immunomodulation, oxidative stress, and genotoxicity. Notably, adenocarcinoma cell lines exhibit reduced viability when exposed to microplastics, with severity influenced by concentration and exposure duration [197]. Recent studies reveal that microplastics

(MPs) are not effectively excreted after ingestion, with evidence of MPs present in human blood. This necessitates a reevaluation of their health effects and understanding of their metabolism in the body [198]. Additionally, the detection of MPs in human faecal samples points to widespread exposure, underscoring the need to investigate their health impacts further [199].

Pregnant women and infants are particularly vulnerable to microplastic exposure. Studies show that levels of polyethene terephthalate (PET) in infant faeces range from 5700 to 82,000 ng/g, with a median of 36,000 ng/g. This is about ten times higher than in adults. This increased exposure likely comes from behaviours like mouthing objects and immature detoxification systems [200]. The recent discovery of microplastics in human placentas raises concerns about their ability to cross the placental barrier. This suggests possible risks to fetal development [160].

This evidence signifies the need to regard microplastics as developmental toxicants with long-standing health implications. Microplastics (MPs) disrupt biochemical and physiological processes in living organisms. MPs induce oxidative stress by invisibly leaching toxic chemicals and eliciting reactive oxygen species (ROS) during inflammation. Prosthetic implants containing MPs released toxicities and free radicals, contributing to tissue damage and systemic oxidative stress [201]. MP-exposed mice and fish had higher levels of lactate dehydrogenase (LDH). This suggests a shift toward anaerobic metabolism, which may similarly affect humans and disrupt energy balance [202]. Additionally, MPs can weaken immune function, especially in genetically vulnerable individuals, making them more prone to infections and chronic diseases [185].

Researchers have identified a three-stage model for the physiological impacts of microplastic ingestion. The first stage involves mechanical damage to the gastrointestinal tract, while the second sees toxic additives leaching into the body. In the final stage, these toxins are absorbed by vital organs, causing cellular stress and systemic toxicity. Microplastics can enter the body through ingestion, inhalation, and skin contact, accumulating in organs like the liver and kidneys. Studies link microplastic exposure to health issues such as intestinal inflammation and liver disorders. Recent mouse research indicates that chronic exposure may weaken the intestinal barrier and disrupt glucose metabolism, potentially leading to insulin resistance and type 2 diabetes [90].

Humans face risks from consuming microplastic-contaminated fish, which may carry harmful toxins [203]. High-density polyethene and low-density polyethene act as endocrine disruptors, while polyvinyl chloride contains hazardous additives like bisphenol A (BPA) linked to serious health issues, including hormone-related cancers. The presence of BPA in breast milk raises concerns about early-life exposure [204]. Microplastics threaten human health through contaminated seafood, drinking water, and inhalation [205], leading to inflammation, cellular damage, and endocrine disruption. Filter-feeding organisms, like shellfish, can expose humans to toxins when consumed whole, resulting in oxidative stress and immune dysregulation [206]. Additionally, microplastics can transport Persistent Organic Pollutants (POPs) and heavy metals, further complicating exposure routes [207].

## Remedial Measures for Microplastic Pollution: An Interdisciplinary Framework

There are various strategies for preventing and controlling microplastic pollution, increasingly emphasizing interdisciplinary approaches. These strategies encompass policy and regulations, education and awareness, and collaborative industrial innovation [208].

### *Policy and Regulations*

Addressing microplastic pollution needs clear policies and regulations that consider the entire life cycle of plastics. This highlights the need for a strong system for the management of the production, use, and disposal of plastic materials. Using life-cycle-based guidance and standards, policy officials can address the key drivers of microplastic production and minimize their environmental effects and risks to human health. An integrated system is a foundational step to achieve a global response to concerns of microplastic contamination and promote responsible plastic governance.

#### Policy-Driven Solutions for Microplastic Control

Preventing and minimizing microplastic pollution will require a comprehensive strategy that entails policy reform, technological advancement, and community involvement. Part of the strategy may involve policies to prohibit the use of microplastics in consumer products, incentives and/or requirements for recycling, prohibiting plastic product inputs, or encouraging the use of plastic products in permissible and economically viable ways to lessen microplastic entry into water-related environments [209]. Regulatory measures regarding plastic pollution are increasing and are intended to shift the regulation of plastics from the production to the disposal aspect of the plastic life-cycle. These initiatives aim to lessen environmental damage and protect human health, even though a unified global response still lacks coherence. Effective regulatory frameworks must be strong, based on evidence, and easy to implement. Policies should encourage principles of circular economy, sustainable materials, and accountability for producers while raising public awareness. Timely actions are crucial to tackle microplastic pollution and promote a healthier future [210].

Political action against microplastic pollution is gaining global attention, but is still behind initiatives like plastic bag bans. A significant step was the European Union's Regulation (EU) 2023/2055, passed in September 2023, which bans synthetic microparticles in consumer products, including plastic glitter. Many current policies focus on restrictions at the end of a product's life, neglecting issues like overconsumption and the effects throughout a product's life cycle. The EU is beginning to take a more comprehensive approach by banning single-use items and pushing for waste reduction [211]. For policies to be effective, they should consider all stages of product life, including preventing problems and properly disposing of older plastics [212]. Moreover, addressing the political and economic factors that allow unsustainable practices is crucial for reducing microplastic release into ecosystems [213].

#### The Role of International Partnerships

Microplastic pollution is a serious environmental issue that needs cooperation from around the world. Its widespread impact demands unified strategies that surpass individual national policies. Key developments, such as the 2014 United Nations Environment Assembly (UNEA) and the 2017 extension of the Basel Convention to include plastic waste, are vital for international governance. Organizations like the United Nations and the UN Environment Programme (UNEP) are crucial in facilitating global dialogue and collaboration to combat microplastic contamination. Their efforts help set consistent international standards for plastic management [167]. The need for integrated policies grows during environmental emergencies, which can cause sudden increases in microplastics in marine ecosystems. Tackling these issues requires coordinated responses [214, 215]. Therefore, international collaboration is crucial for addressing the rising challenge of microplastic pollution.

### *Education and Awareness*

Engaging the public, especially students, is vital for encouraging responsible consumption and effective waste management in the fight against microplastic pollution [216]. Education is important for raising awareness, prompting behavioural change, and promoting civic engagement. By learning about the sources and effects of microplastics, individuals can develop more sustainable habits. Incorporating microplastic education into school curricula allows students to better comprehend the complexities of this issue. Hands-on experiments and discussions illuminate the ecological and health consequences while promoting stewardship [217–219]. Participatory approaches, such as citizen science, further enhance understanding and strengthen connections to local ecosystems [220]. This collaboration between education and public engagement is crucial for effective environmental governance.

### *Collaborative Industrial Innovation*

Addressing microplastic pollution requires collaboration across industries and technological innovation. This complex issue mandates coordinated efforts among sectors to create meaningful change. Industries can both contribute to and help mitigate this crisis by partnering with manufacturers, research institutions, and regulatory bodies. Innovation is crucial for developing eco-friendly materials and enhancing waste treatment technologies. By redesigning products for sustainability, we can decrease microplastic generation [211]. Effective remediation techniques, such as filtration and biodegradation, are vital for managing microplastics from wastewater treatment plants [221]. However, the rise of bioplastics raises concerns about their potential contributions to microplastic pollution, necessitating scrutiny of their degradation [125, 222].

A comprehensive approach that combines sustainable innovation, advanced waste processing, and behavioural change is essential for transitioning to circular industrial systems [223]. Key strategies include implementing Extended Producer Responsibility (EPR) to hold manufacturers accountable and reducing single-use plastics. The proposed Global Plastics Treaty emphasizes international collaboration in regulating production and waste management [224, 225]. Research and advancements in materials science are critical for creating biodegradable or recyclable materials, and technologies like magnetic nano-adsorbents show promise for capturing microplastics in Wastewater Treatment plants (WWTPs) [226]. Community-led cleanup initiatives and ongoing studies are vital for understanding and addressing the microplastic crisis [167, 184]. Mitigating micro-pollutants requires reducing hazardous chemicals, advancing treatment technologies, and fostering community engagement. Improving wastewater treatment infrastructure is crucial for eliminating trace contaminants, and public awareness campaigns can promote responsible consumer habits [156, 157].

## Conclusion

Microplastics, which stem from sources such as cosmetics and synthetic materials, are widespread contaminants in marine, freshwater, and terrestrial environments. They pose serious risks to human health and freshwater ecosystems. In aquatic systems, marine organisms ingest microplastics, disrupting their physiological functions and reproductive abilities, ultimately leading to increased mortality and negatively affecting industries like fisheries and tourism. In humans, the impact of microplastics has been associated with a range of health issues, including respiratory disease, gastrointestinal health, obesity, diabetes, and sleep issues. The accumulation of microplastics in human tissue raises questions around human inflammation and reproductive health. Collective action around the globe is necessary to address this issue, ranging from limiting microplastic emissions, improving public health campaigns, such as limiting single-use plastics, and increasing awareness of pollution. Many nations are implementing regulations to limit the use of plastics and improve recycling. When these practices are enacted, regulations can substantially reduce microplastics within the ecosystem as well as improve sustainability. A comprehensive strategy is needed that focuses on reducing plastics in the environment, improving waste management, and developing effective removal technologies. Ongoing research to better understand the health effects of microplastics is key. Using the knowledge gained from scientific evidence, as well as policy actions and action at the community level, to protect freshwater environments and aquatic diversity for generations to come is a principled approach.

**Acknowledgement** No conflicts of interest are disclosed by the author.

## References


1. Clunies-Ross, P., Smith, G., Gordon, K., & Gaw, S. (2016). Synthetic shorelines in New Zealand? Quantification and characterisation of microplastic pollution on Canterbury's coastlines. *New Zealand Journal of Marine and Freshwater Research, 50*, 317. https://doi.org/10.1080/00288330.2015.1132747
2. De Bhowmick, G., Sarmah, A. K., & Dubey, B. (2020). Microplastics in the NZ environment: Current status and future directions. *Case Studies in Chemical and Environmental Engineering, 3*, 100076. https://doi.org/10.1016/j.cscee.2020.100076
3. Thompson, R. C., Swan, S. H., Moore, C. J., & Saal, F. S. V. (2009). Our plastic age. *Philosophical Transactions of the Royal Society B: Biological Sciences, 364*, 1973. https://doi.org/10.1098/rstb.2009.0054
4. Ilyas, M., Ahmad, W., Khan, H., Yousaf, S., Khan, K., & Nazir, S. (2018). Plastic waste as a significant threat to environment—A systematic literature review. *Reviews on Environmental Health, 33*, 383–406. https://doi.org/10.1515/reveh-2017-0035
5. Pan, D., Su, F., Liu, C., & Guo, Z. (2020). Research progress for plastic waste management and manufacture of value-added products. *Advanced Composites and Hybrid Materials, 3*, 443. https://doi.org/10.1007/s42114-020-00190-0
6. Zeenat, N., Elahi, A., Bukhari, D. A., Shamim, S., & Rehman, A. (2021). Plastics degradation by microbes: A sustainable approach. *Journal of King Saud University - Science, 33*, 101538. https://doi.org/10.1016/j.jksus.2021.101538
7. Anik, A. H., Hossain, S., Alam, M., Sultan, M. B., Hasnine, M. T., & Rahman, M. M. (2021). Microplastics pollution: A comprehensive review on the sources, fates, effects, and potential remediation. *Environmental Nanotechnology, Monitoring & Management, 16*, 100530. https://doi.org/10.1016/j.enmm.2021.100530
8. Akdogan, Z., & Guven, B. (2019). Microplastics in the environment: A critical review of current understanding and identification of future research needs. *Environmental Pollution, 254*, 113011. https://doi.org/10.1016/j.enmm.2021.100530
9. Cole, M., Lindeque, P., Halsband, C., & Galloway, T. S. (2011). Microplastics as contaminants in the marine environment: A review. *Marine Pollution Bulletin, 62*, 2588. https://doi.org/10.1016/j.marpolbul.2011.09.025
10. Browne, M. A. (2015). Sources and pathways of microplastics to habitats. In *Marine anthropogenic litter* (pp. 229–244). Springer. https://doi.org/10.1007/978-3-319-16510-3_9
11. Gall, S. C., & Thompson, R. C. (2015). The impact of debris on marine life. *Marine Pollution Bulletin, 92*, 170. https://doi.org/10.1016/j.marpolbul.2014.12.041

12. Yuan, Z., Nag, R., & Cummins, E. (2022). Human health concerns regarding microplastics in the aquatic environment—From marine to food systems. *Science of the Total Environment, 823*, 153730. https://doi.org/10.1016/j.scitotenv.2022.153730
13. Rahman, A., Sarkar, A., Yadav, O. P., Achari, G., & Slobodnik, J. (2020). Potential human health risks due to environmental exposure to nano- and microplastics and knowledge gaps: A scoping review. *Science of the Total Environment, 757*, 143872. https://doi.org/10.1016/j.scitotenv.2020.143872
14. Galgani, F., Hanke, G., Werner, S., & De Vrees, L. (2013). Marine litter within the European Marine Strategy Framework Directive. *ICES Journal of Marine Science, 70*, 1055. https://doi.org/10.1093/icesjms/fst122
15. Eerkes-Medrano, D., Thompson, R. C., & Aldridge, D. C. (2015). Microplastics in freshwater systems: A review of the emerging threats, identification of knowledge gaps and prioritisation of research needs. *Water Research, 75*, 63. https://doi.org/10.1016/j.watres.2015.02.012
16. Eriksen, M., Mason, S., Wilson, S., Box, C., Zellers, A., Edwards, W., Farley, H., & Amato, S. (2013). Microplastic pollution in the surface waters of the Laurentian Great Lakes. *Marine Pollution Bulletin, 77*, 177. https://doi.org/10.1016/j.marpolbul.2013.10.007
17. Pérez-Méndez, M. A., Fraga-Cruz, G. S., Domínguez-García, S., Pérez-Méndez, M. L., Bocanegra-Díaz, C. I., & Nápoles-Rivera, F. (2025). Microplastic pollution in soil and water and the potential effects on human health: A review. *Processes, 13*, 502. https://doi.org/10.3390/pr13020502
18. Amobonye, A., Bhagwat, P., Raveendran, S., Singh, S., & Pillai, S. (2021). Environmental impacts of microplastics and nanoplastics: A current overview. *Frontiers in Microbiology, 12*, 768297. https://doi.org/10.3389/fmicb.2021.768297
19. Haque, F., & Fan, C. (2023). Fate and impacts of microplastics in the environment: Hydrosphere, pedosphere, and atmosphere. *Environments, 10*, 70. https://doi.org/10.3390/environments10050070
20. Lamichhane, G., Acharya, A., Marahatha, R., Modi, B., Paudel, R., Adhikari, A., Raut, B. K., Aryal, S., & Parajuli, N. (2022). Microplastics in environment: Global concern, challenges, and controlling measures. *International Journal of Environmental Science and Technology, 20*, 4673. https://doi.org/10.1007/s13762-022-04261-1
21. Pilapitiya, P. G. C. N. T., & Ratnayake, A. S. (2024). The world of plastic waste: A review. *Cleaner Materials, 11*, 100220. https://doi.org/10.1016/j.clema.2024.100220
22. Ozumchelouei, E. J., Hamidian, A. H., Zhang, Y., & Yang, M. (2019). Physicochemical properties of antibiotics: A review with an emphasis on detection in the aquatic environment. *Water Environment Research, 92*, 177. https://doi.org/10.1002/wer.1237
23. Wright, S. L., Thompson, R. C., & Galloway, T. S. (2013). The physical impacts of microplastics on marine organisms: A review. *Environmental Pollution, 178*, 483. https://doi.org/10.1016/j.envpol.2013.02.031
24. Bakir, A., Rowland, S. J., & Thompson, R. C. (2012). Competitive sorption of persistent organic pollutants onto microplastics in the marine environment. *Marine Pollution Bulletin, 64*, 2782. https://doi.org/10.1016/j.marpolbul.2012.09.010
25. Dekiff, J. H., Remy, D., Klasmeier, J., & Fries, E. (2014). Occurrence and spatial distribution of microplastics in sediments from Norderney. *Environmental Pollution, 186*, 248. https://doi.org/10.1016/j.envpol.2013.11.019
26. Zarfl, C., & Matthies, M. (2010). Are marine plastic particles transport vectors for organic pollutants to the Arctic? *Marine Pollution Bulletin, 60*, 1810. https://doi.org/10.1016/j.marpolbul.2010.05.026
27. Bakir, A., O'Connor, I. A., Rowland, S. J., Hendriks, A. J., & Thompson, R. C. (2016). Relative importance of microplastics as a pathway for the transfer of hydrophobic organic chemicals to marine life. *Environmental Pollution, 219*, 56. https://doi.org/10.1016/j.envpol.2016.09.046

28. Syberg, K., Khan, F. R., Selck, H., Palmqvist, A., Banta, G. T., Daley, J., Sano, L., & Duhaime, M. B. (2015). Microplastics: Addressing ecological risk through lessons learned. *Environmental Toxicology and Chemistry, 34*, 945. https://doi.org/10.1002/etc.2914
29. Klein, S., Dimzon, I. K., Eubeler, J., & Knepper, T. P. (2017). Analysis, occurrence, and degradation of microplastics in the aqueous environment. In *The handbook of environmental chemistry* (pp. 51–67). Springer. https://doi.org/10.1007/978-3-319-61615-5_3
30. Orb, A., Eisenhauer, L., & Wynaden, D. (2001). Ethics in qualitative research. *Journal of Nursing Scholarship, 33*, 93. https://doi.org/10.1111/j.1547-5069.2001.00093.x
31. Rist, S., & Hartmann, N. B. (2017). Aquatic ecotoxicity of microplastics and nanoplastics: Lessons learned from engineered nanomaterials. In *The handbook of environmental chemistry* (pp. 25–49). Springer. https://doi.org/10.1007/978-3-319-61615-5_2
32. Harrison, J. P., Hoellein, T. J., Sapp, M., Tagg, A. S., Ju-Nam, Y., & Ojeda, J. J. (2017). Microplastic-associated biofilms: A comparison of freshwater and marine environments. In *The handbook of environmental chemistry* (pp. 181–201). Springer. https://doi.org/10.1007/978-3-319-61615-5_9
33. Zhao, J., Ran, W., Teng, J., Liu, Y., Liu, H., Yin, X., Cao, R., & Wang, Q. (2018). Microplastic pollution in sediments from the Bohai Sea and the Yellow Sea, China. *Science of the Total Environment, 640–641*, 637. https://doi.org/10.1016/j.scitotenv.2018.05.346
34. Jiang, J.-Q. (2017). Occurrence of microplastics and its pollution in the environment: A review. *Sustainable Production and Consumption, 13*, 16. https://doi.org/10.1016/j.spc.2017.11.003
35. Rusidi, R. S., et al. (2024). Microplastics pollution mitigation in wastewater treatment: Current practices, challenges, and future perspectives. *Malaysian Journal of Analytical Sciences, 28*, 79–96.
36. Cesa, F. S., Turra, A., & Baruque-Ramos, J. (2017). Synthetic fibers as microplastics in the marine environment: A review from textile perspective with a focus on domestic washings. *Science of the Total Environment, 598*, 1116. https://doi.org/10.1016/j.scitotenv.2017.04.172
37. Singh, R., Kumar, R., & Sharma, P. (2021). Microplastic in the subsurface system: Extraction and characterization from sediments of River Ganga near Patna, Bihar. In *Advances in remediation techniques for polluted soils and groundwater* (pp. 191–217). Elsevier. https://doi.org/10.1016/B978-0-12-823830-1.00013-4
38. Zheng, K., Fan, Y., Zhu, Z., Chen, G., Tang, C., & Peng, X. (2019). Occurrence and species-specific distribution of plastic debris in wild freshwater fish from the Pearl River Catchment, China. *Environmental Toxicology and Chemistry, 38*, 1504. https://doi.org/10.1002/etc.4437
39. Kumar, R., Sharma, P., Manna, C., & Jain, M. (2021). Abundance, interaction, ingestion, ecological concerns, and mitigation policies of microplastic pollution in riverine ecosystem: A review. *Science of the Total Environment, 782*, 146695. https://doi.org/10.1016/j.scitotenv.2021.146695
40. Horton, A. A., Svendsen, C., Williams, R. J., Spurgeon, D. J., & Lahive, E. (2016). Large microplastic particles in sediments of tributaries of the River Thames, UK—Abundance, sources and methods for effective quantification. *Marine Pollution Bulletin, 114*, 218. https://doi.org/10.1016/j.marpolbul.2016.09.004
41. Rezania, S., Park, J., Din, M. F. M., Taib, S. M., Talaiekhozani, A., Yadav, K. K., & Kamyab, H. (2018). Microplastics pollution in different aquatic environments and biota: A review of recent studies. *Marine Pollution Bulletin, 133*, 191. https://doi.org/10.1016/j.marpolbul.2018.05.022
42. Raza, M., Lee, J.-Y., & Cha, J. (2022). Microplastics in soil and freshwater: Understanding sources, distribution, potential impacts, and regulations for management. *Science Progress, 105*, 00368504221126676. https://doi.org/10.1177/00368504221126676
43. Mintenig, S. M., Int-Veen, I., Loder, M. G. J., Primpke, S., & Gerdts, G. (2016). Identification of microplastics in effluents of wastewater treatment plants using focal plane array-based micro-Fourier-transform infrared imaging. *Water Research, 108*, 365. https://doi.org/10.1016/j.watres.2016.11.015

44. Oßmann, B. E., Sarau, G., Holtmannspotter, H., Pischetsrieder, M., Christiansen, S. H., & Dicke, W. (2018). Small-sized microplastics and pigmented particles in bottled mineral water. *Water Research, 141*, 307. https://doi.org/10.1016/j.watres.2018.05.027
45. Leslie, H. A., Brandsma, S. H., Van Velzen, M. J. M., & Vethaak, A. D. (2017). Microplastics en route: Field measurements in the Dutch river delta and Amsterdam canals, wastewater treatment plants, North Sea sediments and biota. *Environment International, 101*, 133. https://doi.org/10.1016/j.envint.2017.01.018
46. Gatidou, G., Arvaniti, O. S., & Stasinakis, A. S. (2018). Review on the occurrence and fate of microplastics in sewage treatment plants. *Journal of Hazardous Materials, 367*, 504. https://doi.org/10.1016/j.jhazmat.2018.12.081
47. Sun, J., Dai, X., Wang, Q., Van Loosdrecht, M. C. M., & Ni, B.-J. (2019). Microplastics in wastewater treatment plants: Detection, occurrence and removal. *Water Research, 152*, 21. https://doi.org/10.1016/j.watres.2018.12.050
48. Carr, S. A., Liu, J., & Tesoro, A. G. (2016). Transport and fate of microplastic particles in wastewater treatment plants. *Water Research, 91*, 174. https://doi.org/10.1016/j.watres.2016.01.002
49. Murphy, F., Ewins, C., Carbonnier, F., & Quinn, B. (2016). Wastewater treatment works (WWTW) as a source of microplastics in the aquatic environment. *Environmental Science & Technology, 50*, 5800. https://doi.org/10.1021/acs.est.5b05416
50. Rolsky, C., Kelkar, V., Driver, E., & Halden, R. U. (2019). Municipal sewage sludge as a source of microplastics in the environment. *Current Opinion in Environmental Science & Health, 14*, 16. https://doi.org/10.1016/j.coesh.2019.12.001
51. Hale, R. C., Seeley, M. E., La Guardia, M. J., Mai, L., & Zeng, E. Y. (2020). A global perspective on microplastics. *Journal of Geophysical Research: Oceans, 125*, e2018JC014719. https://doi.org/10.1029/2018jc014719
52. Hee, Y. Y., Hanif, N. M., Weston, K., Latif, M. T., Suratman, S., Rusli, M. U., & Mayes, A. G. (2023). Atmospheric microplastic transport and deposition to urban and pristine tropical locations in Southeast Asia. *Science of the Total Environment, 902*, 166153. https://doi.org/10.1016/j.scitotenv.2023.166153
53. Wright, S. L., & Kelly, F. J. (2017). Plastic and human health: A micro issue? *Environmental Science & Technology, 51*, 6634. https://doi.org/10.1021/acs.est.7b00423
54. Abbasi, S., Keshavarzi, B., Moore, F., Turner, A., Kelly, F. J., Dominguez, A. O., & Jaafarzadeh, N. (2018). Distribution and potential health impacts of microplastics and micro rubbers in air and street dusts from Asaluyeh County, Iran. *Environmental Pollution, 244*, 153. https://doi.org/10.1016/j.envpol.2018.10.039
55. Bhardwaj, L. K., Rath, P., Yadav, P., & Gupta, U. (2024). Microplastic contamination, an emerging threat to the freshwater environment: A systematic review. *Environmental Systems Research, 13*. https://doi.org/10.1186/s40068-024-00338-7
56. Boucher, J., Friot, D., & International Union for Conservation of Nature. (2017). *Primary microplastics in the oceans: A global evaluation of sources*. IUCN.
57. Horton, A. A., Walton, A., Spurgeon, D. J., Lahive, E., & Svendsen, C. (2017). Microplastics in freshwater and terrestrial environments: Evaluating the current understanding to identify the knowledge gaps and future research priorities. *Science of the Total Environment, 586*, 127. https://doi.org/10.1016/j.scitotenv.2017.01.190
58. Duis, K., & Coors, A. (2016). Microplastics in the aquatic and terrestrial environment: Sources (with a specific focus on personal care products), fate and effects. *Environmental Sciences Europe, 28*, 2. https://doi.org/10.1186/s12302-015-0069-y
59. Culin, J., & Bielic, T. (2016). Plastic pollution from ships. *Journal of Maritime & Transportation Science, 51*, 57. https://doi.org/10.18048/2016.51.04
60. Alomar, C., Estarellas, F., & Deudero, S. (2016). Microplastics in the Mediterranean Sea: Deposition in coastal shallow sediments, spatial variation and preferential grain size. *Marine Environmental Research, 115*, 1. https://doi.org/10.1016/j.marenvres.2016.01.005

61. Rochman, C. M. (2018). Microplastics research—From sink to source. *Science, 360*, 28. https://doi.org/10.1126/science.aar7734
62. Guerranti, C., Martellini, T., Perra, G., Scopetani, C., & Cincinelli, A. (2019). Microplastics in cosmetics: Environmental issues and needs for global bans. *Environmental Toxicology and Pharmacology, 68*, 75. https://doi.org/10.1016/j.etap.2019.03.007
63. Almroth, B. M. C., Åström, L., Roslund, S., Petersson, H., Johansson, M., & Persson, N.-K. (2017). Quantifying shedding of synthetic fibres from textiles; a source of microplastics released into the environment. *Environmental Science and Pollution Research, 25*, 1191. https://doi.org/10.1007/s11356-017-0528-7
64. Kole, P. J., Löhr, A. J., Van Belleghem, F., & Ragas, A. (2017). Wear and tear of tyres: A stealthy source of microplastics in the environment. *International Journal of Environmental Research and Public Health, 14*, 1265. https://doi.org/10.3390/ijerph14101265
65. Fadare, O. O., Wan, B., Guo, L.-H., & Zhao, L. (2020). Microplastics from consumer plastic food containers: Are we consuming it? *Chemosphere, 253*, 126787. https://doi.org/10.1016/j.chemosphere.2020.126787
66. Fadare, O. O., & Okoffo, E. D. (2020). Covid-19 face masks: A potential source of microplastic fibres in the environment. *Science of the Total Environment, 737*, 140279. https://doi.org/10.1016/j.scitotenv.2020.140279
67. Verschoor, A., De Poorter, L., Droge, R., Kuenen, J., De Valk, E., RIVM, & TNO. (2016). *Emission of microplastics and potential mitigation measures*. National Institute for Public Health and the Environment.
68. O'Brien, S., Rauert, C., Ribeiro, F., Okoffo, E. D., Burrows, S. D., O'Brien, J. W., Wang, X., Wright, S. L., & Thomas, K. V. (2023). There's something in the air: A review of sources, prevalence and behaviour of microplastics in the atmosphere. *Science of the Total Environment, 874*, 162193. https://doi.org/10.1016/j.scitotenv.2023.162193
69. Strojny, W., Gruca-Rokosz, R., & Cieśla, M. (2023). Preliminary study of the occurrence of microplastics in the sediments of the Rzeszów reservoir using the laser direct infrared (LDIR) method. *Sustainability, 15*, 16653. https://doi.org/10.3390/su152416653
70. Wu, Y., Wang, S., Wu, L., Yang, Y., Yu, X., Liu, Q., Liu, X., Li, Y., & Wang, X. (2022). Vertical distribution and river-sea transport of microplastics with tidal fluctuation in a subtropical estuary, China. *Science of the Total Environment, 822*, 153603. https://doi.org/10.1016/j.scitotenv.2022.153603
71. Idowu, G. A., Oriji, A. Y., Olorunfemi, K. O., Sunday, M. O., Sogbanmu, T. O., Bodunwa, O. K., Shokunbi, O. S., & Aiyesanmi, A. F. (2024). Why Nigeria should ban single-use plastics: Excessive microplastic pollution of the water, sediments and fish species in Osun River, Nigeria. *Journal of Hazardous Materials Advances, 13*, 100409. https://doi.org/10.1016/j.hazadv.2024.100409
72. Xu, B., et al. (2019). Microplastics in the soil environment: Occurrence, risks, interactions and fate—A review. *Critical Reviews in Environmental Science and Technology, 50*, 2175. https://doi.org/10.1080/10643389.2019.1694822
73. Ekvall, M. T., Lundqvist, M., Kelpsiene, E., Sileikis, E., Gunnarsson, S. B., & Cedervall, T. (2018). Nanoplastics formed during the mechanical breakdown of daily-use polystyrene products. *Nanoscale Advances, 1*, 1055. https://doi.org/10.1039/C8NA00210J
74. Zhang, K., Hamidian, A. H., Tubic, A., Zhang, Y., Fang, J. K. H., Wu, C., & Lam, P. K. S. (2021). Understanding plastic degradation and microplastic formation in the environment: A review. *Environmental Pollution, 274*, 116554. https://doi.org/10.1016/j.envpol.2021.116554
75. Napper, I. E., & Thompson, R. C. (2016). Release of synthetic microplastic plastic fibres from domestic washing machines: Effects of fabric type and washing conditions. *Marine Pollution Bulletin, 112*, 39. https://doi.org/10.1016/j.marpolbul.2016.09.025
76. Minghua, Z., Xiumin, F., Rovetta, A., Qichang, H., Vicentini, F., Bingkai, L., Giusti, A., & Yi, L. (2008). Municipal solid waste management in Pudong New Area, China. *Waste Management, 29*, 1227. https://doi.org/10.1016/j.wasman.2008.07.016

77. Vassilenko, E., Watkins, M., Chastain, S., Mertens, J., Posacka, A. M., Patankar, S., & Ross, P. S. (2021). Domestic laundry and microfiber pollution: Exploring fiber shedding from consumer apparel textiles. *PLoS One, 16*, e0250346. https://doi.org/10.1371/journal.pone.0250346
78. Ozkan, I., & Gündoğdu, S. (2020). Investigation on the microfiber release under controlled washings from the knitted fabrics produced by recycled and virgin polyester yarns. *The Journal of the Textile Institute, 112*, 264. https://doi.org/10.1080/00405000.2020.1741760
79. Li, J., Zhang, K., & Zhang, H. (2018). Adsorption of antibiotics on microplastics. *Environmental Pollution, 237*, 460–467.
80. Karbalaei, S., Golieskardi, A., Hamzah, H. B., Abdulwahid, S., Hanachi, P., Walker, T. R., & Karami, A. (2019). Abundance and characteristics of microplastics in commercial marine fish from Malaysia. *Marine Pollution Bulletin, 148*, 5. https://doi.org/10.1016/j.marpolbul.2019.07.072
81. Naji, A., Esmaili, Z., & Khan, F. R. (2016). Plastic debris and microplastics along the beaches of the Strait of Hormuz, Persian Gulf. *Marine Pollution Bulletin, 114*, 1057. https://doi.org/10.1016/j.marpolbul.2016.11.032
82. Peng, G., Xu, P., Zhu, B., Bai, M., & Li, D. (2017). Microplastics in freshwater river sediments in Shanghai, China: A case study of risk assessment in mega-cities. *Environmental Pollution, 234*, 448. https://doi.org/10.1016/j.envpol.2017.11.034
83. Calero, M., Godoy, V., Quesada, L., & Martín-Lara, M. A. (2021). Green strategies for microplastics reduction. *Current Opinion in Green and Sustainable Chemistry, 28*, 100442. https://doi.org/10.1016/j.cogsc.2020.100442
84. Park, T. J., Lee, S.-H., Lee, M.-S., Lee, J.-K., Lee, S.-H., & Zoh, K.-D. (2019). Occurrence of microplastics in the Han River and riverine fish in South Korea. *Science of the Total Environment, 708*, 134535.
85. Bhattacharya, R. R. N. S., Chandrasekhar, K., Roy, P., & Khan, A. (n.d.) Challenges and opportunities: Plastic waste management in India. http://hdl.handle.net/2451/42242
86. Anderson, P. J., Warrack, S., Langen, V., Challis, J. K., Hanson, M. L., & Rennie, M. D. (2017). Microplastic contamination in Lake Winnipeg, Canada. *Environmental Pollution, 225*, 223. https://doi.org/10.1016/j.envpol.2017.02.072
87. He, S., et al. (2021). Biofilm on microplastics in aqueous environment: Physicochemical properties and environmental implications. *Journal of Hazardous Materials, 424*, 127286. https://doi.org/10.1016/j.jhazmat.2021.127286
88. Mishra, A. K., Singh, J., & Mishra, P. P. (2022). Microplastics in freshwater ecosystem: A serious threat for freshwater environment. *International Journal of Environmental Science and Technology, 20*, 9189. https://doi.org/10.1016/j.scitotenv.2019.134535
89. Osman, A. I., et al. (2023). Microplastic sources, formation, toxicity and remediation: A review. *Environmental Chemistry Letters, 21*, 2129. https://doi.org/10.1007/s10311-023-01593-3
90. Ziani, K., Ioniță-Mîndrican, C.-B., Mititelu, M., Neacșu, S. M., Negrei, C., Moroșan, E., Draganescu, D., & Preda, O.-T. (2023). Microplastics: A real global threat for environment and food safety: A state-of-the-art review. *Nutrients, 15*, 617. https://doi.org/10.3390/nu15030617
91. Wagner, M., et al. (2014). Microplastics in freshwater ecosystems: What we know and what we need to know. *Environmental Sciences Europe, 26*, 12. https://doi.org/10.1186/s12302-014-0012-7
92. Estahbanati, S., & Fahrenfeld, N. L. (2016). Influence of wastewater treatment plant discharges on microplastic concentrations in surface water. *Chemosphere, 162*, 277. https://doi.org/10.1016/j.chemosphere.2016.07.083
93. Shruti, V. C., Jonathan, M. P., Rodriguez-Espinosa, P. F., & Rodríguez-González, F. (2018). Microplastics in freshwater sediments of Atoyac River basin, Puebla City, Mexico. *Science of the Total Environment, 654*, 154. https://doi.org/10.1016/j.scitotenv.2018.11.054

94. Andrady, A. L. (2017). The plastic in microplastics: A review. *Marine Pollution Bulletin, 119*, 12. https://doi.org/10.1016/j.marpolbul.2017.01.082
95. Sharma, S., & Chatterjee, S. (2017). Microplastic pollution, a threat to marine ecosystem and human health: A short review. *Environmental Science and Pollution Research, 24*, 21530. https://doi.org/10.1007/s11356-017-9910-8
96. Rodrigues, S. M., Almeida, C. M. R., Silva, D., Cunha, J., Antunes, C., Freitas, V., & Ramos, S. (2018). Microplastic contamination in an urban estuary: Abundance and distribution of microplastics and fish larvae in the Douro estuary. *Science of the Total Environment, 659*, 1071. https://doi.org/10.1007/s11356-017-9910-8
97. Sadri, S. S., & Thompson, R. C. (2014). On the quantity and composition of floating plastic debris entering and leaving the Tamar Estuary, Southwest England. *Marine Pollution Bulletin, 81*, 55. https://doi.org/10.1016/j.marpolbul.2014.02.020
98. Blair, R. M., Waldron, S., Phoenix, V. R., & Gauchotte-Lindsay, C. (2019). Microscopy and elemental analysis characterisation of microplastics in sediment of a freshwater urban river in Scotland, UK. *Environmental Science and Pollution Research, 26*, 12491. https://doi.org/10.1007/s11356-019-04678-1
99. Browne, M. A., Galloway, T., & Thompson, R. (2007). Microplastic—An emerging contaminant of potential concern? *Integrated Environmental Assessment and Management, 3*, 559. https://doi.org/10.1002/ieam.5630030412
100. Weinstein, J. E., Crocker, B. K., & Gray, A. D. (2016). From macroplastic to microplastic: Degradation of high-density polyethylene, polypropylene, and polystyrene in a salt marsh habitat. *Environmental Toxicology and Chemistry, 35*, 1632. https://doi.org/10.1002/etc.3432
101. Benítez, A., Sánchez, J. J., Arnal, M. L., Müller, A. J., Rodríguez, O., & Morales, G. (2012). Abiotic degradation of LDPE and LLDPE formulated with a pro-oxidant additive. *Polymer Degradation and Stability, 98*, 490. https://doi.org/10.1016/j.polymdegradstab.2012.12.011
102. Cooper, D. A., & Corcoran, P. L. (2010). Effects of mechanical and chemical processes on the degradation of plastic beach debris on the Island of Kauai, Hawaii. *Marine Pollution Bulletin, 60*, 650. https://doi.org/10.1016/j.marpolbul.2009.12.026
103. Campanale, C., Stock, F., Massarelli, C., Kochleus, C., Bagnuolo, G., Reifferscheid, G., & Uricchio, V. F. (2019). Microplastics and their possible sources: The example of Ofanto river in Southeast Italy. *Environmental Pollution, 258*, 113284. https://doi.org/10.1016/j.envpol.2019.113284
104. An, L., Liu, Q., Deng, Y., Wu, W., Gao, Y., & Ling, W. (2020). Sources of microplastic in the environment. In *The handbook of environmental chemistry* (pp. 143–159). Springer. https://doi.org/10.1007/698_2020_449
105. Zhang, K., Gong, W., Lv, J., Xiong, X., & Wu, C. (2015). Accumulation of floating microplastics behind the Three Gorges Dam. *Environmental Pollution, 204*, 117. https://doi.org/10.1016/j.envpol.2015.04.023
106. Mojoudi, F., Hamidian, A. H., Goodarzian, N., & Eagderi, S. (2018). Effective removal of heavy metals from aqueous solution by porous activated carbon/thiol functionalized graphene oxide composite. *Desalination and Water Treatment, 124*, 106. https://doi.org/10.5004/dwt.2018.22695
107. Mojoudi, F., Hamidian, A. H., Zhang, Y., & Yang, M. (2019). Synthesis and evaluation of activated carbon/nanoclay/thiolated graphene oxide nanocomposite for lead (II) removal from aqueous solution. *Water Science and Technology, 79*, 466. https://doi.org/10.2166/wst.2019.071
108. Lahens, L., Strady, E., Kieu-Le, T.-C., Dris, R., Boukerma, K., Rinnert, E., Gasperi, J., & Tassin, B. (2018). Macroplastic and microplastic contamination assessment of a tropical river (Saigon River, Vietnam) transversed by a developing megacity. *Environmental Pollution, 236*, 661. https://doi.org/10.1016/j.envpol.2018.02.005
109. Peng, G., Zhu, B., Yang, D., Su, L., Shi, H., & Li, D. (2017). Microplastics in sediments of the Changjiang Estuary, China. *Environmental Pollution, 225*, 283. https://doi.org/10.1016/j.envpol.2016.12.064

110. Alam, F. C., Sembiring, E., Muntalif, B. S., & Suendo, V. (2019). Microplastic distribution in surface water and sediment river around slum and industrial area (case study: Ciwalengke River, Majalaya district, Indonesia). *Chemosphere, 224*, 637. https://doi.org/10.1016/j.chemosphere.2019.02.188
111. Fan, Y., Zheng, K., Zhu, Z., Chen, G., & Peng, X. (2019). Distribution, sedimentary record, and persistence of microplastics in the Pearl River catchment, China. *Environmental Pollution, 251*, 862. https://doi.org/10.1016/j.envpol.2019.05.056
112. Zhang, G. S., & Liu, Y. F. (2018). The distribution of microplastics in soil aggregate fractions in southwestern China. *Science of the Total Environment, 642*, 12. https://doi.org/10.1016/j.scitotenv.2018.06.004
113. Hitchcock, J. N., & Mitrovic, S. M. (2019). Microplastic pollution in estuaries across a gradient of human impact. *Environmental Pollution, 247*, 457. https://doi.org/10.1016/j.envpol.2019.01.069
114. Hwang, J., Choi, D., Han, S., Jung, S. Y., Choi, J., & Hong, J. (2020). Potential toxicity of polystyrene microplastic particles. *Scientific Reports, 10*, 7391. https://doi.org/10.1038/s41598-020-64464-9
115. Wu, P., Tang, Y., Dang, M., Wang, S., Jin, H., Liu, Y., Jing, H., Zheng, C., Yi, S., & Cai, Z. (2019). Spatial-temporal distribution of microplastics in surface water and sediments of Maozhou River within Guangdong-Hong Kong-Macao Greater Bay Area. *Science of the Total Environment, 717*, 135187. https://doi.org/10.1016/j.scitotenv.2019.135187
116. Xu, Q., Xing, R., Sun, M., Gao, Y., & An, L. (2020). Microplastics in sediments from an interconnected river-estuary region. *Science of the Total Environment, 729*, 139025. https://doi.org/10.1016/j.scitotenv.2020.139025
117. Naidoo, T., Glassom, D., & Smit, A. J. (2015). Plastic pollution in five urban estuaries of KwaZulu-Natal, South Africa. *Marine Pollution Bulletin, 101*, 473. https://doi.org/10.1016/j.marpolbul.2015.09.044
118. Wen, X., et al. (2018). Microplastic pollution in surface sediments of urban water areas in Changsha, China: Abundance, composition, surface textures. *Marine Pollution Bulletin, 136*, 414. https://doi.org/10.1016/j.marpolbul.2018.09.043
119. Ding, L., Mao, R. F., Guo, X., Yang, X., Zhang, Q., & Yang, C. (2019). Microplastics in surface waters and sediments of the Wei River, in the northwest of China. *Science of the Total Environment, 667*, 427. https://doi.org/10.1016/j.scitotenv.2019.02.332
120. Dikareva, N., & Simon, K. S. (2019). Microplastic pollution in streams spanning an urbanisation gradient. *Environmental Pollution, 250*, 292. https://doi.org/10.1016/j.envpol.2019.03.105
121. Fok, L., & Cheung, P. K. (2015). Hong Kong at the Pearl River Estuary: A hotspot of microplastic pollution. *Marine Pollution Bulletin, 99*, 112. https://doi.org/10.1016/j.marpolbul.2015.07.050
122. Mao, Y., Li, H., Gu, W., Yang, G., Liu, Y., & He, Q. (2020). Distribution and characteristics of microplastics in the Yulin River, China: Role of environmental and spatial factors. *Environmental Pollution, 265*, 115033. https://doi.org/10.1016/j.envpol.2020.115033
123. Free, C. M., Jensen, O. P., Mason, S. A., Eriksen, M., Williamson, N. J., & Boldgiv, B. (2014). High levels of microplastic pollution in a large, remote, mountain lake. *Marine Pollution Bulletin, 85*, 156. https://doi.org/10.1016/j.marpolbul.2014.06.001
124. Kay, P., Hiscoe, R., Moberley, I., Bajic, L., & McKenna, N. (2018). Wastewater treatment plants as a source of microplastics in river catchments. *Environmental Science and Pollution Research, 25*, 20264. https://doi.org/10.1007/s11356-018-2070-7
125. Shruti, V. C., & Kutralam-Muniasamy, G. (2019). Bioplastics: Missing link in the era of microplastics. *Science of the Total Environment, 697*, 134139. https://doi.org/10.1016/j.scitotenv.2019.134139
126. Mani, T., Hauk, A., Walter, U., & Burkhardt-Holm, P. (2015). Microplastics profile along the Rhine River. *Scientific Reports, 5*. https://doi.org/10.1038/srep17988

127. Ramirez, M. M. B., Caamal, R. D., & Von Osten, J. R. (2019). Occurrence and seasonal distribution of microplastics and phthalates in sediments from the urban channel of the Ria and coast of Campeche, Mexico. *Science of the Total Environment, 672*, 97. https://doi.org/10.1016/j.scitotenv.2019.03.472
128. Bulannga, R. B., & Schmidt, S. (2024). Micro- and nanoplastics in freshwater ecosystems: Interaction with and impact upon bacterivorous ciliates. *Frontiers in Earth Science, 12*, 1349865. https://doi.org/10.3389/feart.2024.1349865
129. Bellasi, A., Binda, G., Pozzi, A., Galafassi, S., Volta, P., & Bettinetti, R. (2020). Microplastic contamination in freshwater environments: A review, focusing on interactions with sediments and benthic organisms. *Environments, 7*, 30. https://doi.org/10.3390/environments7040030
130. Leiser, R., Schumann, M., Dadi, T., & Wendt-Potthoff, K. (2021). Burial of microplastics in freshwater sediments facilitated by iron-organ flocs. *Scientific Reports, 11*, 24072. https://doi.org/10.1038/s41598-021-02748-4
131. Viitala, M., Steinmetz, Z., Sillanpää, M., Manttari, M., & Sillanpää, M. (2022). Historical and current occurrence of microplastics in water and sediment of a Finnish lake affected by WWTP effluents. *Environmental Pollution, 314*, 120298. https://doi.org/10.1016/j.envpol.2022.120298
132. Issac, M. N., & Kandasubramanian, B. (2021). Effect of microplastics in water and aquatic systems. *Environmental Science and Pollution Research, 28*, 19544. https://doi.org/10.1007/s11356-021-13184-2
133. Nasser, F., Constantinou, J., & Lynch, I. (2019). Nanomaterials in the environment acquire an "eco-Corona" impacting their toxicity to daphnia magna: A call for updating toxicity testing policies. *Proteomics, 20*, e1800412. https://doi.org/10.1002/pmic.201800412
134. Ekvall, M. T., Hedberg, J., Wallinder, I. O., Malmendal, A., Hansson, L.-A., & Cedervall, T. (2021). Adsorption of bio-organic eco-corona molecules reduces the toxic response to metallic nanoparticles in Daphnia magna. *Scientific Reports, 11*. https://doi.org/10.1038/s41598-021-90053-5
135. Giri, S., & Mukherjee, A. (2021). Ageing with algal EPS reduces the toxic effects of polystyrene nanoplastics in freshwater microalgae Scenedesmus obliquus. *Journal of Environmental Chemical Engineering, 9*, 105978. https://doi.org/10.1016/j.jece.2021.105978
136. Liu, S., Junaid, M., Liao, H., Liu, X., Wu, Y., & Wang, J. (2022). Eco-corona formation and associated ecotoxicological impacts of nanoplastics in the environment. *Science of the Total Environment, 836*, 155703. https://doi.org/10.1016/j.scitotenv.2022.155703
137. Scircle, A., Cizdziel, J. V., Missling, K., Li, L., & Vianello, A. (2020). Single-pot method for the collection and preparation of natural water for microplastic analyses: Microplastics in the Mississippi River System during and after historic flooding. *Environmental Toxicology and Chemistry, 39*, 986. https://doi.org/10.1002/etc.4698
138. Mishra, A. K., Singh, J., & Mishra, P. P. (2021). Microplastics in polar regions: An early warning to the world's pristine ecosystem. *Science of the Total Environment, 784*, 147149. https://doi.org/10.1016/j.scitotenv.2021.147149
139. Wong, J. K. H., Lee, K. K., Tang, K. H. D., & Yap, P.-S. (2020). Microplastics in the freshwater and terrestrial environments: Prevalence, fates, impacts and sustainable solutions. *Science of the Total Environment, 719*, 137512. https://doi.org/10.1016/j.scitotenv.2020.137512
140. Vaid, M., Sarma, K., & Gupta, A. (2021). Microplastic pollution in aquatic environments with special emphasis on riverine systems: Current understanding and way forward. *Journal of Environmental Management, 293*, 112860. https://doi.org/10.1016/j.jenvman.2021.112860
141. Watkins, L., McGrattan, S., Sullivan, P. J., & Walter, M. T. (2019). The effect of dams on river transport of microplastic pollution. *Science of the Total Environment, 664*, 834. https://doi.org/10.1016/j.scitotenv.2019.02.028
142. Lima, A. R. A., Costa, M. F., & Barletta, M. (2014). Distribution patterns of microplastics within the plankton of a tropical estuary. *Environmental Research, 132*, 146. https://doi.org/10.1016/j.envres.2014.03.031

143. Klein, S., Worch, E., & Knepper, T. P. (2015). Occurrence and spatial distribution of microplastics in river shore sediments of the Rhine-Main area in Germany. *Environmental Science & Technology, 49*, 6070. https://doi.org/10.1021/acs.est.5b00492
144. He, B., Goonetilleke, A., Ayoko, G. A., & Rintoul, L. (2019). Abundance, distribution patterns, and identification of microplastics in Brisbane River sediments, Australia. *Science of the Total Environment, 700*, 134467. https://doi.org/10.1016/j.scitotenv.2019.134467
145. Luo, W., Su, L., Craig, N. J., Du, F., Wu, C., & Shi, H. (2018). Comparison of microplastic pollution in different water bodies from urban creeks to coastal waters. *Environmental Pollution, 246*, 174. https://doi.org/10.1016/j.envpol.2018.11.081
146. Gray, A. D., Wertz, H., Leads, R. R., & Weinstein, J. E. (2018). Microplastic in two South Carolina estuaries: Occurrence, distribution, and composition. *Marine Pollution Bulletin, 128*, 223. https://doi.org/10.1016/j.marpolbul.2018.01.030
147. Hurley, R., Woodward, J., & Rothwell, J. J. (2018). Microplastic contamination of river beds significantly reduced by catchment-wide flooding. *Nature Geoscience, 11*, 251. https://doi.org/10.1038/s41561-018-0080-1
148. Scherer, C., et al. (2020). Comparative assessment of microplastics in water and sediment of a large European river. *Science of the Total Environment, 738*, 139866. https://doi.org/10.1016/j.scitotenv.2020.139866
149. Zhang, K., Xiong, X., Hu, H., Wu, C., Bi, Y., Wu, Y., Zhou, B., Lam, P. K. S., & Liu, J. (2017). Occurrence and characteristics of microplastic pollution in Xiangxi Bay of Three Gorges Reservoir, China. *Environmental Science & Technology, 51*, 3794. https://doi.org/10.1021/acs.est.7b00369
150. Eo, S., Hong, S. H., Song, Y. K., Han, G. M., & Shim, W. J. (2019). Spatiotemporal distribution and annual load of microplastics in the Nakdong River, South Korea. *Water Research, 160*, 228. https://doi.org/10.1016/j.watres.2019.05.053
151. Yan, M., Nie, H., Xu, K., He, Y., Hu, Y., Huang, Y., & Wang, J. (2018). Microplastic abundance, distribution and composition in the Pearl River along Guangzhou city and Pearl River estuary, China. *Chemosphere, 217*, 879. https://doi.org/10.1016/j.watres.2019.05.053
152. Bond, T., Ferrandiz-Mas, V., Felipe-Sotelo, M., & Van Sebille, E. (2018). The occurrence and degradation of aquatic plastic litter based on polymer physicochemical properties: A review. *Critical Reviews in Environmental Science and Technology, 48*, 685. https://doi.org/10.1080/10643389.2018.1483155
153. Wang, W., Ndungu, A. W., Li, Z., & Wang, J. (2016). Microplastics pollution in inland freshwaters of China: A case study in urban surface waters of Wuhan, China. *Science of the Total Environment, 575*, 1369. https://doi.org/10.1016/j.scitotenv.2016.09.213
154. Corcoran, P. L., Norris, T., Ceccanese, T., Walzak, M. J., Helm, P. A., & Marvin, C. H. (2015). Hidden plastics of Lake Ontario, Canada and their potential preservation in the sediment record. *Environmental Pollution, 204*, 17. https://doi.org/10.1016/j.envpol.2015.04.009
155. Pan, T., Liao, H., Yang, F., Sun, F., Guo, Y., Yang, H., Feng, D., Zhou, X., & Wang, Q. (2023). Review of microplastics in lakes: Sources, distribution characteristics, and environmental effects. *Carbon Research, 2*, 25. https://doi.org/10.1007/s44246-023-00057-1
156. Prabakaran, C. (2023). Micro pollutant on the environment: Causes, impacts, and solutions. *The Pharma Innovation, 12*, 2009.
157. Roman, L., Schuyler, Q., Wilcox, C., & Hardesty, B. D. (2021). Plastic pollution is killing marine megafauna, but how do we prioritize policies to reduce mortality? *Conservation Letters, 14*, e12781. https://doi.org/10.1111/conl.12781
158. Bottari, T., et al. (2021). Microplastics in the Bogue, Boops boops: A snapshot of the past from the southern Tyrrhenian Sea. *Journal of Hazardous Materials, 424*, 127669. https://doi.org/10.1016/j.jhazmat.2021.127669
159. McCormick, A., Hoellein, T. J., Mason, S. A., Schluep, J., & Kelly, J. J. (2014). Microplastic is an abundant and distinct microbial habitat in an Urban River. *Environmental Science & Technology, 48*, 11863. https://doi.org/10.1021/es503610r

160. Li, Y., Tao, L., Wang, Q., Wang, F., Li, G., & Song, M. (2023). Potential health impact of microplastics: A review of environmental distribution, human exposure, and toxic effects. *Environment & Health, 1*, 249. https://doi.org/10.1021/envhealth.3c00052
161. Zhao, X., Gao, S., Ouyang, D., Chen, S., Qiu, C., Qiu, H., & Chen, Z. (2023). Advances on micro/nanoplastics and their effects on the living organisms: A review. *Science of the Total Environment, 904*, 166722. https://doi.org/10.1016/j.scitotenv.2023.166722
162. Franzellitti, S., Canesi, L., Auguste, M., Wathsala, R. H. G. R., & Fabbri, E. (2019). Microplastic exposure and effects in aquatic organisms: A physiological perspective. *Environmental Toxicology and Pharmacology, 68*, 37. https://doi.org/10.1016/j.etap.2019.03.009
163. Pastorino, P., & Barcelo, D. (2023). Microplastics and their environmental effects. *Environmental Toxicology and Pharmacology, 104*, 104324. https://doi.org/10.1016/j.etap.2023.104324
164. Monique, M., et al. (2022). Investigating the effects of microplastic ingestion in Scyliorhinus canicula from the South of Sicily. *Science of the Total Environment, 850*, 157875. https://doi.org/10.1016/j.scitotenv.2022.157875
165. Hermabessiere, L., Dehaut, A., Paul-Pont, I., Lacroix, C., Jezequel, R., Soudant, P., & Duflos, G. (2017). Occurrence and effects of plastic additives on marine environments and organisms: A review. *Chemosphere, 182*, 781. https://doi.org/10.1016/j.chemosphere.2017.05.096
166. Yong, C., Valiyaveettil, S., & Tang, B. (2020). Toxicity of microplastics and nanoplastics in mammalian systems. *International Journal of Environmental Research and Public Health, 17*, 1509. https://doi.org/10.3390/ijerph17051509
167. Ghosh, S., Sinha, J. K., Ghosh, S., Vashisth, K., Han, S., & Bhaskar, R. (2023). Microplastics as an emerging threat to the global environment and human health. *Sustainability, 15*, 10821. https://doi.org/10.3390/su151410821
168. Yang, X., Man, Y. B., Wong, M. H., Owen, R. B., & Chow, K. L. (2022). Environmental health impacts of microplastics exposure on structural organization levels in the human body. *Science of the Total Environment, 825*, 154025. https://doi.org/10.1016/j.scitotenv.2022.154025
169. Brennecke, D., Duarte, B., Paiva, F., Caçador, I., & Canning-Clode, J. (2016). Microplastics as vector for heavy metal contamination from the marine environment. *Estuarine, Coastal and Shelf Science, 178*, 189. https://doi.org/10.1016/j.ecss.2015.12.003
170. Harsha, M. S. V., Prasad, P. A. S., & Prakash, D. B. (2024). Theoretical review on microplastic pollution: A multifaceted threat to marine ecosystems, human health, and environment. *IgMin Research, 2*, 460. https://doi.org/10.61927/igmin203
171. Jabeen, K., Su, L., Li, J., Yang, D., Tong, C., Mu, J., & Shi, H. (2016). Microplastics and mesoplastics in fish from coastal and fresh waters of China. *Environmental Pollution, 221*, 141. https://doi.org/10.1016/j.envpol.2016.11.055
172. Kasamesiri, P., & Thaimuangphol, W. (2020). Microplastics ingestion by freshwater fish in the Chi River, Thailand. *International Journal of Geomate, 18*. https://doi.org/10.21660/2020.67.9110
173. Horton, A. A., & Dixon, S. J. (2017). Microplastics: An introduction to environmental transport processes. *Wiley Interdisciplinary Reviews: Water, 5*, e1268. https://doi.org/10.1002/wat2.1268
174. Carbery, M., O'Connor, W., & Palanisami, T. (2018). Trophic transfer of microplastics and mixed contaminants in the marine food web and implications for human health. *Environment International, 115*, 400. https://doi.org/10.1016/j.envint.2018.03.007
175. Gallo, F., Fossi, C., Weber, R., Santillo, D., Sousa, J., Ingram, I., Nadal, A., & Romano, D. (2018). Marine litter plastics and microplastics and their toxic chemical components: The need for urgent preventive measures. *Environmental Sciences Europe, 30*, 13. https://doi.org/10.1186/s12302-018-0139-z
176. Ehlers, S., Manz, W., & Koop, J. (2019). Microplastics of different characteristics are incorporated into the larval cases of the freshwater caddisfly Lepidostoma basale. *Aquatic Biology, 28*, 67. https://doi.org/10.3354/ab00711

177. Parker, B., Andreou, D., Green, I. D., & Britton, J. R. (2021). Microplastics in freshwater fishes: Occurrence, impacts and future perspectives. *Fish and Fisheries, 22*, 467. https://doi.org/10.1111/faf.12528
178. Roch, S., Friedrich, C., & Brinker, A. (2020). Uptake routes of microplastics in fishes: Practical and theoretical approaches to test existing theories. *Scientific Reports, 10*. https://doi.org/10.1038/s41598-020-60630-1
179. Sarijan, S., Azman, S., Said, M. I. M., & Jamal, M. H. (2020). Microplastics in freshwater ecosystems: A recent review of occurrence, analysis, potential impacts, and research needs. *Environmental Science and Pollution Research, 28*, 1341. https://doi.org/10.1007/s11356-020-11171-7
180. Mintenig, S. M., Loder, M. G. J., Primpke, S., & Gerdts, G. (2018). Low numbers of microplastics detected in drinking water from groundwater sources. *Science of the Total Environment, 648*, 631. https://doi.org/10.1016/j.scitotenv.2018.08.178
181. Bhuyan, M. S. (2022). Effects of microplastics on fish and in human health. *Frontiers in Environmental Science, 10*, 1–17. https://doi.org/10.3389/fenvs.2022.827289
182. Akhbarizadeh, R., Moore, F., & Keshavarzi, B. (2017). Investigating a probable relationship between microplastics and potentially toxic elements in fish muscles from northeast of Persian Gulf. *Environmental Pollution, 232*, 154. https://doi.org/10.1016/j.envpol.2017.09.028
183. Windsor, F. M., Tilley, R. M., Tyler, C. R., & Ormerod, S. J. (2018). Microplastic ingestion by riverine macroinvertebrates. *Science of the Total Environment, 646*, 68. https://doi.org/10.1016/j.scitotenv.2018.07.271
184. Campanale, C., Massarelli, C., Savino, I., Locaputo, V., & Uricchio, V. F. (2020). A detailed review study on potential effects of microplastics and additives of concern on human health. *International Journal of Environmental Research and Public Health, 17*, 1212. https://doi.org/10.3390/ijerph17041212
185. Prata, J. C., Da Costa, J. P., Lopes, I., Duarte, A. C., & Rocha-Santos, T. (2019). Environmental exposure to microplastics: An overview on possible human health effects. *Science of the Total Environment, 702*, 134455. https://doi.org/10.1016/j.scitotenv.2019.134455
186. Rillig, M. C., & Lehmann, A. (2020). Microplastic in terrestrial ecosystems. *Science, 368*, 1430. https://doi.org/10.1126/science.abb5979
187. Panebianco, A., Nalbone, L., Giarratana, F., & Ziino, G. (2019). First discoveries of microplastics in terrestrial snails. *Food Control, 106*, 106722. https://doi.org/10.1016/j.foodcont.2019.106722
188. Hu, X., Yu, Q., Waigi, M. G., Ling, W., Qin, C., Wang, J., & Gao, Y. (2022). Microplastics-sorbed phenanthrene and its derivatives are highly bioaccessible and may induce human cancer risks. *Environment International, 168*, 107459. https://doi.org/10.1016/j.envint.2022.107459
189. Sun, A., & Wang, W.-X. (2023). Human exposure to microplastics and its associated health risks. *Environment & Health, 1*, 139. https://doi.org/10.1021/envhealth.3c00053
190. Prietl, B., Meindl, C., Roblegg, E., Pieber, T. R., Lanzer, G., & Fröhlich, E. (2013). Nano-sized and micro-sized polystyrene particles affect phagocyte function. *Cell Biology and Toxicology, 30*(1), 1–16. https://doi.org/10.1007/s10565-013-9265-y
191. Kwon, J.-H., et al. (2020). Microplastics in food: A review on analytical methods and challenges. *International Journal of Environmental Research and Public Health, 17*, 6710. https://doi.org/10.3390/ijerph17186710
192. Choi, D., Hwang, J., Bang, J., Han, S., Kim, T., Oh, Y., Hwang, Y., Choi, J., & Hong, J. (2020). In vitro toxicity from a physical perspective of polyethylene microplastics based on statistical curvature change analysis. *Science of the Total Environment, 752*, 142242. https://doi.org/10.1016/j.scitotenv.2020.142242
193. Han, S., Bang, J., Choi, D., Hwang, J., Kim, T., Oh, Y., Hwang, Y., Choi, J., & Hong, J. (2020). Surface pattern analysis of microplastics and their impact on human-derived cells. *ACS Applied Polymer Materials, 2*, 4541. https://doi.org/10.1021/acsapm.0c00645

194. Stock, V., Laurisch, C., Franke, J., Dönmez, M. H., Voss, L., Bohmert, L., Braeuning, A., & Sieg, H. (2020). Uptake and cellular effects of PE, PP, PET and PVC microplastic particles. *Toxicology In Vitro, 70*, 105021. https://doi.org/10.1016/j.tiv.2020.105021
195. Oliviero, M., Tato, T., Schiavo, S., Fernández, V., Manzo, S., & Beiras, R. (2019). Leachates of micronized plastic toys provoke embryotoxic effects upon sea urchin Paracentrotus lividus. *Environmental Pollution, 247*, 706. https://doi.org/10.1016/j.envpol.2019.01.098
196. Li, C., Busquets, R., & Campos, L. C. (2019). Assessment of microplastics in freshwater systems: A review. *Science of the Total Environment, 707*, 135578. https://doi.org/10.1016/j.scitotenv.2019.135578
197. Danopoulos, E., Twiddy, M., West, R., & Rotchell, J. M. (2021). A rapid review and meta-regression analyses of the toxicological impacts of microplastic exposure in human cells. *Journal of Hazardous Materials, 427*, 127861. https://doi.org/10.1016/j.jhazmat.2021.127861
198. Leslie, H. A., Van Velzen, M. J. M., Brandsma, S. H., Vethaak, A. D., Garcia-Vallejo, J. J., & Lamoree, M. H. (2022). Discovery and quantification of plastic particle pollution in human blood. *Environment International, 163*, 107199. https://doi.org/10.1016/j.envint.2022.107199
199. Othman, A. R., Hasan, H. A., Muhamad, M. H., 'Izzati Ismail, N., & Abdullah, S. R. S. (2021). Microbial degradation of microplastics by enzymatic processes: A review. *Environmental Chemistry Letters, 19*, 3057. https://doi.org/10.1007/s10311-021-01197-9
200. Zhang, J., Wang, L., Trasande, L., & Kannan, K. (2021). Occurrence of polyethylene terephthalate and polycarbonate microplastics in infant and adult feces. *Environmental Science & Technology Letters, 8*, 989. https://doi.org/10.1021/acs.estlett.1c00559
201. Valavanidis, A., Vlachogianni, T., Fiotakis, K., & Loridas, S. (2013). Pulmonary oxidative stress, inflammation and cancer: Respirable particulate matter, fibrous dusts and ozone as major causes of lung carcinogenesis through reactive oxygen species mechanisms. *International Journal of Environmental Research and Public Health, 10*, 3886. https://doi.org/10.3390/ijerph10093886
202. Wen, B., Zhang, N., Jin, S.-R., Chen, Z.-Z., Gao, J.-Z., Liu, Y., Liu, H.-P., & Xu, Z. (2017). Microplastics have a more profound impact than elevated temperatures on the predatory performance, digestion and energy metabolism of an Amazonian cichlid. *Aquatic Toxicology, 195*, 67. https://doi.org/10.1016/j.aquatox.2017.12.010
203. Rao, B. M. (2019). Microplastics in the aquatic environment: Implications for post-harvest fish quality. *Indian Journal of Fisheries, 66*. https://doi.org/10.21077/ijf.2019.66.1.83125-20
204. Anbumani, S., & Kakkar, P. (2018). Ecotoxicological effects of microplastics on biota: A review. *Environmental Science and Pollution Research, 25*, 14373. https://doi.org/10.1007/s11356-018-1999-x
205. Smith, M., Love, D. C., Rochman, C. M., & Neff, R. A. (2018). Microplastics in seafood and the implications for human health. *Current Environmental Health Reports, 5*, 375. https://doi.org/10.1007/s40572-018-0206-z
206. Walkinshaw, C., Lindeque, P. K., Thompson, R., Tolhurst, T., & Cole, M. (2019). Microplastics and seafood: Lower trophic organisms at highest risk of contamination. *Ecotoxicology and Environmental Safety, 190*, 110066. https://doi.org/10.1016/j.ecoenv.2019.110066
207. Zhang, Q., Xu, E. G., Li, J., Chen, Q., Ma, L., Zeng, E. Y., & Shi, H. (2020). A review of microplastics in table salt, drinking water, and air: Direct human exposure. *Environmental Science & Technology, 54*, 3740. https://doi.org/10.1021/acs.est.9b04535
208. Giri, S., Lamichhane, G., Khadka, D., & Devkota, H. P. (2024). Microplastics contamination in food products: Occurrence, analytical techniques and potential impacts on human health. *Current Research in Biotechnology, 7*, 100190. https://doi.org/10.1016/j.crbiot.2024.100190
209. Sarkar, B., et al. (2021). Challenges and opportunities in sustainable management of microplastics and nanoplastics in the environment. *Environmental Research, 207*, 112179. https://doi.org/10.1016/j.envres.2021.112179
210. Kentin, E., & Kaarto, H. (2018). An EU ban on microplastics in cosmetic products and the right to regulate. *Review of European, Comparative & International Environmental Law, 27*, 254. https://doi.org/10.1111/reel.12269

211. Nikpay, M., & Roodsari, S. T. (2024). Crafting a scientific framework to mitigate microplastic impact on ecosystems. *Microplastics, 3*, 165. https://doi.org/10.3390/microplastics3010010
212. Hettiarachchi, H., & Meegoda, J. N. (2023). Microplastic pollution prevention: The need for robust policy interventions to close the loopholes in current waste management practices. *International Journal of Environmental Research and Public Health, 20*, 6434. https://doi.org/10.3390/ijerph20146434
213. Nielsen, T. D., Hasselbalch, J., Holmberg, K., & Stripple, J. (2019). Politics and the plastic crisis: A review throughout the plastic life cycle. *Wiley Interdisciplinary Reviews: Energy and Environment, 9*, e360. https://doi.org/10.1002/wene.360
214. Wan, S., Yang, X., Chen, X., Qu, Z., An, C., Zhang, B., Lee, K., & Bi, H. (2022). Emerging marine pollution from container ship accidents: Risk characteristics, response strategies, and regulation advancements. *Journal of Cleaner Production, 376*, 134266. https://doi.org/10.1016/j.jclepro.2022.134266
215. Zhang, C., Wang, Q., Zhao, J., Zhao, Y., & Shan, E. (2022). Observational and model studies on transport and inventory of microplastics from a leak accident on the beaches of Yantai. *Marine Pollution Bulletin, 182*, 113976. https://doi.org/10.1016/j.marpolbul.2022.113976
216. Kabir, M. S., Wang, H., Luster-Teasley, S., Zhang, L., & Zhao, R. (2023). Microplastics in landfill leachate: Sources, detection, occurrence, and removal. *Environmental Science and Ecotechnology, 16*, 100256. https://doi.org/10.1016/j.ese.2023.100256
217. Raab, P., & Bogner, F. X. (2020). Microplastics in the environment: Raising awareness in primary education. *The American Biology Teacher, 82*, 478. https://doi.org/10.1525/abt.2020.82.7.478
218. Araújo, J. L., Morais, C., & Paiva, J. C. (2023). Students' attitudes towards the environment and marine litter in the context of a coastal water quality educational citizen science project. *Australian Journal of Environmental Education, 39*, 522. https://doi.org/10.1017/aee.2023.14
219. Lindemann-Matthies, P., Hoyer, E., & Remmele, M. (2021). Collective public commitment: Young people on the path to a more sustainable lifestyle. *Sustainability, 13*, 11349. https://doi.org/10.3390/su132011349
220. Nelms, S. E., et al. (2021). The role of citizen science in addressing plastic pollution: Challenges and opportunities. *Environmental Science & Policy, 128*, 14. https://doi.org/10.1016/j.envsci.2021.11.002
221. Gao, W., Zhang, Y., Mo, A., Jiang, J., Liang, Y., Cao, X., & He, D. (2022). Removal of microplastics in water: Technology progress and green strategies. *Green Analytical Chemistry, 3*, 100042. https://doi.org/10.1016/j.greeac.2022.100042
222. Jiang, T., et al. (2024). A potential threat from biodegradable microplastics: Mechanism of cadmium adsorption and desorption in the simulated gastrointestinal environment. *Frontiers of Environmental Science & Engineering, 18*, 19. https://doi.org/10.1007/s11783-024-1779-4
223. Cordier, M., & Uehara, T. (2019). How much innovation is needed to protect the ocean from plastic contamination? *Science of the Total Environment, 670*, 789. https://doi.org/10.1016/j.scitotenv.2019.03.258
224. Diggle, A., & Walker, T. R. (2020). Implementation of harmonized extended producer responsibility strategies to incentivize recovery of single-use plastic packaging waste in Canada. *Waste Management, 110*, 20. https://doi.org/10.1016/j.wasman.2020.05.013
225. Bezerra, J. C., Walker, T. R., Clayton, C. A., & Adam, I. (2021). Single-use plastic bag policies in the Southern African development community. *Environmental Challenges, 3*, 100029. https://doi.org/10.1016/j.envc.2021.100029
226. Tang, K. H. D., & Hadibarata, T. (2021). Microplastics removal through water treatment plants: Its feasibility, efficiency, future prospects and enhancement by proper waste management. *Environmental Challenges, 5*, 100264. https://doi.org/10.1016/j.scitotenv.2022.15778

# Chapter 13
# Application of AI in Microplastics Environmental Impact Assessment and Remediation

**Jayesh Baldota, Ashish Katke, Tanishka Sharma, and Mahek Pathan**

## Introduction

### *The Global Challenge of Microplastic Pollution*

Synthetic polymers first emerged in the late nineteenth century, around the 1860s, but the significant "plastics boom" occurred only after World War II. Since their initial development as phenol-formaldehyde resins, plastics have become one of the most widely used materials [1]. Initially created to enhance human living standards, plastics have now turned into a major threat to environmental integrity and planetary health [2, 3].

At present, plastics are pervasive across all environmental compartments—air, water, and soil. This is particularly concerning as food packaging for products such as dairy, meat, fish, and beverages, including bottled water, is largely made from plastic [4]. The interaction between food and its plastic packaging almost always results in material transfer from the container to the contents. Consequently, the quality of food can be compromised, as contaminants from packaging materials alter nutritional properties and pose risks to consumer safety [5].

Microplastics have been identified in a variety of ecosystems, including soils, surface waters, coastal sediments, shoreline sands, freshwater sediments, and

J. Baldota (✉)
Department of Pharmaceutics, School of Pharmacy, Vishwakarma University, Pune, Maharashtra, India
e-mail: jayesh.baldota@vupune.ac.in

A. Katke · T. Sharma
School of Pharmacy, Vishwakarma University, Pune, Maharashtra, India

M. Pathan
Northern State Medical University, Arkhangelsk, Russia


S. Kulkarni, S. Das (eds.), *Microplastics and Nanoplastics*, Lecture Notes in Nanoscale Science and Technology 49,
https://doi.org/10.1007/978-3-032-17963-0_13

deep-sea environments; moreover, even precipitation such as rain and snow has been found to contain significant levels of microplastics, many of which are invisible to the naked eye [6]. Indeed, the excessive use of plastic, combined with inefficient waste management systems, particularly concerning collection and disposal, has led to a massive accumulation of plastic waste in the environment. The release of plastic materials is now recognized as a major environmental pollution issue [7].

## *Limitations of Conventional Monitoring and Assessment Methods*

Conventional micro plastics monitoring is often hindered by high cost, intensive labor, and low efficiency. These manual methods provide only a sample in time and space, hence the dynamic distribution and long-term environmental effects of microplastics are hard to track accurately.

### Visual Inspection/Microscopy

Some of the first methods to separate microplastics from other matrix components were by visual inspection using optical microscopies like fluorescence, polarized light, and stereo. Samples are observed under the microscope after pre-treatment procedures such as density separation and organic matter digestion [8]. The size, color, and shape of suspected microplastics are set manually. In the dye-based fluorescence microscopy technique, plastic particles are selectively stained for better contrast. The method is arbitrary, time-consuming, and prone to human error. It cannot detect microplastics below sizes of 100 μm and is unable to identify the chemical composition of the particles. Manual sorting requires immense time and labor with complex and contaminated samples. Also, contamination through lab equipment or air-borne fibers often puts distortion on the result [9].

### Fourier Transform Infrared (FTIR) Spectroscopy

FTIR spectroscopy determines the chemical makeup of suspected microplastics by providing their infrared absorption spectrum. Sample preparation and purification are needed. Subsequent to this, suspected particles are loaded into the FTIR instrument, and its spectral fingerprint is compared with that of a reference polymer [10]. Particles with a detection range of 10–20 μm can be identified using micro-FTIR. Although FTIR is an excellent method for confirming polymers, it is usually impractical for small, opaque, or dark particles. In addition, it takes a great deal of time to analyze every particle, as well as costly equipment and knowledgeable staff.

Accuracy may be reduced by interference in spectral clarity caused by additives or residue [11].

#### Raman Spectroscopy

Raman spectroscopy employs laser light to examine inelastic scattering. The mechanism allows the material to yield a specific spectrograph which characterizes a polymer based on its vibrational modes. When the microscope is fitted (Micro-Raman), it is capable of detecting particles of approximately 1 μm or less in size. As far as sensitivity and specificity are concerned, fluorescence is an issue because it has the ability to particularly spoil the results of environmental samples. Additionally, costly apparatus and trained personnel are required to perform the procedure. In other instances, laser heating damages thermo-sensitive plastics, while black or opaque particles will yield very weak or blocked signals [12]. Similar to FTIR, it is some time before large quantities of single particles can be analyzed, thus limiting its usability for mass monitoring [13].

#### Pyrolysis-Gas Chromatography/Mass Spectrometry (Py-GC/MS)

Py-GC/MS is among the thermo-analytical techniques where polymers of plastics are broken down by pyrolysis, the pyrolysate are later separated by gas chromatography, and mass spectrometry followed their identification. It offers a detailed chemical profile that may be used to ascertain the types of polymers and their amounts. However, the technique is destructive and cannot allow for any future examination of the sample [14]. It simply quantifies plastics on a mass basis and provides no information on the particle size, shape, or number. Sample preparation is an important consideration in that residual organic matter could interfere with the products of pyrolysis. Furthermore, Py-GC/MS needed very sophisticated instruments, skilled operators, and complicated data interpretation, and so it would not be readily employed for routine environmental monitoring [14].

#### Thermogravimetric Analysis (TGA)

By measuring the abrupt mass losses of a sample during gradual heating, TGA aids in the identification of plastics based on their respective decomposition temperatures. Alternatively, if identifying the decomposition gases with a mass spectrometer (TGA-MS) instead is necessary, that could be done. Although TGA is instrumental in ascertaining microplastics; however, the technique is nonspecific and has a lack of resolution for mixed or complex matrices. It is a destructive technique, does not delineate particle size, particle count, and has difficulty distinguishing polymers sharing similar thermal degradation profiles. This can be further confounded by the presence of interfering organic or inorganic materials [15] (Table 13.1 and Fig. 13.1).

**Table 13.1** Studies utilizing microplastic analysis with PY-GC–MS, FTIR, microscopic imaging, hyperspectral imaging, and Raman spectroscopy [16]

| Analytical method | Sample type | Key findings |
|---|---|---|
| Pyrolysis–GC–MS (PY-GC-MS) | Freshwater river sediment | Quantified >700 microplastic particles per kg of dry sediment, predominantly polyethylene and polypropylene |
| | Agricultural soil | Detected PET and PS microplastics within the size range of 200–400 μm |
| | Marine sediment | Demonstrated a high prevalence of polyethylene and polypropylene |
| | Ocean water | Identified polystyrene and polyethylene as the most common polymers |
| Fourier transform infrared spectroscopy (FTIR) | Marine water | Verified the presence of multiple microplastic types, with polypropylene and polyethylene as dominant forms |
| Microscopic imaging | Drinking water, mussels | Confirmed microplastic contamination, mainly PVC and polypropylene |
| Raman spectroscopy | Urban dust, air samples | Characterized a diverse polymer profile, including PP, PE, and PA |
| Hyperspectral imaging | Marine ecosystem | Detected four major polymers: PP, PE, PET, and PS |

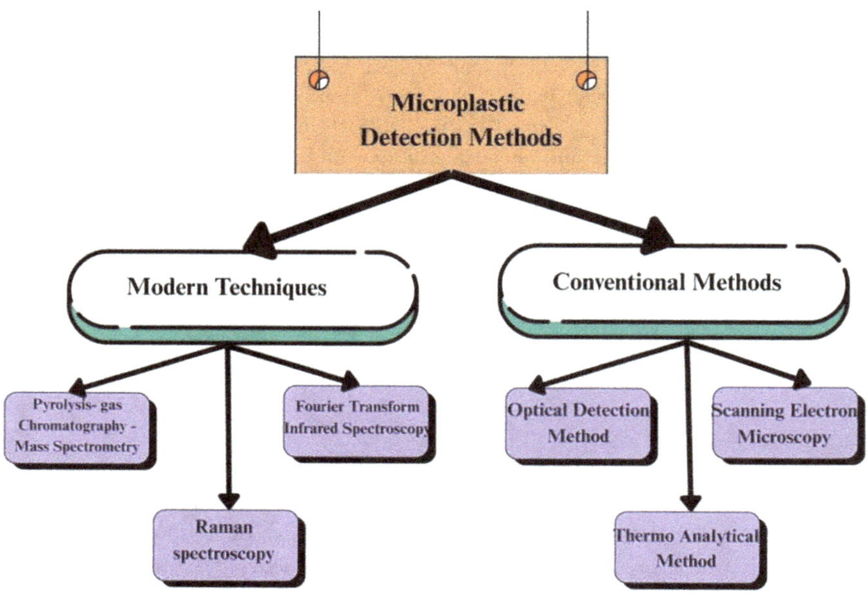


**Fig. 13.1** Flowchart of microplastic detection methods

**Table 13.2** Review of AI-powered microplastic detection techniques. The table details AI technologies, methodologies, polymer types identified, and references for studies utilizing machine learning and deep learning for improved detection accuracy [16]

| AI technology | Study | Polymers identified | Methodology |
|---|---|---|---|
| Machine learning | AI-driven FTIR analysis | PE, PP, PS, PVC, PET | FTIR spectra interpreted using Random Forest (RF) classification |
| | AI in soil microplastic studies | PP, PET, PVC | Soil microplastic characterization via FTIR data analyzed with RF models |
| | Real-time detection with AI | PE, PET, PP, PS | Portable FTIR integrated with ML for real-time microplastic detection |
| Deep learning | Image-based detection (CNN) | PE, PET, PA, PS | Microscopic particle images processed using convolutional neural networks (CNNs) |
| | AI for water sample analysis | PE, PP, PS, PET, PVC | Water sample data evaluated through CNN-based analytical pipelines |
| Support vector machine | Hyperspectral imaging with AI | PE, PP, PS, PA, PET | Hyperspectral imaging datasets classified using SVM algorithms |

## *Introduction to Artificial Intelligence (AI) and Relevant Subfields (ML, DL, CV)*

Artificial Intelligence is a relatively new, transformative arm of computer science concerned with intelligent systems-making machines that can perform tasks that require human intelligence on a lateral: learn, reason, recognize patterns, and make decisions.

- **Machine Learning or ML**: It uses algorithms to learn patterns and make predictions or decisions from data without being explicitly programmed. In microplastics research, ML models are trained to distinguish microplastics from natural particles using labeled training datasets.
- **Deep Learning (DL)**: a branch of ML, applies artificial neural networks, especially convolutional neural networks (CNNs), for complex image data. DL-models can detect microplastic [13, 17] (Table 13.2).

## Background: Microplastics and AI Fundamentals

Conventional microplastics monitoring is often hindered by the high cost and labor-intensive work involved, as low efficiency is another reason. These manual methods offer only a sample in time and space; hence, the dynamic distribution and long-term effects of micro plastics are difficult to track with precision.

## *Microplastics: Definition, Sources, Types, and Distribution*

Microplastic are small tiny plastic particles which are less than 5 mm in size this particles have become largely abundant in Marine fresh water the source of microplastic is twofold primary which are manufactured to be small example microbeads in cosmetics and secondary which constitute of any microplastic which may be originated from the fragmentation of large plastic items in environment due to ultraviolet radiation or abrasion microfibers come in many shapes such as fibre fragments and pallets and polymer composition such as polythene or polypropylene etc. allow for a variety of classification the fact that this microplastic are found everywhere drifting on ocean surface shrinking into deep sea sediments and floating in air which we breathe rises serious concern about that effects on environment and human health [13].

## *Know Environmental and Ecological Impacts of Microplastics*

Micro plastic in environment and ecological impacts of microplastic are very much seen everywhere they take their position in freshwater Marine and terrestrial environments due to the white spread of diverse plastic items and their bad disposal V due to the small size these particles cannot be removed from the environment once released into the environment they pose a long term danger or threat to biodiversity and ecosystem health [17].

Micro plastic also disturb the ecosystem by it means to function as a vector for microbial pathogens and invasion species habitat degradation biodiversity loss and disturbance in ecological balance are the result of the combined effects of these hence tackling the issues of microplastic pollution would be an imperative to the sustainability of ecological and human health across the world [18, 19].

## *Key AI Concepts for Environmental Applications*

There are still many unresolved issues with AI technical limitations, data sharing, data complexity from massive volumes of imagery, depth and interpretability of algorithmic models, cooperation features, etc., even though a lot of data has been published regarding the automatic quantification and interaction with the ecology of anthropogenic plastic debris. There are still a number of issues that need to be resolved in order to close these technological and knowledge gaps and enable a broader application of sophisticated machine learning techniques in microplastic imaging domains. The most crucial and urgent of these is investigating effective and affordable automatic quantification of microplastics in photos based on their physical characterization characteristics [20].

### Machine Learning Algorithms (e.g., SVM, Random Forests, Clustering)

Applications of mL machine learning discovers solutions for environmental data analyse support vector machines are employed in classification based task for instance to classify dirty and clean water samples the random forest method is applied for regression and classified being a suitable and strong ensemble learning tool by which pollutants are precisely predicted or land changes are made more accurately combining the algorithms such as k mean record pollution or hotspot or pollution areas are basically and supervised methods for pattern finding grouping similar environmental points [21].

### Deep Learning Architectures (e.g., CNNs, RNNs)

Deep learning architectures structure and force automated processing of big intricate environment data CNN are famous in image application as microplastic detection satellite imaging classification of the images or vegetation covered identification while RRN do wonders in sequencing data like time series whether condition or pollutant disperse overtime this kind of architectures minimise manual error engineering tasking and improve forecasting potential [22].

### Computer Vision Techniques

In computer vision, AI systems create input for symbols like microscopes, drones, or satellites, which issue an input from various visual sources. Image processing techniques like segmentation or object detection can track plastic particles to monitor wildlife or deforestation. CV processes microscopy images created for the microplastics study, decreasing human interaction in picture analyses and increasing the accuracy and throughput of sample analysis [23].

### Data Preprocessing and Feature Engineering

Processing and featuring and engineering are the main task that can lead to better artificial intelligence model performance preprocessing is basically cleaning data replacing values which are missing scaling the data and removing unwanted noise etc. on environmental data set that ensures the data is of good quality and consisting included under the term feature engineering on the other hand refers to the selection of design of variable which are features that are relevant for the task and thus improving model performance in the detection of microplastic feature parameters describe ML or DL model [24] (Fig. 13.2).

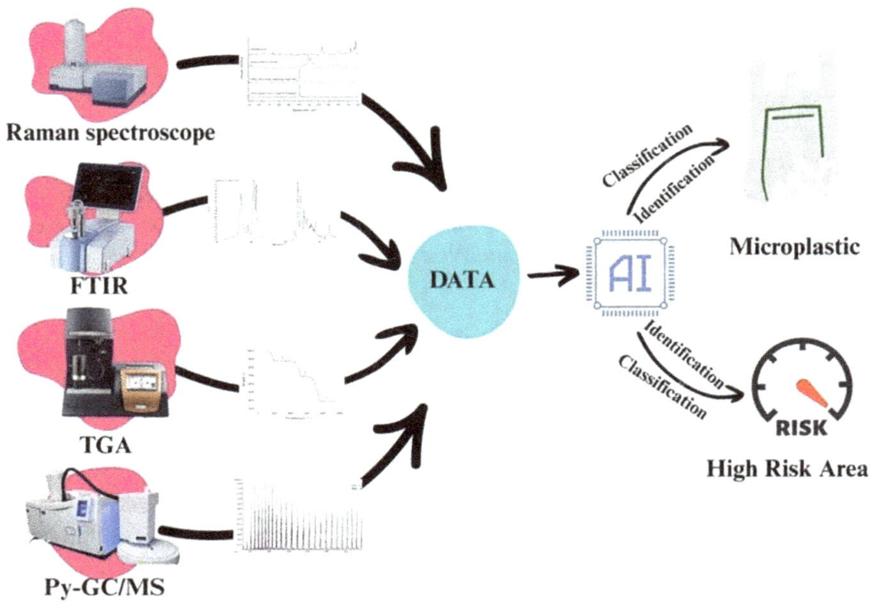


**Fig. 13.2** AI application in microplastic assessment and remediation

## AI Applications in Microplastic Detection and Quantification

The integration of microplastic detection and quantification in artificial intelligence microplastic research has enhanced the speed and accuracy of the detection and qualification process in old and traditional methods manual examination usually takes plenty of time and some cases it also depends over observer bias Anand Les by dint of image analysis and spectral data integration AI methods are available that can scale up and speed up microplastic recognition in a complex environment in this integration approach deep learning machine learning and computer vision will help researchers automobiles detection distinguish different polymer types and quantification of microplastic with little human interventions this will ensure that the data collected is easy to understand [25].

### *Automated Image Analysis for Microplastic Identification*

In comparison to hand identification, computer-aided and automated image analysis methods using AI have facilitated microplastic ID to achieve labor reduction associated with the process and become more consistent. With the use of datasets obtained via multiple microscopy methods, AI systems, especially CNNs, can be trained to recognize microplastic particles by their shape, size, and texture [26].

### Optical Microscopy Image Processing

In general, optical microscopy is used in the first stages of microplastic screening because it is accessible and cheap. Various AI algorithms, those involving image segmentation and classification models, are employed with microscopic images to discriminate microplastics from organic matter and sediments. Features such as color, transparency, and morphology are extracted to train the models for proper identification of microplastics in mixed environmental samples [27].

### Electron Microscopy Image Analysis

Detection of nano plastics and microplastics is beyond the resolving power of light microscopes. The AI methods can be applied on details of surface and polymers in SEM or TEM images. Deep learning methods trained on electron micrographs can tell microplastics from similarly sized non-plastic particles with a high degree of confidence [28].

### Hyperspectral and Raman Spectroscopy Data Analysis

Hyperspector imaging and Raman spectroscopic and provide detailed chemical fingerprints of materials and thrust surface volume analysis polymer and its type in microplastic studies which can help AI models such as support vector system SPM and random forest are applied to classify spectral data deep learning approaches can also be used to manage high dimensional hyperspectoral data automatic the detection and categorization of different types of microplastic based on the spectral signals thurs and answering reliability of chemical characterization in environmental real time samples [29].

## *AI-Enhanced Sensor Technologies for In-Situ Monitoring*

AI enhanced sensor technology's which provide a better monitoring for in situ monitoring sense these instrument provide a very fine chemical finger print of the material analysed hyperspector imaging and Raman spectroscopy would be opt for a polymer type analysis in microplastic studies AI model support vectors SPM and the random forest have been applied to classify this spectral data deep learning methods may also be applied to handle high dimensional hyperspectoral data this kind of data will help us classified different types of microplastic based on spectral signals this will enhance the speed and reliability of chemical characterization in environmental samples [30].

### *Machine Learning for Characterization (Size, Shape, Polymer Type)*

For automation characterization using machine learning to detect its shape size and polymer type automation of these characterization of particle attributes of micro-blasting machine learning is a crucial the modes are instructed to identify micro plastic in terms of the size sheet basis Inspector data conviction neutral networks derive the morphological attributes microscopy images where as SVM and RF are employed to classify Polymers using Raman and ft. IR Spectroscopy data aside from speeding of analysis these technique also enhances the accuracy and reproducibility of results and open avenues for large scale environmental investigation for detection [31, 32].

## AI Applications in Microplastic Environmental Impact Assessment

The application of AI in microplastic environmental impact assessment is important and essential crucial that AI to be applied in forecasting microplastic environmental repercussion by way of advance casting models and risk assessment algorithms traditionally methods heavy dependency on field Sampling and lab testing which takes too much time and present very limited scope of study the artificial intelligence helps in large scale stimulation detection and alternate ecological predictions so that it will have better predictions of micro plastic behaviour in environment bio accumulation bears risk of a various kind of organism in a ecosystem [33].

### *Modeling Fate and Transport of Microplastics in Ecosystems*

AI-driven models take into account site-scale simulations as well as transport of microplastics between environmental compartments such as water bodies, sediments, and atmosphere. Machine learning algorithms in theory would take in different datasets including hydrodynamic conditions of the region, particle properties, and environmental factors to forecast microplastic behavior-movement with time, settling, or resurfacing. This would be excellent assistance in hotspot identification of pollution, transport routes, and cleanup operations. The AI-driven ocean current model, for instance, might forecast areas where microplastics tend to accumulate or even the distant floating of oceanic systems [34] (Fig. 13.3).

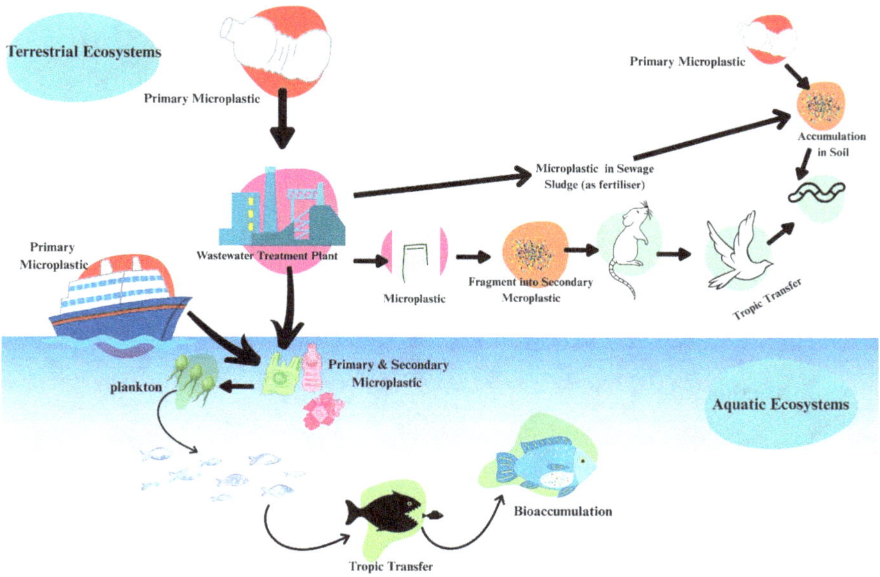


**Fig. 13.3** Microplastic pathways and fate in terrestrial and aquatic ecosystems

## *Predicting Bioaccumulation and Trophic Transfer*

AI will also be used to estimate microplastic bioaccumulation into organisms and trophic transfer through the food chain. Based on species-specific feeding habits, particle size, and digestive retention time, AI models estimate the uptake of microplastics into organisms and then their transfer from one trophic level to another. With deep learning and probabilistic models, in fact, the accumulation process of microplastics can be simulated in aquatic animals, birds, and mammals. These determinations would lead to understanding the long-term health impacts on wildlife and the risks posed to humans as consumers in the food chain [35].

## *AI-Driven Ecological Risk Assessment (ERA)*

Ecological risk assessment ERA identify the potential adverse effect of microplastic on health and Ecosystem exposure data and effect data of toxicity arput Together by Era models to draw probably stick risk assessment uncertainty it can risk to be analysed by Bayesian networks and decision trees which are simultaneously predicted the threshold level at which the environment is secured also the effect of AI based

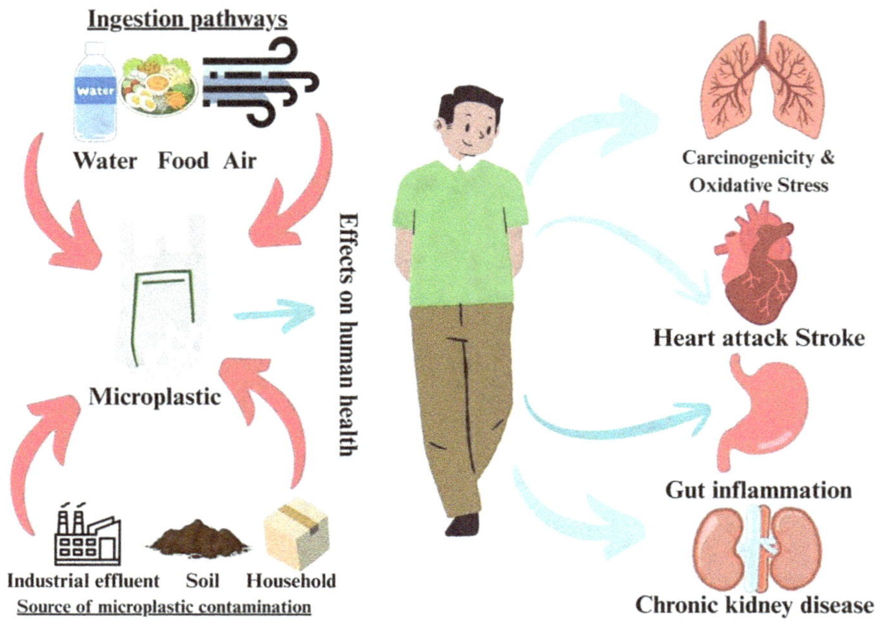


**Fig. 13.4** Sources and health effects of microplastic contamination in humans

risk mapping would be quiet dramatic by incorporation of data from new folds and lab work therefore enabling regulatory bodies in making worthwhile decisions from the data collected from ERA [36] (Fig. 13.4).

## *Identifying Patterns in Toxicity Data and Predicting Biological Effects*

microplastic toxicity and adverse biological effects on the environment are evaluated based on the study of large and complicated data set from laboratories bioasis and field observation experiment meaning full pattern and multidimensional toxicity data set are detected by a techniques including and supervised learning techniques such as clustering and PCA and supervised models such as random forest and neural networks [37].

**The usual procedures involve**:

- **Data acquisition**: Data obtained either experimentally or from the literature describes organism exposure to various microplastics differing in size, shape, polymer type, and concentration.
- **Data preprocessing**: Data cleaning, feature normalization (exposure dose, duration), and encoding categorical variables.
- **Feature selection**: Choose which variables (particle size, surface charge, etc.) contribute most to toxicity.

- **Model training**: ML algorithms are used to classify toxicity levels or predict biological endpoints of interest (mortality, inhibition of growth, oxidative stress, gene expression).
- **Recognition of patterns**: Identifying correlations between properties of particles and specific biological effects. AI thus enables in silico toxicity screening, significantly reducing the need for extensive animal testing, and helps predict sub-lethal and chronic impacts that might not be immediately evident through conventional assays [38].

## *Source Apportionment and Tracking Pollution Hotspots*

AI stands behind source apportionments—the process of determining where microplastic pollution originates—and pollution hotspot tracking through space and time. By integrating data arising from remote sensing, weather systems, hydrological models, industrial activity logs, citizen reports, and more, AI models can determine trends and behaviors and thereby forecast zones of accumulation. AI has certain key applications and steps [39]:

- **Real-time data integration**: Sensor data (e.g., from smart buoys, drones, satellites) are injected into AI systems to monitor the presence of microplastics dynamically.
- **Spatiotemporal modeling**: Deep learning models (e.g., LSTM networks) utilize time-series and spatial data to predict microplastic accumulation in places such as river mouths, coastlines, or urban outflows [40] (Fig. 13.5).

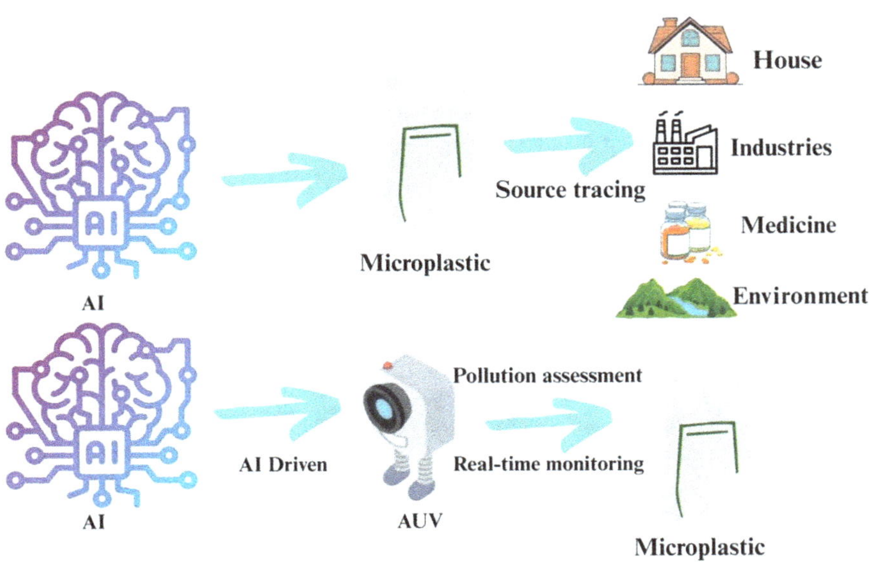


**Fig. 13.5** AI-driven microplastic monitoring and source tracing

# AI Applications in Microplastic Remediation Strategies

From the standpoint of AI, enhancing microplastic removal would mean machine-learning optimization of filtration systems, automated sorting via computer vision, and deployment of AI robots for targeted removal. In these form-fit applications, cleanup efforts are put together for greater efficiency and accuracy, increasing scalability for a global environmental problem.

## *Optimization of Existing Remediation Technologies*

The focus of microplastic remediation is shifting from simple removal to optimizing existing technologies to be more efficient, cost-effective, and sustainable. This involves enhancing the performance of established methods to address the unique challenges posed by microplastics, such as their small size, diverse forms, and widespread distribution [41].

### Filtration and Separation Processes

Filtration and separation are the two main processes that remove microplastics from water bodies and wastewater treatment plants. The optimization of these processes focuses on improving their ability to trap microplastics in terms of the range of sizes, shapes, and densities.

- **Advanced Membrane Filtration**: Alongside separation, filtration using the traditional filter medium often does not effectively trap the small microplastics. Optimizing the material used in membrane filters focuses on creating membranes with smaller pores and altering the membrane surfaces. Research has been undertaken with membranes and using nanofibers and graphene oxide coating to increase the block rate of microplastics waterfalls flat without seriously fouling the filter medium.
- **Advanced Coagulation-Flocculation**: This process is the enhancement of coagulation-flocculation with the use of chemicals (coagulants) to aggregate the small microplastic particles, so that they may more easily settle out or be filtered. Optimization focuses on creating more effective and environmentally friendly coagulants like alginate-based biopolymers, which play a more role in aggregating as broad as possible with coagulants to aggregate more varieties of microplastic particles with fibers and other fragments as well.
- **Magnetic Separation**: In a new emerging process filtration-separation method, some studies suggest adding magnetic materials that bind to the microfibers and other microplastic particles themselves. Optimization focuses on generating highly effective and biodegradable magnetic nanoparticles to recover and reuse

again. This is potentially a highly effective way of filtering separation of microplastic pollution compressively from big volumes.
- **Air Flotation Systems**: Dissolved air flotation (DAF) systems are used to add fine air bubbles into water to adhere and lift microplastic elements to the water surface for skimming. Design optimization has centered on altering components of the DAF system and special surfactants added to improve buoyancy and removal efficiencies for microplastic [42].

#### Biodegradation and Chemical Degradation Enhancement

Biodegradation and chemical degradation enhancement off micro plastic the microplastic is notorious for its stability made off micro plastic through biological and chemical processes it is specifically vital for the removal of microplastic that already present in environment

- **Biodegradation**: the designing of enzyme or microorganism that are capable to digest polymeric chain of very common plastic like PET and PE into their monomeric state making of such procedure to work towards optimization of more powerful and effective enzymes and bioreactors where the enzyme can carry out it works search a setting will accelerate microbial enzymatic breakdown off micro plastic to harmless monomers.
- **Chemical degradation**: while the process of chemical recycling is a method that break down polymer into their monomer from which a new plastic can be made the challenge is making the chemical process of solvolysis and pyrolysis efficient and selective for Mishra of micro plastic these are multiple in kinds and contaminated with each other repairing in the environment micro plastic waste is treated as a resource in this chemical process plants there by closing the loop of plastic pollution [43, 44].

### *AI in Designing Novel Removal and Degradation Materials/Methods*

Artificial Intelligence (AI) is transforming the field of microplastic remediation by accelerating the discovery and design of innovative solutions. By leveraging AI, scientists can move beyond traditional trial-and-error methods to create more effective and efficient materials and processes.

- **Materials Discovery**: AI algorithms can sift through vast databases of chemical compounds and material properties to predict which materials are likely to be best at capturing or degrading microplastics. For example, AI can be used to develop next-generation adsorbents with targeted surface chemistries to selectively bind with microplastic polymers. This work promotes and encourages the design of advanced filters and membranes.

- **Process Optimization**: AI can simulate complex chemical and physical processes to optimize remediation strategies. For instance, in AOPs, AI models can predict the ideal combination of catalysts, light, and reaction times to enhance microplastic degradation efficiency while minimizing energy use [44, 45].

### *Predictive Modeling for Remediation Efficiency and Strategy Selection*

AI facilitates easier forecasting of the success of remediation projects and forecasting the most effective remediation strategy for a specific scenario.

- **Real-time Feedback and Adjustment**: Integrating AI with in-situ sensors (as mentioned in the previous section) can provide real-time user feedback on remediation system success. If a remediation system detects reduced regeneration, the AI can notify the operator, and if linked to the control sensor can also make real-time changes to maintain maximum efficiency level and adjust any given parameter in a remediation loop situation [46, 47] (Fig. 13.6).

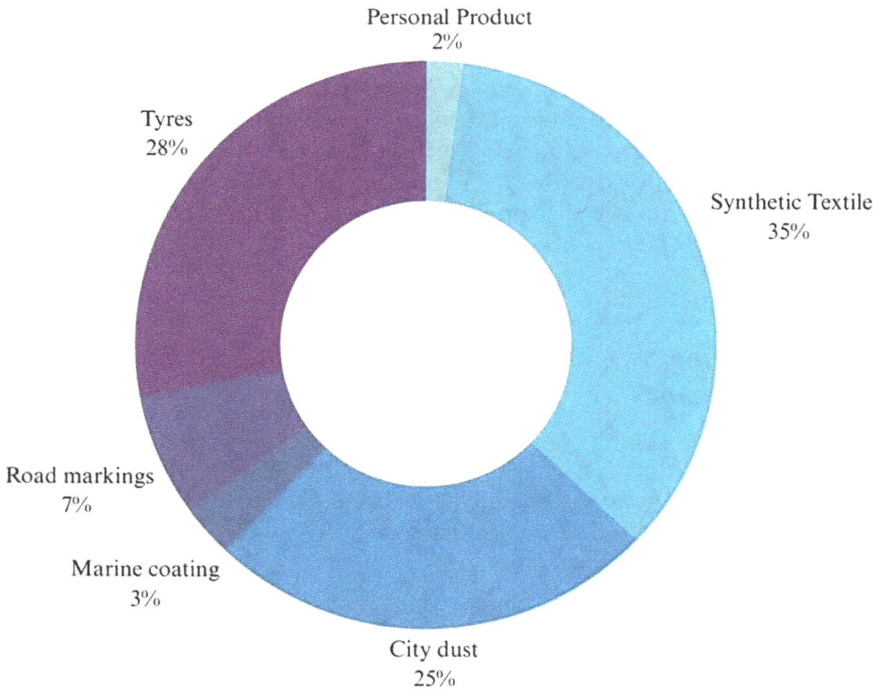


**Fig. 13.6** Major sources of microplastic pollution

### *Potential Role of AI in Robotics for Microplastic Collection*

The combination of robotics and artificial intelligence is powerful for automating the detection and collection of microplastics, especially for remote locations and larger scale situations.

- **Autonomous Collection Robots**: AI drones and robotic boats could autonomously scan and collect microplastics directly from oceans, rivers, and wastewater treatment plants operatively. These collection robots can have sensors that identify hotspots of microplastics and can easily navigate to these locations for collection.
- **Smart suction and filtration robots**: Robots can be built with suction and filtering systems. The robots can collect only microplastics while also considering the impact on marine organisms through AI controls and computer vision algorithms which can recognize the difference between microplastics and small organisms.
- **Sorting and recycling**: After microplastics are collected, they are a heterogeneous material that is made up of many varied materials. AI sorting robots can use spectral analysis and machine learning to sort and identify different polymer types and thus for effective chemical recycling and valorization of waste being collected [48, 49].

## Challenges, Limitations, and Future Directions

The integration of AI into microplastic remediation is a rapidly evolving field, but it is not without significant challenges. Addressing these limitations will be crucial for realizing the full potential of AI in tackling the global microplastic crisis.

### *Data Availability, Quality, and Standardization Issues*

The effectiveness of AI models is heavily dependent on the data on which they are trained. In the context of microplastic remediation, this presents several hurdles [50].

- **Absence of Standardized Data**: There is no unified, standardized method of microplastic sampling, analysis, and data reporting. Previously this has caused variability in one dataset to the next making it hard to train robust, generalizable AI methods.
- **Limited High-Quality Datasets**: It is difficult to obtain high-quality datasets on microplastics, including datasets describing their physical and chemical characteristics, their spatial and temporal distributions in a variety of environments, and the performance of varying types of remediation technologies in situ. These datasets are often insufficient, unoriginal, or skewed.

- **Cost and Labor**: Manual methods of microplastic analysis are laborious and time intensive. The expense of generating the mass of well-labeled datasets necessary to train deep learning models is cumbersome in the entire process and creates a bottleneck [51].

## *Model Interpretability and Explainability (XAI)*

While AI models can be highly effective, their decision-making processes are often opaque, a problem known as the "black box" dilemma. This is a significant limitation, especially in environmental science where trust and understanding are paramount.

- **Trust and Validation**: If a model suggests a remediation option, environmental managers should want to know why the model made that suggestion. Otherwise, it is difficult to validate the model's output or develop public trust in the technology.
- **Scientific Insight**: Many AI models are black-box models, which means we lose the opportunity to develop new scientific insights. For example, an AI model may produce a suggestion for a successful biodegradation method. If that model were explainable, it is possible it would also provide information on the critical environmental factors or microbial interactions that contributed to that success [52].

## *Computational Costs and Infrastructure Requirements*

Developing and deploying sophisticated AI models for microplastic remediation requires significant computational resources and infrastructure, which can be a major barrier.

- **Computational Demands**: The training of deep learning models on large environmental datasets is computationally expensive and requires considerable hardware, such as GPUs, and energy resources. These costs can be prohibitive for research institutions or organizations, particularly in developing nations.
- **Infrastructure for Applications**: For researchers to deploy artificial intelligence systems that require real-time monitoring or robotic collection, they require substantial infrastructure for data transfer, processing, and communications. This is especially challenging in remote or harsh conditions, such as the open ocean.
- **Maintenance and Scaling**: Maintenance, updating actors and models, and most tasks associated with artificial intelligence are always ongoing. Transforming these systems from a small-scale pilot project to a large-scale, regional, or global, solution is a significant technical and financial effort [53, 54].

### *Integration of Multi-modal Data (Imagery, Sensors, Chemical Data)*

- **Enhanced Understanding**: Using data from different sources helps AI understand microplastic pollution better. For instance, an AI system could use satellite images to find plastic waste, chemical data from sensors to identify types of plastics, and microscope images to sort out the different kinds of microplastics being seen.
- **Improved Prediction**: Integrating data across these streams could allow improved predictions, which would improve predictive models for microplastic transport, fate, and impact, thereby identifying sources and help in predicting a microplastic plume and its movement for remediation efforts [55, 56].

### *Development of Real-Time Monitoring and Early Warning Systems*

The ability to monitor microplastic pollution in real time is essential for effective management and for preventing new contamination events. AI is central to this effort.

- **Real-Time Data Processing**: Traditional microplastic analysis is slow and retrospective. AI, when combined with smart sensors, can process data in real time to obtain immediate insight into pollution levels, trends and triggers that can lead to immediate responses to spills or pollution events.
- **Predictive Early Warning**: AI models can be trained to recognize conditions that occur prior to pollution events. For example, while looking at water quality data and corresponding meteorological conditions, follow-on AI could predict the likelihood of a microplastic influx from a river to a coastal area which could mitigate the effects using timely preventative action.
- **IoT and AI**: IoT will be instrumental in creating the network of connected sensors and monitoring stations, with AI being the "brain" that will receive data from all of the IoT sensors, and provide a real time aggregative microplastic situation and connect it to its IoT connection with all other monitoring environments [57, 58].

### *AI for Policy Making and Environmental Management*

AI can move beyond technical applications to inform and improve environmental policy and management decisions related to microplastics.

- **Evidence-Based Policy**: AI models can provide policy makers with evidence-based information about the sources, transport, and effects of microplastics to better inform regulations, such as advertising and regulating certain plastic products, and regulating industrial sectors.

- **Allocating resources**: AI can help make better decisions about where to allocate resources to remediation efforts. Using AI can help identify areas of maximum risk and allow to model various remediation efforts for success to be employed in high-risk areas.
- **Assessing impacts of microplastics**: Models can use AI to evaluate ecological and economic long-term effects of microplastic pollution associated with different policy scenarios. This feature could allow the policymakers to compare interventions and start to assess benefits and trade-offs of each option before the intervention is put in place [59, 60].

## Case Studies

Several research studies highlight the growing use of AI and machine learning (ML) to improve the detection and identification of microplastics. These methods offer significant improvements in traditional techniques by increasing both efficiency and accuracy.

**Spectroscopic Analysis and Machine Learning**:

- **Raman Spectroscopy**: Cowger et al. [61] utilized deep learning models to identify polymers like polyethylene and polypropylene from a large dataset of Raman bands [62]. This approach demonstrated how AI can improve on traditional error minimization methods and save time [63, 16].
- **FTIR Spectroscopy**: Singh et al. [64] developed a rapid and portable microplastic identification system that combines machine learning with Fourier-transform infrared (FTIR) spectroscopy. Their research focused on identifying common polymers such as polyethylene, polystyrene, and polyethylene terephthalate.
- **Adaptable AI Systems**: Boyle and Ormeci [65] created an adaptable AI system for microplastic identification and detection. The system uses image capturing and analysis to examine variables like quantity, color, and size. This small and user-friendly technology, driven by a machine learning technique, is particularly effective for identifying larger ($\geq$0.5 mm) colored plastic particles and is suitable for both scientific research and citizen science projects [66].
- **Hyperspectral Imaging (HSI)**: Yang et al. [67] explored the potential of combining hyperspectral imaging (HSI) with machine learning to identify a variety of microplastics, including polyamide, polystyrene, and polyethylene, within environmental samples. This method allows for the detection of multiple polymers in a single analysis [16].

## Applications of Artificial Intelligence in Detection of MP

Artificial Intelligence and Machine Learning is being applied to detecting and classifying plastic particles in food products. Deep Learning specifically can be trained on bigger datasets of images to optically recognize, identify, and classify various microplastic types. Machine Learning techniques can also process the spectral analysis data from infrared analysis to classify the polymer compounds in food samples automatically. AI and Machine Learning can be adapted to monitor food samples for microplastics, in real time. Uncertainty models can be built based on the types of food, specific handling practices, and risks of microplastic contamination in each of the types of food considered [68].

Along with artificial intelligence, machine learning applications to support a transparent and traceable supply chain involve blockchain technologies. Data on microplastic contamination could be recorded on a blockchain at any stage of production and distribution to locate the sources of risk and contamination, and act accordingly [69].

Deep learning and machine learning can deliver more reliable and accurate results than traditional computer vision, and manual classification systems, which are quite common have manual and repetitive sampling and classification tasks fraught with human error. Importantly, these new methods and techniques are fast, even with most local computers which do not possess high powered graphics processing units (GPUs) [70].

## Comparative Analysis of Artificial Intelligence-Based Modern Techniques with Conventional Techniques

Artificial intelligence gives cutting-edge solutions to microplastic pollution, representing a great danger to nature and human life. Sometimes, an AI approach can outperform the conventional methods of locating, identifying, and removing microplastics. Microplastics are tiny pieces of plastic that can harm individuals, animals, and ecosystems, bringing forth long-term health problems. Traditional approaches to analyze microplastics include microscope observation and infrared and ultraviolet-visible spectroscopy. There are classical approaches where microplastics are visually or chemically identified under microscopes and infrared and UV-Vis spectroscopies, which rightly can lead to accurate results. The drawback is that these techniques can be quite time-consuming and could become unreliable if operated by untrained personnel. Newer techniques such as Raman spectroscopy and FTIR provide solutions for the differential identification of plastic types in complex samples but require costly instruments and extensive sample preparation work.

AI can identify microplastics with a higher speed and accuracy, especially when dealing with huge amounts of data. Although AI can face problems such as the availability of big datasets, the need for maximum computing power, and the requirement of a transparent explanation [71, 72].

## Challenges in Application of AI in Microplastic

Identification of microplastics in food through AI is a large challenge due to the fact that there are not many good, correctly labeled datasets that illustrate the types of foods and how contaminated they can be. In addition, samples of food are complicated, and microplastics are made of various structures and compositions, making it challenging for AI to identify and label them correctly. Although these methods can provide a significant amount of useful information, applying them with AI requires extensive technical effort and collaboration among various disciplines. In addition, AI systems typically require high-performance computers and expert skills, which may render them difficult to apply in most research and government institutions [73].

AI may be utilized to make detection of microplastics more effective, but it is never a substitute for such primary tools as spectroscopy and microscopes that are utilized to analyze and identify material. Rather, AI assists by improving the data from images and spectra, which can make results more targeted, eliminate errors, and enable quicker analysis [74].

## Conclusion

These requirement future investigation in the area of micro plastic will emphasize an inclusive AI enabled Framework for environment monitoring and decision support an inter disciplinary efforts must be set against AI research environmental scientist policy makers and ethics to be included in the process the socio inclination of Technology the prime focus should be on low price field messengers that are rug and AI smart to predict micro plastic in internet of thing is a real time early warning system they must also work on improving data in fusion and debasing algorithms to take heterogeneous data set as input and enhancing model robustness similarly the first possible pilot projects and field verification test will greatly helping determining whether the technology's are needed ready for fields in the future lightweight edge AI models will actually release into production cloud infrastructure and AI specific hardware it will be these solutions easy and accessible to all the people enhance so field ready XAI methods conversely will become at most important in model interpretability enhancement of microplastic.

## Future Directions

With microplastics entering food webs, MP particles thus threaten environmental integrity while constituting a final potential health risk to humans. Consumers are thus engulfed in the so-called buzzword being familiar with this new contaminant. So, an urgent need arises for the development of fast and efficient methods to treat this. Analytical instruments, in days past, suffered from drawbacks in sensitivity, specificity and throughput. So these institutions have evolved in the direction of some technologies which deserve further probing. Hence, it can be distinguished from each other under a detectable gadget concerning them; i.e., they provide new possibilities for rapid screening of all corresponding MPS and their accurate discovery in complex food matrices. The advances in regulatory surveillance and detection are greatly benefited by the use of artificial intelligence. AI-powered analytical tools can provide the data on which decisions are made to improve policy options and awareness programs. The monitoring of microplastics by AI is an attitudinal change. It signals a better coordinated worldwide response to the pollution of MPs in food chains international and a scalable approach.

## References

1. Issac, M. N., & Kandasubramanian, B. (2021). Effect of microplastics in water and aquatic systems. *Environmental Science and Pollution Research, 28*(16), 19544–19562. https://link.springer.com/article/10.1007/s11356-021-13184-2
2. Acharya, S., Rumi, S. S., Hu, Y., & Abidi, N. (2021). Microfibers from synthetic textiles as a major source of microplastics in the environment: A review. *Textile Research Journal, 91*(13–18), 004051352199124.
3. Altunışık, A. (2023). Prevalence of microplastics in commercially sold soft drinks and human risk assessment. *Journal of Environmental Management, 336*, 113720.
4. An, L., Liu, Q., Deng, Y., Wu, W., Gao, Y., & Ling, W. (2020). Sources of microplastic in the environment. In *The handbook of environmental chemistry* (Vol. 95(95), pp. 143–159). Springer.
5. Dauvergne, P. (2018). Why is the global governance of plastic failing the oceans? *Global Environmental Change, 51*, 22–31.
6. Anderson, P. J., Warrack, S., Langen, V., Challis, J. K., Hanson, M. L., & Rennie, M. D. (2013). Microplastic contamination in Lake Winnipeg, Canada. *Environmental Pollution, 225*, 223–231.
7. Anuar, S. T., Altarawnah, R. S., Mohd Ali, A. A., Lee, B. Q., Khalik, W. M. A. W. M., Yusof, K. M. K. K., et al. (2022). Utilizing pyrolysis–gas chromatography/mass spectrometry for monitoring and analytical characterization of microplastics in polychaete worms. *Polymers, 14*(15), 3054. https://www.ncbi.nlm.nih.gov/pmc/articles/PMC9370765/#
8. Picó, Y., & Barceló, D. (2020). Pyrolysis gas chromatography-mass spectrometry in environmental analysis: Focus on organic matter and microplastics. *TrAC Trends in Analytical Chemistry, 130*, 115964. https://www.sciencedirect.com/science/article/pii/S016599362030193X

9. Astray, G., Soria-Lopez, A., Barreiro, E., Mejuto, J. C., & Cid-Samamed, A. (2023). Machine learning to predict the adsorption capacity of microplastics. *Nanomaterials, 13*(6), 1061. https://www.ncbi.nlm.nih.gov/pmc/articles/PMC10051191/
10. Aydin, R., Yozukmaz, A., Şener, İ., Temiz, F., & Giannetto, D. (2023). Occurrence of microplastics in most consumed fruits and vegetables from turkey and public risk assessment for consumers. *Life, 13*(8), 1686.
11. Azooz, E. A., Abduladheem, N. A., & El, K. (2024). An overview of the green synthesis of coinage metallic nanoparticles: Protocols, challenges, and cancer treatments. *Nano Life, 15*(5).
12. Baruah, A., Sharma, A., Sharma, S., & Nagraik, R. (2021). An insight into different microplastic detection methods. *International Journal of Environmental Science and Technology, 19*(6), 5721–5730. https://link.springer.com/article/10.1007/s13762-021-03384-1
13. Bhagat, J., Zang, L., Nishimura, N., & Shimada, Y. (2020). Zebrafish: An emerging model to study microplastic and nanoplastic toxicity. *Science of the Total Environment, 728*, 138707. https://www.sciencedirect.com/science/article/abs/pii/S0048969720322245
14. Bhardwaj, A., Sam, L., Akanksha, Martín-Torres, F. J., & Kumar, R. (2016). UAVs as remote sensing platform in glaciology: Present applications and future prospects. *Remote Sensing of Environment, 135*, 196–204.
15. Bradney, L., Wijesekara, H., Palansooriya, K. N., Obadamudalige, N., Bolan, N. S., Ok, Y. S., et al. (2019). Particulate plastics as a vector for toxic trace-element uptake by aquatic and terrestrial organisms and human health risk. *Environment International, 131*, 104937.
16. Burghardt, T. E., & Pashkevich, A. (2023). Road markings and microplastics—A critical literature review. *Transportation Research Part D: Transport and Environment, 119*, 103740.
17. Cincinelli, A., Scopetani, C., Chelazzi, D., Lombardini, E., Martellini, T., Katsoyiannis, A., et al. (2013). Microplastic in the surface waters of the Ross Sea (Antarctica): Occurrence, distribution and characterization by FTIR. *Chemosphere, 135*, 391–400.
18. Cowger, W., Roscher, L., Jebens, H., Chamas, A., Maurer, B., Gehrke, L., et al. (2024). Generation of macro- and microplastic databases by high-throughput FTIR analysis with microplate readers. *Analytical and Bioanalytical Chemistry, 416*(6), 1311–1320.
19. Dekiff, J. H., Remy, D., Klasmeier, J., & Fries, E. (2014). Occurrence and spatial distribution of microplastics in sediments from Norderney. *Environmental Pollution, 186*, 248–256.
20. Deshoulles, Q., Le Gall, M., Dreanno, C., Arhant, M., Priour, D., & Le Gac, P. Y. (2022). Chemical coupling between oxidation and hydrolysis in polyamide 6—A key aspect in the understanding of microplastic formation. *Polymer Degradation and Stability, 197*, 109851.
21. Di Fiore, C., Carriera, F., Iannone, A., Paris, E., Gallucci, F., & Avino, P. (2024). First approach for defining an analytical protocol for the determination of microplastics in cheese using pyrolysis–gas chromatography–mass spectrometry. *Applied Sciences, 14*(13), 5621. https://www.mdpi.com/2076-3413/14/13/5621
22. Kurniawan, S. B., Said, N. S. M., Imron, M. F., & Abdullah, S. R. S. (2021). Microplastic pollution in the environment: Insights into emerging sources and potential threats. *Environmental Technology & Innovation, 23*, 101390.
23. Steffen, W., Richardson, K., Rockstrom, J., Cornell, S. E., Fetzer, I., Bennett, E. M., et al. (2015). Planetary boundaries: Guiding human development on a changing planet. *Science, 347*(6223) https://www.science.org/doi/10.1126/science.1259855
24. Henry, B., Laitala, K., & Klepp, I. G. (2019). Microfibres from apparel and home textiles: Prospects for including microplastics in environmental sustainability assessment. *Science of the Total Environment, 652*(652), 483–494. https://www.sciencedirect.com/science/article/pii/S004896971834049X
25. Vitali, C., Peters, R., Janssen, H. G., & Nielen, M. W. F. (2022). Microplastics and nanoplastics in food, water, and beverages; part I. Occurrence. *TrAC Trends in Analytical Chemistry, 159*, 116670.
26. McCormick, A., Hoellein, T. J., Mason, S. A., Schluep, J., & Kelly, J. J. (2014). Microplastic is an abundant and distinct microbial habitat in an Urban River. *Environmental Science & Technology, 48*(20), 11863–11871.

27. La Nasa, J., Biale, G., Fabbri, D., & Modugno, F. (2020). A review on challenges and developments of analytical pyrolysis and other thermoanalytical techniques for the quali-quantitative determination of microplastics. *Journal of Analytical and Applied Pyrolysis, 149*, 104841.
28. Fries, E., Dekiff, J. H., Willmeyer, J., Nuelle, M. T., Ebert, M., & Remy, D. (2013). Identification of polymer types and additives in marine microplastic particles using pyrolysis-GC/MS and scanning electron microscopy. *Environmental Science Processes & Impacts, 15*(10), 1949–1956.
29. Ugwu, K., Herrera, A., & Gómez, M. (2021). Microplastics in marine biota: A review. *Marine Pollution Bulletin, 169*, 112540.
30. Li, J., Liu, H., & Paul Chen, J. (2018). Microplastics in freshwater systems: A review on occurrence, environmental effects, and methods for microplastics detection. *Water Research, 137*(137), 362–374.
31. Li, Q., Ma, C., Zhang, Q., & Shi, H. (2021). Microplastics in shellfish and implications for food safety. *Current Opinion in Food Science, 40*, 192–197.
32. Lin, J., Cheng, Q., Kumar, A., Zhang, W., Yu, Z., Hui, D., et al. (2024). Effect of degradable microplastics, biochar and their coexistence on soil organic matter decomposition: A critical review. *TrAC Trends in Analytical Chemistry, 183*, 118082. https://www.sciencedirect.com/science/article/abs/pii/S016599362400565X
33. Mühlschlegel, P., Hauk, A., Walter, U., & Sieber, R. (2013). Lack of evidence for microplastic contamination in honey. *Food Additives and Contaminants - Part A Chemistry, Analysis, Control, Exposure and Risk Assessment, 34*(11), 1982–1989.
34. Nel, H. A., Sambrook Smith, G. H., Harmer, R., Sykes, R., Schneidewind, U., Lynch, I., et al. (2020). Citizen science reveals microplastic hotspots within tidal estuaries and the remote Scilly Islands, United Kingdom. *Marine Pollution Bulletin, 161*, 111376.
35. Peixoto, D., Pinheiro, C., Amorim, J., Oliva-Teles, L., Guilhermino, L., & Vieira, M. N. (2019). Microplastic pollution in commercial salt for human consumption: A review. *Estuarine, Coastal and Shelf Science, 219*, 161–168.
36. Periyasamy, A. P., & Tehrani-Bagha, A. (2022). A review on microplastic emission from textile materials and its reduction techniques. *Polymer Degradation and Stability, 199*, 109901.
37. Petersen, M., Yu, Z., & Lu, X. (2021). Application of Raman spectroscopic methods in food safety: A review. *Biosensors, 11*(6), 187.
38. Shi, Y., Liu, P., Wu, X., Shi, H., Huang, H., Wang, H., et al. (2021). Insight into chain scission and release profiles from photodegradation of polycarbonate microplastics. *Water Research, 195*, 116980.
39. Sommer, F., Dietze, V., Baum, A., Sauer, J., Gilge, S., Maschowski, C., et al. (2018). Tire abrasion as a major source of microplastics in the environment. *Aerosol and Air Quality Research, 18*(8), 2014–2028.
40. Sridhar, A., Kannan, D., Kapoor, A., & Prabhakar, S. (2022). Extraction and detection methods of microplastics in food and marine systems: A critical review. *Chemosphere, 286*, 131653. https://www.sciencedirect.com/science/article/pii/S004565352102125l#
41. Sun, Q., Ren, S. Y., & Ni, H. G. (2020). Incidence of microplastics in personal care products: An appreciable part of plastic pollution. *Science of the Total Environment, 742*, 140218.
42. Tagorti, G., & Kaya, B. (2022). Genotoxic effect of microplastics and COVID-19: The hidden threat. *Chemosphere, 286*, 131898.
43. Baduge, S. K., Thilakarathna, S., Perera, J. S., Arashpour, M., Sharafi, P., Teodosio, B., et al. (2022). Artificial intelligence and smart vision for building and construction 4.0: Machine and deep learning methods and applications. *Automation in Construction, 141*(2), 104440. https://www.sciencedirect.com/science/article/pii/S0926580522003132
44. Wang, W., Zhang, J., Qiu, Z., Cui, Z., Li, N., Li, X., et al. (2022). Effects of polyethylene microplastics on cell membranes: A combined study of experiments and molecular dynamics simulations. *Journal of Hazardous Materials, 429*, 128323.

45. Zada, L., Leslie, H. A., Vethaak, A. D., Tinnevelt, G. H., Jansen, J. J., de Boer, J. F., et al. (2018). Fast microplastics identification with stimulated Raman scattering microscopy. *Journal of Raman Spectroscopy, 49*(7), 1136–1144. https://onlinelibrary.wiley.com/doi/pdf/10.1002/jrs.5367
46. Xiang, Y., Jiang, L., Zhou, Y., Luo, Z., Zhi, D., Yang, J., et al. (2022). Microplastics and environmental pollutants: Key interaction and toxicology in aquatic and soil environments. *Journal of Hazardous Materials, 422*, 126843. https://www.sciencedirect.com/science/article/abs/pii/S0304389421018112?via%3Dihub
47. Yakovenko, N., Carvalho, A., & ter Halle, A. (2020). Emerging use thermo-analytical method coupled with mass spectrometry for the quantification of micro(nano)plastics in environmental samples. *TrAC - Trends in Analytical Chemistry, 131*, 115979.
48. Yang, C., Xie, J., Gowen, A., & Xu, J. L. (2024). Machine learning driven methodology for enhanced nylon microplastic detection and characterization. *Scientific Reports, 14*(1), 3464.
49. Yaseen, T., Sun, D. W., & Cheng, J. H. (2013). Raman imaging for food quality and safety evaluation: Fundamentals and applications. *Trends in Food Science & Technology, 62*, 137–189.
50. Zhang, Y., Zhang, D., & Zhang, Z. (2023). A critical review on artificial intelligence-based microplastics imaging technology: Recent advances, hot-spots and challenges. *International Journal of Environmental Research and Public Health, 20*(2), 1150. https://www.mdpi.com/1660-4601/20/2/1150
51. Zhao, M., Zou, G., Li, Y., Pan, B., Wang, X., Zhang, J., et al. (2025). Biodegradable microplastics coupled with biochar enhance Cd chelation and reduce Cd accumulation in Chinese cabbage. *Biochar, 7*(1), 31.
52. Frias, J. P. G. L., & Nash, R. (2019). Microplastics: Finding a consensus on the definition. *Marine Pollution Bulletin, 138*, 145–147.
53. Alimi, O. S., Fadare, O. O., & Okoffo, E. D. (2021). Microplastics in African ecosystems: Current knowledge, abundance, associated contaminants, techniques, and research needs. *Science of the Total Environment, 755*, 142422.
54. Aljaibachi, R., & Callaghan, A. (2018). Impact of polystyrene microplastics on Daphnia magna mortality and reproduction in relation to food availability. *PeerJ, 6*, e4601.
55. Andrades, R., dos Santos, R. A., Martins, A. S., Teles, D., & Santos, R. G. (2019). Scavenging as a pathway for plastic ingestion by marine animals. *Environmental Pollution, 248*, 159–165. https://www.sciencedirect.com/science/article/pii/S026974911833344X
56. Banaee, M., Gholamhosseini, A., Sureda, A., Soltanian, S., Fereidouni, M., & Ibrahim, A. (2021). Effects of microplastic exposure on the blood biochemical parameters in the pond turtle (Emys orbicularis). *Environmental Science and Pollution Research, 28*(8), 9221–9234. https://www.semanticscholar.org/paper/Effects-of-microplastic-exposure-on-the-blood-in-Banaee-Gholamhosseini/50702e0463d09ceab3a2b8565bdd805d027b2d59
57. Barboza, L. G. A., Dick Vethaak, A., Lavorante, B. R. B. O., Lundebye, A. K., & Guilhermino, L. (2018). Marine microplastic debris: An emerging issue for food security, food safety and human health. *Marine Pollution Bulletin, 133*, 336–348. https://www.sciencedirect.com/science/article/pii/S0025326X1830376X?via%3Dihub
58. Berglund, E., Fogelberg, V., Nilsson, P. A., & Hollander, J. (2019). Microplastics in a freshwater mussel (Anodonta anatina) in Northern Europe. *Science of the Total Environment, 697*, 134192.
59. Bessa, F., Ratcliffe, N., Otero, V., Sobral, P., Marques, J. C., Waluda, C. M., et al. (2019). Microplastics in gentoo penguins from the Antarctic region. *Scientific Reports, 9*(1), 14191. https://www.nature.com/articles/s41598-019-50621-2.pdf
60. Besseling, E., Foekema, E. M., van den Heuvel-Greve, M. J., & Koelmans, A. A. (2013). The effect of microplastic on the uptake of chemicals by the Lugworm Arenicola marina (L.) under environmentally relevant exposure conditions. *Environmental Science & Technology, 51*(15), 8795–8804.

61. Cowger, W., Willis, K. A., Bullock, S., Conlon, K., Emmanuel, J., Erdle, L. M., Eriksen, M., Farrelly, T. A., Hardesty, B. D., Kerge, K., Li, N., Li, Y., Liebman, A., Tangri, N., Thiel, M., Villarrubia-Gómez, P., Walker, T. R., & Wang, M. (2024). Global producer responsibility for plastic pollution. Science Advances, 10(17). https://doi.org/10.1126/sciadv.adj8275
62. Ivleva, N. P. (2021). Chemical Analysis of Microplastics and Nanoplastics: Challenges, Advanced Methods, and Perspectives. Chemical Reviews, 121(19), 11886–11936. https://doi.org/10.1021/acs.chemrev.1c00178
63. Bayo, J., Olmos, S., & López-Castellanos, J. (2020). Microplastics in an urban wastewater treatment plant: The influence of physicochemical parameters and environmental factors. Chemosphere, 238, 124593. https://doi.org/10.1016/j.chemosphere.2019.124593
64. Singh, S., Chakma, S., Alawa, B., Kalyanasundaram, M., & Diwan, V. (2023). Identification, characterization, and implications of microplastics in soil – A case study of Bhopal, central India. Journal of Hazardous Materials Advances, 9, 100225. https://doi.org/10.1016/j.hazadv.2022.100225
65. Boyle, K., & Banu Örmeci. (2024). Beta Testing an AI-Based Physical Analysis Technology for Microplastic Quantification and Characterization. Water, 16(17), 2518–2518. https://doi.org/10.3390/w16172518
66. Wang, F., Wang, B., Duan, L., Zhang, Y., Zhou, Y., Sui, Q., Xu, D., Qu, H., & Yu, G. (2020). Occurrence and distribution of microplastics in domestic, industrial, agricultural and aquacultural wastewater sources: A case study in Changzhou, China. Water Research, 182, 115956. https://doi.org/10.1016/j.watres.2020.115956
67. Yang, Z., Bu, J., Li, Z., Li, C., Yi, Y., Wang, X., & Liu, Q. (2023). A novel index-based method associated with aquatic ecosystem for evaluating river longitudinal connectivity: A case study for cascade dams in the Yalong River, China. Ecological Indicators, 154, 110903–110903. https://doi.org/10.1016/j.ecolind.2023.110903
68. Choy, C. A., Robison, B. H., Gagne, T. O., Erwin, B., Firl, E., Halden, R. U., et al. (2019). The vertical distribution and biological transport of marine microplastics across the epipelagic and mesopelagic water column. *Scientific Reports, 9*(1), 7843.
69. Eo, S., Hong, S. H., Song, Y. K., Lee, J., Lee, J., & Shim, W. J. (2018). Abundance, composition, and distribution of microplastics larger than 20 μm in sand beaches of South Korea. *Environmental Pollution, 238*, 894–902.
70. Fadare, O. O., Wan, B., Guo, L. H., & Zhao, L. (2020). Microplastics from consumer plastic food containers: Are we consuming it? *Chemosphere, 253*, 126787.
71. Furtado, R., Menezes, D., Santos, C. J., & Catry, P. (2016). White-faced storm-petrels Pelagodroma marina predated by gulls as biological monitors of plastic pollution in the pelagic subtropical Northeast Atlantic. *Marine Pollution Bulletin, 112*(1–2), 117–122.
72. Galloway, T. S., & Lewis, C. N. (2016). Marine microplastics spell big problems for future generations. *Proceedings of the National Academy of Sciences of the United States of America, 113*(9), 2331–2333.
73. Gandara e Silva, P. P., Nobre, C. R., Resaffe, P., Pereira, C. D. S., & Gusmão, F. (2016). Leachate from microplastics impairs larval development in brown mussels. *Water Research, 106*, 364–370.
74. Gautam, A., Gore, P. M., & Kandasubramanian, B. (2020). Nanocluster materials in photosynthetic machines. *Chemical Engineering Journal, 385*, 123951.